回顾历史，纪念先贤，开拓视野，启迪后人

Review history, commemorate our predecessors,
broaden the mind, enlighten posterity

中国植物分类学纪事

A Chronicle of Plant Taxonomy in China

主编　马金双

编者　马金双　胡宗刚　廖　帅　叶　文　鲍棣伟

Editor in Chief: Jin-Shuang MA

Authors: Jin-Shuang MA, Zong-Gang HU, Shuai LIAO, Wen YE, David E. Boufford

河南科学技术出版社

·郑州·

内容提要

本书以编年纪事方式记载当代，特别是过去百年间，中国植物分类学的主要研究机构，主要植物分类学家及其成就，植物分类学图书、期刊及其他重要论著等，全国性与国际性植物分类学学术会议，以及重要的植物学采集内容。

Introduction

This chronicle provides an account of modern plant taxonomy in China. It covers the major taxonomic institutions, key taxonomists and their major achievements, taxonomic books, journals and other important publications, national and international congresses, and important botanical expeditions and collections, particularly during the past hundred years.

图书在版编目（CIP）数据

中国植物分类学纪事 / 马金双主编 . —郑州 : 河南科学技术出版社，2020.1（2021.1 重印）

ISBN 978-7-5349-9539-2

Ⅰ . ①中… Ⅱ . ①马… Ⅲ . ①植物分类学—中国 Ⅳ . ① Q949

中国版本图书馆 CIP 数据核字 (2019) 第 194642 号

出版发行：河南科学技术出版社
地址：郑州市郑东新区祥盛街 27 号　邮编：450016
电话：（0371）65737028　65788613
网址：www.hnstp.cn
策划编辑：陈淑芹
责任编辑：田　伟
责任校对：司丽艳
整体设计：张　伟
责任印制：张　巍
地图审图号：GS（2019）5470 号
地图编制：西安地图出版社
封面绘图：李爱莉
印　　刷：河南省邮电印刷厂
经　　销：全国新华书店
开　　本：889mm × 1194mm　1/16　印张：43.25　插页：9　字数：1200 千字
版　　次：2020 年 1 月第 1 版　2021 年 1 月第 2 次印刷
定　　价：650.00 元

王文采序

我国古代有四大发明：指南针、火药、造纸术、印刷术。此外，在数学、天文、医药、水利等方面也有杰出的专家做出重要贡献。这些为促进世界科技的发展发挥了重要作用。到了18世纪欧洲工业革命兴起，同时，其自然科学各学科蓬勃发展。这时，在这些方面我国却大大落后了。科技落后就要挨打。1842年，腐败的清政府战败，签订丧权辱国的《南京条约》，割地赔款，五口通商，国门大开。随着外国商人的来临，也有不少欧洲的植物采集人员涌入。他们的足迹遍及我国29个省（区），采走大量植物标本。根据这些标本，欧洲有关国家的植物学家发表了万余新种，以及不少新属和一些新科。这种情况一直延续到20世纪初才发生了转变。

1916年和以后几年，留学美国的钱崇澍、胡先骕、陈焕镛3位教授先后学成回国，在南京、北京、广州建立了植物学机构。1929年，留学法国的刘慎谔、林镕2位教授学成回国，在北京建立了植物学研究机构。这些先生筹建研究所，建立图书馆、实验室，派遣人员到全国各地采集植物标本，建立植物标本馆。他们自己也积极投身植物分类学研究，并做出大量的研究成果。

胡老发现了7个新属：木兰科的合果含笑属 *Paramichelia*、拟单性木兰属 *Parakmeria*（与郑万钧合作）、桑科梨桑属 *Smithiodendron*、野茉莉科秤锤树属 *Sinojackia*、木瓜红属 *Rehderodendron*、紫草科车前紫草属 *Sinojohnstonia*，他与郑万钧教授合作发表的活化石水杉 *Metasequoia glyptostroboides* 轰动了国际植物学界。此外，他还发表了壳斗科、桦木科、桑科、肉豆蔻科、豆科、山茶科、海桐科等多科新种。

陈老返国后，在 Elmer D. Merrill 的帮助下对海南、广东的植物区系进行研究，对樟科和壳斗科有深入的研究，发表了 5 个新属：豆科任豆属 *Zenia*、苦苣苔科世纬苣苔属 *Tengia*、漏斗苣苔属 *Raphiocarpus*、扁蒴苣苔属 *Cathayanthe*，以及他与匡可仁教授合作发表的又一裸子植物活化石银杉 *Cathaya argyrophylla*，又一次轰动了国际植物学界。此外，他还发表了小檗科、山茶科、椴树科、绣球花科、紫薇科等多科的新种。

钱老发表了兰科的一新属：独花兰属 *Changnienia*，以及豆科、荨麻科、山茱萸科、山茶科、报春花科、忍冬科等多科新种。

刘老主持编著了中国北部的蓼科、龙胆科、旋花科、忍冬科图谱，主持编写了东北木本植物图志、东北草本植物志和东北植物检索表；刘老注重植物标本的采集工作，足迹遍及我国北部、西部及西南部，撰写了数篇植物地理学论文。

林老是菊科专家，发现了2个新属：重羽菊属 *Diplazoptilon*、紊蒿属 *Elachanthemum*（和林有润合作），此外，还发表了菊科、龙胆科、壳斗科、槭树科、梧桐科等科的新种。

上述诸老的分类学研究成果的发表宣告了西方植物学家掌握中国植物分类学研究局面的结束。此后，上述诸老建立的研究所开展的分类学研究工作在内战、外战的不利环境下却仍能不间断地进行，新的研究人员逐渐增加，各地的标本采集不断进行，诸专科、专属的研究工作不断增加和完成，终于在进入 21 世纪不久，对我国植物学研究和我国经济建设有重大意义的中国植物志和英文版中国植物志两部著作完成出版了。看到这两部著作，我们大家都十分高兴。对二巨著的顺利完成，我们应该向诸老创建我国植物分类学研究事业所做出的贡献表示崇高的敬意！

最近我看到马金双教授等编著的中国植物分类学纪事书稿。此书掌握的文献广泛、全面；记录了中国植物分类学研究历史中发表的全部著作。要了解中国植物分类学的历史，或想了解上述诸老的工作，把此书一翻，就能很快达到目的。正如本书开始时所说的“回顾历史，纪念先贤，开拓视野，启迪后人”。鉴于本书将会起到的重要作用，我应对作者们编写本书付出的劳动表示崇高的敬意！

王文采
中国科学院植物研究所研究员
中国科学院院士
2017 年 10 月 9 日

Foreword by Wen-Tsai WANG

There were four great inventions in ancient China: the compass, gunpowder, paper making and printing. Many outstanding experts in mathematics, astronomy, medicine, water conservancy and in other fields played an important role in promoting the development of science and technology in the world. By the eighteenth century, however, with the advent of the industrial revolution in Europe, the study of natural sciences was flourishing, but had moved to the West. At that time, China lagged far behind in those areas. Lagging behind in sciences and technology leaves a nation vulnerable to attacks. In 1842, the corrupt Qing government was defeated and signed the 'Nanking Treaty' that humiliated the country. It divided up the land for compensation and set up five trading exchanges, then the national door opened widely. With the arrival of foreign businessmen, there was also an influx of European plant collectors. Their footprints covered twenty-nine provinces and autonomous regions throughout the country. A large number of plant specimens were taken away. Based on those specimens, about 10,000 new species of plants, many new genera and a few new families were published by botanists from European countries. This situation lasted until the beginning of the twentieth century.

In 1916 and in the following years, Sung-Shu CHIEN, Hsen-Hsu HU and Woon-Young CHUN, who studied in the United States, successively returned to China and established botanical institutes in Nanjing, Beijing and Guangzhou. In 1929, Professors Tchen-Ngo LIOU and Yong LING, after studying in France, returned to China to set up a botanical research institute in Beijing. These gentlemen set up research institutes, set up libraries and laboratories, and dispatched personnel to collect plant specimens throughout the country and establish herbaria. They were also actively involved in research on plant taxonomy and did a great deal of work of lasting importance.

Professor HU discovered seven new genera: *Paramichelia* (Magnoliaceae), *Parakmeria* (Magnoliaceae; in collaboration with Wan-Chun CHENG), *Smithiodendron* (Moraceae), *Sinojackia* and *Rehderodendron* (Styracaceae), *Sinojohnstonia* (Boraginaceae) and, in collaboration with Professor Wan-Chun CHENG, *Metasequoia glyptostroboides* (Cupressaceae), a living fossil that caused a sensation in the botanical world. Additionally, Hu also published new species of Fagaceae, Betulaceae, Moraceae, Myristicaceae, Fabaceae, Theaceae and Pittosporaceae.

After returning from his studies abroad, Professor CHUN, with the help of Elmer D. Merrill, studied the floras of Hainan and Guangdong. He studied the Lauraceae and Fagaceae in depth and published five new genera: *Zenia* (Fabaceae), *Tengia*, *Raphiocarpus* and *Cathayanthe* (Gesneriaceae), and *Cathaya argyrophylla* (Pinaceae), another living gymnosperm fossil, co-published by CHUN and Professor Ke-Zen KUANG, which again caused a sensation in the international botanical community. In addition, he also published new species of Berberidaceae, Theaceae, Tiliaceae, Hydrangeaceae, Lythraceae and many other families.

Professor CHIEN published the new genus *Changnienia* (Orchidaceae), as well as some new species of Fabaceae, Urticaceae, Cornaceae, Theaceae, Primulaceae, Caprifoliaceae and many other families.

Professor LIOU presided over the compilation of *Flore Illustree du nord de la Chine*: *Hopei (Chihli) et ses provinces voisines* with Polygonaceae, Gentianaceae, Convolvulaceae and Caprifoliaceae published. He also presided over the compilation of *Illustrated Flora of Ligneous Plants of Northeast China*, *Flora Plantarum Herbacearum Chinae Boreali-Orientalis* and *Claves Plantarum Chinae Boreali-Orientalis*. Professor LIOU focused on the collection of plant specimens, covering all of northern, western and southwestern China and wrote several articles on plant geography.

Professor LING was an expert on Asteraceae. He discovered two new genera: *Diplazoptilon* and *Elachanthemum* (in collaboration with Yuou-Ruen LING), as well as other new species of Asteraceae, Gentianaceae, Fagaceae, Aceraceae, Sterculiaceae and other families.

These publications based on the results of taxonomic research by the older generation marked the end of domination by western botanists of plant taxonomy in China. Since then, taxonomic research was carried out continuously and uninterruptedly by the older generation in the research institutes mentioned above, even during the unfavorable environment of civil and foreign wars. The number of new researchers is gradually increasing. The collection of specimens in various places throughout China continues. Research has increased and major works have been completed. Soon after entering the 21st century, two major publications, *Flora Reipublicae Popularis Sinicae* and *Flora of China* in English, of great significance to botanical research in China and for economic construction in our country, have been published. We are all very happy to see these two works. On the successful completion of these two great achievements, we should pay tribute to the contributions of the older generation for establishing the foundations of plant taxonomic research in our country!

I recently read the manuscript of *A Chronicle of Plant Taxonomy in China* by Professor Jin-Shuang MA and others. The book introduces a wide range of historical literature published on plant taxonomy in China. To understand the history of plant taxonomy in China, or to understand the work of the older generation, just open this book and you can quickly reach your goal. As the book begins "View history, commemorate our predecessors, broaden the mind, enlighten posterity". For the important role played by this book, I should pay tribute to the efforts of the authors with my highest respect!

Wen-Tsai Wang

Research Professor, Institute of Botany, Chinese Academy of Sciences, Beijing, China
Academician, the Chinese Academy of Science
October 9, 2017

胡启明序

中国是世界园林之母！中国的植物种类多达30 000多种，高居北温带榜首，同时也名列世界前茅！然而，自18世纪以来，欧美等发达国家对中国植物种类进行了长期而又大规模的采集与研究。这些资料，特别是原始文献和模式标本以及记录等，基本都散落在国外。自林奈1753年以来的这260多年中国植物的研究历史，可谓历尽沧桑，惨不忍睹。一百年前，中国留美的学者归来之后才逐渐开始有中国人研究中国自己的植物，并逐渐掌握研究的主动权。然而，这些历史由于种种原因，至今不是十分清楚，特别是前一个半世纪我们基本没有发言权，一百年前自己开始工作之后又面临外敌入侵以及后来的内乱，加之后来的各类运动，以致出现了不重视科学档案管理，有的单位领导只顾彰显自身的功绩而忽视甚至否认前任所取得之成就等不良风气，使得各类资料丧失殆尽，收集起来格外困难。我很高兴有机会看到了中国植物分类学纪事这本书的全部稿件。作为20世纪50年代就入行的中国植物分类学者，很高兴在此与海内外同仁共享阅后感想与体会。

首先，这本书的内容可谓是中国植物分类学历史上十分全面、非常详细、令人信服的纪事。作者们在收集资料、挖掘整理、细节考证、深入研究等方面值得称赞。其次，该书的设计格式以及方式可谓独具匠心；每一条目都以简明的格式呈现，使人一目了然；而且全书还配了诸多珍贵的历史照片，同时还有3个附录。这些对读者以及使用者无疑是十分重要的。再次，我国植物分类学历史上至今还没有这样的书，特别是中英文对照。这应归功于该书的团队，不仅作者们是国内植物学分类和历史方面的著名专家，更有国际上研究我国植物的权威学者参加。这使得该书出版后必将在学术界产生应有的影响。再次，“回顾历史，纪念先贤，开拓视野，启迪后人”，书前的铭记无疑是该书的宗旨所在。最后，作为分类学者，我十分欣赏作者们在考证历史史实上所付出的努力，特别是对中国首位哈佛大学女博士陈秀英、留学法国的刘厚博士以及留学美国的刘汝强（刘毅然）博士等所做的记载，这些历史事实过去未曾报道。分类学历史考证是一项十分花费时间与精力，而且进展十分缓慢的工作；特别是在中国，作者们能够广泛收集海内外的各类信息，挖掘出如此翔实而又丰富的内容，确实非常值得称赞。

早期研究中国植物主要是西方人主导，后期我们又经历了战乱以及各类运动，所以留给后人的资料十分有限。也正是如此，书中还有很多有待改进或者说可以深入研究的地方。比如采集史方面，采集人记载方面，海外的俄语、德语、法语等资料的收集以及挖掘方面，还有一定的空间与距离。书中列举大量历史事实，若适当评论会使其更有可读性与感染力。但这些毫不影响作为中国植物分类学首部历史纪事的价值与权威，并将永久载入历史。

是为序！

胡启明

中国科学院华南植物园研究员

2017年12月7日

Foreword by Chi-Ming HU

China is the mother of the world gardens! As many as thirty thousand species of plants occur in China, which ranks the country at the top of plant diversity in the North Temperate zone, and also near the top worldwide! However, since the 18th century, developed countries from Europe and the United States conducted long-term and large-scale acquisition of specimens and living plants and carried out research on Chinese plants. These materials, especially the original documents and type specimens as well as the records, are basically scattered abroad. The history of plant research in China for more than 260 years since Linnaeus in 1753 can be described as the vicissitudes of life and so miserable to view. Around one hundred years ago, Chinese scholars began to study in the United States, then returned to China where they gradually began to study their own plants and grasp the initiative of research. The history of the early activities has, however, been unclear, especially the early one and a half centuries, since none of the milestones were recorded by Chinese historians. Nearly one hundred years ago, at the start of modern taxonomic research by Chinese botanists, we faced invasion by foreign enemies and an ensuing civil war. In addition, the various movements that followed resulted in a lack of attention to scientific archives. The leaders of some institutions manifested only their own merits, while neglecting or even denying the achievements of their predecessors. These bad habits made it extremely difficult to gather information as well to reconstruct past events and contributions. I am very happy to have the opportunity to read the full text of *A Chronicle of Plant Taxonomy in*

China. As a Chinese plant taxonomist since the 1950s, I am very happy to share my thoughts and experiences with colleagues both at home and abroad.

First of all, the content of this book can be described as a very comprehensive, very detailed and convincing chronicle on the history of Chinese plant taxonomy. The authors should be lauded for their efforts to collect materials, analyze old documents, study in detail the old texts and to report in-depth their findings. Second, the book's design format and approach are distinctive; each entry is presented with a concise format for easy consultation; and the book is accompanied by many precious historical photographs, along with three appendices. These are undoubtedly very important for the readers and users. Third, so far there is no similar book on the history of Chinese plant taxonomy, not to mention its presentation in both English and Chinese. The results can be attributed to the book's team, not only the famous experts who are authoritative in the botanical taxonomy and history of the country, but also internationally authoritative in the study of Chinese plants. The book will surely have an impact in the academic community and with readers around the world. Fourthly, the inscription in the opening pages, "Review history, commemorate our predecessors, broaden the mind, enlighten posterity" is undoubtedly the purpose of the book; and finally, as a plant taxonomic scholar, I very much appreciate the efforts of the authors to research the historical facts, especially the records of Dr. Hsiu-Ying CHEN, the first Chinese female recipient of a Ph. D. degree in botany from Harvard University, of Dr. Hou LIOU, who studied in France, and Dr. Ju-Chiang LIU (Yi-Ran LIU), who studied in the United States. These historical facts have not been reported previously. Compiling taxonomic history is a very time consuming and painstaking task, and progress is very slow, especially in China. The authors have searched widely, both at home and abroad, to collect all kinds of information. To uncover and present such a rich trove of information is indeed worthy of praise.

Early studies of Chinese plants were mainly dominated by Westerners. In China, we later experienced war and all kinds of movements. The information left for future generations is very limited. Because of this, there are still many places in the book to be improved or to be studied in further detail, such as the history of collecting, collector's records, overseas publications in Russian, German, French and other languages. Although the book presents numerous historical facts, it could be made more readable and contagious if a appropriate comments could be added. However, this does not affect the value and authority of the book as this first chronicle of Chinese plant taxonomy, and it certainly will be permanently embedded in the botanical history of China.

Chi Ming Hu

Research Professor, South China Botanical Garden, Chinese Academy of Sciences, Guangzhou

December 7, 2017

编写说明

一、本纪事采用公历纪年。为方便读者，在附录一中列出与公历纪年相对应的天干地支，以及 1912 年前的朝代年号和 1912—1949 年的中华民国年号。正文按年、月、日次序编排；凡无日可考则以月计，无月可考则以年计，且按作者或名称之字母顺序，列于相应之月或年之末。如果某一项工作目前没有完成，或连续出版物，起始年代之后加号“+”表示，如“1998+”。

二、本纪事记载内容的起始时间为当代植物分类学的命名起点（1753 年），特别是涵盖中国植物分类学百年，以第一篇近代植物分类学文献的发表为起点（钱崇澍，1916 年），截止时间为 2017 年 7 月 31 日（深圳第十九届国际植物学大会闭幕）。本书的目的是记载中国植物分类学的研究历史，包括相关的机构、出版物，主要国际和国内学术与采集活动，关键人物等，以及与植物分类学相关的其他内容。

三、本纪事收载的出版物中，凡涉及的地理范围以省（市、区）为起点，出版物包括志书、名录、检索表、图鉴或图志、手册、工具书等，但是不包括具体分类群的专著。

四、采集则只关注原始采自于中国的，其中中国采集人（队）选取标准为一万号以上；国际植物采集活动相关历史资料贫乏，本纪事尽可能对此进行了记载。附录二详细列出了国内外收藏中国植物的主要标本馆（或室）。

五、本纪事中所有的中外人名为罗马表达方式且一律名前姓后；中文人名采用姓全拼大写，名首字母大写，其余小写，如 Cheng-Yih WU。如一位学者的姓名有两种或两种以上的拼写方式，则全部予以记载，如吴征镒，既有 Cheng-Yih WU 又有 Zheng-Yi WU。为方便读者，将汉语拼音实行之前的老一代姓名拼写方式（如韦氏等）与拼音对照以及中文人名列于附录三。

六、本纪事收载的期刊以分类学为主，包括其变迁历史等；但不包括各类高校的自然科学版等刊物。专业出版物则以分类学命名法规范畴为主，同时兼收一些中国没有正规书号的各类“内部资料”或“内部印制”，以及相关的各类参考资料。

由于历史原因，早年标志性教科书适当收录，但当代教科书以及研究生教材等一般不在考虑之列。

七、对于留学或研修者，主要记载获得博士学位的人员及其毕业院校、出归国时间、出国前和归国后的具体工作与服务机构。出国前或归国后从事植物分类学研究的人员均予以记载。本纪事基本遵循东亚高等植物分类学文献概览（马金双，2011）的原则，收录人员一般以目前已故或已退休者为主。1990 年之后的留学或研修者一般不做收录。

八、本纪事以两种语言编辑，即英文（包括其他西方语言转译为英文，以及极少为拉丁文表述）和中文（极少为日文表述）；所有词条均以两种语言同时列出。排列方式以原文在前，另一种语言于后（包括译文）。如某文献以中文出版，则中文在前，英文于后；反之亦然。若原始无英文标题，则以单引号‘标题’以示与原作者的标题不同。

本纪事以历史为依据而记载，对于无据可考的不予收录，除非具有重要意义。

Introduction to the format and contents

The western calendar is used in the Chronicle. As an aid to readers outside of China, a comparison with the Chinese calendar is in Appendix I. The calendar gives the dates for both the Chinese empire (pre 1912) and the Republic (1912 to 1949). Major content is arranged by year, month, and day. Events without detailed dates are listed last. If a work is incomplete or is part of a continuing series, a plus sign (+) follows the starting year, for example, 1988+.

The chronicle extends from Carl Linnaeus (i.e. 1753) through to the Nineteenth International Botanical Congress held in Shenzhen, China, in 31 July 2017. In particular, the Chronicle mostly covers the past 100 years, since the publication of the first modern taxonomic paper written by a Chinese botanist (Chien, 1916). The objective has been to document the history of study of Chinese plants, including the institutions, publications, major academic events and activities, key scholars in plant taxonomy, as well as related subjects.

The publications reported in the Chronicle pertain only to the province level and above. The publications include floras, checklists, keys, icones and other works of illustrations, and manuals (including handbooks), but excluding various monographs of special taxa.

Only collections of specimens from China are recorded, and only the collections by Chinese botanists numbering more than 10,000 specimens. International activities are covered to the extent possible, since details of those activities and events are often lacking. The important herbaria related to Chinese specimens both within China and abroad are provided in Appendix II.

All personal names, for both Chinese and non Chinese persons, are recorded as given name followed by family name, with the family name written in upper case letters for Chinese persons and given name(s) with only the first letter in upper case, for example, Zheng-Yi WU. If more than one transliteration was used by the same person during her or his lifetime, both are provided, with the original spelling first, for example Cheng-Yih WU and Zheng-Yi WU. As an aid to readers, Appendix III gives the modern pinyin spelling for some of the older scholars who romanized their names in one of the older systems (such as the Wade-Giles system) before the advent of the pinyin system.

The journals that are cited are those that mainly deal with taxonomy, even though they may have changed their focus over the course of their history. Various publications from universities and colleges are not included. Specialty publications are mainly limited to those that are affected by the International Code of Nomenclature. Some unofficial publications, including materials and prints meant only for use in China, as well as various references without an ISBN code, are also recorded. For historical reasons, a few landmark textbooks from the early days of Chinese taxonomy are also included, but modern textbooks as well as works for graduate students are not.

Entries for Chinese students or scholars who went abroad for further study are mainly limited to those who earned a Ph.D. degree. Included is the university or institution they attended, dates of study, research focus and affiliated organizations. Whether their research work was taxonomic, even before leaving China or after returning, is included. The Chronicle mainly records those who have passed away or have retired, as in *The Outline of Taxonomic Literature of Higher Plants from East Asia* (MA, 2011). Graduate students and scholars who went abroad after 1990 are generally not included.

The Chronicle is in two languages, English (including various translations into English, and rarely Latin) and Chinese (with very few in Japanese). Items listed in both languages are arranged in order of the original language followed by the translation into the other language. For example, if the original was in Chinese, the Chinese entry is placed first, and vice versa. Original publications lacking a title in a western language are indicated by the 'translated title' in single quotes.

The chronicle is based on verifiable historic facts; unverified accounts are omitted without further explanation unless deemed to be of significant relevance.

致谢

自2015年元旦开始撰写，本书得到海内外无数朋友与同仁的帮助，没有他们本书不可能完成（按单位与姓氏顺序）：北京林业大学张玉钧，北京师范大学胡晓江、刘全儒，大连自然博物馆张淑梅，东北林业大学王洪峰、郑宝江，复旦大学陈家宽，广西植物研究所刘演、唐赛春，贵阳市科技干部培训中心韩国营，杭州师范大学吴玉环，华东师范大学朱瑞良，江苏省中国科学院植物研究所 南京中山植物园李梅，兰州大学冯虎元，南京大学田兴军，内蒙古师范大学哈斯巴根，内蒙古大学曹瑞、马平、赵东平、赵利清，上海师范大学曹同，深圳市中国科学院仙湖植物园张力、张寿洲，四川大学何兴金、粟和毅，云南大学何兆荣、王焕冲、朱维明，台湾成功大学蒋镇宇，台湾师范大学黄生、王振哲，台湾农委会特有生物保育中心许再文，台湾自然科学博物馆吴声华、杨宗愈，台湾中山大学江友中、刘和义，台湾大学谢长富、胡哲明、王国雄，台湾中央研究院生物多样性研究中心彭镜毅，武汉大学郭友好、汪小凡，西安植物园卢元、寻路路，西北大学刘培亮，西北农林科技大学王辉、吴振海、徐朗然，西藏自治区高原生物研究所土艳丽，浙江大学傅承新、刘军、赵云鹏，中国科学院成都生物研究所彭玉兰、印开蒲，中国科学院华南植物园陈忠毅、邓云飞、董仕勇、胡启明、黄观程、黄向旭、王瑞江、邢福武、张奠湘，中国科学院昆明植物研究所方瑞征、贾颖、李恒、李嵘、康珠永初、吴曙光、杨云珊、乐霁培、周卓、朱卫东，中国科学院南京地质古生物研究所郭双星、李春香，中国科学院沈阳应用生态研究所曹伟、姬兰柱、李维，中国科学院西北高原生物研究所陈世龙、吴玉虎，中国科学院武汉植物园张燕君、梁琼，中国科学院新疆生态与地理研究所管开云、李文军、潘伯荣、张道远，中国科学院西双版纳热带植物园刘红梅、Harald Schneider、杨玺、朱华、朱仁斌，中国科学院植物研究所曹子余、陈之端、高天刚、贾渝、靳淑英、景新明、孔宏智、李爱莉、李良千、李敏、林祁、刘方谱、马克平、孙久琼、王文采、吴鹏程、杨永、于宁宁、张宪春、张志耘、赵星武，中山大学李植华、廖文波，中国科学院上海辰山植物科学研究中心/上海辰山植物园邓玲丽、杜诚、李慧茹、刘夙、田代科、王樟华、汪远、严靖、闫小玲、严岳鸿、周雅萍、左云娟；特别是英国皇家植物园邱园图书馆和档案馆 Julia Buckley，美国康奈尔大学 Edward Cobb，瑞典乌普萨拉大学 Stefan Ekman，美国哈佛大学阿诺德树木园 William (Ned) Friedman 和 Lisa Pearson，俄罗斯科学院马洛夫植物研究所 Dmitry Geltman，美国 Oberlin College 的 Ken Grossi，中国科学院植物研究所图书馆韩芳桥，美国密苏里植物园何思和张丽兵，新加坡植物园何文钏，荷兰国立标本馆 Karien Lahaise，日本东京大学小石川植物园邑田仁，意大利佛罗伦萨大学自然历史博物馆 Chiara Nepi，美国俄勒冈州立大学档案研究中心 Chris Petersen，奥地利维也纳大学 Heimo Rainer，美国夏威夷大学 Tom Ranker，美国伊利诺伊大学校友联合会 Elaine Schaufele，苏格兰爱丁堡植物园 Lesley Scott 和 Mark F. Watson，德国柏林植物园与博物馆 Nicholas Turland，美国圣路易斯华盛顿大学中央图书馆 Melissa Vetter，美国哈佛大学植物图书馆 Gretchen Wade 和 Judith Warnement 提供相关资料。

该项目得到上海市绿化和市容管理局辰山科研专项（G152433）资助。

我们诚挚感谢许多海内外同仁和机构慷慨地提供了照片，并在相应的照片下予以注明；特别感谢中国科学院植物研究所王文采先生和中国科学院华南植物园胡启明先生百忙中斧正并撰序。

编者

Acknowledgements

Since the project began in January 1, 2015, this compilation has received considerable help from friends and colleagues both within and outside China. It could not have been finished without assistance of the following, arranged alphabetically by institution and individual.

Chinese sources of information: Beijing Forestry University, Yu-Jun ZHANG; Beijing Normal University, Xiao-Jiang HU and Quan-Ru LIU; Biodiversity Research Center, Academia Sinica, Ching-I PENG; Chengdu Institute of Biology, Chinese Academy of Sciences, Yu-Lan PENG and Kai-Pu YIN; Dalian Natural History Museum, Shu-Mei ZHANG; East China Normal University, Rui-Liang ZHU; Fairy Lake Botanical Garden, Shenzhen and Chinese Academy of Sciences, Li ZHANG and Shou-Zhou ZHANG; Fudan University, Jia-Kuan CHEN; Guangxi Institute of Botany, Yan LIU and Sai-Chun TANG; Guizhou Science and Technology Training Center, Guo-Ying HAN; Hangzhou Normal University, Yu-Huan WU; Inner Mongolia Normal University, Khasbagan; Inner Mongolia University, Rui CAO, Ping MA, Dong-Ping ZHAO and Li-Qing ZHAO; Institute of Applied Ecology, Chinese Academy of Sciences, Wei CAO, Lan-Zhu JI and Wei LI; Institute of Botany, Beijing, Chinese Academy of Sciences, Zi-Yu CAO, Zhi-Duan CHEN, Tian-Gang GAO, Yu JIA, Shu-Ying JIN, Xin-Ming JING, Hong-Zhi KONG, Ai-Li LI, Liang-Qian LI, Min LI, Qi LIN, Fang-Pu LIU, Ke-Ping MA, Jiu-Qiong SUN, Wen-Tsai WANG, Pan-Cheng WU, Yong YANG, Ning-Ning YU, Xian-Chun ZHANG, Zhi-Yun ZHANG and Xing-Wu ZHAO; Institute of Botany, Jiangsu Province and Chinese Academy of Science, Nanjing Botanical Garden Memorial Sun Yat-Sen, Mei LI; Kunming Institute of Botany, Chinese Academy of Sciences, Rhui-Cheng FANG, Ying JIA, Kangzhu YONGCHU, Heng LI, Rong LI, Shu-Guang WU, Yun-Shan YANG, Ji-Pei YUE, Zuo ZHOU and Wei-Dong ZHU; Lanzhou University, Hu-Yuan FENG; Library of the Institute of Botany, Chinese Academy of Sciences, China, Fang-Qiao HAN; Nanjing Institute of Geology and Palaeontology, Chinese Academy of Sciences, Shuang-Xing GUO and Chun-Xiang LI; Nanjing University, Xing-Jun TIAN; Northeast Forestry University, Hong-Feng WANG and Bao-Jiang ZHENG; Northwest A&F University, Hui WANG, Zhen-Hai WU, Lang-Rang XU; Northwest Institute of Plateau Biology, Chinese Academy of Sciences, Shi-Long CHEN and Yu-Hu WU; Northwest University, Pei-Liang LIU; Shanghai Chenshan Plant Science Research Center, Chinese Academy of Sciences, Ling-Li DENG, Cheng DU, Hui-Ru LI, Su LIU, Dai-Ke TIAN, Zhang-Hua WANG, Yuan WANG, Jing YAN,

Xiao-Ling YAN, Yue-Hong YAN, Ya-Ping ZHOU and Yun-Juan ZUO; Shanghai Normal University, Tong CAO; Sichuan University, Xing-Jin HE and Ho-Yi SU; South China Botanical Garden, Chinese Academy of Sciences, Zhong-Yi CHEN, Yun-Fei DENG, Shi-Yong DONG, Chi-Ming HU, Guan-Cheng HUANG, Xiang-Xu HUANG, Rui-Jiang WANG, Fu-Wu XING and Dian-Xiang ZHANG; Sun Yat-Sen University, Zhi-Hua LI and Wen-Bo LIAO; Taiwan Cheng Kung University, Tzen-Yuh CHIANG; Taiwan Endemic Species Research Institute, Tsai-Wen HSU; Taiwan Museum of Natural Science, Sheng-Hua WU and Tsung-Yu Aleck YANG; Taiwan Normal University, Shong HUANG and Jenn-Che WANG; Taiwan Sun Yat-Sen University, Yu-Chung CHIANG, Ho-Yih LIU; Taiwan University, Chang-Fu HSIEH, Jer-Ming HU, Kuo-Hsiung WANG; Wuhan Botanical Garden, Chinese Academy of Sciences, Yan-Jun ZHANG and Qiong LIANG; Wuhan University, You-Hao GUO and Xiao-Fan WANG; Xi'an Botanical Garden, Yuan LU and Lu-Lu XUN; Xinjiang Institute of Ecology and Geography, Chinese Academy of Sciences, Kai-Yun GUAN, Wen-Jun LI, Bo-Rong PAN and Dao-Yuan ZHANG; Xishuangbanna Tropical Botanical Garden, Chinese Academy of Sciences, Hong-Mei LIU, Harald Schneider, Xi YANG, Hua ZHU and Ren-Bin ZHU; Xizang Institute of Plateau Biology, Yan-Li TU; Yunnan University, Zhao-Rong HE, Huan-Chong WANG and Wei-Ming CHU; Zhejiang University, Cheng-Xin FU, Jun LIU and Yun-Peng ZHAO.

Non-Chinese sources of information: Archives Research Center, Oregon State University, U.S.A., Chris Petersen; Arnold Arboretum of Harvard University, U.S.A., William (Ned) Friedman and Lisa Pearson; Botanischer Garten und Botanisches Museum, Germany, Nicholas Turland; Botany Libraries, Harvard University, U.S.A., Gretchen Wade and Jude Warnement; Cornell University, U.S.A., Edward Cobb; Illinois University Alumni Association, U.S.A., Elaine Schaufele; John M. Olin Library, Washington University, St. Louis, U.S.A., Melissa Vetter; Koishikawa Botanical Garden, University of Tokyo, Japan, Jin Murata; Komarov Botanical Institute, Russian Academy of Sciences, Russia, Dmitry Geltman; Library and Archives, Royal Botanic Gardens, Kew, England, U.K., Julia Buckley; Missouri Botanical Garden, U.S.A., Si HE and Li-Bing ZHANG; Natural History Museum, University of Florence, Italy, Chiara Nepi; Naturalis, Nationaal Herbarium Nederland, Karien Lahaise; Oberlin College, U.S.A., Ken Grossi; Royal Botanic Garden, Edinburgh, Scotland, U.K. Lesley Scott and Mark F. Watson; Singapore Botanic Garden, Singapore, Boon-Chuan HO; Universität Wien, Austria, Heimo Rainer; University of Hawaii, U.S.A., Tom Ranker; Uppsala University, Sweden, Stefan Ekman.

The project received financial support from the Special Fund for Scientific Research of Chenshan, Shanghai Landscaping and City Appearance Administrative Bureau (G152433).

Our sincere thanks are due to the numerous colleagues and institutions who generously provided photos, credited in the caption below each contribution; and particularly to professors Wen-Tsai WANG (PE) and Chi-Ming HU (IBSC) for expressing their thoughts in the forewords.

Authors

目录

Contents

'On Zenia', a poem by Hsen-Hsu HU on the color painting of *Zenia insignis* drawn by Cheng-Ru FENG (Jeng-Ru FENG) (around Spring, 1940), a new genus and a new species of Chinese legume named by Woon-Yung CHUN in honor of Hung-Chun JEN (H. C. ZEN) in 1946 (Photo provided by Institute of Botany)

胡先骕为冯澄如绘制的任豆彩图所题的诗——《任公豆歌》(约 1940 年春)；任豆（*Zenia insignis* Chun）是陈焕镛 1946 年为纪念任鸿隽而命名的中国豆科植物新属和新种（相片提供者：植物研究所）

正文
Text

1700s

1740-1757, Père Nicolas le Cheron d'Incarville (1706-1757) was the first of the French missionaries to collect plants in China. More than 149 specimens from Beijing and 144 specimens from Macau were collected by him.[1] It was largely through his specimens that were sent to Paris, France, that Europeans became aware of the richness of the Chinese flora. The missionaries were soon followed by Western scientists seeking to know more about the Chinese flora and explorers seeking ornamental plants for the gardens of America and Europe.[2] d'Incarville's specimens are in France (P).[3]

1740—1757 年，Père Nicolas le Cheron d'Incarville（1706—1757）为法国派往中国采集植物的传教士中的第一人，采集的标本包括从北京采集的 149 份和从澳门采集的 144 份。正是这些送往法国巴黎的标本，令欧洲人意识到中国植物的丰富程度，引发了西方学者寻找更多中国植物的兴趣，并掀起了采集人员为欧美植物园寻求更多观赏植物的热潮。d'Incarville 的标本现存于法国（P）。

1753, Carl Linnaeus[4], ***Species Plantarum***, with about 100 species from China, including some collected from Whampoa [Huangpu], near Canton [Guangzhou], by his student, Peter Osbeck (1723-1805), between August 23, 1751 and January 4, 1752.[5]

1753 年，Carl Linnaeus，**植物种志**，记载了大约百余种中国植物，包括他的学生 Peter Osbeck（1723—1805）于 1751 年 8 月 23 日至 1752 年 1 月 4 日在广州黄埔采集的一些物种。[6]

---

1 http://plants.jstor.org/stable/10.5555/al.ap.person.bm000366455, accessed 19 September 2017.

2 Jane Kilpatrick, 2014, *Fathers of Botany*. Kew Publishing, Royal Botanic Gardens, Kew and the University of Chicago Press, Chicago. An easily readable and fascinating account of the French missionaries who were among the first to send massive collections of natural history objects, especially plants, to European institutions.

3 Adrien R. Franchet, 1882, Les Plantes de Père d'Incarville dans l'Herbier du Muséum d'Histoire Naturelle de Paris, *Bulletin de la Société Botanique de France*, 29: 2-13. The genus *Incarvillea* Jussieu was named for him.

4 Swedish botanist, physician and zoologist, as Carl von Linné (1707-1778) after his ennoblement in 1761, but most as Carolus Linnaeus in Latin in the academic world. For details see TL2, 3: 71-111, 1981. And also Gordon McGregor Reid, 2009, Carolus Linnaeus (1707-1778): his life, philosophy and science and its relationship to modern biology and medicine, *Taxon* 58(1): 18-31.

5 Elmer D. Merrill, 1916, Osbeck's Dagbok Ofwer en Ostindsk Resa, *American Journal of Botany* 3(10): 571-588.

6 董洪进、刘恩德、彭华，2011，中国植物分类学编目的过去、现在和将来，植物科学学报 26（6）：755—762。

1800s

1812-1831, John Reeves[7] (1774-1856), Tea Inspector for the East India Company, sent to Europe at least 900 botanical drawings by Chinese artists for the Royal Society of London. He also introduced many famous Chinese flowering plants to the West.[8]

1812—1831 年，John Reeves（1774—1856），东印度公司的茶叶检察官，向英国的皇家园艺学会寄送了至少 900 幅中国画家的植物画，同时引种了很多中国的著名花卉植物。

1831，Alexander Georg von Bunge（Алекса́ндр Андре́евич Бу́нге, 1803-1890) collected around Peiping [Beijing] (and Mongolia). Most collections are in Russia (LE), but his personal collections are in France (P).[9]

1831 年，Alexander Georg von Bunge（1803—1890）在北京附近（以及蒙古等地）采集，标本现存于俄罗斯（LE），但个人所藏存于法国（P）。

1843-1862, Robert Fortune (1812-1880) of Scotland travelled to China five times (1843-1845, 1848-

7 English tea inspector by trade, was appointed Inspector of Tea for the East India Company in 1808. Four years later he was sent to China and spent the next 19 years living in Macao and working in Canton [Guangzhou] during the tea season. A keen amateur naturalist, Reeves documented the animals and plants in and around Canton. He also collected specimens and commissioned talented Chinese artists to paint them in the Western scientific tradition under his supervision. The largest proportion of these drawings are of marine and freshwater fish. He generously shared his collections with other naturalists, and in particular they attracted the interest of Sir John Richardson (1787-1865), who was also a keen naturalist. As a result, Reeves engaged the Chinese artists to make copies of all of his fish drawings so that a complete set could be given to Richardson. Richardson subsequently wrote an important scientific paper on the fishes of Japan and China, describing around 80 new fish species that were based entirely on these drawings. Reeves also sent living specimens of beautiful Chinese flowering plants back to England, and was responsible for the introduction of many attractive garden plants to the West, including chrysanthemums, azaleas and wisteria. Reeves returned to England in 1831. In 1817 he was made a Fellow of both the Linnean Society and of the Royal Society. He was also honored by having the epithet, reevesii, applied to nearly 30 species of animals, and the plant genus, *Reevesia* (https://web.archive.org/web/20070322084232/http://www.nhm.ac.uk/nature-online/online-ex/art-themes/drawingconclusions/more/fish_more_info.htm, accessed 19 September 2017).

8 Kate Bailey, 2010, The Reeves Collection of Chinese botanical drawings, *The Plantsman* 9: 218-225; Judith Magee, 2011, *Image of the Nature: Chinese Art and the Reeves Collection*, 112 p; Natural History Museum, London; Martyn Rix, 2013, Botanical Illustration in China and India, *American Scientist* 101: 300-307.

9 Alexander Georg von Bunge, 1833, Enumeratio Plantarum quas in China Boreali Collegit Dr. A. Bunge anno 1831, *Mémoires présentés a l'Académie impérial des Science de Saint-Petersbourg par divers Savans et dans les Assemblées* II: 75-147; 1835, Bruchstuck ans dem Tagebuche des Professors Dr. A. Bunge auf dessen Reise nach China in den Jahren 1830 and 1831, *Dorpater Jahrbücher für Literatur, Statistik und Kunst, besonders Russlands* 4: 251-262 and 341-356.

1851, 1853-1856, 1858-1859,[10] and 1861-1862). He mostly collected and introduced horticultural and economic plants, particularly living tea plants, from coastal areas of eastern China to Darjeeling, India. His major collections are in England (BM, K).[11]

1843—1862 年，苏格兰人 Robert Fortune（1812—1880）受伦敦皇家园艺学会的派遣，5 次（1843—1845、1848—1851、1853—1856、1858—1859、1861—1862）[12] 来华在华东沿海大规模采集、调查中国植物资源，考察园林植物与经济植物的栽培情况并引种，尤其是从中国东部沿海将茶引种至印度大吉岭。Robert Fortune 著有多部采集游记，个人所采集的标本存于英国（BM、K）。

1844-1886, Henry F. Hance[13](1827-1886), British vice-consul to Whampoa [Huangpu] and consul to Canton [Guangzhou], collected (personally but largely through many others) in Hong Kong, Guangdong, Hainan and surrounding areas. He studied Chinese plants from 1854 through 1886. The major part of his collections is in England (especially BM).

1844—1886 年，Henry F. Hance[14]（1827—1886）1844 年入职香港政府，后任黄埔和广州领事，

---

10 From March 4, 1858 through (before) April 28, 1859, Robert Fortune introduce more than 50,000 tea plants from China into the United States of America for the government of the U.S.A. For more detail, see: Alistair Watt, 2016, *Robert Fortune, A Plant Hunter in the Orient*, 183-196 p; Kew Publishing, Kew.

11 Robert Fortune is best known for introducing tea from China to India. He was employed by the Royal Botanic Garden, Edinburgh, and later by the Horticultural Society's Garden at Chiswick, London. As a result of his success, the British made large profits and were able to distribute tea throughout the world. Fortune's travels resulted in the introduction to Europe of many new, exotic plants with beautiful flowers. His most famous accomplishment was the successful transportation of tea from China to India in 1848 on behalf of the British East India Company. Fortune stayed in China for about two and a half years, from 1848 to 1851. He travelled to some areas of China that had seldom been visited by Europeans, including remote areas of Fujian, Guangdong, Anhui, Jiangxi and Jiangsu provinces. Fortune employed many different means to transport tea plants, seedlings and other botanical discoveries, but he is most well-known for his use of Nathaniel Bagshaw Ward's portable Wardian cases to sustain the plants. Using these small greenhouses, Fortune introduced 20,000 tea plants and seedlings to the Darjeeling region of India. He also brought with him a group of trained Chinese tea workers who would facilitate the production of tea leaves. With the exception of a few plants that survived in established Indian gardens, most of the Chinese tea plants Fortune introduced to India perished. The technology and knowledge that was brought over from China, however, may have been instrumental in the later flourishing of the Indian tea industry. See Alistair Watt, 2016, *Robert Fortune, A Plant Hunter in the Orient*, 420 p; Kew Publishing, Kew; as well as: Sarah Rose, 2009, *For all the Tea in China: Espionage, Empire and the Secret Formula for the World's Favourite Drink*, Random House, UK, and 2010, *For All the Tea in China, How England Stole the World's Favorite Drink and Changed History*, Penguin Group, USA.

12 自 1858 年 3 月 4 日至 1859 年 4 月 28 日（之前），受雇于美国政府从中国引种茶苗 5 万株。这一采集活动在以往的历史中几乎全部被忽略了，直到最近才被系统整理出来。详细参见：Alistair Watt, 2016, *Robert Fortune, A Plant Hunter in the Orient*, 183-196 p; Kew Publishing, Kew。

13 Henry F. Hance devoted his spare time to the study of Chinese plants. Born in London, his first appointment was to Hong Kong in 1844. He later became vice-consul in Whampoa, consul in Canton and finally consul in Amoy [Xiamen], where he died in 1886.

14 他所收集的标本达 22 437 号，送给英国（自然）博物馆（BM）。参见：王印政、覃海宁、傅德志，2004，中国植物采集简史，中国植物志 第 1 卷：第 668 页；科学出版社，北京。

1854—1886 年在香港、广东及海南等地采集，并大规模收集植物标本（包括通过他人采集），工余之暇研究当地植物。他的植物标本主要存于英国（特别是 BM）。

Henry F. Hance (1827-1886) (Photo provided by Royal Botanic Gardens, Kew)
Henry F. Hance（1827—1886）（相片提供者：邱园）

1847, Robert Fortune, ***Three Years' Wanderings in the Northern Provinces of China*** *including a visit to the tea, silk and cotton countries: with an account of the agriculture and horticulture of the Chinese, new plants, etc.*, 420 p; J. Murray, London.

1847 年，Robert Fortune，**中国北部三年之旅**，420 页；J. Murray，伦敦。

1847, John Lindley, ***The Elements of Botany,*** *structural and physiological*: being a fifth edition of *The outline of the first principles of botany, with a sketch of the artificial methods of classification, and a glossary of technical terms*, ed. 5, 142 p; Bradbury and Evans, London. Translated into Chinese in 1858.

1858、2014 年，韦廉臣、艾约瑟辑译，李善兰笔述，**植物学**，8 卷本，1858；黑海书馆，上海。[15] 韦廉臣、艾约瑟辑译，李善兰笔述，**植物学**，影印本[16]，196 页，2014；上海交通大学出版社，上海。

1852, Robert Fortune, ***A Journey to the Tea Countries of China*** *including Sung-Lo and the Bohea hills:*

15 本书的原文并非单一版本，实际上是几个不同版本甚至不同作者的著作摘译编辑而成；参见：潘吉星，1984，谈"植物学"一词在中国和日本的由来，大自然探索 3:167-172。即使是主要的原始版本，目前记载也不一致且学术界也有争议。早年记载为 1841 年的第 4 版，参见：汪子春，1984，我国传播近代植物学知识的第一部译著《植物学》，自然科学史研究 3（1）：90-96；罗桂环，1987，我国早期的两本植物学译著——《植物学》和《植物图说》及其术语，自然科学史研究 6（4）：383-387。近几年的考证则认为是 1847 年的第 5 版，且还有其他学者的专著；详细参见：芦笛，2015，晚清《植物学》一书的外文原本问题，自然辩证法通讯 37（6）：1-8；邢鑫，2015，《植物学》卷八"分科"考，或问 28：95-106。在此主要是参考了后者的意见，故以 1847 年列出。

16 博物学文化丛书，刘华杰主编。

with a short notice of the East India company's tea plantations in the Himalaya Mountains, 398 p; J. Murray, London.

1852 年，Robert Fortune，**中国茶乡之旅**，398 页；J. Murray，伦敦。

1853, Robert Fortune, ***Two Visits to the Tea Countries of China*** *and the British tea plantations in the Himalaya with a narrative of adventures and a full description of the culture of the tea plant, the agriculture, horticulture and botany of China*, 315 p; J. Murray, London. Translated into Chinese in 2016.

2016 年，敖雪岗译，**两访中国茶乡**，415 页；江苏人民出版社，南京。

1854-1864, Carl J. Maximowicz (Карл Иванович Максимович, 1827-1891) made three trips (1854-1857, 1859-1860, 1860-1864) to the Far East, including Amur [Heilongjiang], Manchuria [Northeast China] and Japan. The major portion of his collections are in Russia (LE).[17]

1854—1864 年，Carl J. Maximowicz（1827—1891）3 次（1854—1857、1859—1860、1860—1864）赴远东地区，包括黑龙江流域、中国东北和日本采集；所采植物标本现存于俄罗斯（LE）。

Carl J. Maximowicz (1827-1891) (Photo provided by Komarov Botanical Institute)
Carl J. Maximowicz (1827—1891)（相片提供者：科马洛夫植物研究所）

1855 and 1859, Richard K. Maack (Ричард Карлович Маак, 1825-1886) collected twice in Amur [Heilongjiang] and the Ussuri area. His specimens are in Russia (LE).[18]

1855、1859 年，Richard K. Maack（1825—1886）两次赴黑龙江和乌苏里江流域采集；所采集植物标本现存于俄罗斯（LE）。

17 Audrey L. Lievre, 1997, Carl Johann Maximowicz (1827-1891), *The New Plantsman* 4(3): 131-143.

18 Michie A, Gustav Radde and Richard K. Maack, 1867, *Das Amur-Gebiet and seine Bedeutung. Reisen in Theilen der Mongolei, den angrenzeenden gegenden Ostsibiriens, an Amur and seine Nebenflussen. Nach den neuesten Berichten, vornehmlich nach Aufzeichnungen*, 268 p; Spamer, Leipzig.

1856, 1857, 1861, Robert Swinhoe (1836-1877), English Consul and biologist (particularly ornithologist), collected three times in Taiwan. He collected mainly animals but also plants. Swinhoe (1863) *List of plants of the island of Formosa* accounting for 234 species, including 33 ferns, is considered to be the first catalog of the plants of Taiwan. His specimens are in England (K).

1856、1857、1861年，英国领事、生物学家（特别是鸟类学家）郇和[19]（1836—1877）三度来台湾采集，其采集对象以动物为主，但所采植物亦不少，并于1863年发表台湾植物目录[20]，记录了植物234种，包括蕨类33种，被称为第一部研究台湾植物的文献。他的标本存于英国（K）。

1857, Robert Fortune, ***A Residence Among the Chinese Inland, on the Coast and at Sea****—being a narrative of scenes and adventures during a third visit to China, from 1853-1856*, 440 p; J. Murray, London.

1857年，Robert Fortune，**中国内地、沿海及海上之旅行**，440页；J. Murray，伦敦。

1857, Carl J. Maximowicz, Die ersten botanischen nachrichten über das Amurland, *Mélanges biologiques tirés du Bulletin de l'Académie impériale des sciences de St. Pétersbourg* [1856] 2: 407-442 p, plate; 472-474 p; 513-568 p.

1857年，Carl J. Maximowicz，阿穆尔盆地的首次植物学考察见闻，*Mélanges biologiques tirés du Bulletin de l'Académie impériale des sciences de St. Pétersbourg* [1856] 2：407-442页，图版：472-474页；513-568页。

1859, Charles Darwin, ***On the Origin of Species***, 502 p; J. Murray, London. Translated into Chinese in 1920.

1920年，马君武译，**达尔文物种原始**，628页；中华书局，上海。

1859，Richard K. Maack, ***Journey on the Amur in 1855***, 320 p; Tip. K. Vul'fa, Saint Petersburgh. Translated into Chinese in 1977.

1977年，吉林省哲学社会科学研究所翻译组译，**黑龙江旅行记**，496页；商务印书馆，北京。

1859，Carl J. Maximowicz, Primitiae Florae Amurensis, Versuch einer Flora des Amur-Landes, with Index Florae Pekinensis and Index Florae Mongolicae, *Mémoires de l'Académie impérial des sciences de Saint-Pétersbourg par divers savans et lus dans ses assemblées* 9: 1-504.

1859年，Carl J. Maximowicz，阿穆尔植物志初编——附北京和蒙古植物名录，*Mémoires de l'Académie impérial des sciences de Saint-Pétersbourg par divers savans et lus dans ses assemblées* 9:

19 又译为史温侯或斯文豪。

20 彭镜毅、杨智凯，2008，台湾植物物种多样性研究与现况，台湾物种多样性研究现况研讨会论文集，第1-32页；"国立"自然科学博物馆，台中。

1-504。

1861, George Bentham, ***Flora Hongkongensis***—*A description of the flowering plants and ferns of the Island of Hongkong*, 20 p + 482 p; Reprinted in 1872 with supplement by Henry F. Hance, 59 p.

1861 年，George Bentham，**香港植物志** —— 香港的有花植物和蕨类植物记述，20 页 + 482 页；1872 年重印版附有 Henry F. Hance 的增补，59 页。

1861, Richard K. Maack, ***Journey in the Valley of the Ussuri***, 1: 203 p, 2: 344 p; Tip. V. Bezobrazovai Komp., Saint Petersburgh.

1861 年，Richard K. Maack，**乌苏里流域旅行记**，第 1 卷：203 页，第 2 卷：344 页；Tip. V. Bezobrazovai Komp.，圣彼得堡。

1861, Eduard A. Regel, Tentamen Florae Ussuriensis, oder Versuch einer Flora des Ussuri-Gebietes, nach den von Herrn R. Maack (1859) gesammelten Pflanzen bearbeitet, *Memoires presents a l'Academie imperial des Sciences de Saint Pétersbourg* VII, 4(4): 1-128.

1861 年，Eduard A. Regel，乌苏里植物志初编 —— Herrn R. Maack (1859) 乌苏里采集植物志，*Memoires presents a l'Academie imperial des Sciences de Saint Pétersbourg* VII，4(4): 1-128。

1863, Robert Fortune, ***Yedo and Peking***—*a narrative of a journey to the capitals of Japan and China*, 395 p; J. Murray, London.

1863 年，Robert Fortune，**江户**[21] **与北平** —— 日本和中国首都旅行之叙事，395 页；J. Murray，伦敦。

1863-1870 and 1872-1874, Abbe Jean Pierre Armand David[22](1826-1900), French Catholic missionary and naturalist, collected in north, northwest, east, central and southwest China (1863-1864, 1866, 1868-1870 and 1872-1874). He collected both animals and plants, particularly in Sichuan and Shaanxi. His plant specimens are in France (P), where they were studied and described by Adrien R. Franchet (1834-1900).[23]

1863—1870 年、1872—1874 年，法国遣使会教士，博物学家 Abbe Jean Pierre Armand David（谭卫道，

21 江户为明治维新时期的日本首都，即今日东京之旧称。

22 Bernard Scott CM, PÈRE JEAN PIERRE ARMAND DAVID CM (https://cmglobal.org/en/files/2013/08/ScottBDavid.pdf, accessed 3 October 2017); Jane Kilpatrick, 2014, *Fathers of Botany*. Kew Publishing, Royal Botanic Gardens, Kew and the University of Chicago Press, Chicago.

23 Abbe Jean Pierre Armand David, 1873, *Natural History of North China, with notices of that of the south, west and northeast and of Mongolia and Thibet*, 45 p; translated from Shanghai Nouveliste; Shanghai Evening Courier and Shanghai Budget; 1875, Abbe Jean Pierre Armand David (1826-1900), *Journal de mon troisieme voyage d'exploration dans I'empire Chinois*, 1: 383 p and 2: 348 p; Librairie Hachette et Cie, Paris.

1826—1900），先后数次（1863—1864、1866、1868—1870、1872—1874）来华，在华北、西北、华东以及华中和华西采集大量生物标本，特别是在四川、陕西等地采集大量的动植物标本；植物标本存于法国（P），主要由 Adrien R. Franchet（1834—1900）研究并发表。

Bust of Abbe Jean Pierre Armand David in Deng Chi Gou, Feng Yong Zhai National Reserve, Baoxing, Sichuan, where the first giant panda was discovered by David in 1869 (Photo provided by Bin-Bin LI)
四川省宝兴县（原穆坪县）蜂桶寨国家级自然保护区邓池沟的传教士谭卫道半身像。1869 年大卫首次发现大熊猫的地方（相片提供者：李彬彬）

1865-1901, Père Émile Marie Bodinier (1842-1901), French missionary and collector of plant specimens and seeds, arrived in Guizhou in 1865, where, with the exception of stays in Hong Kong and Beijing, he remained until his death in 1901. His collections, from the vicinity of Beijing, Hong Kong and Macau and especially from Guizhou, were studied primarily by Adrien R. Franchet. His collections are in France (P).[24]

1865—1901 年，法国传教士与植物采集家 Père Émile Marie Bodinier（1842—1901）来华采集，除北京、香港、澳门等地的周边外，主要于 1865 年起在贵州采集标本和种子，直至 1901 年故去；所采标本由 Adrien R. Franchet 研究。他的标本现存于法国（P）。

1866-1883, Emil V. Bretschneider (Эмиль В. Бретшнейдер, 1833-1901), physician to the Imperial Russian Legation to Peking [Beijing], also Sinologist and expert on both Chinese and non-Chinese botanical literature. He collected around Beijing during his stay in China, sent specimens to Russia (LE) for study by Carl. J. Maximowicz, but most important was his relationship with westerners, particularly botanists and plant taxonomists from both Europe and North America. He also collected (including through others) the seeds of trees and shrubs, economic, horticultural and vegetable plants of great value and distributed them among many botanical gardens and arboreta in Europe and North America.

1866—1883 年，俄罗斯帝国驻华使馆医生、汉学家和植物文献学家 Emil V. Bretschneider（1833—1901），在北京期间于附近采集，植物标本现存于俄罗斯（LE）等，主要由 Carl. J. Maximo-

24 Émile Marie Bodinier, https://fr.wikipedia.org/wiki/%C3%89mile-Marie_Bodinier, accessed 10 September 2017; Jane Kilpatrick, 2014, *Fathers of Botany*. Kew Publishing, Royal Botanic Gardens, Kew and the University of Chicago Press, Chicago.

wicz 研究；他尤以与同时代的西方学者密切联系著名，其中主要是欧洲和北美的植物学家和植物分类学家。与此同时，他大量（包括通过他人）采集植物，特别是具有经济价值的乔灌木、经济作物，以及园林植物和蔬菜的种子，并送给欧美等国的植物园和树木园。

1866-1877 and 1877-1893, Carl J. Maximowicz, Diagnoses Plantarum Novarum Japoniae et Mandshurae, parts 1-20, *Bulletin de l'Academie Imperiale des Sciences de Saint-Pétersbourg. St. Petersburg*. vols. 10-22; and Diagnoses Plantarum Novarum Asiaticarum, parts 1-8, *Bulletin de l'Academie Imperiale des Sciences de Saint-Pétersbourg. St. Petersburg*, vols. 23-32.

1866—1877 年、1877—1893 年，Carl J. Maximowicz，日本与满洲植物纪要，1-20 卷，*Bulletin de l'Academie Imperiale des Sciences de Saint-Pétersbourg. St. Petersburg*, vols. 10-22；亚洲新植物纪要，1-8 卷，*Bulletin de l'Academie Imperiale des Sciences de Saint-Pétersbourg. St. Petersburg*, vols. 23-32。

1867, Alphonse de Candolle, ***Lois de la Nomenclature Botanique***, 60 p, V. Masson et fils, Paris. Partly translated into Chinese in 1919.

1919 年，董延禧摘译，万国采用之植物命名法，博物杂志 1:73-78。[25]

1867-1888, Nikolai M. Przewalski[26](Николáй Михáйлович Пржевáльский, 1839-1888), Imperial Russian geographer and explorer of central and eastern Asia; five expeditions (1867-1869, 1870-1873, 1876-1877, 1879-1880, with geographer, Vsevolod I. Roborowski (1856-1910) and 1883-1885, with geographer, Vsevolod I. Roborowski and Peter K. Kozlov (1863-1953), botanist), expeditions to Ussuri, north, northwest and west China as well as

Nikolai M. Przewalski (1839-1888) (Photo provided by Komarov Botanical Institute)
Nikolai M. Przewalski（1839—1888）（相片提供者：科马洛夫植物研究所）

25 此为 1900 年巴黎通过的版本，可能是通过翻译而首次用中文记载的命名法规内容。

26 Daniel Brower, 1994, Imperial Russia and Its Orient: The Renown of Nikolai Przhevalsky, *Russian Review* 53(3): 367-381.

central Asia. His plant specimens are in Russia (LE).[27]

1867—1888 年，俄罗斯帝国地理学家、中亚和东亚考察家 Nikolai M. Przewalski[28]（1839—1888）先后 5 次（1867—1869、1870—1873、1876—1877、1879—1880、1883—1885）来华在乌苏里、华北、西北、西部，以及中亚大规模采集。其中，第 4 次协同地理学家 Vsevolod I. Roborowski（1856—1910），第 5 次协同地理学家 Vsevolod I. Roborowski 和植物学家 Peter K. Kozlov（1863—1953）。所采植物标本现存于俄罗斯（LE）。

1867, 1881-1892, 1893-1895, Abbé Pierre Jean Marie Delavay[29](1834-1895), French missionary, collected and sent to France (P) more than 200,000 specimens from Yunnan between 1881-1895 for study by Adrien R. Franchet (1834-1900).[30] Duplicate specimens are in several herbaria in Europe (B, G, H, K, L, LE) and in the United States (A, F, MO, US).

1867 年、1881—1892 年、1893—1895 年，法国传教士 Abbé Pierre Jean Marie Delavay（赖神甫，1834—1895）在云南采集了 20 多万份标本，并于 1881—1895 年送至法国（P）由 Adrien R. Franchet（1834—1900）研究。他的复份标本存于欧洲（B、G、H、K、L、LE）和美国（A、F、MO、US）。

1870, Emil V. Bretschneider, The study and value of Chinese Botanical Works, *Chinese Recorder and Missionary Journal*, 3: 157-163, 172-178, 218-227, 241-249, 264-272 and 150-152; Reprinted as ***On the Study and Value of Chinese Botanical Works*** *with notes on the History of Plants and Geographical Botany*, 51 p, 1871; Rozerio, Marcal and Co., Foochow. Translated into Chinese in 1935.

1935 年 7 月，石声汉[31]译，胡先骕校，**中国植物学文献评论**，82 页；商务印书馆，上海。

1871-1902, Charles Ford (1844-1927), named Superintendent, Department of Botany and Afforestation, Hong Kong. Hong Kong Herbarium (HK) founded with Charles Ford as the first curator in 1878. Ford

27 Nikolai M. Przewalski, 1870, *Travels in the Ussuri Region 1867-1869*, 297 p; Vtip. N. Nekliudova, S.-Petersburg; 1875 and 1876, *Mongolia and Tangut Country and the solitudes of northern Tibet; being a narrative of three years' travel in eastern High Asia*, vols. 1: 287 p and 2: 320 p; S.-Peter bug; 1876, *From Kuldzh to Tian Shan and Lob-Nor*, 156 p, Gos. Idz-vo Georgr. Lit-ty, Moskva; 1883, *From Zaisan through Hami to Tibet and the Sources of the Yellow River*, 473 p, Tip. V.S. Balaskeva, S. Petersburg; 1888, *From Kiakhta to the Sources of the Yellow River, Exploration of the Northern borderline of Tibet and the Journey through Lobnor along the Basin of the River Tarim*, 536 p; V. S. Balaskeva, S.-Petersburg.

28 Donald Rayfield, 1976, *The Dream of Lhasa, The life of Nikolay Przhevalsky, Explorer of Central Asia*, 221 p; Paul Elek, London.

29 Socièté d'Horticulture de Haute-Savoie, 2013, *Jean-Marie Delavay: botaniste Savoyard des Missions Estrangéres*, 26 p; Societe d'Horticulture de Haute-Savoie, Annecy; Jane Kilpatrick, 2014, *Fathers of Botany*. Kew Publishing, Royal Botanic Gardens, Kew and the University of Chicago Press, Chicago.

30 Frederick C. Stern, 1944, The discoveries of the great French missionaries in central and western China, *Proceedings of the Linnean Society*, 156(1): 16-20.

31 湖南湘潭人（1907—1971），中国农业史学家、农业教育家、植物生理学家。详细参见：武汉大学生命科学院院史编纂委员会，2012，*奋进岁月 铸就辉煌——武汉大学生命科学学院院史*，481 页；武汉大学出版社，武汉。

collected in Guangdong, Hainan, Taiwan and Fujian and especially employed local native collectors. His specimens are in Hong Kong (HK) and England (K).

1871—1902 年，Charles Ford (1844—1927) 任香港植物与造林部主管，1878 年香港植物标本室成立（HK），Charles Ford 任首任馆长。他长期在广东、海南，以及台湾和福建等地采集，特别是雇用当地人员采集。他的植物标本现存于香港（HK）和英国（K）。

Charles Ford (1844-1927) (Photo provided by Royal Botanic Gardens, Kew)
Charles Ford（1844—1927）（相片提供者：邱园）

1874-1882, Francis B. Forbes (1839-1908), American businessman, collected in Shanghai and Chefoo [Yantai], Shandong; also employed local native collectors. Forbes studied Chinese plants in Europe from 1875-1876 and collaborated with William B. Hemsley (1843-1924) at K, England, on ***Index Florae Sinensis***; his private collections are in England (BM).

William B. Hemsley (1843-1924) (Photo provided by Royal Botanic Gardens, Kew)
William B. Hemsley（1843—1924）（相片提供者：邱园）

1874—1882 年，美国商人 Francis B. Forbes（1839—1908）长期（1857—1882）在中国经商；1874 年开始对植物感兴趣并在上海和烟台采集，同时雇用当地人为他采集。1875—1876 年 Francis B. Forbes 赴欧洲研究中国植物，并与邱园的 William B. Hemsley（1843—1924）合作编写**中国植物名录**。他个人的植物标本则存放在英国（BM）。

1876 年，傅兰雅[32]（1839—1928，英国人）在上海编辑发行了中国最早的自然科学专刊**格致汇编**（1876—1877, 1879—1892；共发行 7 卷 60 册），包括动植物学等文章。[33]

1876, John Fryer (1839-1928, British) issued ***Chinese Scientific Magazine*** (60 parts in 7 volumes) in 1876-1877, 1879-1892. This was the earliest scientific natural history journal, including zoological and botanical papers, published in Shanghai, China.

1876-1891, Michael V. Pevtsov (Михаил В. Певцов, 1843-1902), explored inland Asia several times (1876, 1878-1879, 1889-1891, 1893, 1895, 1896); his plant specimens are in Russia (LE)[34].

1876—1891 年，Michael V. Pevtsov（1843—1902）数次（1876、1878—1879、1889—1891、1893、1895、1896）到亚洲内陆考察；所采植物标本均存于俄罗斯（LE）。

Grigory N. Potanin (1835-1920) (Photo provided by Komarov Botanical Institute)
Grigory N. Potanin（1835—1920）（相片提供者：科马洛夫植物研究所）

1876-1895, Grigorii N. Potanin (Григорий Николаевич Потанин, 1835-1920), Imperial Russian, undertook

32 傅兰雅，1839 年 8 月 6 日生于英国，1928 年 7 月 2 日卒于美国；圣公会教徒，翻译家，单独翻译或与人合译西方书籍 129 部（绝大多数为科学技术性质），是在华外国人中翻译西方书籍最多的人。清政府曾授予其三品官衔和勋章。傅兰雅大学毕业后于清咸丰十一年（1861 年）到香港就任圣保罗书院院长。两年后受聘任京师同文馆英语教习，清同治四年（1865 年）转任上海英华学堂校长，并主编字林洋行的中文报纸《上海新报》。同治七年（1868 年）受雇任上海江南制造局翻译馆译员达 28 年，翻译科学技术书籍。清光绪二年（1876 年）创办格致书院，自费创刊科学杂志格致汇编，所载多为科学常识，带有新闻性，设有“互相问答”一栏，从创刊号至停刊，差不多期期都有，共刊出了 322 条，交流了 500 个问题。光绪三年（1877 年）被举为上海益智书会总编辑，从事科学普及工作。光绪二十二年（1896 年）去美国担任加州大学东方文学语言教授，后加入美国籍。参见：王扬宗，2000，傅兰雅与近代中国的科学启蒙，科学出版社，北京。王渝生，2000，中国近代科学的先驱——李善兰，84 页，科学出版社，北京。

33 王铁军，2004，傅兰雅与格致汇编，中国科学院自然科学研究所官网（http://www1.ihns.ac.cn/readers/2004/wangtiejun.htm, accessed 7 December 2017）。

34 Michael V. Pevtsov, 1883, *Sketch of a Journey in Mongolia*; 1892-1896, *The Labours of the Tibet Expedition under Michael V. Pevtsov*, vols. 1, 1895, 2, 1892, 3, 1896 (by Vsevolod I. Roborovski and Peter K. Kozlov).

exploratory expeditions in Asia (1876-1878, 1879-1880, 1884-1886, 1891-1894, 1899). The first two expeditions were in northern China and particularly in Mongolia; the second two were in central China, particularly in Gansu, Hebei, Henan, Shanxi, Shaanxi, Sichuan, and last time to east Mongolian area. His plant specimens are in Russia (LE).[35]

1876—1895 年，俄罗斯帝国亚洲考察家[36]Grigorii N. Potanin（1835—1920），5 次（1876—1878、1879—1880、1884—1886、1891—1894、1899）在中国北部大规模采集；前两次主要在蒙古，后两次则主要在中国中部（甘肃、河北、河南、山西、陕西、四川等），最后一次到蒙古东部。他所采集的植物标本现存于俄罗斯（LE）。

1880, Emil V. Bretschneider, Early European researches into the flora of China, *Journal of the North China, Branch of the Royal Asiatic Society, new series* 15: 1-192.

1880 年，Emil V. Bretschneider，欧洲人早期对中国植物的研究，*Journal of the North China*, *Branch of the Royal Asiatic Society*，*new series* 15: 1-192。

1882, 1893, 1896, Emil V. Bretschneider, ***Botanicon Sinicum***, Notes on Chinese botany from native and western sources, I, General Part, Literature, *Journal of the North China, Branch of the Royal Asiatic Society, new series* 16: 18-230, 1882; II, The Botany of the Chinese Classics, *Journal of the North China, Branch of the Royal Asiatic Society, new series* 25: 1-468, 1893; III, Botanical Investigations into the Materia Medica of the Ancient Chinese, *Journal of the North China, Branch of the Royal Asiatic Society, new series* 29: 1-623, 1896; Reprinted in three parts, dated 1881, 1892 and 1895, with subtitles: part I, General Introduction and Bibliography; part II, The botany of the Chinese classics; part III, Botanical investigations into the Materia Medica of the ancient Chinese.

1882 年、1893 年、1896 年，Emil V. Bretschneider，**中国植物**，基于本土和西方的中国植物学注释，I：综论，文献；*Journal of the North China, Branch of the Royal Asiatic Society, new series* 16: 18-230，1882；II：中国的经典植物学，*Journal of the North China, Branch of the Royal Asiatic Society*，*new series* 25: 1-468，1893；III：古代中国中草药的植物学研究，*Journal of the North China, Branch of the Royal Asiatic Society, new series* 29: 1-623，1896；先行分三部分印刷时间是 1881 年、1892 年、1895 年。

1883-1888, Adrien R. Franchet, ***Plantae Davidianae ex Sinarum Imperio***, vols. 1: 1-390 p, *Plantes de*

35 Grigorii N. Potanin, 1881 and 1883, *Sketches of North-Western Mongolia*, 1: 1-426 p, 2: 1-181 p, 3: 1-372 p, 4: 1-1,026 p; 1893, *The Tanguto-Tibetan Border Regions of China and Central Mongolia*, 1: 1-567, 2: 1-437; 1899, Sketches of a journey to Szechuan and the eastern frontier of Tibet in 1892-1893, *Bulletin of the Russian Imperial Geographical Society* 35: 368-418; Emil V. Bretschneider, 1899, A list of stations of Potanin's expedition to Szechuan and eastern Tibet in 1893, *Bulletin of the Russian Imperial Geographical Society* 35: 427-436. 吴吉康、吴立珺译，2013，蒙古纪行，473 页；兰州大学出版社，兰州。

36 他所采集的标本上万号，多达 1 000 种。参见：王印政、覃海宁、傅德志，2004，中国植物采集简史，中国植物志 第 1 卷：第 666-667 页；科学出版社，北京。

Mongolie du Nord et du Centre de la Chine; 2: 1-334 p, *Plantes du Thibet Oriental (Province de Moupine)*; originally published in five parts in *Nouvelles Archives du Museum d'Histoire Naturelle, Paris* 5: 153-272, 1883; 6: 1-126, 1883; 7: 55-200, 1884; 8: 183-254, 1885; 10: 33-198, 1887-1888; based on specimens collected by Abbe Jean Pierre Armand David in China from 1863-1870 and 1872-1874, includes many taxa described as new by Adrien R. Franchet.

1883—1888 年，Adrien R. Franchet，**Abbe Jean Pierre Armand David 中国采集植物志**，第1卷：390 页，1883，蒙古、华北、华中植物；第2卷：334 页，1888，藏东（宝兴）植物；基于 1863—1870 年和 1872—1874 年的采集，包括很多新类群。原文分 5 部分发表于 *Nouvelles Archives du Muséum d'Histoire Naturelle*，*Paris* 5: 153-272, 1883；6: 1-126, 1883；7: 55-200, 1884；8: 183-254, 1885；10: 33-198, 1887-1888。

Sven Hedin (1868-1952) (Photo Credit: https://commons.wikimedia.org/wiki/Sven_Hedin#/media/File:Sven_Hedin_in_McClure%27s_Magazine_December_1897.jpg, accessed December 2018)

Sven Hedin（1868—1852）（相片来源：https://commons.wikimedia.org/wiki/Sven_Hedin#/media/File:Sven_Hedin_in_McClure%27s_Magazine_December_1897.jpg，2018 年 12 月进入）

1884-1935, Sven A. Hedin (1868-1952), Sweden Explorer to central Asia: 1884-1885, 1890-1891, 1893-1897, 1899-1902, 1905-1908, 1923, 1927-1935, especially in northern and western China. Hedin's specimens are in Sweden (S).[37]

1884—1935 年，瑞典探险家 Sven A. Hedin（1868—1952）先后数次（1884— 1885、1890—1891、1893—1897、1899—1902、1905—1908、1923、

37 Sven A. Hedin, 1898, *En färd genom Asien, 1893 and 1897*; *Through Asia*, 1: 1-664 p, 2: 667-1,279 p; Methuen, London; *Through Asia with nearly three hundred illustrations from sketches and photographs by the author Sven A. Hedin and John T. Bealby*, 1899, 1,255 p; Harper and Brothers, New York / London; 1903, *Central Asia and Tibet Towards the Holy City of Lassa*, 1: 613 p and 2: 617 p; Hurst and Blackett, Ltd., London, / C. Scribners sons, New York; 1904, *Adventures in Tibet*, 487 p; Hurst and Blackett, London; 1904-1907, *Scientific Results of a Journey in Central Asia 1899-1902*, vols. 1-6; Lithographic Institute of the General Staff of the Swedish Army, Stockholm; 1909, *Trans-Himalaya–Discoveries and Adventures in Tibet*, vols. 1: 436 p, 2: 414 p, 3: 426 p; The MacMillan Co, New York; 1917-1922, *Southern Tibet: discoveries in former times compared with my own researches in 1906-1908*, vols. 1-11; Lithographic Institute of the General Staff of the Swedish Army, Stockholm; 1924, *Von Peking nach Moskau*, 321 p; F. A. Brockhaus, Leipzig; 1932, *Ratsel der Gobi, die Fortsetzung der grossen Fahrt durch Innerasien in den Jahren 1928-1930*, 335 pp, F. A. Brockhaus, Leipzig; 1927-1935, 1943, *History of the expedition in Asia, 1927-1935, Reports from the scientific expedition to the north-western provinces of China under the leadership of Dr. Sven A. Hedin*, The Sino-Swedish expedition, publication 23-25: Part I, 1927-1928, Part II, 1928-1933, Part III, 1933-1935, Part IV, General reports of travels and field-work, by Folke Bergman, Gerhard Bexell, Birger Bohlin and Gösta Montell, 1943; Elanders Boktryckeri Aktiebolag, Stockholm Göteborg. 罗桂环著，2009，中国西北科学考查团综论，278 页；中国科学技术出版社，北京。

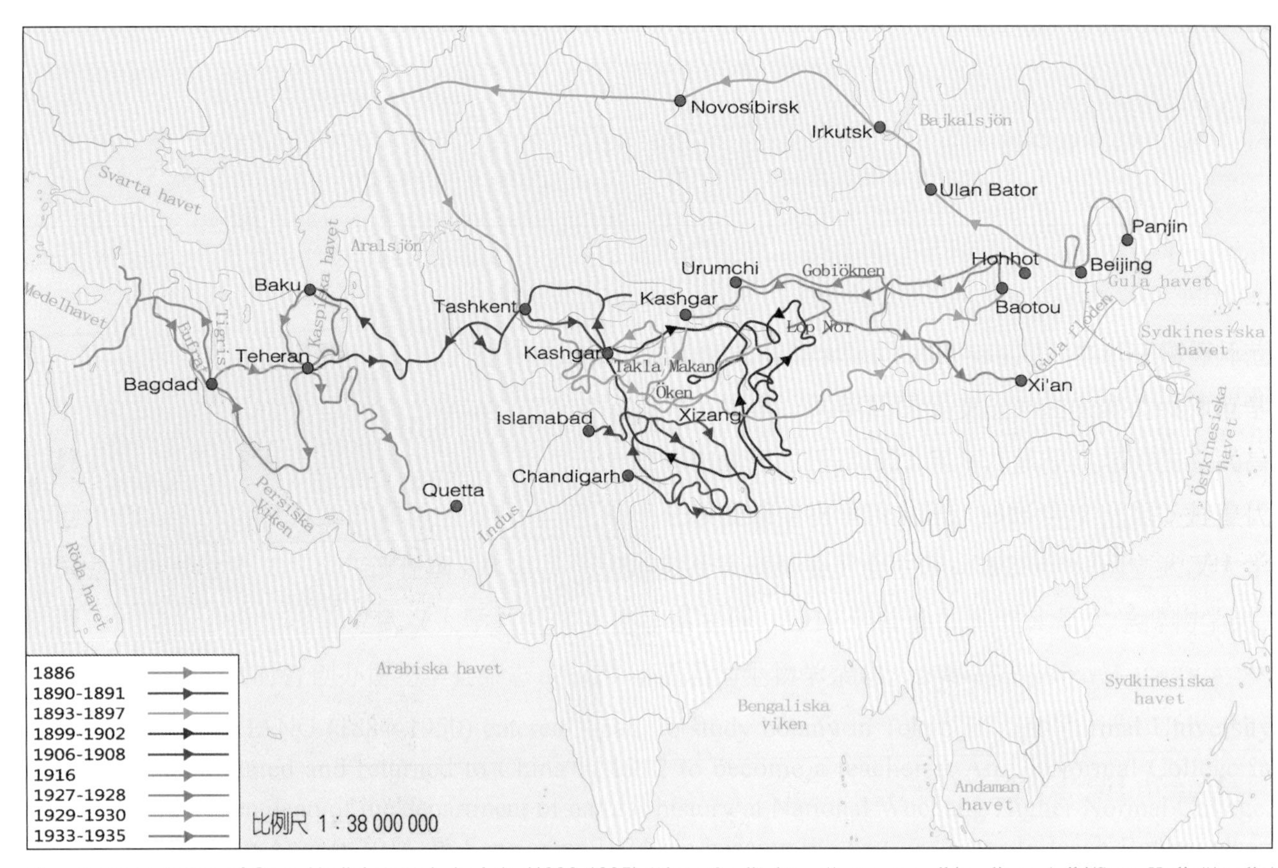

Map showing routes of Sven Hedin's travels in Asia (1886-1935) (Photo Credit: https://commons.wikimedia.org/wiki/Sven_Hedin#/media/File:Exploring_expeditions_of_Sven_Hedin_1886-1935.jpg, accessed December 2018)

1886—1935 年间 Sven Hedin 的中亚考察路线（相片来源：https://commons.wikimedia.org/wiki/Sven_Hedin#/media/File:Exploring_expeditions_of_Sven_Hedin_1886-1935.jpg，2018 年 12 月进入）

1927—1935）在中亚，特别是在中国北部和西部进行过大规模考察采集，所采植物标本现存于瑞典（S）。从 1937 年起该考察团的研究报告以 Sven A. Hedin 探险队在中国西北地区学术探险报告的形式陆续刊行，至 1980 年已出版了 50 卷，内容极为丰富，其中有 4 卷专门报道了植物学方面的成果，即第 13、22、31、33 卷。[38]

1885-1900, Augustine Henry (1857-1930), Chinese Customs officer from Ireland, collected largely in Hubei (and eastern Sichuan [now Chongqing]), Hainan, Taiwan and Yunnan, not only personally but also through local native collectors. Most of his collections are in England (K) but others were distributed widely[39]. His own specimens are in Ireland (DBN)[40] and partly in the United States (NY).

1885—1900 年，Augustine Henry（1857—1930），爱尔兰人，中国海关官员，先后在湖北（和四川东部，今重庆）、海南、台湾、云南大规模雇用当地人员采集。他所采集的植物标本主要存于英国

38 George Kishi, 1984, *To the Heart of Asia—The Life of Sven Hedin*, 153 p; The University of Michigan Press, Ann Arbor, MI.

39 T. Y. Aleck Yang and Chang-Fu Hsieh, 2013, Revision of Dr. Augustine Henry's A list of plant from Formosa (1896), *Collection and Research* 26: 43-49.

40 Seamus O'Brien, 2011, *In the Footsteps of Augustine Henry and His Chinese Plant Collectors*, 367 p; Garden Art Press, Woodbridge, Suffolk.

（K），但世界各地的其他标本馆也有存放。他个人的植物标本存于爱尔兰（DBN）[41]，也有一部分存于美国（NY）。

1886-1905, Francis B. Forbes and William B. Hemsley, An enumeration of all the Plants known from China proper, Formosa, Hainan, Corea, the Luchu Archipelago and the Island of Hong Kong, together with their distribution and synonymy[42], *Journal of the Linnean Society, Botany* 23: 1-521, 1886-1888, 26: 1-592, 1889-1902, 36: 1-686, 1903-1905.

1886—1905 年，Francis B. Forbes、William B. Hemsley，中国内地、台湾、海南、朝鲜、琉球群岛及香港岛的植物名录[43]，*Journal of the Linnean Society. Botany* 23：1–521 页，1886—1888 年，26：1–592 页，1889—1902 年，36：1–686 页，1903—1905 年。

Augustine Henry (1857-1930) (Photo provided by Royal Botanic Gardens, Kew)
Augustine Henry（1857—1930）（相片提供者：邱园）

1888-1905, Père Jean André Soulié (1858-1905), French missionary based in western Sichuan and eastern Xizang (Tibet), collected more than 7,000 numbers of plant specimens, initially around Kangding and Dong'eluo (Tongelo on labels), Sichuan, but later farther west around Batang and in eastern Xizang (Tibet) and Cigu (Tsekou on labels), Yunnan. His specimens are in France (P).[44]

1888—1905 年，法国传教士 Père Jean André Soulié（1858—1905）于四川西部和西藏东部采集 7 000 多号标本。采集开始主要是在四川康定及东俄洛一带，后延伸至四川巴塘以西直至西藏东部及云南的德钦（一带）。他的标本现存于法国（P）。

41 马金双、叶文，2013，书评：In the Footsteps of AUGUSTINE HENRY and His Chinese Plant Collectors，植物分类与资源学报 35（2）：216–218。

42 Reprint with title *Index Florae Sinensis*.

43 单行本印刷时以中国植物索引为题。

44 Jean Soulié—Archives des Missions Étrangères de Paris. http://archives.mepasie.org/fr/notices/notices-biographiques/soulia (accessed 26 September 2017); Jane Kilpatrick, 2014, *Fathers of Botany*. Kew Publishing, Royal Botanic Gardens, Kew and the University of Chicago Press, Chicago.

1889-1890, Adrien R. Franchet, ***Plantae Delavayanae****—Plantes de Chine recueillis au Yun-nan par l'Abbé Delavay*, vols. 1: 1-80 p, 1889; 2: 81-160 p, 1889, 3: 161-240 p, 1890; Paul Klineksieck, Paris.

1889—1890 年，Adrien R. Franchet，**Delavay 采集植物志**—— Abbé Pierre Jean M. Delavay 中国云南采集植物志，第 1 卷：1-80 页，1889；第 2 卷：81-160 页，1889；第 3 卷：161-240 页，1890；Paul Klineksieck，巴黎。

1889, Carl J. Maximowicz, ***Flora Tangutica*** *sive Enumeratio Plantarum regionis Tangut (Amdo) Provinciae Kansu, nec non Tibetiae praesertim orientali-borealis atque Tsaidam, ex collectionibus N. M. Przewalski atqut G. N. Potanin*, vol. 1: 1-110 p, Thalamiflorae et Disciflorae.

1889 年，Carl J. Maximowicz，**唐古特植物志**，N. M. Przewalski 和 G. N. Potanin 采自唐古特地区的植物名录，第 1 卷：110 页；Thalamiflorae et Disciflorae。

1889, Carl J. Maximowicz, ***Enumeratio Plantarum Hucusque in Mongolia nec non adjacente parte Turkestaniae Sinensis Lectarum***, vol. 1: 138 p, Thalamiflorae et Disciflorae.

1889 年，Carl J. Maximowicz，**蒙古及其临近的中国新疆的植物名录**，第 1 卷：138 页；Thalamiflorae et Disciflorae。

1890, Carl J. Maximowicz, Plantae Chinenses Potanianae nec non Piasezkianae, *Trudy Imp. S.-Peterburgsk. Bot. Sada* 11(1): 1-112.[45]

1890 年，Carl J. Maximowicz，Potanin 和 Piasezki 采自中国的植物，*Trudy Imp. S.-Peterburgsk. Bot. Sada* 11(1): 1-112。

1890-1901, Père Giuseppe Giraldi (1848-1901)[46], Italian missionary, at the request of Antonio Biondi [47](1849-1929), collected more than 8,000 plant specimens, especially non-flowering plants, in more than 10 years in Shaanxi, particularly around Huxian in the Qinling Mountains. His specimens are in Italy (FI), but duplicates were distributed widely (A, B, H, K, LE, W).

45 i.e. *Acta Horti Petropolitani.*

46 Faustino Ghilardi, 1921, *Il P. Giuseppe Giraldi, missionario in Cina: 1848-1901*, 112 p; Collegio S. Bonaventura.

47 Italian estate owner and plant collector, although he never published any botanical works, he developed a rich personal herbarium from his own and others collections. Besides from Italy, another large part of this collection is a set of Chinese specimens he obtained from two Franciscan Fathers, Giuseppe Giraldi in Shaanxi and Cipriano Silvestri (1872−1955) in Hubei (Xiangyang), both of whom he had trained in the art of botanical collecting. The pair were missionaries working in China and the plants he received from them, along with his personal examples from Italy, were donated to the Central Herbarium in Florence (FI) between 1903 and 1920. http://plants.jstor.org/stable/10.5555/al.ap.person.bm000375483 (accessed 8 November 2017).

1890—1901 年，意大利传教士 Père Giuseppe Giraldi（1848—1901）受 Antonio Biondi（1849—1929）的委托，在陕西以秦岭的户县为基地，10 年间采集至少 8 000 份植物标本，且侧重隐花植物。他的标本现存于意大利（FI），复份标本遍布世界各地（A、B、H、K、LE、W）。[48]

1891, Père Paul Guillaume Farges (1844-1912), French missionary, traveled to China in 1867, but did not make collections until he received a request from Adrien R. Franchet[49] in 1891 to collect plant specimens in northeastern Sichuan where he was based an area previously unexplored. Over the next nine years Farges sent numerous specimens to Paris, of which many were new to science, more than 100 species were named in his honor. Farges specimens are in France (P, with duplicates at A, B, F, G, L, MO and likely elsewhere).

1891 年，法国传教士 Père Paul Guillaume Farges（1844—1912）于 1867 年进入中国，1891 年受 Adrien R. Franchet 委托前往未被采集的四川北部进行采集，之后的 9 年间向巴黎发送回很多标本，包括 100 多个以他的名字命名的新种。他的标本现存于法国（P，复份标本遍布世界各地 A、B、F、G、L、MO 等）。

1893, Augustine Henry, ***Notes on Economic Botany of China***, 68 p; Presbyterian Mission Press, Shanghai; Reprinted, 1986, with introduction by E. Charles Nelson (1951-); Boethius, Kilkenny.

1893 年，Augustine Henry，**中国经济植物注释**，68 页；长老会出版社，上海；1986 年重印，Boethius，Kilkenny，并附有 E. Charles Nelson（1951—）所写的介绍。

1894, Thomas H. Huxley, *Evolution and Ethics and Other Essays*, 396 p; The MacMillan Co., London. Translated into Chinese in 1897-1898.

1897—1898 年，严复译，天演论，国闻汇编，第 2 期、第 4–6 期；国文报馆，天津。

1894-1920, Père Pierre Julien Cavalerie[50](1869-1927) collected around 8,400 specimens, mostly in Guizhou. Over 225 species of plants were named for Cavalerie. The main portion of his collections is in France (P) with large sets of duplicates in the Netherlands (L), Austria (W) and England (K).

1894—1920 年，法国传教士 Père Pierre Julien Cavalerie（1869—1927）在中国采集 8 400 份标本，主要采自贵州；超过 225 个新种以他的名字命名。他的标本主要存于法国（P），但大量的复份存于

48 崔友文、李培元，1964，P. Giraldi 在陕西采集植物地点的考证，植物分类学报 9（3）：308-312。

49 Jane Kilpatrick, 2014, *Fathers of Botany*. Kew Publishing, Royal Botanic Gardens, Kew and the University of Chicago Press, Chicago; Paul Guillaume Farges. https://es.wikipedia.org/wiki/Paul_Guillaume_Farges (Accessed 29 September 2017).

50 Jane Kilpatrick, 2014, *Fathers of Botany*. Kew Publishing, Royal Botanic Gardens, Kew and the University of Chicago Press, Chicago; Père Pierre Julien Cavalerie, 1908, Note sur les renonculacées du Kouy-Tchéou, *Bulletin de l'Académie internationale de géographie botanique*, n° 219: VII-VIII; 1911, Les aurantiacées du Kouy-Tchéou, *Bulletin de géographie botanique*, n° 261: 210-211; 1911, La flore de Gan-Chouen-Fou (Kouy-Tchéou, Chine), *Bulletin de géographie botanique*, n° 262: 231; 1911, Les liliacées au Kouytchéou, *Bulletin de géographie botanique*, n° 263: 243-248.

荷兰（L）、奥地利（W）和英国（K）。

1895, Yasusada Tashiro (1856-1928), Jinzo Matsumura (1856-1928), Bunzo Hayata[51](1874-1934) and Takiya Kawakami (1871-1915) from Japan, collected plant specimens throughout Taiwan. Most of their specimens were sent to Tokyo Imperial University (now the University of Tokyo, TI) for study, resulting in a series of accounts of the plants of Taiwan.

1895 年，日本人田代安定（1856—1928）、松村任三（1856—1928）、早田文藏（1874—1934）、川上泷弥（1871—1915）等在我国台湾大规模采集，标本大多数送回东京帝国大学（今东京大学，TI）研究，并发表一系列有关台湾植物的论文。

1895-1897, Vladimir L. Komarov (Влади́мир Лео́нтьевич Комаро́в, 1869-1945) collected more than 6,000 specimens representing about 1,300 species in Manchuria [northeast China] and Korea in 1896. His collections are in Russia (LE).[52]

1895—1897 年，Vladimir L. Komarov（1869—1945）在中国东北和朝鲜半岛采集超过 6 000 份植物标本，约 1 300 种，所采标本现存于俄罗斯（LE）。

Vladimir L. Komarov (1869-1945) (Photo provided by Komarov Botanical Institute)

Vladimir L. Komarov（1869—1945）（相片提供者：科马洛夫植物研究所）

1895-1934, Pèrc Joseph Henri Esquirol[53](1870-1934), French missionary, linguist and collector of botanical specimens, primarily in Guizhou. More than 200 species of plants were named for him, most by Augustin Abel Hector Léveillé (1863-1918). Esquirol's specimens are in France (P), with duplicates in Europe (B, G) and the United States (A).

---

51 Hiroyoshi Ohashi, 2009, Bunzo Hayata and his contributions to the flora of Taiwan, *Taiwania* 54(1): 1-27.

52 Vladimir L. Komarov, 1898, The Manchurian Expedition of 1896, *Bulletin of Russian Imperial Geographical Society* 34: 117-184.

53 Jane Kilpatrick, 2014, *Fathers of Botany*. Kew Publishing, Royal Botanic Gardens, Kew and the University of Chicago Press, Chicago.

1895—1934 年，法国传教士、语言学家兼标本采集员 Père Joseph Henri Esquirol（1870—1934）在中国采集大量植物标本，主要采集地在贵州；200 多个新种以他的名字命名，其中多数是 Augustin Abel Hector Léveillé（1863—1918）发表的。他的标本主要存于法国（P），复份标本存于欧洲（B、G）和美国（A）。

1896 年，日本人在台北设立苗圃，1911 年改称林业试验场，1921 年正式称为台北植物园。此为中国最早的植物园。

1896, A nursery founded in Taipei by the Japanese government became the Forest Experiential Farm in 1911, then was officially named the Taipei Botanical Garden in 1921. This is the oldest botanical garden in China.

1896, Augustine Henry, A list of Plants from Formosa, with some preliminary remarks on the Geography, Nature of the Flora and Economic Botany of the Island, *Transactions of the Asiatic Society of Japan*, 24 Supplement: 1-118.[54]

1896 年，Augustine Henry，台湾植物名录，以及该岛的地理、植被性质和经济植物初步观察，*Transactions of the Asiatic Society of Japan*，24 Supplement: 1-118。

1897 年，**农学报**创刊，转载从日本引入的中文植物学术语，并逐渐开始由日文翻译各类有关教科书、参考书以及词典等。[55]

1897, ***Agricultural Papers*** issued and botanical terms in Chinese introduced from Japan were published; various textbooks, references and dictionaries were gradually translated from Japanese into Chinese.

1898, Emil V. Bretschneider, ***History of European Botanical Discoveries in China***, vols. 1 and 2: 1-1,167 p; Press of the Imperial Russian Academy of Sciences, St. Petersburg; Reprinted 1962 and 1981; Zentral-Antiquariat der deutschen demokratischen Republik, Leipzig. Describes botanical contributions of some 500 authors and collectors: itineraries, bibliographies, lists of discoveries; numerous biographical notes; the index contains about 1,000 personal names. This work remains an important document in the botanical history of China.

1898 年，Emil V. Bretschneider，**欧洲人在中国的植物学发现历史**，2 卷本：1、2：1-1,167 页；沙俄帝国科学院出版社，圣彼得堡；1962 年和 1981 年分别由德国重印；包括约 500 位作者和采集者的日程、文献目录、发现记载，以及一些传记注释，人名索引有千余人。该著作为中国植物学历史上的重要文献。

---

54 Tsung-Yu YANG and Chang-Fu HSIEH, 2013, Revision of Dr. Augustine Henry's A list of Plant from Formosa (1896), *Collection and Research* 26: 43-49.

55 朱京伟，2002，中国における日本製植物学用語の受容，明海日本语 第 7 号，71-100。

1899[56], Manabu Miyoshi, '*Botany Textbook*', 701 p, Toyama, Tokyo (三好学著[57], **植物学讲义**[58], 701 页；富山房，東京). Translated into Chinese in 1935.

1935 年，沙俊译，**植物分类**，155 页；商务印书馆，上海。[59]

1899-1905, Père Jean-Théodore Monbeig[60](1875-1914), French missionary made botanical collections in northwestern Yunnan and southeastern Xizang (Tibet). His collections are in France (P), with duplicates elsewhere (A, BM, K).

1899—1905 年，法国传教士 Père Jean-Théodore Monbeig (1875—1914) 在云南西北部和西藏东南部采集大量标本；他的标本今日主要存于法国 (P)，其他欧美国家也有复份标本 (A、BM、K)。

Emil V. Bretschneider (1883-1901) (Photo provided by Hunt Institute for Botanical Documentation)

Emil V. Bretschneider (1883-1901) (相片来源：Hunt Institute for Botanical Documentation)

1899-1918, Ernest H. Wilson[61] (1876-1930), also called Chinese Wilson, made five botanical expeditions in China: two Veitch Expeditions to Hubei and Sichuan in 1899-1902 and 1903-1905, two Arnold Arboretum of Harvard University Expeditions to Hubei and Sichuan, 1907-1909 and 1910-1911 and a third Arnold Arboretum of Harvard University expedition to Taiwan 1918. His collections of over 65,000 plant specimens and more than 5,000 species, are mainly in the United States (A) and England (K).

56 1899 年为首版。翻译版没有标明具体的翻译版本信息。

57 三好学 (1861—1939)，日本当代植物学奠基人。

58 在此引用的页码为 1902 年第 5 版。

59 分别作为万有文库和自然科学丛书中的一本于 1935 和 1936 年出版发行。

60 Jane Kilpatrick, 2014, *Fathers of Botany*. Kew Publishing, Royal Botanic Gardens, Kew and the University of Chicago Press, Chicago.

61 Richard A. Howard, 1980, E. H. Wilson as a botanist, *Arnoldia* 40(4): 54-193; Shiu-Ying HU, 1980, Mapping the collecting localities of E. H. Wilson in China, *Arnoldia* 40(3): 139-145.

1899—1918 年，Ernest H. Wilson (威尔逊[62]，1876—1930)，被称为中国的威尔逊，曾 5 次来华采集，其中两次受英国的园艺公司派遣赴湖北和四川，两次受美国哈佛大学阿诺德树木园派遣赴湖北和四川，第三次代表哈佛大学阿诺德树木园曾到达台湾。采集植物标本超过 65 000 份[63]，约 5 000 种，主要藏于美国（A）和英国（K）。

62 原译名威理森，今作威尔逊。参见：胡启明译，2015，中国 —— 园林之母，305 页；广东科技出版社，广州。

63 参见：王印政、覃海宁、傅德志，2004，中国植物采集简史，中国植物志 第 1 卷：第 685 页；科学出版社，北京。

Ernest H. Wilson (1876-1930) (Photo provided by Arnold Arboretum of Harvard University)

Ernest H. Wilson（1876—1930）（相片提供者：哈佛大学阿诺德树木园）

Ernest H.Wilson, his wife Helen G. Wilson and daughter Muriel Q. Wilson (Photo provided by Arnold Arboretum of Harvard University)

威尔逊与夫人 Helen 和女儿 Muriel（相片提供者：哈佛大学阿诺德树木园）

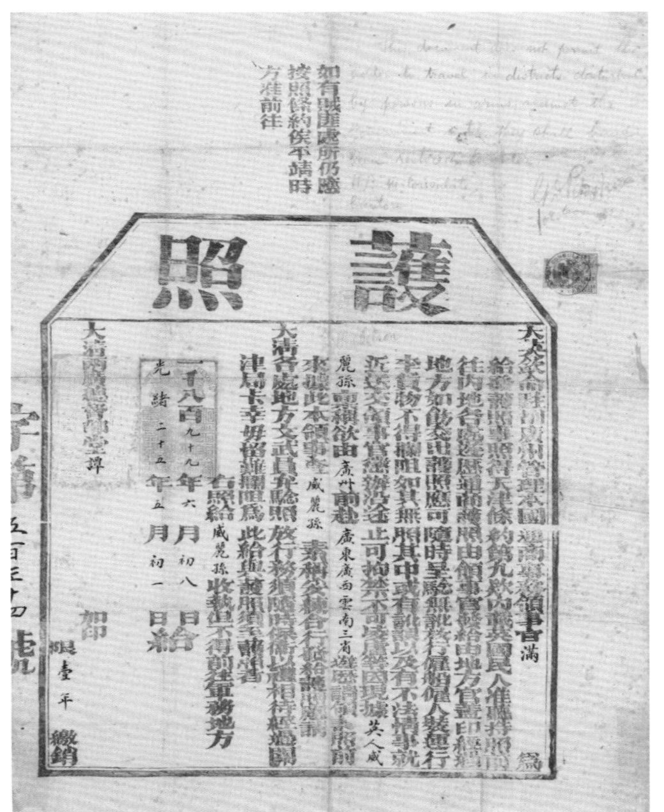

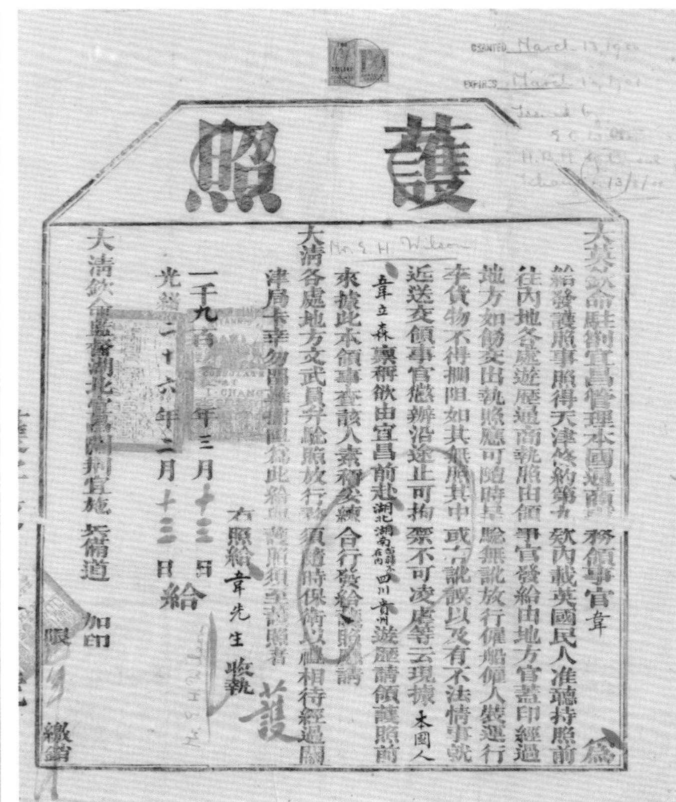

Passports of Ernest H. Wilson (1899, 1900, 1908) (Photo provided by Arnold Arboretum of Harvard University)
Wilson 的护照（1899,1900,1908）（相片提供者：哈佛大学阿诺德树木园）

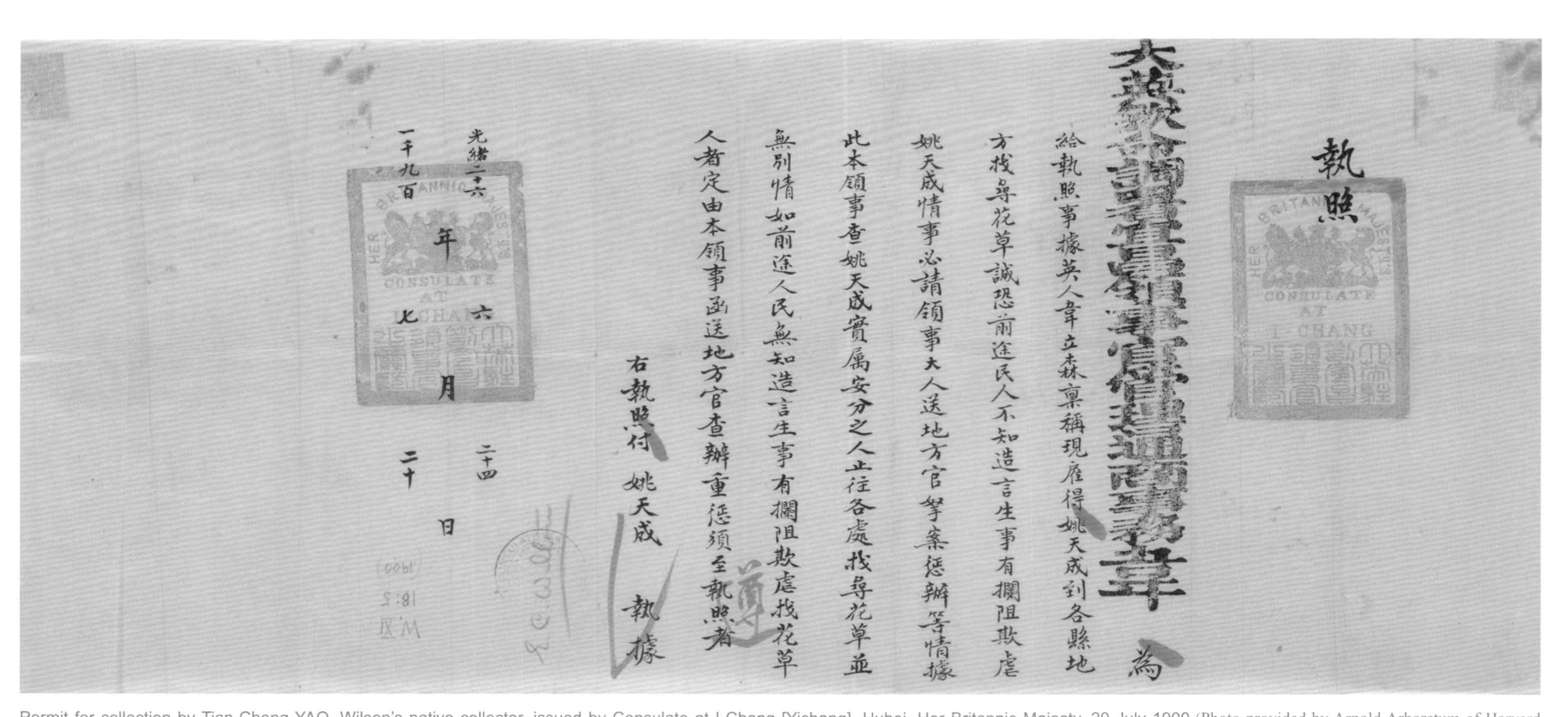

執照

為

給執照事據英人韋立森稟稱現雇得姚天成到各縣地
方找尋花草誠恐前途民人不知造言生事有攔阻欺虐
姚天成情事必請領事大人送地方官拏案懲辦等情據
此本領事查姚天成實屬安分之人止往各處找尋花草並
無別情如前途人民無知造言生事有攔阻欺虐找花草
人者定由本領事函送地方官查辦重懲須至執照者

右執照付姚天成執據

光緒二十六年六月二十四日

一千九百年七月二十日

Permit for collection by Tian-Cheng YAO, Wilson's native collector, issued by Consulate at I-Chang [Yichang], Hubei, Her Britannic Majesty, 20 July 1900 (Photo provided by Arnold Arboretum of Harvard University)

1900 年 7 月 20 日，英国驻鄂宜昌领事官颁发给 Wilson 的中国采集者姚天成之采集许可（相片提供者：哈佛大学阿诺德树木园）

1. 2. 3. Ernest H. Wilson's collecting teams (Photo provided by Arnold Arboretum of Harvard University)

1. 2. 3. Wilson 的采集队（相片提供者：哈佛大学阿诺德树木园）

4. Preparation for one of Ernest H. Wilson's expeditions (Photo provided by Arnold Arboretum of Harvard University)

4. Wilson 考察采集之一的准备（相片提供者：哈佛大学阿诺德树木园）

Arnold Arboretum of Harvard University, 2017 (Photo provided by Jin-Shuang MA)
2017 年的哈佛大学阿诺德树木园（相片提供者：马金双）

1900

1900-1901, Friedrich Ludwig. E. Diels, Die Flora von Central-China, Nach der vorhandenen Litteratur and neu mitgeteilten Original-Materiale, *Botanische Jahrbucher für Systematik, Pflanzengeschichte und Pflanzengeographie* 29: 169-576, 1900 and 577-659, 1901. Translated into Chinese in 1931-1934.

1931—1934年，董爽秋译，中国中部植物，自然科学（国立中山大学理工学院）3（2）：305- 320，3（3）：435-474，1931；3（4）：684-723，4（1）：85-132，4（2）：221-281，1932；4（3）：427-479，4（4）：619-671，5（2）：267-316，1933；5（3）：455-506，6（1）：25-60，6（2）：277-316，1934。

Friedrich Ludwig E. Diels (1874-1945) (Photo provided by Botanischer Garten und Botanisches Museum Berlin-Dahlem)

Friedrich Ludwig E. Diels（1874—1945）（相片提供者：柏林达莱植物园与植物博物馆）

1901

1901 and 1905, John M. Coulter, ***Plant Studies—****An Elementary Botany*, 392 p; revised edition, 392 p; D. Appleton and Company, New York. Translated into Chinese in 1907.

1907 年，奚若、蒋维乔译，**胡尔德氏植物学教科书**，400 页；商务印书馆，上海。

1901-1907, Vladimir L. Komarov, Flora Manshuriae, *Trudy Imp. S.-Peterburgsk. Bot. Sada* 20: 1-559, 1901, 22: 1-452, 1903 and 453-787, 1904; 25: 1-334, 1905 and 335-853, 1907. Translated into Japanese in 1927 to 1933.

1901—1907 年，Vladimir L. Komarov，满洲植物志，*Trudy Imp. S.-Peterburgsk. Bot. Sada* 20：1-559，1901，22：1-452，1903，453-787，1904；25：1-334，1905，335-853，1907。日文版，**满洲植物志**，7 卷本，1927—1933；南满洲铁路株式会社广务部调查课编；每日新闻社，大阪。

1902

1902, William B. Hemsley and Henry H. W. Pearson, The Flora of Tibet or High Asia being a consolidated account of the various Tibetan Botanical Collections in the Herbarium of the Royal Gardens, Kew, together with an exposition of what is known of the Flora of Tibet, *Journal of the Linnean Society Botany* 35: 124-265.

1902 年，William B. Hemsley、Henry H. W. Pearson，西藏植物志，*Journal of the Linnean Society Botany* 35：124-265。

1903

1903年，张珽[64]（1884—1950）赴日本留学；先入宏文书院学习，1906年考入日本东京高等师范大学博物专科学习植物学；1912年毕业于东京高等师范大学，同年回国到安庆，在安徽优级师范学校教书；1914年8月受聘为国立武昌高等师范学校博物学部主任；1923年9月该校改名为国立武昌师范大学，1924年9月又改名国立武昌大学，期间任教授兼生物学系主任；1926年2月至5月曾代理国立武昌大学校长，同年改任国立武昌中山大学教授兼理科委员会主席；1928年任国立武汉大学教授，1929—1949年任国立武汉大学教授兼生物学系主任。1919年曾参加国立武昌大学博物学会，为创建人之一；当时曾创办博物学会杂志，1924年更名为生物学杂志。

1903, Ting CHANG (1884-1950) entered Japan, to study botany in Tokyo Higher Normal University in 1906. He graduated and returned to China in 1912 to become a teacher at Anhui Normal College in Anqing, Anhui, then dean of the department of natural history at National Wuchang Higher Normal College, Wuchang, Hubei, in August 1914. In September 1924, he became the first professor to teach Botany when National Wuchang Higher Normal College was changed to National Wuchang University; then National Wuchang University changed to National Wuhan University in 1928. He was dean of the department of biology, National Wuhan University from 1929-1949. In 1950, he became director of the department of biology, Wuhan University.

64 字镜澄，又名张肇，安徽桐城人；1903年桐城中学毕业。

1904

1904 年 6 月 18 日，黄以仁[65]（1874—1944）在四川峨眉山采集。目前在国内能找到他所采集的最早的一份标本现存于第二军医大学植物标本室[66]（SMMU-BH-002081）[67]。黄以仁最晚在 1911 年之前于江苏（惠山、无锡、常州、虞山、常熟、苏州、南京）、上海和山东（烟台）等地采集，所采标本由日本植物学家研究[68]。这些植物标本现存于日本（TI）。这可能是国人最早的用于植物学研究的采集之一。

1904.06.18, I-Jin WANG (I-Yen WONG, 1874-1944) collected on Emei Shan, Sichuan. The earliest known specimens collected by I-Jin WANG in China are in the botanical herbarium (SMMU-BH-002081) of the Second Military Medical University, Shanghai. Wang also collected at least before 1911 in Jiangsu (Huishan, Wuxi, Changzhou, Yushan, Changshu, Suzhou and Nanjing), Shanghai and Shandong (Yantai). Those collections are in Japan (TI), and were studied by Japanese botanists[69]. They may be one of the earliest specimens collected by a Chinese botanist and used in botanical research.

1904 年，日本人在台湾成立第一个植物标本室，1924 年改名为中央研究所林业部腊叶馆；1945 年台湾光复改为台湾省林业试验所植物标本室（TAIF）；现有 40 多万藏量，为今日台湾最早并且最大的植物标本馆。[70]

---

65 江苏无锡人；字子彦，1902 年曾跟随钟观光学习；参见：洪余庆，1985，我国著名植物学前辈钟观光传略，镇海文史资料 1：53-60。1906—1909 年在日本东京大学学习植物学；参见：小倉谦编，1940，东京帝国大学理学部植物学教研室沿革，342 页；东京帝国大学理学部植物学教研室，东京。黄以仁归国后（具体归国时间不详）先后任北京大学农学院和河南大学农学院教授；1944 年抗战时受日军所逼，年已古稀时与河南大学其他师生被迫流浪，病逝于河南大学临时校址荆紫关。

66 孟世勇、刘慧圆、余梦婷、刘全儒、马金双，2018，中国植物采集先行者钟观光的采集考证，生物多样性 26（1）：79-88。

67 黄以仁 1933 年在平汉沿线李家寨（河南）采集的标本 20 余份，存于中国科学院植物研究所（PE；http://www.cvh.ac.cn/search/%E9%BB%84%E4%BB%A5%E4%BB%81?page=1&searchtype=1&n=3，2017 年 10 月 22 日查阅）。

68 松田定久，1911，黄以仁氏惠山虞山无锡常熟采集，植物学杂志 25(299)：237-250；黄以仁氏惠山虞山等地采集品二，植物学杂志 25(298): 440-441；许（虔孙）张二氏采集浙江省植物，植物学杂志 25(294)：281-284；1914，黄以仁采集的中国有颖的植物，植物学杂志 28(334，335)：296-300，316-322。

69 Sadahisa Matsuda, 1911, A list of plants collected by Wang-i-jin in the Wai-shan, Yu-shuan, Mou-sek, Shong-Shuk and other places, *Botanical Magazine Tokyo* 25(299): 237-250; On the specimens collected by Wang I Jen in Hwei Shan and Yu Shan, Kiangsu, *Botanical Magazine Tokyo* 25(298): 440-441; List of some plants collected by Hsu and Chang in Chekiang, *Botanical Magazine Tokyo* 25(294): 281-284; 1914, A list of some Chinese glumaceous plants collected by Hwang-yi-jen, *Botanical Magazine Tokyo* 28(334, 335): 296-300, 316-322.

70 陈建帆、张艺翰、陈建文，2014，林业试验所植物标本馆的历史、现状与展望，林业研究专讯 21（3）：4-10。

The earliest known specimen collected by a Chinese botanist (I-Jin WANG, 18 June 1904, Emei Shan, Sichuan, SMMU)
(Photo provided by Min JIA)
目前已知中国植物学家采集的最早标本（黄以仁，1904 年 6 月 18 日，四川峨眉山，SMMU）（相片提供者：贾敏）

1904, The first herbarium in Taiwan was founded by the Japanese government. In 1924, it became the herbarium of the Research Institute of Central Forestry, then in 1945, the herbarium of Taiwan Forestry Research Institute (TAIF). It houses more than 400,000 specimens, making it not only the earliest but also the largest herbarium in Taiwan today.

1904 and 1925, Alfred B. Rendle, ***The Classification of Flowering Plants***, vols. 1: 403 p, 1904, Gymnosperms and monocotyledons, 2: 644 p, 1925, Dicotyledons; 2nd edition in 1953 and 1952; University Press, Cambridge. Translated into Chinese in 1958 and 1965.

1958、1965年，钟补求（第1册），钟补求、杨永执合译（第2册），**有花植物分类学**，第1册，裸子植物、单子叶植物，377页；1958，第2册，双子叶植物，581页，1965；科学出版社，北京。[71]

1904-1932, George Forrest[72] (1873-1932), from Scotland, made seven plant collecting trips to China (1904-1906, 1910, 1912-1914, 1917-1919, 1921-1923, 1924-1925 and 1930-1932). He explored in Yunnan over a period of 28 years, for both private gardens and for the Royal Botanic Garden, Edinburgh, especially concentrating on the alpine plants, including *Rhododendron* and *Primula*. His collection numbers exceed 31,000, the main set of which is in

George Forrest and his native collector (Photo provided by Royal Botanic Garden, Edinburgh)

Goerge Forrest 和他的当地采集者（相片提供者：爱丁堡植物园）

71 译自第2版；但该书第2版实际上是第1版的修正（未改版）。

72 Forrest was born in Falkirk, Scotland on 13 March 1873 and died in Tengyueh, Yunnan, China, January 5, 1932. Forrest entered the Herbarium of the Royal Botanic Garden, Edinburgh in 1902. A year later, he was recommended to Liverpool horticulturist and cotton broker Arthur K. Bulley, who was sponsoring an expedition to southwestern China in search of exotic plants, particularly species of rhododendrons. Forrest made his first expedition to Yunnan in 1904, and until his last trip in 1930-1932, total he made seven trips to Yunnan in collecting samples and seeds. In total, he brought back about 31,000 plant specimens.

Scotland (E)[73]; duplicates were widely distributed, even to India (CAL).[74]

1904—1932 年，苏格兰 George Forrest（傅礼士，1873—1932）先后 7 次（1904—1906 年、1910 年、1912—1914 年、1917—1919 年、1921—1923 年、1924—1925 年、1930—1932 年），前后达 28 年，在中国云南为私家园林和爱丁堡植物园等采集，采集时特别注重高山植物，如杜鹃花属和报春花属。他的植物标本多达 31 000 余号[75]，完整地存放在苏格兰（E），但复份标本散落在世界各地，甚至包括印度（CAL）。

73 Royal Botanic Garden Edinburgh, 1911-1934, Plantae Chinenses Forrestianae, Plants discovered and collected by George Forrest during his first exploration of Yunnan and eastern Tibet in the years 1904, 1905, 1906; *Notes from the Royal Botanic Garden, Edinburgh* 5: 65-148, 1911-1912; 8: 1-45, 1913; Catalogue of the plants (excluding Rhododendron) collected by George Forrest during his fifth exploration of Yunnan and Eastern Tibet in the years 1921-1922, 14: 75-393 (nos. 19334-23258), 1924; Catalogue of all the plants collected by George Forrest during his fourth exploration of Yunnan and Eastern Tibet in the years 1917-1919, 17: 1-406 (nos 13599-19333 and corrections), 1929-1930; 18: 119-158 and 275-276 (Conifers), 1934; and Friedrich Ludwig E. Diels, New and Imperfectly known species, *Notes from the Royal Botanic Garden, Edinburgh* 5: 161-308, 1912; Catalogue of all the plants collected by George Forrest during his first exploration of Yunnan and eastern Tibet in the years 1904, 1905, 1906, *Notes from the Royal Botanic Garden, Edinburgh* 7: 1-298 (nos. 1-5099), 299-333 (natural order), 321-411 (alphabetized), 1912-1913.

74 Kandasamy Ramamurthy and Uma P. Samamadar, 1985, *Types of Taxa based on Forrest's Collections at CAL, Type Collections in the Central National Herbarium*, 26 p; Botanical Survey of India, Howrah.

75 参见：王印政、覃海宁、傅德志，2004，中国植物采集简史，中国植物志 第 1 卷：第 686 页；科学出版社，北京。

1905

1905.06.11-18, Second International Botanical Congress held in Vienna. The Great Qing Government sent as China's representative the secretary (Tscheng-Fan-Tsching) of the Qing Empire Legation in Vienna.[76]

1905 年 6 月 11 日至 18 日，第二届国际植物学大会在维也纳举行，大清帝国派驻维也纳公使秘书

272 NATURE [JULY 20, 1905

tumuli which in all probability represent more recent additions to the original scheme of observation, as we have found at Stenness; and show that Trowlesworthy was for long one of the chief centres of worship on Dartmoor. Their azimuths are S. 64° E. and S. 49° W., dealing, therefore, with the May year sunrises in November and February and the solstitial sunset in December. It is probable that, as at the Hurlers, tumuli were used instead of stones not earlier than 1900 B.C.

Stalldon Moor (lat. 50° 27′ 45″) I have already incidentally referred to. The azimuth of the stone row as it leaves the circle, *not* from its centre as I read the 6-inch map, is N. 3° E.; as the azimuth gradually increases for a time, we may be dealing with Arcturus, but local observation is necessary.

The differences between the Cornish and Dartmoor monuments give much food for thought, and it is to be hoped that they will be carefully studied by future students of orientation, as so many questions are suggested. I will refer to some of them.

(1) Are the avenues, chiefly consisting of two rows of stones, a reflection of the sphinx avenues of Egypt? and, if so, how can the intensification of them on Dartmoor be explained?

nothing more recondite than an inspection of a precessional globe to have been precisely the stars, the "morning stars," wanted by the priest-astronomers who wished to be prepared for the instant of sunrise at the critical points of the May or solstitial year.

NORMAN LOCKYER.

THE BOTANICAL CONGRESS AT VIENNA.

THE International Botanical Congress, held at Vienna on June 11–18, was an impressive demonstration of the activity of botany as a science, and of the enthusiasm of its adherents. Vienna is not the most central town for a meeting-place, but, nevertheless, more than six hundred botanists, men and women, representing nearly all the important, and many of the less important, botanical institutions of the world, met together there. As might have been expected, the central European element predominated, but there were a goodly number of Americans representing the southern and far western as well as the eastern States, while from the Far East came a deputation of two Chinese.

On the first day of the Congress, members were invited to be present at the opening of the Botanical

274 NATURE [JULY 20, 1905

congress is the meeting together and getting to know one's fellow-workers; and an expression of thanks is due to the organising committee under the joint presidentship of Profs. Wiesner and von Wettstein, with Dr. Zahlbruckner as the energetic secretary, to Prof. Flahault, the firm and genial president of the conference on nomenclature, and finally to Dr. Briquet, whose name must always be associated with the latest attempt to solve the vexed question of plant-nomenclature.

At the final meeting, in response to an invitation from the Belgian Government voiced by Prof. Errera, Brussels was selected as the place of meeting for the third congress, which will be held in 1910.

A. B. RENDLE.

NOTES.

AN important step in the direction of the adoption by this country of a decimal system of weights and measures has been taken by the Board of Trade. In reply to a resolution sent to the Board of Trade by the secretary of the Association of Chambers of Commerce, in which the Board was asked to authorise weights of 20 lb., 10 lb., and 5 lb. as aliquot parts of the cental, Lord Salisbury has written :—" With reference to your letter of March 14 last, in which you suggest that new denominations of weights of 20 lb., 10 lb., and 5 lb. should be legalised for use in trade, the Board of Trade have given careful consideration to the representations which have been made, and they are prepared to assent to the application. Steps will, therefore, be taken for the preparation of

Report by Alfred B. Rendle (1865–1938), 20 July 1905, of two (unnamed) Chinese delegates who attended the International Botanical Congress in Vienna in 1905 (The Botanical Congress in Vienna, *Nature* vol.72, no, 1864, 272–274)

1905 年 7 月 20 日，Alfred B. Rendle (1865—1938) 报道两位中国人出席 1905 年维也纳国际植物学大会（维也纳植物学大会，*Nature* 72 卷，1864 期，第 272–274 页）

76 Richard von Wettstein, Julius von Wiesner and Alexander Zahlbruckner, 1906, Verhandlungen des Internationalen Botanischen Kongresses in Wien 1905, 260 p; Verlag von Gustav Fischer, Jena.

（程范清[77]）与会[78]。

1905-1907, Peter K. Kozlov, ***Transactions of the Expedition of the Imperial Russian Geographical Society made in 1899-1901 under direction of Peter K. Kozlov***, 1: 1-256 p, 1905, 2: 257-734 p, 1907; *Peter K. Kozlov*, 1947, ***Mongolia and Kam Three years' travels in Mongolia and Tibet (1899-1901),*** 1-437; Gos. Izd-vo geogr. lit-ry, Moskva.[79]

1905—1907 年，Peter K. Kozlov, **1899—1901 年俄罗斯帝国地理学会 Peter K. Kozlov 领导的中国、蒙古和康区考察**，第 1 卷：1-256 页，1905；第 2 卷：257-734 页，1907；Peter K. Kozlov，1947，**蒙古和康区 —— 在蒙古和西藏的 3 年旅行（1899—1901 年）**，1-437 页；Gos. Izd-vo geogr. lit-ry，Moskva。

1905-1918, Frank Meyer[80] (1875-1918) was sent to China by the United States Department of Agriculture (USDA) and by the Arnold Arboretum of Harvard University to collect economic plants and to introduce Chinese trees and shrubs to the United States. His four lengthy expeditions (1905-1908, 1909-1911, 1912-1915, 1916-1918) across China, mainly in the north, northeast and northwest, resulted in the introduction of approximately 2,500 plants to the United States. His herbarium specimens are mainly in the United States (A, MO, NA).

1905—1918 年，美国农业部和哈佛大学阿诺德树木园联合资助 Frank Meyer（1875—1918）赴中国采集经济植物并引种乔灌木，前后共 4 次（1905—1908 年、1909—1911 年、1912—1915 年、1916—1918 年）横跨中国，特别是"三北"地区，引入约 2 500 种有用植物。他的植物标本主要存于美国（A、MO、NA）。

---

77 或曾范清（目前尚未获得准确的中文表述）。

78 Alfred Barton Rendle (1865-1938) was an English botanist, FRS, and Keeper of Botany at the Natural History Museum from 1906 to 1930. In 1905, Rendle attended the International Botanical Congress in Vienna, where he was appointed on to the editorial committee for the International Rules of Botanical Nomenclature (now superseded by the International Code of Nomenclature for algae, fungi, and plants) a role which he continued in until 1935. Rendle published a number of books. Perhaps the best known of these was The Classification of Flowering Plants, which saw a gap of over 20 years between the publication of its two volumes-the first was published in 1904, but readers had to wait until 1925 for volume two. He was also botany editor for the *Encyclopædia* Britannica Eleventh Edition, published in 1911. Rendle was president of the Quekett Microscopical Club from 1916 to 1921 and president of the Linnean Society from 1923 to 1927.

79 Peter K. Kozlov, 1951, *Journeys through China and Mongolia*, 1-284; Gos. Izd-vo geogr. lit-ry, Moskva; 1897, Russian Expedition in Tibet, *The Geographical Journal* 9(5): 546-555 p, 1899-1901, Vsevolod I. Roborovski and Peter K. Kozlov (also as Peter K. Kozloff), *Works of the Expedition of the Imperial Russian Geographical Society to Central Asia carried out in 1893-1895 under command of Vsevolod I. Roborovski*, 1(1-2): 1-388 p, 1900, 1(3): 1-389-610 p, 1901, 2(1-2): 1-296 p, 1899, 3(1-6): 1-19 p, 1-55 p and 1-43 p, 1899.

80 As Fans Nicholaas Meijer in origin, after 1908 as Fank Nicholas Meyer (from TL2 3: 402). For details see: David Fairchild, 1919, A Hunter of Plants, *The National Geographic Magazine* 36: 57-77.

1906

1906年，陈嵘[81]（1888—1971）赴日本留学，1909—1913年在北海道帝国大学林科学习，回国后于1913年创办并任浙江省甲种农业学校校长，1915—1922年任江苏省立第一农业学校林科主任；1916年发起组织中华农学会，1917年支持发起林学会；1923—1924年赴美国哈佛大学学习树木学，获硕士学位，1925年赴德国撒克逊林学院进修，同年回国任金陵大学森林系教授、主任，直至1952年院系调整，转赴北京任中央林业部林业科学研究所所长。[82]

Yung CHEN (1888-1971) (Photo provided by Zong-Gang HU)
陈嵘（1888—1971）（相片提供者：胡宗刚）

1906, Yung CHEN (1888-1971) went to Japan, and studied forestry in Hokkaido Imperial University between 1909 and 1913. He returned to China in 1913 to establish Zhejiang Forest Agriculture School. He served as principal, then dean of forestry, in the First Agriculture School of Jiangsu between 1915 and 1922; sponsored the Agricultural Association of China in 1916 and advocated for the Chinese Forestry Society in 1917. In 1923, he traveled to the United States to study dendrology at Harvard University, where he received a masters's degree in 1924. He then went to Germany in 1925 to study forestry at Saxon Forestry College and returned to China in the same year to serve as professor at the University of Nanking [Nanjing] until the reorganization of colleges and universities in China in 1952. Soon after that he was appointed director of the Institute of Forestry, China Central Forestry Administration, in Beijing.

1906, Jinzo Matsumura and Bunzo Hayata, Enumeratio Plantarum in Insula Formosa sponte cerescentium hucusque rite cognitarum adjectis descriptionibus et figuris specierum pro regione novarum,

81 原名正嵘，字宗一，浙江安吉人；1905年至平阳县高等学堂学习。

82 陈嵘，1937、1953，中国树木分类学，1 086页，1937；中国农学会，南京；修订版，1 191页，1953；中国图书，南京；补订版，1 191页，1957；科学技术出版社，上海。

Journal of the College of Science, Imperial University of Tokyo 22: 1-702.

1906 年，松村任三、早田文藏，台湾植物名录，*Journal of the College of Science*，*Imperial University of Tokyo* 22: 1-702。

1906 年，钟观光[83]（1868—1940）和 7 位老师率领浙江宁波师范学堂学生 70 余人到附近的天童寺

Kuan-Kuang TSOONG (front row, first from right) with the sponsoring members of Shanghai Scientific Instruments Company (Photo provided by Ren-Jian ZHONG/TSOONG, the grandson of Kuan-Kuang TSOONG, and his wife, Jin-Zheng CHEN)
钟观光（前右一）与上海科学仪器馆发起成员合影（相片提供者：钟观光之孙钟任建与妻陈锦正）

83 字宪鬯，浙江镇海人，1899 年创办“四明实学会”，1901 年创办上海科学仪器馆，1902—1903 年任教于江阴南菁书院（后改为江苏高等学堂），1904 年创办上海科学仪器馆理科传习所并讲授理化博物知识，1904—1908 年任教于宁波师范学校，1908 年任教于上海理科传习所，1908—1909 年在杭州西湖疗养，1909—1911 年任教于宁波旅沪公学自然博物室，1911 年创设实学通艺馆，1912—1915 年任南京临时政府教育部参事，1916—1917 年任湖南高等师范学校博物教授，1918—1926 年任北京大学副教授，1927—1930 年任国立浙江大学农学院教授，兼浙江省博物馆自然部主任，1931—1932 年于南京任中央研究院自然历史博物馆植物编辑员，1933—1940 年任北平研究院植物学研究所专任研究员，1940 年 9 月 30 日于故乡病逝。

太白山采集[84]。此为目前记载的最早的教师带领学生进行的野外实习。

1906, Kuan-Kuang TSOONG (1868–1940) and seven teachers lead more than 70 students from Ningbo Teachers College to Taibai Shan, Tiantongsi, Ningbo, Zhejiang. This is the first known field excursion led by teachers with students in China.

84 梅季　点辑，1984，八指头陀诗文集，第 489–490 页；光绪丙午闰四月望前一日，宁波师范学堂教务长兼理科教员钟君宪鬯、庶务长冯君友笙、监学员张君申之、东文兼图画教员顾君麟如、体操教员应君惠吉、算学教员叶君德之、育德学堂监督陈君屺怀、体操教员林君莲村，偕学生 70 余人，入太白山（即天童寺，位于浙江省宁波市东 25 千米的太白山麓）采集植物。敬安率监院僧候山门，则见龙旗飘飏于青松翠竹之间。龙骧虎步，整队而来，若临大敌，因之欢喜赞叹，得未曾有。虽禅悦法喜，无此乐也。岳麓书社，长沙。释敬安撰，梅季　点校，2007，八指头陀诗文集，402–403 页；岳麓书社，长沙。

1908

1908, Bunzo Hayata, Flora Montana Formosae, an enumeration of the plants found on Mt. Morrison, the Central Chain and other mountainous regions of Formosa at altitudes of 3,000–13,000 ft., *Journal of the College of Science, Imperial University of Tokyo* 25: 1–260.

1908 年，早田文藏，台湾山地植物志[85]，*Journal of the College of Science*，*Imperial University of Tokyo* 25: 1–260。

85 又作：台湾高地带植物志。参见：大场秀章，2017，早田文藏，台湾林业试验所，台北。

1909

1909 年，陈焕镛[86]（1890—1971）入麻州[87]农学院学习森林学和昆虫学，1912 年转入纽约州立大学林学院，1915 年毕业获学士学位，并进入哈佛大学学习树木学，1919 年获硕士学位并返国；1929 年于国立中山大学农学院创建农林植物研究所并设立植物标本室（IBSC），任主任，1930 年创办 *Sunyatsenia*；1935 年与广西大学在广西梧州合办广西大学植物研究所，兼任所长（1935—1938），1935 年当选中央研究院评议会评议员；1954 年中山大学农林植物研究所更名为中国科学院华南植物研究所，仍任所长（1954—1971），并兼广西分所所长（1954—1958、1958—1961）；1955 年当选中国科学院首届学部委员；1959 年中国植物志编辑委员会成立，任第一届编辑委员会，并与钱崇澍（1883—1965）共同担任主编（1959—1971）。

1909, Woon-Young CHUN (1890-1971) enrolled in Massachusetts Agriculture College, Amherst, to study forestry and entomology, then transferred to the Forestry Department, New York State University at Syracuse, in 1912. He graduated with a bachelor's degree in 1915, then enrolled as a graduate student in the Bussey Institution, Harvard University. He graduated with a masters's degree in 1919 and returned to China. He established the Botanical Institute and the herbarium (IBSC) in the Agriculture College of National Sun Yat-Sen University, Guangzhou, Guangdong and became its director in 1929. As director, he initiated the journal *Sunyatsenia*[88] in 1930. He served as director (concurrently, 1935-1938) of the Guangxi Institute of Botany, Guangxi University, in Wuzhou, Guangxi, co-founded by the Botanical Institute of Sun Yat-Sen University and Guangxi University in 1935. In the same year, 1935, he was appointed to the council of Academia Sinica. In 1954, the name of the Botanical Institute, Sun Yat-Sen University, was changed to the South China Institute of Botany, Chinese Academy of Sciences. He served there as director (1954-1971) and also as the director (concurrently, 1954-1958, 1958-1961) of the Guangxi Branch. In 1955, he was elected an academician of the Chinese Academy of Sciences. He and Sung-Shu CHIEN (1883-1965) were co-editors in chief of *Flora Reipublicae Popularis Sinicae* (1959-1971).

1909 年，张之铭[89]（1872—1937）自浙江宁波采集标本，交由日本人研究，植物标本存于日本（TI）[90]。这可能是国人采集的较早用于植物学研究的记录之一。

---

86 字文农，号韶钟；祖籍广东，生于香港；1905 年赴美国华盛顿州西雅图读中学，1909 年毕业后才转入美国东部的大学。

87 马萨诸塞州的简称为麻州或麻省。

88 Wan-Yi ZHAO, Ke-Wang XU, Qiang FAN and Wen-Bo LIAO, 2016, Contributions to the botanical journal *Sunyatsenia* from 1930 to 1948, *Phytotaxa* 269(4): 237-270.

89 号伯岸，晚号豚翁，浙江鄞县（今宁波）人；民国藏书家；从小经商，然嗜书如命；少时在上海与学校诸友建立实学通艺馆，专储藏仪器以待求者；曾在日本东京横滨侨居多年。

90 松田定久，1909、1913、1914，张之铭氏寄赠之植物标品，植物学杂志 23(267)：150-154，27(321)：187-192，27(322)：205-212，27(323)：234-242，27(324)：271-276，28(325)：5-19。

1909, Chih-Ming CHANG (1872–1937) collected in Ning-po [Ningbo], Cheh-kiang [Zhejiang]. His collections were studied by Japanese botanists and are in Japan (TI)[91]. This may be one of the earlier examples of Chinese plants collected by a Chinese person and used in research.

91 Sadahisa Matsuda, 1909, 1913, 1914, A list of plants collected from Ning-po, Cheh-kiang, *Botanical Magazine, Tokyo* 23(267): 150–154; 27(321): 187–192, 27(322): 205–212, 27(323): 234–242, 27(324): 271–276, 28(325): 5–19.

1910

1910年9月，钱崇澍[92]（1883—1965）赴美留学，1914年获美国伊利诺伊大学学士学位，后入芝加哥大学、哈佛大学学习，1916年回国，先后受聘于江苏甲种农业学校、南京金陵大学、国立东南大学、国立清华大学、北京大学、厦门大学、国立四川大学和复旦大学，并长期在中国科学社生物研究所任教授兼植物部主任，参加编写我国第一本大学教科书（高等植物学，1923）；1916年发表的宾夕法尼亚毛茛的两个亚洲近缘种（*Rhodora* 18：189-190）是中国人用拉丁文为植物命名和分类的第一篇文献；1948年当选为中央研究院院士；1949年被任命为中国科学院植物分类研究所研究员兼第一任所长（1950—1965）；1955年当选为中国科学院首批学部委员；1959—1971年与陈焕镛共同担任中国植物志编辑委员会第一届主编。

Sung-Shu CHIEN (1883-1965) (Photo provided by Zong-Gang HU)
钱崇澍（1883—1965）（相片提供者：胡宗刚）

1910.09, Sung-Shu CHIEN (1883-1965) studied at University of Illinois in the United States, received Bachelor Degree in 1914, then to the University of Chicago and Harvard University. He returned to China in 1916. He became a teacher, then professor in the First Agricultural School of Jiangsu, University of Nanking [Nanjing], National Southeast University, National Tsinghua University, Peking University, Amoy University, National Sichuan University and Fudan University. He was a professor and dean of botany in the Biological Laboratory, Science Society of China for many years and co-author of the first Chinese textbook of advanced botany (1923). His *Two Asiatic allies of Ranunculus pensylvanicus* (*Rhodora* 18: 189-190, 1916) is the first paper describing a new species in Latin and the first modern taxonomic paper written by a Chinese botanist. He was elected an academician of Academia Sinica in 1948. He was appointed research professor[93] and the first director of the Institute of Plant Taxonomy (Institute of Botany from 1953), Chinese Academy of Sciences (1950-1965). He was elected an

92 浙江海宁人，号雨农，1905年入南洋公学，1909年毕业，并转入唐山路矿学堂（即唐山交通大学，今西南交通大学）。

93 i.e. Research Scientist at professor level, similar to curator in the West.

academician of the Chinese Academy of Sciences in 1955. From 1959 to 1971, he was the Co-Editor in Chief (with Woon-Young CHUN) of *Flora Reipublicae Popularis Sinicae*.

1910 年，江苏南京的金陵大学相继设立农学科、林学科以及生物学科。

1910, Three plant related departments, Agriculture, Forestry and Biology, were established in the University of Nanking, Nanjing, Jiangsu.

1910 年，川上泷弥，**台湾植物名录**，165 页、119 页；台湾总督府民政部殖产局，台北。

1910, Takiya Kawakami, ***A List of Plants of Formosa***, 165 + 119 p; Bureau of Productive Industry, Government of Formosa, Taihoku [Taibei].

1911

1911, Bunzo Hayata, Materials for a Flora of Formosa, Supplementary notes to the Enumeratio Plantarum Formosanarum and Flora Montana Formosae, based on a study of the collections of the Botanical Survey of the Government of Formosa, principally in the herbarium of the Royal Botanic Gardens, Kew, *Journal of the College of Science, Imperial University of Tokyo* 30: 1-471.

1911 年，早田文藏，台湾植物志资料，*Journal of the College of Science*，*Imperial University of Tokyo* 30: 1-471。

1911 年，金陵大学植物标本馆（今南京大学植物标本馆，N）于南京成立[94]。

1911, The herbarium of the University of Nanking (now herbarium of Nanjing University, N) was established in Nanking [Nanjing].

1911, Frank N. Meyer, ***Chinese Plant Names***, 40 p; Division of Foreign Plant Introduction, Bureau of Plant Industry, USDA, Washington, D. C..

1911 年，Frank N. Meyer，**中国植物名称**，40 页；Division of Foreign Plant Introduction，Bureau of Plant Industry，美国农业部，华盛顿特区。

1911, Frank N. Meyer, Agricultural explorations in the fruit and nut orchards of China, *USDA Bureau of Plant Industry Bulletin* 204: 1-62.

1911 年 , Frank N. Meyer, 中国的果树和果品之农业考察，*USDA Bureau of Plant Industry Bulletin* 204: 1-62。

1911-1917, Charles S. Sargent[95], ***Plantae Wilsonianae***—An Enumeration of the Woody Plants Collected in Western China for the Arnold Arboretum during the years 1907, 1908 and 1910 by E. H. Wilson, 1(1): 1-144 p, 1911, 1(2): 145-312 p, 1912, 1(3): 313-611 p, 1913; 2(1): 1-262 p, 1914, 2(2): 263-422 p, 1915, 2(3): 423-661 p, 1916; 3(1): 1-188 p, 1916, 3(2): 189-419 p, 1916, 3(3): 421-666 p, 1917.

1911—1917 年，Charles S. Sargent, **威尔逊采集植物志**——1907、1908 和 1910 年威尔逊在华西

94 焦启源，1936，金陵大学植物标本室之概况，科学 11（2）：139-243。

95 Charles S. Sargent (1841-1927) was an American botanist. He was appointed as the first director of the Arnold Arboretum of Harvard University in Boston, MA, in 1872. Sargent held the post until his death in 1927. He traveled China in 1903 via Siberia, visiting Manchuria, Peiping, Shanghai, and Hong Kong from August through October. For more details, see Alfred Rehder, 1927, Charles Sprague Sargent, *Journal of Arnold Arboretum* 8(2): 69-86.

为阿诺德树木园采集的中国木本植物名录，1（1）：1–144 页，1911，1（2）：145–312 页，1912，1（3）：313–611 页，1913；2（1）：1–262 页，1914，2（2）：263–422 页，1915，2（3）：423–661 页，1916；3（1）：1–188 页，1916，3（2）：189–419 页，1916，3（3）：421–666 页，1917。

1911-1921, Bunzo Hayata, ***Icones Plantarum Formosanarum*** nec non et contributiones ad floram Formosanam, or Icones of the plants of Formosa and materials for a flora of the island, based on a study of the collection of the botanical survey of the Government of Formosa, vols. 1-10[96], 1911-1921; Bureau of Productive Industry, Government of Formosa, Taihoku [Taibei].

1911—1921 年，早田文藏，**台湾植物图谱**——台湾植物志资料，10 卷本；台湾总督府民政部殖产局，台北。

1911-1935, Francis Kingdon-Ward[97](1885-1958) collected in the border area of China-Burma [Myanmar] and in the Himalaya more than ten times, including many times in Yunnan and Tibet (and Sichuan) (1911, 1913, 1921, 1922-1923, 1924-1926, 1926-1928, 1930, 1932-1933 and 1935). His plant collections are in England (BM, K) with duplicates widely distributed.

1911—1935 年，Francis Kingdon-Ward（1885—1958）先后十多次到中缅边境及喜马拉雅地区采集，其中数次（1911、1913、1921、1922—1923、1924—1926、1926—1928、1930、1932—1933 和 1935）到达中国云南和西藏（以及四川）等地。他的植物标本主要存于英国（BM, K），复份散落在世界各地。

Francis Kingdon Ward (1885-1958) (Photo provided by Royal Botanic Gardens, Kew)
Francis Kingdon Ward（1885—1958）（相片提供者：邱园）

96 Bunzo Hayata, 1917, *Supplement to Icones Plantarum Formosanarum VI*, General Index to the Flora of Formosa, as recorded in all literature up to the publication of *Icones Plantarum Formosanarum VI*, 155 p; Tokyo. 台湾植物总目录，1917 年（大正六年）；台湾总督府民政部殖产局；晃文馆，台北。

97 Born as Frank Kingdon Ward, also used in some occasions.

1911—1944 年，台湾博物学会，**台湾博物学会汇报**，第 1–34 卷。

1911-1944, Natural History Society of Taiwan, ***Journal of the Natural History Society of Taiwan***, vols. 1-3, 1911-1913; ***Transactions of the Natural History Society of Formosa***, vols. 4-32, 1914-1942; ***Transactions of the Natural History Society of Taiwan***, vols. 33-34, 1943-1944.

1912

1912 年，京师优级师范学堂改称国立北京高等师范学校，并设立博物部；1912—1916 年彭世芳（1886—1940）[98] 任博物馆主任。

1912, Peiping Senior Teaching College changed to Peiping High Teaching College, with Department of Natural History, and Shih-Fang PENG (1886-1940) as director from 1912 to 1916.

1912, Stephan T. Dunn and William J. Tutcher, Flora of Kwangtung [Guangdong] and Hongkong (China): being an account of the flowering plants, ferns and fern allies together with keys for their determination preceded by a map and introduction, *Bulletin of Miscellaneous Information, Royal Botanic Gardens, Kew, Additional Series* 10: 1-370.

1912 年，Stephan T. Dunn、William J. Tutcher，（中国）广东和香港植物志，*Bulletin of Miscellaneous Information*，*Royal Botanic Gardens*，*Kew, Additional Series* 10：1-370。

1912 年，祁天赐（1876—1937）在东吴大学创建我国第一个生物系，成立了全国唯一的生物材料处。该处制作生物制品供应全国教学，如标本等。尽管祁天赐的专长是动物，但生物材料处同时也供应相关的植物学材料。

1912, Nathaniel G. Gee[99](1876-1937) established the first department of biology in China at Soochow [Suzhou] University, and a department of biological materials, the only one in China. The department of biological materials, as the name implies, provided biological materials, such as specimens for teaching, nationwide. Even through his specialty was mainly zoology, the department of biological materials also provided plant material[100].

1912, William R. Price[101](1886-1975) collected botanical specimens in Taiwan, initially with Henry J. Elwes[102](1846-1922). At least 1,133 of his herbarium specimens are in England (K), Taiwan, China (TAI) and

98 号型伯，江苏吴县（今苏州）人，早年留学日本东京高等师范学校，曾任北京高等师范学校博物部首任教务主任，北京师范大学、北京女子高师植物学教授，编写了大量生物学教科书及专著。

99 Nathaniel G. Gee arrived in China in 1901 as a professor of natural sciences in Soochow University. From 1920 to 1932 he was an advisor on premedical education in China for the Rockefeller Foundation, then a fundraiser in the United States for Yenching University in Peking [Beijing] for 3 years before returning to the United States as professor of biology at Lander College in Greenwood, South Carolina, his home state.

100 William J. Haas, 1996, *China Voyager: Gist Gee's Life in Science*, 345 p; M. E. Sharpe, Armonk, NY / London.

101 https://www.kew.org/blogs/library-art-and-archives/early-20th-century-plant-hunting-taiwan (Accessed 17 October 2017).

102 https://plants.jstor.org/stable/10.5555/al.ap.visual.kdcas4599?searchUri (Accessed 17 October 2017).

Japan (TI).

1912 年，William R. Price（1886—1975）在台湾采集植物标本，最初与 Henry J. Elwes（1846—1922），目前至少有 1 133 份标本藏在英国（K）、中国台湾（TAI）和日本（TI）。

1912，Ernest H. Wilson, ***Vegetation of Western China***—a series of 500 photographs with index, with an introduction by Charles S. Sargent, 19 p; printed for the subscribers, London.

1912 年, Ernest H. Wilson, **华西植被** ——500 张照片及索引, 以及 Charles S. Sargent 的 19 页导读; Printed for the subscribers，伦敦 。

1913

1913 年 2 月至 1916 年 11 月，胡先骕[103]（1894—1968）赴美国加州大学伯克利分校学习，获植物学学士学位。1918 年任教于南京高等师范学校，1921 年任教于国立东南大学；1920 年 7 月赴浙江东部大举采集植物标本，1921 年又赴江西采集植物标本。另参见 1923 年条目。

1913.02-1916.11, Hsen-Hsu HU (1894-1968) studied in the United States, earning a bachelor's degree in botany from the University of California, Berkeley, California. He returned to become a professor in Nanking Higher Teachers College in 1918, then professor in National Southeast University in 1921. He largely collected plant specimens in east Zhejiang in 1920 and in Jiangxi in 1921. See also 1923.

1913 年 12 月，刘厚[104]（1894—1984）赴法勤工俭学，先入翁特农林学校学农，1921 年在里昂中法大学学习农业植物学的同时兼职学校行政事务。1923 年考取理科硕士学位，然后入巴黎天然博物馆（即自然历史博物馆）；1929 年辞去中法大学行政职务专门从事植物分类学研究。1933 年获巴黎大学理科博士学位，同年 12 月归国。[105]

1913.12, Hou LIU (Ho LIOU, 1894-1984) studied agriculture in France, initially at the École d›agriculture et de foresterie de Wentong, then studied agricultural botany at the Université de Lyon de Chine et de France in 1921. At the same time he was also involved in administrative work at the Université de Lyon de Chine et de France; enrolled as a master's student at the Institut de Botanique de Paris (i. e. Musèum national d' Hisfoire Naturelle) in 1923, then studied plant taxonomy, without administrative duties, starting in

103 字步曾，号忏庵，江西新建人，1909 年考入京师大学堂预科，1912 年秋参加江西省留学考试，被录取为西洋留学生。

104 字大悲，四川古宋（今兴文）人。

105 刘厚留学 20 年后归国，1934—1935 年任职南京国民政府行政院实业部简任技正，1936 年任广州实业部商品检验局局长，1937 年任复旦大学（重庆北碚）教授，1938 年任贵州省农业改进所森林系主任并筹办创立西南垦殖公司，1939 年任西南垦殖公司理事长兼总经理，同时在平坝大规模栽培桐、茶和樟等 4 000 多亩；1946 年任农林部湖北金水农场场长，1947 年任农林部河北垦业农场场长。1949 年初赴台后任士林园艺试验所技正、罗东山林管分所所长、花莲山林管所所长、台北区农业改良场场长（1963 年，70 岁，从政府退休），1963—1973 年兼职政治大学、淡江文理学院、辅仁大学和政治作战学校法语教授；1978 年移居新西兰，1983 年 9 月初回到成都，与旅法同学、分离 35 年的妻子、著名桑蚕教育家张振华（1893—2006）团聚，并与她一道共同担任四川省成都市市政府参事；次年 3 月过世。详细参见：张振华，1989，共产党政策好，人心思留思归 —— 记个人经历中三次大的转折，蜀都春晓 —— 蓉参史料专辑，第 191-197 页；成都市人民政府参事室编，成都。2016 年 9 月 12 日，刘厚台北的大女儿刘良玉来到贵州平坝，将父亲收藏的资料捐献给平坝监狱（即其父 1939 年在贵州建立的西南垦殖公司），并捐赠其父的铜像，同时参观当年父亲组织栽培的樟树等（自贵州省平坝监狱 2016 年 9 月 12 日微博，题目：华侨的乡愁（http://weibo.com/3557356894/E7WoulEWW?-type=comment#_rnd1503860191675, accessed 18 August 2017）。另参见刘厚，1978，刘大悲先生年谱，33+57 页；台北（自印）。刘厚女儿刘良玉赠送年谱，特此致谢。

1. Hou LIOU (1894-1984) (Photo provided by Shirley LIU / Jin-Peng LAI)

1. 刘厚（1894—1984）（相片提供者：刘良玉 / 来金朋）

2. Hou LIOU and colleagues (Photo provided by Shirley LIU / Jin-Peng LAI)

2. 刘厚与同事（相片提供者：刘良玉 / 来金朋）

3. Bronze bust of Hou LIOU (Photo provided by Shirley LIU / Jin-Peng LAI)

3. 刘厚铜像（相片提供者：刘良玉 / 来金朋）

4. Pingba Leaders with Shirley LIU (Photo provided by Shirley LIU / Jin-Peng LAI)

4. 平坝领导与刘良玉（相片提供者：刘良玉 / 来金朋）

1929. He received his Ph. D. degree in 1933[106] and returned to China in December of the same year.

1913.12-1915.07.04, Urbain Jean Faurie (1847-1915) collected in Taiwan for the second time. Faurie collected extensively in Japan, South Korea and Taiwan. His collections are in France (P), England (BM, K), Japan (TI) and the United States (A, UC).[107, 108]

1913 年 12 月至 1915 年 7 月 4 日，法国传教士 Urbain Jean Faurie（1847—1915）第 2 次[109]来中国台湾采集[110]。他曾经在日本、韩国和中国台湾采集很多植物标本，他的标本今日主要存于法国（P）、英国（BM、K）、日本（TI）和美国（A、UC）。

1913, Francis Kingdon Ward, ***The Land of the Blue Poppy***—*Travels of a naturalist in eastern Tibet,* 283 p; University Press, Cambridge.

1913 年，Francis Kingdon Ward, **绿绒蒿之乡** —— 一个博物学家在西藏东部的旅行，283 页；大学出版社，剑桥。

1913, Ernest H. Wilson, ***A Naturalist in Western China***, with vasculum, camera and gun: being some account of eleven years' travel, exploration and observation in the more remote parts of the flowery kingdom, with an introduction by Charles S. Sargent, 1: 251 p, 2: 229 p. Methuen and Co. Ltd., London; Reprinted in 2011 by Cambridge University Press (digital version), Cambridge. Translated into Chinese between 1917 and 1920.

1917—1920 年，胡先骕译，中国西部植物志，科学 3（10）：1 079-1 092，1917；农商公报 52：7-14，1917（转载）；东方杂志 15（8）：104-114，1918（转载）；中国西部果品志，科学 4（10）：1 010-1 019，1919；农商公报 68：7-12，1919（转载）；序言，中美木本植物之比较，科学 5（5）：478-491，5（6）：623-638，1920。

1913 年，钟心煊[111]（1893—1961）赴美国留学伊利诺伊大学，次年转入哈佛大学，1917 年入研究生，毕业获硕士学位；1920 年回国创办南开大学生物系，1922 年创办厦门大学植物系（并建立植物标本馆，

106 Ph.D. dissertation: Ho LIOU, 1934, Lauracées de Chine et d'Indochine—Contribution à l'étude systématique et phytogéographique, 226 p; Advisor: (Marie Antoine) Alexandre Guilliermond (1876-1945); Librairie Hermann, 6 rue de la Sorbonne, Paris. 参见：刘慎谔，1934，国立北平研究院植物学研究所丛刊第 2 卷第 4 期第 115-116 页的专门介绍。

107 http://archives.mepasie.org/fr/notices/notices-biographiques/faurie-1 (Accessed 17 October 2017).

108 https://moussons.revues.org/2281 (Accessed 17 October 2017).

109 第一次来台湾时间为 1903 年 4-7 月，采集有限；李瑞宗著，佛里神父，194 页，2017；行政院农业委员会林业试验研究所，台北。

110 台湾林业试验所于 2017 年 9 月 30 日在台北植物园落成 Urbain Jean Faurie 的铜像，以示纪念，参见：http://www.cna.com.tw/news/ahel/201709300280-1.aspx (Accessed 24 October 2017).

111 字仲襄，江西南昌人，1910 年夏考取清华学校首届留美预备班。

AU)，1931 年后一直在国立武汉大学任教。

1913, Hsin-Hsuan CHUNG (1893-1961) traveled to the United States to study at the University of Illinois, then transferred to Harvard University in 1914; enrolled as a graduate student in 1917 and earned a masters's degree in botany in 1920. He returned to China to establish the department of biology and to became a professor at Nankai University, Tientsin [Tianjin], then established the department of biology and became a professor at Amoy [now Xiamen] University in 1922; moved to National Wuhan University in 1931, where he stayed for the remainder of his career.

1914

1914年2月，由吴家煦[112]组织，经教育部批准，中华博物研究会在上海成立。该会所有同志均为东西洋留学生，且从事自然科学研究，其中植物科主任为彭世芳；并在苏、浙、鄂、鲁、粤等省设立支部。

1914.02, Natural Science Association of China[113] founded in Shanghai, organized by Chia-Hsun WU and approved by the Bureau of Education. The members were trained both in the West or Japan, mainly in the natural sciences. Shih-Fang PENG was in charge of botany. Branches were established in Guangdong, Hubei, Jiangsu, Shandong and Zhejiang.

1914年10月至1928年10月，**博物学杂志**由中华博物研究会于上海创刊，共出版2卷8期（不规则）。[114]

1914.10–1928.10, ***Journal of Natural Science*** issued by the Natural Science Association of China in Shanghai. Two volumes in eight parts were published (irregularly).

1914年，武昌高等师范学校设立博物部，植物学家张珽加入，植物学家辛树帜为其学生之一。

1914, Department of Natural History was founded in Wuchang Higher Teaching College. Botanist Ting CHANG was one of its faculty members; botanist Shu-Chih HSIN was one of his graduates.

1914–1915, Reginald J. Farrer[115](1880–1920) and a companion, Kew-trained William Purdom (1880–1921), set out on an ambitious expedition to Gansu, Qinghai and Xizang. Their collections are in England (K).

1914—1915年，Reginald J. Farrer (1880—1920) 和邱园培训的同仁 William Purdom（1880—1921）到甘肃、青海和西藏考察。他们采集的植物标本存放在英国（K）。

1914–1915, Augustin Abel Hector Léveillé, ***Flore du Kouy-Tchéou***, autographée en partie par l'auteur,

---

112 字和士，江苏苏州人（生卒不详）。曾留学日本："曩昔留学东瀛，蹴居小石川区，日夕徜徉于植物园内，于分类学颇有所得，返国以还，惟以采制标本为事，足迹几遍全国，每得一植物，先之以对照写生，继之以剖花实，然后记载其形态，检索其学名，明定其科属，积年既多，得腊叶标本3 500种，各附以彩色图谱。"吴家煦，1943，中国博物学变迁史，文化汇刊3（1）：21-29。

113 英文翻译参见：吴国盛，2016，自然史还是博物学？读书1：89-95。

114 李楠、姚远，2011，《博物学杂志》办刊思想探源，编辑学报23（5）：398-400。

115 Traveler, plant collector, plantsman and writer, born on 17 February, 1880, London, but died on the way to botanize in Myanmar on 17 October 1920.

Reginald J. Farrer (1880-1920) (Photo provided by Royal Botanic Gardens, Kew)
Reginald J. Farrer（1880—1920）（相片提供者：邱园）

William Purdom (1880-1921) (Photo provided by Royal Botanic Gardens, Kew)
William Purdom（1880—1921）（相片提供者：邱园）

532 p; Le Mans.

1914—1915 年，Augustin Abel Hector Léveillé，**贵州植物志**，532 页；Le Mans。[116]

1914—1915 年，吴家煦，江苏植物志略，博物学杂志（中华博物研究会）1（1）：135-144，1（2）：149-152。[117]

1914-1915, Chia-Hsun WU, Jiangsu Plants, *Journal of Natural Sciences (Natural Science Association of China)* 1(1): 135-144, 1(2): 149-152.

1914, 1915, 1919-1921, Nathaniel G. Gee, List of plants found in Kiangsu province, *National Review (China)* 15: 667-679, 739-742, 809-812, 1914; A preliminary list of the plants of Kiangsu province, *A Text-book of Botany*, 361 p, appendix 1-70 p, 1915; Commercial Press, Shanghai; 1919-1921, A catalogue of plants of Kiangsu, *Science (China)* 4: 1,117-1,124, 1919; 5: 207-212, 603-622, 729-748, 800-814, 1,157-

116 熊源新、曹威，2017，贵州植物分类学研究概述，山地农业生物学报 36（1）：1-11。

117 除期刊的正规（序数）页码之外，该文还有两个大写表示的序数词页码，分别为一至十和一至四。

1,165, 1920; 6: 211-229, 318-385, 417-434, 622-637, 720-733, 1921; Reprinted 1-177 p, 1921.

1914 年、1915 年、1919—1921 年，祁天锡，江苏发现的植物名录，中国公论西报 15：667-679，739-742，809-812，1914；1915，江苏省植物名录初编，植物学教科书，361 页，附录 70 页；商务印书馆，上海；1919—1921，江苏植物名录，科学 4：1 117-1 124，1919；5：207-212，603-622，729-748，800-814，1 157-1 165，1920；6：211-229，318-385，417-434[118]，622-637，720-733，1921；重印 1-177 页，1921。

1914-1918, Heinrich R. E. Handel-Mazzetti[119](1862-1940), Austrian botanist, explored in Guizhou, Hunan, Sichuan and Yunnan. He collected more than 13,000 numbers of plant specimens and more than 30,000 sheets. Most specimens are in Austria (W, original set in WU), but many duplicates are in other herbaria. He described around 35 new genera and more than 1,300 new species, many published in his *Symbolae Sinicae* (see

Portrait of Heinrich R. E. Handel-Mazzetti (1862-1940) (Photo provided by Universität Wien)

Heinrich R. E. Handel-Mazzetti（1862—1940）的肖像（相片提供者：维也纳大学）

118 Collaborated with Sung-Shu CHIEN；与钱崇澍合作；其中科学所刊部分由后者翻译单独成册。

119 Heinrich Raphael Eduard Freiherr von Handel-Mazzetti (usually as Heinrich R. E. Handel-Mazzetti in the botanical world) studied botany at the University of Vienna, obtaining doctorate in 1907. From 1905 he served as an assistant at the Botanical Institute in Vienna. On behalf of the Austrian Academy of Sciences, he traveled to China in 1914 (with Camillo Schneider, 1876-1951) and conducted botanical research in the provinces of Yunnan (1914, 1915, 1916), Sichuan (1914), Guizhou (1917) and Hunan (1917, 1918). In China he also undertook cartographic surveys. He returned to Vienna in 1919, where he devoted his energies toward studies of the Chinese flora, and published a series reports on Chinese plants (entitled as *Plantae Novae Sinensis, diagnosibus brevibus descriptae a Dre Henr. Handel-Mazzetti*, in Akademie der Wissenschatten in Wien, Sitzung der mathematisch-naturwissenschaftlichen Klass, 1920-1926, 1-298 p; Bundled Offprints is at RBGE Library). He died after being hit by a vehicle near the Botanical Institute on 1 February 1940. Heinrich R. E. Handel-Mazzetti, 1927, *Naturbilder aus Südwest-China*, 380 p; Österreichischer bundesverlag für Unterricht, Wissenschaft und Kunst, Wien and Leipzig (*A botanical pioneer in South West China*, 1996, 192 p; David Winstanley, Brentwood). Heinrich R. E. Handel-Mazzetti et al., 1929-1937, *Symbolae Sinicae*, Botanische Ergebnise der Expedition der Akademie der Wissenschaften in Wien nach Sudwest-China, 1914-1918, Heinrich R. E. Handel-Mazzetti, Teil. 7. See also 1929-1937.

1929-1937).

1914—1918 年，Heinrich R. E. Handel-Mazzetti（1862—1940），奥地利植物学家，在云南及四川等地大规模采集，经贵州赴湖南后回国，采集植物标本约 1.3 万号（达 3 万份，全套存于 WU，部分存于 W）[120]，复份标本其他标本馆也有存放。他描述了 35 个新属，约 1 300 个新种，多数发表在他的中国植物纪要中（参见 1929—1937 年条目）。

1914-1937, Père Emile Licent (1876-1952), French Jesuit priest, paleontologist, archaeologist, explored largely in northern China, including Gansu, Hebei, Henan, Inner Mongolia, Shaanxi, Shandong and Shanxi;

Père Emile Licent (1876-1952) (Photo provided by Yong LI)
Père Emile Licent（1876—1952）（相片提供者：李勇）

120 苔藓部分在芬兰赫尔辛基；参见：Qing-Hua WANG, 2015, An introduction to the bryophyte herbarium of Natural History Museum in Finland, *Chenia* 12: 184-185。

established Le Musée Haongho Paiho de Tientsin (i.e. today's Tianjin Natural History Museum)[121] and its herbarium (TIE) in 1922-1923. His plant specimens are mainly in Tientsin (TIE), but also in other countries (A, BM, C, GH, K, P, PR and W).

1914—1937 年，法国耶稣会传教士，古生物学家、考古学家 Père Emile Licent（桑志华，1876—1952），对中国北方进行了广泛和深入的考察（包括山东、河北、山西、河南、陕西、甘肃、内蒙古等）[122]，1922—1923 年创建了天津北疆博物馆（今天津自然博物馆前身）及其标本馆（TIE）[123]。他所采集的植物标本主要在天津（TIE），但海外也有存放（A、BM、C、GH、K、P、PR、W）。

121 Tracey L.D. LU, 2014, *Museums in China—Power, politics and identities*, 235 p; Routledge, Abingdon, Oxon, UK.

122 天津自然博物馆，陈锡欣主编，1994，天津自然博物馆八十周年，342 页；天津科学技术出版社，天津。

123 天津自然博物馆，孙景云主编，2004，天津自然博物馆建馆九十（1914—2004）周年文集，239 页；天津科学技术出版社，天津。

1915

1915 年 10 月 25 日，由中国留美学生组织的中国科学社于美国纽约州康奈尔大学成立，主要发起人为任鸿隽（1886—1961）、秉志（1886—1965）等 9 人，任鸿隽为社长；1918 年中国科学社从美国迁回上海，次年转至南京（社址设于南京成贤街文德里），1928 年在上海法租界兴建新社址（亚尔培路 309 号，今陕西南路），遂又迁至上海。1960 年，中国科学社停止活动。

1915.10.25, Science Society of China founded at Cornell University, Ithaca, New York, by Chinese students. Among the nine main sponsors were Hung-Chun JEN (Hung-Chun ZEN, 1886-1961), and Chi PING (1886-1965). Hung-Chun JEN was chosen to be the president. In 1918, the Science Society of China was transferred from the United States to Shanghai, China, then to Nanjing the following year, and back to Shanghai in 1928. It ceased existence in 1960.

1915 年，江苏省立第一农校植物标本室成立（今南京林业大学植物标本馆，NF）。[124]

1915, The herbarium of the First Agricultural School of Jiangsu Province established (today's herbarium of Nanjing Forestry University, NF).

1915，Nathaniel G. Gee, ***A Text Book of Botany***, 361 p; with a preliminary list of the plants of Kiangsu [Jiangsu] Province, 77 p, index i-vii; Commercial Press, Shanghai.

1915 年，祁天赐著，**植物学教科书**，361 页；附“江苏植物名录”，77 页；商务印书馆，上海。

1915, Frank N. Meyer, *China—a fruitful field for plant exploration*, Yearbook of the United States Department of Agriculture, 205-224 p.

1915 年，Frank N. Meyer，中国——植物考察的乐园，美国农业部年鉴，第 205-224 页。

1915 年，熊岳树木园由日本人于辽宁省熊岳镇建立。这是中国大陆最早的树木园。[125]

1915, Xiongyue Arboretum in Xiongyue, Liaoning Province was founded by Japanese. This is the earliest arboretum in mainland China.

1915 年，吴家煦，植物采集法、植物标本制作法、植物记载法，中华学生界，1(1): 1-9, 1(3): 1-10,

---

124 傅立国主编，1993，中国植物标本馆索引，458 页；中国科学技术出版社，北京。

125 1909 年成立熊岳诚苗圃，1913 年改为农事试验场分场（http://www.byqxww.com/2015/1113/29876.shtml；2017 年进入）。另外参见：草间正庆，1977，熊岳城随想 —— 草间正庆遗稿，66 页；熊岳城会，日本（日本语）。

1(4): 1-10, (12): 11-22。

1915, Chia-Hsun WU, 'Plant Collecting Method, Specimen Making Method and Plant Recording Method', *Chinese Students* 1(1): 1-9, 1(3): 1-10, 1(4): 1-10, (12): 11-22.

1915-1917, Augustin Abel Hector Léveillé, ***Catalogue des Plantes du Yun-nan*** avec renvoi aux diagnoses originales, observations et descriptions d'espèces nouvelles, 299 p; Le Mans.

1915—1917 年，Augustin Abel Hector Léveillé，**云南植物名录**，299 页；Le Mans。

1915—1950 年，**科学**在国内发行，月刊；共发行 32 卷 400 多期。

1915-1950, ***Science (China)***, monthly, was issued in China. Thirty two volumes and over 400 issues were published.

1916

1916, Sung-Shu CHIEN, Two Asiatic allies of *Ranunculus pensylvanicus*, *Rhodora* 18, 213: 189-190. This is the first modern taxonomic paper in which species from China were described as new by a Chinese taxonomist.

1916 年，钱崇澍，宾夕法尼亚毛茛的两个亚洲近缘种，*Rhodora* 18，213：189-190。此为第一篇由中国植物分类学家描述中国植物新种的当代植物分类学文章。

1916, George W. Groff [126](1884-1954) was appointed the first dean of the department of agriculture, Canton Christian College, where he organized the plant collections. *Lingnan Agricultural Journal* issued in 1921, then changed to *Lingnan Science Journal* in 1922, when department of agriculture, Canton Christian College, was changed to department of agriculture, Lingnan University. Groff was the first dean (1927-1932) when the department of agriculture, Lingnan University, was changed to Agricultural College, Lingnan University in 1927.

1916 年，岭南学堂农学部成立，George W. Groff（高鲁甫，1884—1954）出任主任 [127]，并主持采集植物标本 [128]。1921 年岭南学堂农学部改为岭南大学农科，发行岭南农业杂志，1922 年改名岭南科学杂志。1927 年农科又改为农学院，高鲁甫任首任院长（1927—1932）。

126 American agriculturalist, botanist and missionary teacher who devoted his life to Chinese agriculture and botany. Arrived in Canton [Guangzhou] in 1908, began teaching in the middle school, then taught at Canton Christian College; from 1912-1917 served as a science and agriculture tutor; between 1918 and 1920 split his time between serving the USDA and Canton Christian College in China, where he established a herbarium and inaugurated an extensive plant collecting program. His first plant collections date from 1910-1911 in Guangxi; professor of Agriculture at Canton Christian College, where he helped to found the Lingnan Agricultural College, of which he was made dean. Groff remained in this role until 1935. During his time at Lingnan Groff built up the college and established the English language *Lingnan Agricultural Journal* (later *Lingnan Science Journal*). He was appointed director of Lingnan Economic Plant Receiving Station; in 1937 he led a National Geographic expedition in Guangxi; remained at Lingnan after the Japanese invasion of 1937, but moved back to the U.S. in 1941 due to ill health. Here he established a plant exchange station at Sarasota, Florida, collecting and propagating promising plants for shipment to China. He remained in Florida for the rest of his days, save for a year working for the United Nations as an agricultural rehabilitation officer in southern China in 1946-1947; travelled widely in China and also collected plants in the Philippines, Malaysia, Thailand and Vietnam. Additionally, he made a significant collection of bamboos from Southeast Asia and imported livestock from the U.S. for breeding in China (https://libraries.psu.edu/findingaids/409.htm, Accessed 19 September 2017).

127 著有荔枝和龙眼（1921），以及广东植物等，参见 1921—1924 年条目。

128 1931 年年底至 1932 年 2 月，高鲁甫、莫古礼、邵尧年、冯钦组成植物采集旅行团，前往广东台山、新会、开平、恩平、高州、雷州、北海、茂名，广西郁林及广西与越南毗邻地区考察与采集植物。

1916, Elmer D. Merrill[129](1876-1956), American botanist, director of the Philippine Bureau of Science, visited Canton Christian College at the invitation of American Christian School, helped in founding the herbarium, which became the herbarium of Lingnan University, with collections from Guangzhou[130].

1916 年，时任菲律宾科学局长的美国植物学家 Elmer D. Merrill（梅尔，1876—1956）受美国教会学校岭南学堂的邀请来广州，协助组建植物标本室，并于广州附近采集标本。后来在该标本室基础上建立了岭南大学标本室。

Elmer D. Merrill (1876-1956) (Photo provided by Zong-Gang HU)
Elmer D. Merrill（1876—1956）（相片提供者：胡宗刚）

1916 年，岭南学堂自然历史博物馆成立植物标本室，1918 年改为岭南大学植物标本馆（LU），1952 年并入中山大学植物标本馆（SYS）。

1916, Herbarium of Natural History Museum of Canton Christian College founded. In 1918, it became the herbarium (LU) of Lingnan University. In 1952, it merged with the herbarium (SYS) of Sun Yat-Sen University.

1916-1935, Musee Hoangho Paiho de Tientsin, ***Publications du Musée Hoangho Paiho de Tien Tsin***, 1-111 (published irregularly).

1916—1935 年，天津北疆博物院，**天津北疆博物院丛刊（黄河白河博物馆丛刊）**，1-111（不定期出版）。

---

129 Elmer D. Merrill spent 22 years in the Philippines (1902-1924) where he became a recognized authority on the flora of the Asia-Pacific region. He was an accomplished administrator. He served as dean of the College of Agriculture and director of the Agricultural Experiment Station, University of California, Berkeley, CA (1924-1929), director of the New York Botanical Garden (1929-1935) and administrator of the botanical collections of Harvard University and director of the Arnold Arboretum of Harvard University (1935-1946). Over the course of his career he authored nearly 500 publications, described approximately 3,000 new species of plants and amassed over one million herbarium specimens for the institutions under his administration. He contributed greatly to the development of Chinese botany by helping Chinese students and by collaborating with Chinese botanists throughout his life (https://arboretum.harvard.edu/wp-content/uploads/I_B_EDM_2012.pdf, accessed Sept. 2017).

130 George W. Groff, Edward DING and Elizabeth H. Groff, 1923, An enumeration of the McClare collection of Hainan plants, *Lingnan Agricultural Review* 1(2): 27-86.

1917

1917, Reginald J. Farrer, ***On the Eaves of the World***, 1: 311 p, 2: 328 p; Edward Arnold and Co., London.

1917 年，Reginald J. Farrer，**在世界屋檐上**，第 1 卷：311 页，第 2 卷：328 页；Edward Arnold and Co.，伦敦。

1917 年，凌道扬[131]（1888—1993）、陈嵘于南京发起创建中华森林会，1921 年创办林学杂志森林[132]，1928 年更名为中华林学会，1951 年又更名为中国林学会。

1917, Dao-Yang LING (Dau-Yang LIN, 1888-1993) and Yung CHEN organized the Chinese Forest Association in Nanjing; *Journal of Forestry* issued in 1921[133]. The Chinese Forest Association in Nanjing was renamed Chinese Society of Forestry in 1928; renamed Chinese Society of Forestry in 1951.

1917, Boris V. Skvortzov[134] (Борис Васильевич Скворцов, 1896-1980), Russian botanist, especially phycologist studying diatoms, worked on diatoms in Harbin, Heilongjiang, for nearly half a century. He made extensive botanical expeditions, particularly in northeastern China and published many accounts of diatoms of northeast China as well as other botanical papers during his stay in China.

1917 年，Boris V. Skvortzov[135] (1896—1980)，植物学家、硅藻植物专家，俄罗斯人，在黑龙江哈尔滨研究硅藻等近半个世纪，在东北等地进行了大规模采集；在华期间，发表了很多有关中国东北硅藻类以及其他与植物有关的文章。

---

131 广东新安（今深圳）人；中国近代著名林学家、农学家、教育家、水土保持专家。

132 中国林学会，2008；中国林学会史，294 页；上海交通大学出版社，上海。

133 Chinese Society of Forestry, 2008, A History of Chinese Society of Forestry, 294 p; Shanghai Jiao Tong University Press, Shanghai.

134 Also Boris W. Skvortsow or Boris V. Skvortsov. More details see David M. Williams and Geraldine Reid, 2001, A bibliography of the scientific work of Boris V. Skvortzov (1896-1980) with commentary on publications concerning diatoms (Bacillariophyta), *Bulletin of the Natural History Museum London (Botany)* 31(2): 89-106. David M. Williams, Maria Gololobova & Elena Glebova, 2016, Boris Vasil'evich Skvortzov (1896-1980): notes on his life, family and scientific studies, *Diatom Research* 21(3): 313-321.

135 俄罗斯人，也作为 Boris W. Skvortsow 或 Boris V. Skvortsov；1902 年随俄罗斯帝国法官家庭移居中国，1914 年高中毕业后进入俄罗斯帝国圣彼得堡大学专攻藻类学，1917 年开始发表中国藻类的第一篇文章，1919 年于哈尔滨商业学校任教，1922 年参与创办东省文物研究会（今黑龙江省博物馆前身）并一直在那里从事植物学等研究，特别是硅藻；新中国成立后先后任东北林业研究所（筹备处）研究员和东北林学院一级教授；1962—1970 年受雇于巴西圣保罗植物研究所，并在国家公园及附近采集；与此同时继续研究苏联贝加尔湖和日本木崎湖的藻类；一生发表论著 430 多篇（部），其中包括很多中国东北的硅藻新类群。

1918

1918 年 2 月，杜亚泉主编，**植物学大辞典**，1 590 页；商务印书馆，上海。杜亚泉[136]（1873—1933）为主持者[137]，黄以仁为 13 位作者中唯一的植物学者。

1918.02, Ya-Chuan TU[138] (editor in chief), ***Botanical Nomenclature—A Complete Dictionary of Botanical Terms***, 1,590 p; Commercial Press, Shanghai. Ya-Chuan TU (1873-1933) was in charge of the whole work. I-Jen WANG was the only one of 13 authors who was trained in botany.

1918 年，刘汝强[139]（刘毅然，1895—1987）赴美留学，1918 年 9 月至 1919 年 3 月于俄亥俄的 Oberlin 学院读书，1919 年 3 月至 1921 年 3 月任法国华工青年会干事，1921 年 4 月至 1924 年 12 月于费城药专和威斯康星大学麦迪逊分校分别获得本科和硕士学位。另参见 1931 年条目。

1918, Ju-Chiang LIU (Yi-Ran LIU, 1895-1987) went to the United States to study at Oberlin College in Oberlin, Ohio, then to France to work as coordinator of the Chinese Association of Youth Work between March 1919 and March 1921. He attended the Pharmaceutical School in Philadelphia, Pennsylvania, April 1921, the University of Wisconsin, Madison, Wisconsin, December 1924, from where he obtained a bachelor's degree in 1924 and a master's degree in 1927. See also 1931.

1918, Augustin Abel Hector Léveillé, ***Catalogue Illustrè et alphabetique des plantes du Seu Tchouen***, 221 p; Le Mans.

1918 年，Augustin Abel Hector Léveillé，**四川植物图录与名录**，221 页；Le Mans。

1918—1919 年，韩旅尘，广东植物名录，博物学会杂志 2: 26-28, 3: 25-27。

1918-1919, Lü-Chen HAN, 'A checklist of Guangdong plants', *Journal of Natural History* 2: 26-28, 3: 25-27.

1918—1921 年，钟观光在中国大规模采集[140]，东起福建沿海，西至滇川山地，北至山西，南至海

136 作者 13 人（按原书繁体字笔画排序）：孔庆莱、吴德亮、李祥麟、杜亚泉、杜就田、周越然、周藩、陈学郢、莫叔略、许家庆、黄以仁、凌昌焕、严保诚。

137 浙江山阴人，中国科学界先驱者之一。详细参见：谢振声，1988，杜亚泉传略，中国科技史料 9（3）：8-14。

138 Aslo as Ya-T'siun TU.

139 北京人，字毅然（后以字行），1903—1909 年北京基督教小学、1909—1913 年北京潞河中学、1913 年 9 月至 1917 年 6 月清华预科、1917 年 9 月至 1918 年 6 月清华预科教员。

140 钟观光，1919 年 6 月 18 日至 1919 年 12 月 6 日，旅行采集记，北京大学日报；1920—1922 年旅行采集记，地学杂志 11（7）-12（5）。

南，足迹遍及福建、广东、广西、云南、浙江、安徽、江西、湖北、四川、河南、山西等 11 个省区，采得植物标本 15 000 余号（BNU、PE、PEY、SYS）[141]，1924 年在北京大学建立植物标本室(PEY)[142]；他可能早在 1904 年[143]于上海创办科学仪器馆时便开始采集，标本很可能用于科学仪器馆的教学。[144]

1918-1921, Kuan-Kuang TSOONG began collecting plants in China. Particularly large collections were made while he was a teacher in Peking University between 1918 and 1921. He collected more than 15,000 numbers of plant specimens in Anhui, Fujian, Guangdong, Guangxi, Henan, Hubei, Jiangxi, Shanxi, Sichuan, Yunnan and Zhejiang (specimens at BNU, PE, PEY, SYS) and established the herbarium (PEY) at Peking University in 1924. His earliest collections may have been made as early as 1904, particularly to obtain teaching materials for his Shanghai Scientific Instruments Company.

Kuan-Kuang TSOONG (1868-1940) (Photo provided by Jin-Zheng CHEN and Ren-Jian ZHONG/TSOONG)
钟观光（1868—1940）(相片提供者：陈锦正与钟任建)

1918—1924 年，**博物学会杂志**由武昌高等师范学校博物学系于武昌创刊，共出版 5 卷 15 期（不定期；后博物学系改为生物系，杂志从第 6 卷起更名为**生物学杂志**，且卷期号重排）。

1918-1924, ***The Journal of Natural History*** issued by the Department of Natural History Society of Wuchang Teacher's College in Wuchang, Hubei. Five volumes and fifteen parts were published (irregularly). The title was changed to ***Biology Magazine*** from volume 6 part 1 as volume 1 part 1, when the Department of Natural History was changed to the Department of Biology.

141 孟世勇、刘慧圆、余梦婷、刘全儒、马金双，2018，中国植物采集先行者钟观光的采集考证，生物多样性 26（1）：79-88；朱宗元、梁存柱，2005，钟观光先生的植物采集工作 —— 兼记我国第一个植物标本室的建立，北京大学学报（自然科学版）41（6）：825-832；浙江植物志编辑委员会，1993，浙江植物志总论，第 2 页。

142 北京大学生命科学学院，2015，北京大学生命科学九十年，第 8 页；北京大学生命科学学院，北京（内部印制）。早期记载 1905 年的依据与出处未知。参见傅立国主编，1993，中国植物标本馆索引，第 19 页；中国科学技术出版社，北京。

143 姚远，2010，《科学世界》及其物理学和化学知识传播，西北大学学报（自然科学版）40（5）：934-940；王勇忠，2010，民国时期中小学科学仪器制造概况，教育史研究 1：43-49；谢振声，1989，上海科学仪器与《科学世界》，中国科技史杂志 10（2）：61-66；谢振声，1990，上海科学仪器馆述略，科学 42（1）：70-71。

144 孟世勇、刘慧圆、余梦婷、刘全儒、马金双，2018，中国植物采集先行者钟观光的采集考证，生物多样性 26（1）：79-88。

1919

1919年8月16日，上海中华博物研究会与北京博物调查会合并，成立中华博物学会；同时设南北（北京、上海）两个会址，选举袁希涛[145]（1866—1930）为会长，吴家煦、陈宝泉[146]（1874—1937）为副会长，并决定继续进行发行博物杂志、编译博物专书、设立标本陈列馆、推行科学名词审查等工作[147]。

1919.08.16, The Natural Science Association of China in Shanghai and the Survey Society of Natural History in Peiping [Beijing] merged to form the Chinese Society of Natural History with sites in Beijing and Shanghai. Hsi-Tao YUAN (1866-1930) was appointed president, Chia-Hsun WU and Pao-Chuan CHEN (1874-1937) were appointed vice presidents. The society aimed to publish journals, translate and compile books, establish museums to display specimens and issue dictionaries of scientific terminology etc.

1919年，彭世芳，北京栽培植物俗名之研究，博物杂志 1: 30-36。

1919, Shih-Fang PENG, 'On the common names of cultivated plants in Beijing', *Magazine of Natural History* 1: 30-36.

1919年，李顺卿[148]（1894—1972）赴美国留学，先获得耶鲁大学森林学硕士学位，1923年获芝加哥大学植物学博士学位；回国后曾历任国立北京高等师范学校教授、国立北平大学农学院农业生物系教授、国立北平师范大学生物系教授兼系主任并兼任教务长、安徽大学教授兼农学院院长（1934年）及校长（1935年）、国立中央大学农学院教授兼森林系主任、国民政府农林部林业司司长；1932年赴美国哈佛大学及英国皇家植物园邱园从事木本植物分类学研究，1949年赴台湾并任台湾省林产管理局局长，1952—1957年被聘为台湾大学植物系教授兼系主任，1961—1962年再次赴哈佛大学研究中国木本植物，1965年退休后定居美国。

1919, Shun-Ching LEE (Shun-Ching LI, 1894-1972) went to the United States to study. He earned a masters's degree in Forestry at Yale University then a Ph. D. in botany at the University of Chicago in 1923[149]. He returned to China to become a professor in Peiping [Beijing] High Teacher's College, then professor in the department of biology, Agriculture School of Peiping [Beijing] University, professor and dean of the department of biology, Peiping [Beijing] Normal College and councilor of Peiping [Beijing]

145 号观澜，江苏（上海）宝山人，祖籍浙江宁波。清末民初教育家。

146 字筱庄、小庄、肖庄；天津人，中国近代教育家。

147 吴家煦，1943，中国博物学变迁史，文化汇刊3（1）：21-29。

148 字干臣，山东海阳人；1919年金陵大学林科首届毕业生，获森林学学士学位。

149 Ph.D. dissertation: Factors controlling forest successions at Lake Ithaca, Minnesota.

Normal College, chairman of the Agriculture School and president of Anhui University, professor in the Agriculture School and dean of the Department of Forestry, National Central University, director of Forestry Bureau, National Agriculture and Forestry Administration. He studied woody plants at Harvard University and at the Royal Botanic Gardens, Kew in 1932. He went to Taiwan in 1949 as director of the Forestry Production Bureau of Taiwan, and became dean of the department of botany, Taiwan University from 1952 to 1957, and went to Harvard University to study Chinese woody plants again from 1961 to 1962. He retired to live in the United States in 1965.

Woon-Young CHUN on the field trip to Hainan (Photo provided by Zong-Gang HU)
陈焕镛于海南野外（相片提供者：胡宗刚）

1919—1920 年，陈焕镛于哈佛大学获得硕士学位后，回国往海南岛采集，因染病被人抬出大山。陈焕镛在海南所采标本，先后在上海和南京遭火灾被毁。

1919-1920, Woon-Young CHUN returned to China after earning a master's degree from Harvard University. He collected on Hainan, but had to stop because of illness. Unfortunately, all of the collections from this trip were lost in fires in Shanghai and Nanjing.

1919—1925 年，**博物杂志**由北京高等师范学校博物学会创刊，一共出版 8 期（不定期）。

1919-1925, ***Magazine of Natural History*** issued by Natural Historical Society of Peking High Teacher's College; at least 8 parts published irregularly.

1919-1940, Floyd A. McClure (1897-1970) joined Canton Christian College (Lingnan University) as a teacher and specialist of bamboo taxonomy. He collected in Guangdong (including Hong Kong), Guangxi and Hainan, as well as in Vietnam. He collected over 20,000 numbers of specimens (A, SYS, US). During his stay in China, he was also employed by the United States Department of Agriculture to introduce bamboos into the United States.

1919—1940 年，Floyd A. McClure（莫古礼[150]，1897—1970）来岭南学堂（岭南大学）任教并研究竹子；在海南、广东（以及香港）和广西以及越南采集植物标本 20 000 号（A，SYS，US）；在华期间，同时受雇于美国农业部，并向美国大规模引种竹子[151]。

150 出身于俄亥俄州农家，毕业于俄亥俄州立大学，1918 年和 1919 年分别获艺术和哲学学士学位；1927—1928 年获硕士学位，1932—1936 年获博士学位；1919 年进入广州岭南学堂（即后来的岭南大学）任教，除回国攻读学位外，直到 1940 年离开中国，主要从事竹子研究；1940 年返回美国后，继续在美国史密森学会从事美洲的竹子研究，代表性著作为：*The bamboos*—a fresh perspective, xv+347 p, fig. 1-99, 1966; Harvard University Press, Cambridge, Massachusetts。

151 Egbert H. Walker and W. Andrew Archer, 1971, Floyd Alonzo McClure (1897-1970), *Taxon* 20(5/6): 777-784; Frederick G. Meyer, 1972, Floyd Alonzo McClure (1897-1970)—A Tribute, *Economic Botany* 26(1): 1-12; 金文驰，2015，琼岛寻踪——莫古礼两次海南采集纪行，生命世界 5：80-87；胡启明、曾飞燕，2011，广东植物志，10：329 页；广东科技出版社，广州（附录 2，华南植物研究所（园）植物标本采集简史）。

1920

1920, Marcel E. Hardy, ***The Geography of Plants***, 327 p; Clarendon Press, Oxford. Translated into Chinese in 1933.

1933 年，胡先骕译订[152]，**世界植物地理**，213 页；商务印书馆，上海。

1920 年，刘慎谔[153]（1897—1975）赴法国留学，1921—1929 年先后在法国的南锡大学农学院、孟伯里埃农业专科学校、克来孟大学理学院、里昂大学理学院和巴黎大学理学院学习；1926 年在克来孟大学理学院毕业获理科硕士学位，1929 年在巴黎大学毕业获法国国家理学博士学位；同年回国并任北平研究院植物学研究所专职研究员兼所长；1950 年任东北农学院东北植物调查研究所所长，1953 年改任中国科学院林业研究所筹备处副主任，1954 年任中国科学院林业土壤研究所研究员、副所长兼植物研究室主任直至 1975 年过世。[154]

1920, Tchen-Ngo LIOU (1897-1975) studied in France from 1921-1929 at College of Agriculture, University of Nancy, then in the College of Agronomy in Montpellier, University of Clermont-Ferrand, Faculty of Science at University of Lyon, and finally in University of Paris. He received a master's degree in Science at University of Clermont-Ferrand in 1926 and a Ph. D. degree in

Tchen-Ngo LIOU (1897-1975) (Photo provided by Zong-Gang HU)
刘慎谔（1897—1975）（相片提供者：胡宗刚）

152 实为编译，详细参见原书译者序。在此依据原书依然采用译订，而不是编译。

153 山东牟平人，字士林，中学毕业后，考入了保定留法高等工艺学校预备班。

154 刘慎谔文集编辑委员会，1985，刘慎谔文集，342 页；科学出版社，北京。

Science at the University of Paris in 1929[155]. He returned to China and served as a research professor and director of the Institute of Botany, National Academy of Peiping [Beijing], from 1929 to 1949, director of the Institute of Botanical Survey, Northeast Agriculture College, Harbin, from 1950 to 1953, vice director of preparation of the Institute of Forestry, Chinese Academy of Sciences, Shenyang, 1953, research professor and vice director of the Institute of Forestry and Pedology, Chinese Academy of Sciences, and director of botany department, from 1954 to 1975.

1920 年，林镕[156]（1903—1981）赴法国留学，先就读于南锡大学农学院，于 1923 年获学士学位，后入克来孟大学理学院，并于 1927 获得克来孟大学理学院硕士学位；再入巴黎大学理学院，并于 1930 年获得博士学位；1930 年回国后，任教于北平大学农学院，同时兼职于北平研究院植物学研究所；抗战时期辗转于国立西北联合大学、国立西北农学院、中国西北植物调查所、福建省研究院农林植物研究所、厦门大学和福建农学院，1946 年返回北平主持北平研究院植物学研究所的复员并在北平诸多高校兼职，1949 年后任中国科学院植物研究所研究员、副所长、代所长等职，并于 1955 年入选首届中国科学院学部委员，1957 年后兼任生物学部副主任、主任，且任中国植物志第二届（1973—1974 年）和第三届（1975—1976 年）主编。林镕留学专业为真菌学，而回国后才改为研究种子植物，在旋花科、龙胆科，特别是菊科造诣深厚，不但组织完成了中国植物志菊科（74-80 卷）的编写，而且还培养了诸如陈艺林（1930— ）、林有润（1937— ）、刘尚武（1934— ）和石铸（1934—2005）等人才。[157]

Yong LING (1903-1981) (Photo provided by Zong-Gang HU)
林镕（1903—1981）（相片提供者：胡宗刚）

1920, Yong LING (1903-1981) studied

155 Ph.D. dissertation: Tchen-Ngo LIOU, 1929, Études sur la Géographie Botanique des Causses, Archives de Botanique, Tome III, memoire 1, 1-220 p; CAEN; Advisor: Marcel Denis (1897-1929).

156 江苏丹阳人；1919 年高中毕业。

157 陈艺林、林慰慈，2013，林镕文集（1903—1981），945 页；科学出版社，北京。

at College of Agriculture, University of Nancy, France and received a bachelor's degree in 1923, a master's degree at University of Clermont-Ferrand in 1927, and a Ph. D. degree at the University of Paris in 1930[158]. After returning to China he was a teacher at the Agricultural College of Peking University and concurrently a researcher in the Institute of Botany, National Academy of Peiping [Beijing]. During the anti-Japanese war, he moved to National Northwestern Associated University, National Northwest Agricultural College, Botanical Survey of North-Western China, then to the Institute of Agriculture and Forestry, Fujian Academy of Sciences, then to Amoy University and Fujian Agricultural College. After the war, he returned to Peiping [Beijing] where he was in charge of the recovery of the Institute of Botany, National Academy of Sciences. At the same time he taught at various colleges and universities in Beijing. After 1949, he was a research professor, vice director and acting director in the Institute of Botany, Chinese Academy of Sciences. He was elected an academician of the Chinese Academy of Sciences in 1955. He was Editor in Chief of the 2nd and 3rd editorial committee of *Flora Reipublicae Popularis Sinicae* in 1973-1974 and 1975-1976. Yong LING studied fungi abroad, but changed his research interests to seed plants after returning to China. He made major contributions studying Convolvulaceae, Gentianaceae, and particularly Asteraceae. He completed seven parts of volumes 74-80 of *Flora Reipublicae Popularis Sinicae*. He trained students such as Yi-Ling CHEN (1930-), Yuou-Ruen LING (1937-), Shang-Wu LIU (1934-) and Chu SHIH (1934-2005).

1920年，董爽秋[159]（董桂阳，1897—1980）赴法留学，入里昂大学，两年后转赴德国柏林大学专攻植物分类学，1927年获博士学位。1928年易名爽秋，历经艰辛终于返国任安徽大学生物学系教授。1930年任国立中山大学生物学系教授、

Shuang-Chiu TUNG (Kwei-Yang TUNG, 1897-1980) (Photo provided by Zong-Gang HU)
董爽秋（董桂阳，1897—1980）（相片提供者：胡宗刚）

158 Ph.D. dissertation: Yong LING, 1929, Etude des phénomènes de la sexualité chez les Mucorinées—suivi d'un appendice sur les Mucorinées d'Auvergne et spécialement les Mucorinées de Sol, *Revue Générale de Botanique* 42: 1- 202; Advisor: Fernand Moreau (1886-1980).

159 安徽贵池（池州）人；早年考入北京大学预科；1917年入上海商务印书馆当练习生；1919年返回安庆，次年考取公费留学生。参见：吴汉卿，2012年5月11日星期五，著名教育家董爽秋，池州日报B3版（人生驿站·人文池州）。

生物系主任（1933 年 7 月至 1938 年 4 月）兼教务长，后任教于贵州大学、国立西南联合大学；1943 年任国立西北大学理学院生物学系教授；1946 年任国立兰州大学教授兼教务长、文理学院院长、植物系主任[160]，1950 年受聘任湖南大学自然科学院院长兼生物学系主任、教授；1953 年高校调整后于 1953—1962 年和 1964—1972 年任湖南师范学院生物系主任兼教授。

1920, Shuang-Chiu TUNG (Kwei-Yang TUNG or Koe-Yang TONG, 1897-1980) studied at the University of Lyon in France. After two years he transferred to Universität zu Berlin, Germany, where he received a Ph. D. degree in 1927[161], for his studies in plant taxonomy. He returned to Anhui University and changed his name from Kwei-Yang to Shuang-Chiu in 1928. He later went to Sun Yat-Sen University as dean (1933.07-1938.04) of the department of biology, then professors of Kweichow [Guizhou] University, and National Southwest Associated University; then professor of department of biology, School of Science, National Northwest University in 1943. He was professor and provost of National Lanchow (Lanzhou) University, chairman of School of Art and Science and dean of Department of Botany in 1946. In 1950, he was dean of the School of Natural Sciences, Hunan University. He was a professor and dean of the department of biology, Hunan Normal College, from 1953 to 1962 and from 1964 to 1972.

1920—1921 年，由南京高等师范学校、北京大学、北京高等师范学校、广州岭南大学、上海商务印书馆及其他教育机关合作，集资 1.8 万元，由南京高等师范学校植物学教授胡先骕主持调查采集浙江、江西植物，采得标本 3 万份。[162]

1920-1921, Exploration of Zhejiang and Jiangxi co-organized by Nanjing Higher Teaching College, Peking University, Peking High Teacher's College, Lingnan University, Shanghai Commercial Press and other educational organizations, with a budget of 18,000 Chinese Yuan, led by Hsen-Hsu HU, Nanjing Higher Teaching College; 30,000 plant specimens were collected.

1920 and 1928, Vladimir L. Komarov, Botanical itineraries of major Russian expeditions to central Asia (Les itinéraires botaniques des principales expéditions russes en Asie centrale), vol. 1, Itineraries of N. M. Przewalski, *Trudy Imp. S.-Peterburgsk. Bot. Sada* 34: 1-192, 1920; vol. 2, Itineraries of G. N. Potanin, 1876-1890, *Trudy Imp. S.-Peterburgsk. Bot. Sada* 34: 197-404, 1928.

1920 和 1928 年，Vladimir L. Komarov，俄罗斯在中亚重要考察中的植物学行程，第 1 卷：N. M. Przewalski 行 程，*Trudy Imp. S.-Peterburgsk. Bot. Sada* 34：1-192，1920； 第 2 卷，G. N. Potanin 1876—1890 年行程，*Trudy Imp. S.-Peterburgsk. Bot. Sada* 34：197-404，1928。

160 兰州大学生命科学学院院志编撰委员会，2016，兰州大学生命科学学院院志，414 页；兰州大学出版社，兰州。

161 Ph.D. dissertation: Koe-Yang TONG, 1931, Studien über die Familie der Hammamelidaceae, mit besonderer Berücksichtigung der Systematik und Entwicklungsgeschichte von *Corylopsis*, 1-72 p, f. 1-32; *Bulletin of Department of Biology, Sun Yatsen University* 2: 1-71, 1931, and *Lingnan Science Journal* 12(1): 186-189, 1933; Advisor: Friedrich Ludwig E. Diels (1874-1945).

162 胡先骕，1921，江西植物名录，科学 6（11）：1 144-1 171，6（12）：1 232-1 247；胡先骕，1921，江西浙江植物标本鉴定名表 6（12）：1 248-1 254。

1921

1921 年，国立东南大学成立生物系，系主任胡先骕，钱崇澍任教授，其后陈焕镛、陈桢[163]（1894—1957）、张景钺[164]（1895—1975）加入。1925 年生物系分为植物系和动物系。

1921, Department of biology was established at the National Southeast University, with Hsen-Hsu HU as dean, Sung-Shu CHIEN as professor. They were later joined by Woon-Young CHUN, Chen CHEN (1894-1957), Ching-Yueh CHANG (1895-1975). The department of biology was divided into a department of botany and a department of zoology in 1925.

1921，Reginald J. Farrer, ***The Rainbow Bridge***, 383 p; Edward Arnold and Co., London.

1921 年，Reginald J. Farrer，**彩虹桥**，383 页；Edward Arnold and Co.，伦敦。

1921 年，胡先骕，江西植物名录（福建崇安县植物附）I、II，科学 6（11）：1 144-1 171，6（12）：1 232-1 247；胡先骕，江西浙江植物标本鉴定名表，科学 6（12）：1 248-1 254。

1921, Hsen-Hsu HU, A list of Kiangsi [Jiangxi] plants (including Tsung An District of Fukien [Fijian]) I and II, *Science (China)* 6(11): 1,144-1,171, 6(12): 1,232-1,247. Hsen-Hsu HU, A list of identification of plant specimens of Kiangsi [Jiangxi] and Chekiang [Zhejiang], *Science (China)* 6(12): 1,248-1,254.

1921 年，彭世芳、王烈、陈映璜，**博物词典**，12、326、68、4、4、112 页；中华书局，上海。

1921, Shih-Fang PENG, Lieh WANG and Ying-Huang CHEN, '***Dictionary of Natural History***', 12+326+68+4+4+112 p; Chinese Books, Shanghai.

---

163 江西铅山人，1913 年入上海中国公学预科，1914 年毕业后考入金陵大学农林科，1918 年毕业获得农学学士学位，并留校任育种学助教。1919 年考取国立清华大学专科公费赴美留学；先在康奈尔大学农学系进修，1920 年转入哥伦比亚大学动物学系学习，1921 年获硕士学位后随著名遗传学家 Thomas H. Morgan（1866—1945）从事研究。1922 年回国后任南京国立东南大学生物系教授；1926 年任国立清华大学生物系教授，并担任系主任，抗战时期任教于国立西南联合大学，1946 年复任国立清华大学生物系主任；1948 年当选为中央研究院院士和北平研究院评议员；1949 年以后继续担任清华大学生物系主任，1952 年调入北京大学生物系；1953 年起担任中国科学院动物研究室主任，1955 年选聘为中国科学院学部委员；1957 年动物研究室改制为动物研究所，担任所长；中国较早的现代生物学者及教育家，也是中国动物遗传学的创始人。

164 原籍江苏武进，生于湖北光化，1920 年毕业于清华学校（国立清华大学前身），1926 年获美国芝加哥大学科学博士学位，1948 年选聘为中央研究院院士；长期任北京大学生物系教授、系主任，培养了大批生物学人才，特别是植物形态学方面的专家，如严楚江、徐仁、王伏雄、李正理等都是他的学生；著名植物形态学家；1955 年选为中国科学院学部委员。参见：张景钺文集编辑委员会，1995，张景钺文集，319 页；北京大学出版社，北京。

1921, Francis Kingdon Ward, ***In Farthest Burma***—*The record of an arduous journey of exploration and research through the unknown frontier territory of Burma and Tibet*, 311 p; Seeley, Service and Co., Limited, London.

1921 年，Francis Kingdon Ward，**最遥远的缅甸** —— 一次通过缅甸和西藏未知边界的艰险探索和研究之旅，311 页；Seeley, Service and Co., Limited，伦敦。

1921-1924, George W. Groff, ***Plants of Kwangtung Province and Their Chinese Names***, 1: 362 p, 1921; 2: 346 p, 1924; 3: 140 p, 1921 (unpublished manuscript, only five copies prepared).

1921—1924 年，George W. Groff，**广东省植物以及它们的中文名字**，1：362 页，1921；2：346 页，1924；3：140 页，1921（未刊行手稿，只准备了 5 份）。

1921、1924 年，胡先骕，浙江植物名录、增订浙江植物名录，科学 6（1）：70-101，1921，9（7）：818-847，1924。

1921 and 1924, Hsen-Hsu HU, Enumeration of Plants in Chekiang province, China and Supplement to Enumeration of Plants in Chekiang province, China, *Science (China)* 6(1): 70-101, 1921, and 9(7): 818-847, 1924.

1921-1934, Karl August Harald (Harry) Smith (1889-1971) collected in Hebei, Sichuan, Shanxi three times (1921-1922, 1924, 1934). His more than 11,000 specimens are mainly in Sweden (UPS) with duplicates at many institutions (particularly also GB, S and TIE).[165]

1921—1934 年，Karl August Harald (Harry) Smith(1889—1971)，3 次来华(1921—1922 年、1924 年、1934 年)，在河北、四川、山西等地采集。他的 11 000 多份植物标本主要存于瑞典（UPS)，但其他标本馆也存有复份（特别是 GB、S、TIE)。

Karl August Harald (Harry) Smith (1889-1971) (Photo provided by Uppsala University)

Karl August Harald (Harry) Smith（1889—1971）（相片提供者：乌普萨拉大学）

165 Goran Herner, 1988, Harry Smith in China: Routes of his botanical travels, *Taxon* 37(2): 299-308.

1922

1922 年 8 月 18 日，第一个生物学研究机构，中国科学社生物研究所于南京成立；秉志任所长，同时兼任动物部主任，胡先骕任植物部主任（至 1929 年）；植物标本室于次年成立（NAS）[166]。至 1936 年，在该所从事植物分类学者有钱崇澍（自 1929 年起任植物部主任）、裴鉴（1902—1969）、邓叔群[167]（1902—1970）、方文培（1899—1983）、孙雄才[168]（1898—1964）、郑万钧（1904—1983）、吴中伦（1913—1995）、贺贤育等，而曾在该所从事植物分类学的有陈焕镛、秦仁昌[169]（1898—1986）、汪振儒[170]（汪燕杰，1908—2008）、曲仲湘[171]（曲桂龄，1905—1990）、陈长年[172]等。

1922.08.18, The Biological Laboratory of the Science Society of China, the first biological research institute in China, was founded in Nanjing. The first director and head of zoology was Chi PING while head of botany was Hsen-Hsu HU (until 1929). The herbarium (NAS) was founded by the following year. Sung-Shu CHIEN (as head of botany since 1929), Chien PEI (1902-1969), Shu-Chun TENG (1902-1970), Wen-Pei FANG (1899-1983), Hsiong-Tsai SUN (1898-1964), Wan-Chun CHENG (1904-1983), Chung-Lwen

---

166 薛攀皋，1992，中国科学社生物研究所——中国最早的生物学研究机构，中国科技史料 13（2）：47–57。

167 福建福州人，1915—1923 年就读于清华学堂（清华大学），1923 年赴美，1928 年获美国康奈尔大学森林学硕士学位和植物病理学博士学位，1928 年回国后任教于岭南大学、国立中央大学、中央研究院、甘肃水利林牧公司、沈阳农学院等；1948 年选聘为中央研究院院士；中国科学院微生物研究所研究员、副所长。

168 江苏宜兴人，1926 年毕业于南京国立东南大学农科，历任江苏省水产专门学校生物学教师、国立中山大学植物系讲师、英士大学植物学及树木学教授、中国科学社生物研究所研究员、华东药学院植物学教授兼生药系主任、南京药物研究所研究员兼植物研究室主任等职；著名唇形科植物分类学家。参见：编辑部，1964，中国唇形科分类学家孙雄才教授逝世，植物学报 12（2）：212；胡俊铉，朱家壁，1985，怀念孙雄才教授，药学通报 70（1）：55-57。

169 江苏省武进人，1914—1919 年在江苏省第一甲种农校学习林业，1919—1924 年在金陵大学林学系学习林科，获学士学位，1924—1925 年任国立东南大学理学院生物系植物学助教，1926—1927 年任国立中央大学理学院生物系植物学讲师，1927—1932 年任中央研究院自然历史博物馆植物部主任，其中 1930—1932 年赴丹麦哥本哈根大学植物学博物馆研修（期间访问了瑞典、德国、法国、奥地利、捷克和斯洛伐克的标本馆，以及英国皇家植物园标本馆和英国自然历史博物馆），1932—1934 年任静生生物调查所研究员兼植物标本室主任，1934—1938 年创建庐山森林植物园并任主任，1939—1945 年创建庐山森林植物园云南丽江植物工作站并任主任，1941—1944 年任农业部丽江金沙江流域国有林区管理处主任，1945—1948 年任云南大学生物系和森林学系教授兼系主任，1948—1950 年任云南省建设厅农业改进所所长，1950—1954 年任云南省农林厅林业局副局长，1955—1979 年任中国科学院植物研究所研究员兼植物分类与植物地理学研究室主任，1979—1986 年任中国科学院植物研究所顾问；1986 年 7 月 22 日病逝于北京。参见：王中仁，2005，中国蕨类植物系统学创始人——秦仁昌，仙湖　4：32-34。

170 祖籍广西桂林，生于北京；名燕杰，字振儒（以字行）；参见 1935 年条目。

171 又名曲桂龄，河南唐河人，1930 年获国立中央大学生物专业学士学位，毕业后先后任职于中国科学社生物研究所和复旦大学以及西部科学院植物部；1945 年赴加拿大多伦多大学学习，后转入美国明尼苏达大学并于 1948 年获植物生态学硕士学位；回国后先后任教于复旦大学生物系、南京大学生物系、云南大学生物系；著名植物生态学家。参见：曲仲湘著，1990，曲仲湘论文集，338 页；中国环境科学出版社，北京。

172 生辰不详，野外采集时过世。

The Biological Laboratory of the Science Society of China founded, Nanking [Nanjing], 1922 (Photo provided by Zong-Gang HU)

1922 年中国科学社生物研究所于南京成立（相片提供者：胡宗刚）

WU (Chung-Lun WU, 1913-1995) and Xian-Yu HO were also employed as plant taxomists in the Biological Laboratory in 1936. Woon-Young CHUN, Ren-Chang CHING (1898-1986), Chen-Ju WANG (Yen-Chieh WANG, 1908-2008), Chung-Hsiang CHU (Kuei-Ling CHU, 1905-1990) and Chang-Nian CHEN once worked as plant taxomists in the Biological Laboratory.

1922, Woon-Young CHUN, ***Chinese Economic Trees***, 309 p; plates 100; Commercial Press, Shanghai. This is the first taxonomic monograph on Chinese plants written in English by a Chinese author.

1922 年，陈焕镛，**中国经济树木**，309 页，图版：100；商务印书馆，上海。此为国人撰写的第一本英文植物分类学专著。

1922, Elmer D. Merrill visited Albert N. Steward at the University of Nanking to organize the herbarium. Merrill met with Woon-Young CHUN, Hsen-Hsu HU and Sung-Shu CHIEN at National Southeast University to begin a long collaboration with Chinese botanists.

1922 年，Elmer D. Merrill 受金陵大学植物学教授史德蔚邀请来南京，为该校植物标本室进行规划，并与国立东南大学植物学教授陈焕镛、胡先骕、钱崇澍相晤，由此开始与中国植物学家建立长期的合作。

1922, Hans Wolfgang Limpricht[173], Botanische Reisen in den Hochgebirgen Chinas und Ost-Tibets, *Repertorium Specierum Novarum Regni Vegetabilis Beihefte* 12: 1-297.

1922 年，Hans Wolfgang Limpricht，中国高山地带与西藏东部植物之旅，*Repertorium Specierum Novarum Regni Vegetabilis Beihefte* 12：1-297。

1922, Ferdinand A. Pax, Aufzählung der von Dr. Limpricht in Ostasien gesammelten Pflanzen, *Repertorium Specierum Novarum Regni Vegetabilis Beihefte* 12: 298-515.

1922 年，Ferdinand A. Pax，Limpricht 博士在东亚采集的植物名录，*Repertorium Specierum Novarum Regni Vegetabilis Beihefte* 12：298-515。

1922 年，钱崇澍（领队）、陈焕镛、秦仁昌和黄宗组成采集队，赴湖北宜昌采集，并聘请了当年为英国人威尔逊在湖北西部采集的向导，遂沿着威尔逊采集路线，采集其所发现而国内尚未有的标本；采集队经兴山、神农架东侧到达巴东，最高采集地点到达海拔 3 000 米的新店子；采得近千号标本，

173 Hans Wolfgang Limpricht（1877—1954?），德国植物学家，1910 年来华，在上海语言学校任教，同时兼职于同济医学院，后赴天津任教；在华期间于江苏、浙江、福建、秦岭和华北一带采集，特别是 1914 年夏专门赴川西及藏东采集，植物标本送德国学者研究，并存德国（B）和波兰（WRLS）（https://plants.jstor.org/stable/10.5555/al.ap.person.bm000335632，2017 年 7 月进入）。

每号至少 10 份。

1922, Sung-Shu CHIEN (leader), Woon-Young CHUN, Ren-Chang CHING and Chung HUANG, explored around Yichang, Hubei, using Ernest H. Wilson's local guide, with aim to find plants not previously conserved in China yet. They traveled to Badong via Xingshan from the east side of Shennongjia, reaching 3,000 m elevation at Xindianzi. They collected nearly 1,000 numbers of specimens with at least 10 sheets per number.

1922 年，厦门大学成立 1 年后，钟心煊创建动植物学系并建立植物标本室（AU）。[174]

1922, The department of botany as well as it's herbarium (AU), Amoy University (now Xiamen University, Xiamen, Fujian), was established by Hsin-Hsuan CHUNG[175], just one year after the founding of the university.

1922-1927, Lingnan University, ***Lingnan Agricultural Review***, vols. 1-4; 1928-1941, 1945, 1948, 1951, ***Lingnan Science Journal***, vols. 5-20, 21, 22, 23.

1922—1927 年，岭南大学，**岭南农业评论**，第 1-4 卷；1928—1941、1945、1948、1951，**岭南科学杂志**，第 5-20、21、22、23 卷。

1922-1948, Joseph F. K. Rock[176](1884-1962), Austrian American botanist, collected in Gansu, Qinghai (Kokonor Lake [Qinghai Hu]), Sichuan, Xizang, Yunnan and neighboring countries, between 1922-1924, 1924-1927. He was supported by funds from the United States Department of Agriculture, The National Geographic Society, the Arnold Arboretum of Harvard University and the Harvard Yenching Institute. His specimens are widely distributed but are mainly in the United States (A, GH, US). During his later years in China, particularly between 1933 and 1948, he focused on the ethnography of the Naxi (Na-khi) people of

174 厦门大学抗战期间被日本占领，标本等被运往台湾；抗战胜利后标本运回，但标志性的红杉树横切面圆盘（标本馆 1927 年从美国获得）则留在了台湾，至今仍在台湾大学植物研究所大楼一层摆放并附有说明。参见：厦门大学报，2013 年 1 月 5 日，第 1 015 期，厦门大学生物学系系史，第 2 版。谢长富，2019，台湾大学生命科学馆的世界爷树木圆盘横断面来台始末，植物苑电子版（2019 年 5 月号）3：4–22 页（https://www.ntuipb.info/newsletter，2019 年进入）。

175 Hsin-Hsuan CHUNG, 1929, The study of Botany in Fukien [Fujian], *Lingnan Science Journal* 7: 121-130.

176 Joseph Rock, botanist, anthropologist, explorer, linguist, author and world-renowned personality, immigrated to the United States from Austria in 1905. From 1907 to 1920 he lived in Hawaii, where he became a self-taught specialist on the Hawaiian Flora. He joined the faculty of the College of Hawaii in 1911. The college was changed to the University of Hawaii in 1920. He established the first herbarium (HAW) and served as its first curator from 1911 until 1920. From 1920 (from 1922 in China) until 1949, he explored and collected in Asia for various United States institutions and agencies. He introduced many useful plants from Asia to North America and Europe. He was a research fellow at the Harvard Yenching Institute between 1945 and 1950. He returned to Hawaii where he became a professor of oriental studies in 1948. Shortly before his death in 1962, the University of Hawaii awarded him an honorary doctor of science degree. In 2009, the University of Hawaii Manoa herbarium was named the Rock Herbarium in his honor. He has been considered by many to be the father of Hawaiian botany. He died of a heart attack in 1962.

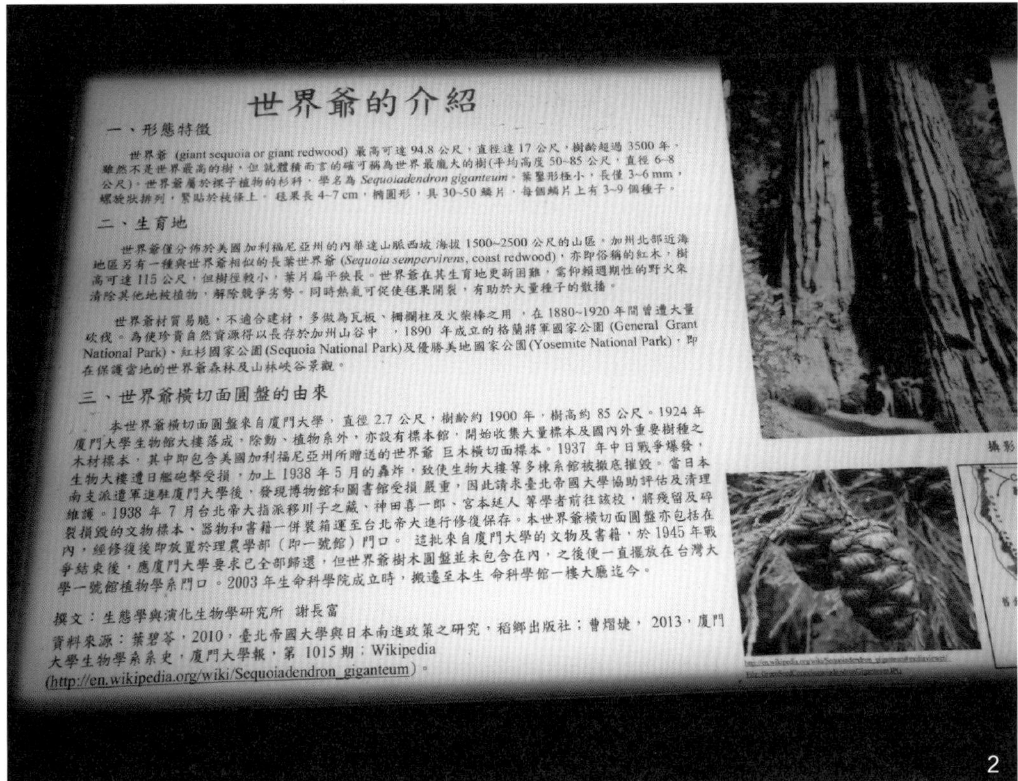

Cross section of trunk of *Sequoiadendron giganteum*, treasure of the herbarium (AU), department of biology, Amoy University, before the war with Japan, now in the Institute of Plant Biology, Taiwan University. Figure I, Jer-Ming HU (left) the current curator of Herbarium (TAI) and Chang-Fu HSIEH (right), former curator of the herbarium (TAI), who wrote the explanation in Figure 2. Figure 2, Explanation (Photo provided by Jin-Shuang MA)

抗战前之厦门大学生物系植物标本馆（AU）镇馆之宝（北美巨杉之年轮）现位于台湾大学植物研究所。图 1: 台湾大学植物标本馆现任馆长胡哲明博士（左）和台湾大学植物标本馆前任馆长、解说词（图 2）的撰写者谢长富博士（右）。图 2: 解说词（相片提供者：马金双）

Lijiang, northwestern Yunnan. His *The ancient Na-khi Kingdom of southwest China* was published by the Harvard-Yenching Institute in 1947.[177]

1922—1948 年，美籍奥地利植物学家 Joseph F. K. Rock（洛克[178],1884—1962）1922—1924 年、1924—1927 年于云南、四川、西藏、甘肃、青海（青海湖）等地在美国农业部、国家地理学会、哈佛大学阿诺德树木园和哈佛燕京学社的资助下大规模采集，他的植物标本现存欧美多地，主要在美国（A、GH、US）。他在华后期，特别是 1933—1948 年，主要居住于云南西北部的丽江并致力于纳西族文化的研究。他所著的**中国西南古纳西王国**[179]于 1947 年在哈佛燕京学社出版。

Joseph F. K. Rock (1884-1962) (Photo provided by University of Hawaii)
Joseph F. K. Rock（1884—1962）（相片提供者：夏威夷大学）

177 Alvin K. Chock, 1963, J. F. Rock, 1884-1962, *Taxon* 12(3): 89-102; Jeffrey Wagner, 1992, The botanical legacy of Joseph Rock, *Arnoldia* 52(2): 29-35.

178 参见：王大卫，2003，寻找天堂——追寻美籍奥地利探险家、学者约瑟夫·洛克在中国 27 年的生死之旅，287 页；中国文联出版社，北京；为剑主编，2004，洛克与香格里拉，283 页；海天出版社，深圳。

179 刘宗岳等，译校本，368 页；1999 年，云南美术出版社，昆明。

1923

1923 年 11 月，胡先骕、邹秉文、钱崇澍编著，**高等植物学**，470 页；商务印书馆，上海。此为国人撰写的第一部大学植物学教科书。

1923.11, Hsen-Hsu HU, Ping-Wen TSOU and Sung-Shu CHIEN[180], ***Advanced Botany***, 470 p; Commercial Press, Shanghai. This is the first textbook for colleges and universities written by Chinese botanists.

1923 年 12 月 12 日，国立东南大学口字房毁于火灾，其中农科生物系损失最大，所藏动植物标本 30 000 余号，付之一炬。

1923.12.12, The Quadrangle Building of National Southeast University in Nanjing burned down with loss of up to 30,000 biological collections in the department of biology.

1923, Peter K. Kozlov, ***Mongolia, Amdo and the Dead City Khara-Khoto,*** Expedition of the Russian Geographical Society of P. K. Kozlov in mountainous Asia, 1907–1909, 678 p; *Mongolie, Amdo und die tote Stadt Chara-Choto,* Die Expedition der Russischen Geographischen Gesellschaft 1907–1909, 1925, 305 p; Neufeld and Henius, Berlin. Translated into Chinese in 2011.

2011 年，陈贵星译，**死城之旅**，361 页；新疆人民出版社，乌鲁木齐。

1923, Thomas S. Lindsay, ***Plant Names***, 93 p; Sheldon Press, London, The MacMillan Co., New York. Translated into Chinese in 1945.

1945 年，盛诚桂编，**植物命名考**，115 页；商务印书馆，重庆。

1923, The National Geographic Society Central China (Gansu) Expedition, led by Frederick R. Wulsin[181] (1891–1961) of the United States, included participants from humanity, zoology and botany. Ren-Chang CHING, assistant teacher from National Southeast University, was the botanist. He collected from spring to autumn in Ningxia, Gansu and Qinghai. The collections of 1, 158 numbers were studied by Egbert H. Walker

---

180 Original spelling is Sien-Siu HU, Ping-Wen CHOU, and Chung-Shu TSIEN.

181 American Anthropologist, graduated from Harvard University, A.B. 1913 and Ph.D. 1929, collected zoological specimens in East Africa and Madagascar, 1914–1915; and in China, Mongolia, Kokonor [Qinghai] and Indo-china, 1921–1924, made archaeological journeys to the Belgian Congo and French Equatorial Africa, 1927–1928; and to Persia, 1930–1931; also served as first lieutenant in the U.S. Army during WWI and Quartermaster Corps during WWII; lecturer at Boston University, 1935–1936 and lecturer and professor at Tufts College (now University), 1945–1957 (http://snaccooperative.org/ark:/99166/w6m3650s, accessed 19 September 2017).

(1899-1981); the results, by Egbert H. Walker, were published in the United States in 1941[182]. Ren-Chang CHING also published an account of the trip in the same year.[183]

1923 年，美国国家地理学会资助 Frederick R. Wulsin（吴立森，1891—1961）组织中国中部（甘肃）科学考察团，成员包括人文、动物和植物；植物学者为国立东南大学助教秦仁昌，由春至秋，自包头出发，经宁夏、甘肃到青海西宁采集；所采植物标本 1 158 号由美国学者 Egbert H. Walker（和嘉，1899—1981）鉴定，并于 1941 年出版（包括秦仁昌的考察报告）；秦仁昌本人也于 1941 年发表植物采集记略。

1923, Francis Kingdon Ward, ***The Mystery Rivers of Tibet****—a description of the little known land where Asia's mightiest rivers gallop in harness through the narrow gateway of Tibet, its peoples, fauna and flora*, 316 p; Seeley, Service and Co. Ltd., London. Translated into Chinese in 2002.

2002 年，李金希、尤永弘译，**神秘的滇藏河流**——横断山脉江河流域的人文与植被，272 页；四川民族出版社，成都、中国社会科学出版社，北京。

1923 年，吴韫珍[184]（1899—1942）赴美留学，在康奈尔大学主修园艺、辅修植物分类学，1927 年获得博士学位，回国后任教于国立清华大学；1933 年利用学术休假赴奥地利维也纳与 Heinrich R. E. Handel-Mazzetti 研究中国植物，1937 年抗战时随校迁滇，任国立西南联合大学教授，1942 年病逝于云南昆明；培养的弟子之一为著名植物学家吴征镒[185]（1916—2013）。

1923, Yun-Chen WU[186] (Wen-Chen WU, 1899-1942) studied horticulture and plant taxonomy at Cornell University, Ithaca, New York, U.S.A.. He earned his Ph. D. degree in 1927[187] and returned to China to become a professor at National Tsinghua University. He spent his sabbatical leave in 1933 with Heinrich R. E. Handel-Mazzetti in Vienna, Austria, studying Chinese plants. In 1937, he travelled from Beijing to

182 Egbert H. Walker, 1941. Plant Collected by R. C. Ching in Southern Mongolia and Kansu Province, China, *Contributions from the United States National Herbarium* 28(4): 563-675; inculding the report of expedition by Ren-Chang CHING, 573-593.

183 Ren-Chang CHING, 1941, A botanical trip in the Ho La Shan, Inner Mongolia, *Bulletin of the Fan Memorial Institute of Biology* 10(5): 257-265.

184 上海青浦人，号振声，1918 年入金陵大学农科，1922 年毕业后任教于安徽省立农校。参见：杨亲二，1990，英年早逝学风永存——记植物分类学家吴韫珍先生，植物杂志 4：44-45。

185 原籍安徽歙县，出生江西九江，1931 年考入江苏省立扬州中学，1933 年考入国立清华大学生物系，1937 年本科毕业留校任教，1940—1942 年入国立西南联合大学理科研究所攻读研究生；1949 年参与组建中国科学院，1950 年任中国科学院植物研究所研究员兼副所长，1955 年选聘为中国科学院学部委员，1958 年任中国科学院昆明植物研究所所长，1979 年兼任中国科学院昆明分院院长，1977 年主编云南植物志、1983 年主编西藏植物志、1986 年主编中国植物志，1989 年联席主编英文版中国植物志。中国当代著名植物分类学家（http://www.cas.cn/ky/kjjl/gjzgkxjsj/2007n/wzy/grjj/200907/t20090729_2282584.shtml，2017 年年底进入）。

186 Zhe-Kun ZHOU and Hang SUN, 2016, Wu Zhengyi and his contributions to plant taxonomy and phytogeography, *Plant Diversity* 38: 259-261.

187 Ph.D. dissertation: Some studies in catalase activity of apple leaf tissue.

Kunming during the war to become a professor at National Southwestern Associated University. He died in 1942 of illness in Kunming, Yunnan. One of the students he trained was Cheng-Yih WU (Zheng-Yi WU, 1916-2013), one of China's most famous botanists.

1923-1924, and 1932, Science Society of China, ***Memoirs of the Science Society of China***, vols. 1(1-3).

1923—1924 年、1932 年，中国科学社，**中国科学社研究丛刊**，第 1 卷第 1-3 号。[188]

1923—1925 年，胡先骕再次赴美研修中国植物，并于 1924 年和 1925 年分别获哈佛大学硕士和博士学位 [189]，成为哈佛大学第一个中国植物分类学博士。胡先骕回国后在植物分类学界做出非凡的成就，被公认为中国植物分类学奠基人。[190]

1923-1925, Hsen-Hsu HU traveled to the United States to study Chinese plants at Harvard University. He earned a masters's of science degree in 1924 and a doctor of science degree in

Hsen-Hsu HU at Harvard University (Photo provided by Zong-Gang HU)
胡先骕于哈佛大学（相片提供者：胡宗刚）

188 第 1 号发表于 1924 年，而第 2 号发表于 1923 年，第 3 号发表于 1932。

189 胡先骕，1925，中国植物志属，两卷本；第 1 卷：1-536 页，第 2 卷：537-1 097 页。博士论文，指导教师：John G. Jack (1861-1949)。原稿现存哈佛大学图书馆。手稿经作者回国之后修订与增补，内容也不完全一致。胡先骕亲自打印之后装订的卷数和页码与哈佛大学图书馆的原稿不同；中国科学院植物研究所图书馆的为 3 卷本，页码分别为 556、475 和 442 页，且还有索引 1 册，69 页。详细参见：王宗训，1985，回忆北平研究院植物研究所，中国科技史料 6（2）：16-20。

190 1928 年与秉志等人在尚志学会和中华教育文化基金会的支持下，于北平创办静生生物调查所，秉志任所长兼动物部主任，胡先骕任植物部主任，并受聘在北京大学和国立北平师范大学讲授植物学；1932 年任所长；1934 年派秦仁昌创办庐山森林植物园；1938 年派俞德浚会同蔡希陶创建云南省农林植物研究所并兼任所长。1940 年赴江西泰和就任国立中正大学首任校长，1946 年返回北平主持静生生物调查所工作，1948 年与国立中央大学森林学系郑万钧教授联合发表有关裸子植物水杉新种的论文，新中国成立后任中国科学院植物分类研究所研究员，1968 年 7 月 16 日忧闷中突发心肌梗死逝世。胡宗刚，2008，胡先骕先生年谱长编，688 页；江西教育出版社，南昌。马金双，2008，新书介绍:《胡先骕先生年谱长编》，植物分类学报 46（5）：793-794。

1925[191]. He was the first Chinese student to earn a doctor of science degree in plant taxonomy at Harvard University. Great contributions in plant taxonomy were made by Hsen-Hsu HU after he returned to China. He is commonly recognized as the founder of plant taxonomy in China.[192]

Franklin P. Metcalf (1892-1955) (Photo provided by Oberlin College)
Franklin P. Metcalf（1892—1955）（相片提供者：Oberlin College）

1923-1929, Franklin P. Metcalf[193](1892-1955) collected in Fujian, Guangdong, Hainan and Indochina. He collected more than ten thousands numbers represented by around sixty thousand specimens that are widely distributed (A, BM, C, FJFC, FNU, M, MICH, MO, MSC, NF, P, S, SYS, US). The Herbarium of the department of biology, Fukien Christian College (today's Fujian Normal University, FNU), was founded by Franklin P. Metcalf.[194]

1923—1929 年，Franklin P. Metcalf（1892—1955）在福建、广东、海南及中南半岛等地采集1万多号6万多份标本(A、BM、C、FJFC、FNU、M、MICH、MO、MSC、NF、P、S、SYS、US）散落世界各地，并建立福建协和学院生物系植物标本室（今福建师范大学植物标本室，FNU）。[195]

191 Ph.D. Dissertation: Hsen-Hsu HU, 1925, Synopsis of Chinese Genera of Phanerogams with Descriptions of Representative Species; vols. 1-2, 1-536 and 537-1,097, Advisor: John G. Jack (1861-1949). The manuscript was reprinted internally as three volumes in the library of Institute of Botany, Chinese Academy of Sciences, after revised and enlarged since back to China.

192 Jin-Shuang MA and Kerry Barringer, 2005, Dr. Hsen-Hsu HU (1894-1968)—A founder of modern plant taxonomy in China, *Taxon*, 54(2): 559-566.

193 Franklin P. Metcalf received his A.B. from Oberlin College, Oberlin, Ohio, in 1913, then worked at Cornell University from 1913 to 1923. From 1923 to 1928 he was a professor of botany at Fukien Christian University. During 1928-1929 he was a China Medical Board Fellow of the Rockefeller Foundation and then a Fellow of Lingnan University, Canton [Guangzhou], China, from 1930-1931. In 1931 he received his Ph.D. degree in systematic botany from Cornell University. From 1932 to 1940, he was the curator of the herbarium of Lingnan Natural Historical Survey and Museum. He returned to the United States from 1940 to 1942 and spent time at the Arnold Arboretum of Harvard University. He was inducted into the United States Army Air Forces where he served as a captain from 1942 to 1947, when he retired (http://oasis.lib.harvard.edu/oasis/deliver/~ajp00028, accessed 19 September 2017).

194 1929, Hsin-Hsuan CHUNG, The study of botany in Fukien, *Lingnan Science Journal* 7: 121-130.

195 Franklin P. Metcalf, 1929, The herbarium at Fukien Christian University, *Proceedings of the Natural History Society of Fukien Christian University, Foochow, China*, 2: 26-27.

1924

1924 年 9 月 1 日，国立广东大学生物学系正式成立[196]，费鸿年（1900—1993）任系主任。1926 年国立广东大学更名为国立中山大学。

1924.09.01, Department of Biology, National Kwangtung University, officially founded, with Hung-Nien PEI (1900-1993) as the department head. National Kwangtung University was renamed National Sun Yat-Sen University in 1926.

1924 年，留学法国学习生物的周太玄[197]（1895—1968）、刘慎谔、汪德耀[198]（1903—2000）、张玺[199]（1897—1967）、林镕、刘厚等 40 余人，在里昂成立中国生物科学学会。这是中国第一个生物学术团体。

1924, The Chinese Society of Biological Sciences, the first academic society of biology of China, was founded by Tai-Hsuan CHOU (1895-1968), Tchen-Ngo LIOU, Te-Yueh WANG (1903-2000), Hsi CHANG (1897-1967), Yong LING and Hou LIU in Lyon, France, with forty Chinese students in France.

1924, Francis Kingdon Ward, ***The Romance of Plant Hunting***, 275 p; Edward Arnold and Co., London.

1924 年，Francis Kingdon Ward，**搜寻植物的传奇**，275 页；Edward Arnold and Co.，伦敦。

1924, Francis Kingdon Ward, ***From China to Hkamti Long***, 317 p; Edward Arnold and Co., London.

1924 年，Francis Kingdon Ward，**从中国到康提垄**，317 页；Edward Arnold and Co.，伦敦。

1924 年，辛树帜[200]（1894—1977）留学英国，后转入德国，1925—1927 年在柏林植物园学习植物分类学，1927 年受邀回国任国立中山大学教授，1929—1931 年兼任生物系主任，1928—1931 年主持广西大瑶山生物考察；1931 年筹建国立编译馆并于 1933 年任馆长，1936 年任国立西北农林专科学校校长，1939 年国立西北农林专科学校易名国立西北农学院任首任院长；期间，联合北平研究院植

---

196 冯双，2011，中山大学生命科学院（生物学系）编年史，8 页；中山大学出版社，广州。

197 原名周焯、号朗宣，后改名周无、号太玄，祖籍河南祥符，出生于四川新都，中国近代知名生物学家、教育家、翻译家、诗人、编辑。

198 江苏灌云人，细胞生物学家，厦门大学原校长。

199 字尔玉，河北邢台人，著名的海洋动物学家、湖沼学家，中国贝类学的开创者和奠基人；曾任中国科学院海洋研究所副所长。

200 湖南临澧人，1915 年入武昌高等师范学校生物系，1919—1924 年在长沙第一师范、长沙明德中学、长郡中学等校任生物教员。

物学研究所刘慎谔创办中国西北植物调查所，1946年出任国立兰州大学首任校长，1950—1967年又调回西北农学院任院长。

1924, Shu-Chih HSIN (1894-1977) studied in the United Kingdom, then transferred to Germany to study plant taxonomy at the Botanische Garten Berlin-Dahlem from 1925 to 1927. He was invited to return to China in 1927 to become a professor in the department of biology, National Sun Yat-Sen University. He served as dean of the department of biology from 1929 to 1931. He organized the largest biological exploration of Dayao Shan, Guangxi, from 1928 to 1931. He established the National Translation Department in 1931 and served as director of the department from 1933 to 1936. He was president of the National Northwest School of Agriculture and Forestry in 1936, then president of the National Northwest Agricultural College in 1939, co-founder with Tchen-Ngo LIOU of the Northwest Institute of Botanical Survey, president of National Lanchow (Lanzhou) University in 1946 and president of Northwest Agricultural College from 1950 to 1967.

Shu-Chih HSIN (1894-1977) (Photo provided by Zong-Gang HU)
辛树帜（1894—1977）（相片提供者：胡宗刚）

1924年，钟心煊，中国木本植物目录，中国科学社研究丛刊，1（1）：1-271页。

1924, Hsin-Hsuan CHUNG, A catalogue of trees and shrubs of China, *Memoirs of the Science Society of China* 1(1): 1-271.

1924—1926年，国立东南大学植物系讲师秦仁昌率采集队历往绥远[201]、甘肃、蒙古、青海、浙江、安徽、福建及江西边界等处采集；派采集员郑万钧等人先后往安徽九华山、浙江天目山采集，共得标本2.5万份。加上旧存及一些机构赠予标本，至1926年植物系标本室收藏已鉴定标本3.18万份。

1924-1926, Ren-Chang CHING, lecturer, department of botany, National Southeast University, led

201 今内蒙古。

collecting expeditions to Anhui, Fujian, Gansu, Jiangxi, Mongolia, Shuiyuan[202], Qinghai and Zhejiang. Wan-Chun CHENG and others were sent to Jiuhua Shan, Anhui and Tianmu Shan, Zhejiang, for further collection. They collected more than 25,000 plant specimens. With donations from other institutions, the number of identified specimens in the herbarium of the department of botany, National Southeast University, reached 31,800 sheets by 1926.

202 Inner Mongolia.

1925

1925 年，北京大学生物系正式成立；谭熙鸿（1891—1956）任主任。

1925, The department of biology of Peking University was officially established, with Hsi-Hung TAN (1891-1956) as dean.

1925, Michiya Miura, ***List of Plants in Manchuria and Mongolia—Vascular Plants***, South Manchurian Railway No 25; 381+35 p; Agricultural Experiment Station, Kungchuling.

1925 年，三浦密成（三浦道哉），**蒙满植物名录**，南满洲铁道株式会社兴业部农务课产业资料第 25 号，381 页 +35 页；农业试验站，公主岭。

1925 年，裴鉴[203]（1902—1969）入美国斯坦福大学植物系，1927 年获学士学位，1928 年获硕士学位，1930 年获博士学位；1931—1944 年任中国科学社生物研究所研究员，并先后兼任中央研究院动植物研究所研究员和国立中央大学、国立药学专科学校、复旦大学、光华大学、金陵大学教授；1944—1949 年任“中央”研究院植物研究所研究员，1950—1954 年任中国科学院植物分类研究所华东工作站主任、研究员，1954—1960 年任中国科学院南京中山植物园主任、研究员，1960—1969 年任中国科学院南京植物研究所所长、研究员。

1925, Chien PEI (1902-1969) went to Stanford University in Stanford, California, United States of America, to study botany. He received a bachelor's degree in 1927, a master's degree in 1928 and a Ph. D. degree in 1930[204]. He returned to China to become a professor in the Biological Laboratory of the Science Society of China from 1931 to 1944 and concurrently research professor and professor at the Institute of Zoology and Botany, Academia Sinica, the National Central University, the National Pharmaceutical School, Fudan University, Kwang Hua University and the University of Nanking. He was a research professor in the Institute of Botany, Academia Sinica, from 1944 to 1949, director and research professor of the East China Station of the Institute of Plant Taxonomy, Chinese Academy of Sciences, from 1950 to 1954, director and research professor of Nanjing Botanical Garden Memorial Sun Yat-Sen, Chinese Academy of Sciences, from 1954 to 1960, and director and research professor of Nanjing Institute of Botany, Chinese Academy of Sciences, from 1960 to 1969.

1925-1932, Yoshimatsu Yamamoto, ***Supplementa Iconum Plantarum Formosanarum***, vols. 1-5; 1: 47

---

203 四川华阳人，1916 年入清华预科，1925 年毕业。

204 Chien PEI, 1932, The Verbenaceae of China, *Memoirs of the Science Society of China*, vol. 1(3): 1-193, pl. 32; Advisor: Elmer D. Merrill.

Chien PEI (1902-1969) (Photo provided by Zong-Gang HU)
裴鉴（1902—1969）（相片提供者：胡宗刚）

p, 1925; 2: 40 p, 1926; 3: 48 p, 1927; 4: 28 p, 1928; 5: 47 p, 1932; Government Research Institute, Taihoku [Taibei].

1925—1932 年，山本由松，**续台湾植物图谱**，5 卷本，第 1 卷：47 页，1925；第 2 卷：40 页，1926；第 3 卷：48 页，1927；第 4 卷：28 页，1928；第 5 卷：47 页，1932；台湾中央研究所，台北。

1925-1948, Science Society of China, ***Contributions from the Biological Laboratory of the Science Society of China***, vols. 1-5 (1925-1929); *Botanical Series* 6-12 (3, 1930-1948).

1925—1948 年，中国科学社，**中国科学社生物研究所汇报**，1-5 卷，1925—1929 年；1930 年动植物分开，植物第 6-12 卷第 3 期，1930—1948 年。自 1933 年中文名称更名为**中国科学社生物研究所论文集**。

1926

1926.08, Fourth International Botanical Congress held in Ithaca, New York, United States of America. Ching-Yueh CHANG, from the Science Society of China, attended.

1926 年 8 月，第四届国际植物学大会于美国伊萨卡举行，中国科学社代表张景钺参加。[205]

1926 年，刘慎谔、林镕、莫定森，审定中国生物名称之商榷，生物科学 1：13-29。

1926, Tchen-Ngo LIOU, Yong LING and Ding-Sen MO, 'Discussion on the examination and approval of Chinese names of biological terminology', *Les Sciences Biologiques* 1: 13-29.

1926 年，国立清华大学成立生物系，钱崇澍任教授兼主任，仅教授 2 人，学生 3 人。1931 年 5 月生物学馆落成，系主任为陈桢。

1926, Department of biology, National Tsinghua University, founded with Sung-Shu CHIEN as dean, composed of two professors and only three students. By May 1931, the biology building was completed and Chen CHEN became dean.

1926, 2001, 2006, Francis Kingdon Ward, ***The Riddle of the Tsangpo Gorges***, 328 p, 1926; Edward Arnold and Co., London; Reprinted, ***Frank Kingdon Ward's Riddle of the Tsangpo Gorges****—retracing the epic journey of 1924-1925 in south-east Tibet*, edited by Kenneth N. E. Cox (1964-), with a new foreward, introduction and more color photos plus new information about the expedition, especially since the 1980s after China opened to the West, 319 p, 2001; and revised and updated, with additional material by Kenneth N. E. Cox, Ken Storm Jr. and Ian Baker, 335 p, 2006; Antique Collectors' Club, Woodbridge, Suffolk. Translated into Chinese in 1933, 1987 and 2003.

1933、1987、2003 年，杨庆鹏译，郑宝善、刘熙审校，**西康之神秘水道记**，边政丛书第 1 种，300 页，1933；蒙藏委员会，南京；杨图南编译，**西康之神秘水道记**，亚洲民族考古丛刊第 5 辑，300 页，1987；南天书局，台北；杨庆鹏译，**西康之神秘水道记**，西南史地文献第 35 卷，中国西南文献丛书第 3 辑，318 页，2003；兰州大学出版社，兰州。

1926 and 1934, John Hutchinson, ***The Families of Flowering Plants*** **Dicotyledons**, 1: 328 p, 1926, ***The Families of Flowering Plants*** **Monocotyledons**, 2: 243 p, 1934; The MacMillan Co, London; ed. 2, 792 p, 1959; Clarendon Press, Oxford; ed. 3, 968 p, 1973; Clarendon Press, Oxford; Translated into Chinese in 1937,1954 and 1955.

205 张景钺，1927，国际植物学会第四届大会，科学 12（3）：433-435。

Ching-Yueh CHANG (back row first person from right) in Morphology and Paleobotany group at Fourth International Botanical Congress, 1926 (Photo provided by library of Cornell University)
张景钺（后排右1）1926年于第四届国际植物学大会形态学和古植物学组（相片提供者：康奈尔大学图书馆）

1937 年，1954—1955 年，黄野萝[206] 译、胡先骕校，**双子叶植物分类**，514 页，1937；商务印书馆，上海；中国科学院植物研究所译，**有花植物科志 I．双子叶植物**，526 页，1954；唐进、汪发缵、关克俭译，**有花植物科志 II．单子叶植物**，495 页，1955；商务印书馆，上海。

206 原名黄在璇（1902—1981），号正仔，江西贵溪人，著名土壤学家、农业教育家。

1927

1927 年，胡先骕，种子植物分类学近来之趋势，科学 13(3): 315-323。

1927, Hsen-Hsu HU, 'Recent trends of taxonomy of seed plants', *Science (China)* 13(3): 315-323.

1927 年，北川政夫[207]，**关东州植物志**，第 1-15 页、第 1-167 页、第 1-16 页，图版 1-4；大连第二中学校博物研究部，大连。

1927, Masao Kitagawa[208], ***Flora of the Territory of Kwangtung***, 1-15 p, 1-167 p, 1-16 p; Plates 1-4; Dairen Daini Chūgakkō Hakubutsu Kenkyūbu, Dairen.

1927 年，刘棠瑞[209]（1911—1997）赴日留学，1929 年考入东京高等师范，1933 年入日本京都帝国大学理学部植物学科，1936 年取得理学学士学位；1937 年回国受聘于日本人在上海所办的自然科学研究所，1939 年任广东省立文理学院生物系教授、生物系主任，并兼任国立中山大学师范学院教授，1948 年任台湾省立博物馆研究员兼研究组长；1955 年任台湾大学农学院森林系教授，1959 年任森林系主任，1972 年任台湾大学农学院院长，同时开办林学博士班，讲授树木学、植物分类学、树木分布学、本地树木等课程，培养出如耿煊（1923—2009）、廖日京[210]（1929—2013）、黄增泉[211]（1931— ）、

---

207 北川政夫（1910—1995），日本植物学家，出生于大连，毕业于东京帝国大学植物学系；曾任大陆科学研究院的研究人员，大量采集中国东北植物标本并出版满洲国植物考（1939）和满洲植物新考（1979）；战后返回日本，1946 年任职千业农业专门学校，1950 年任横滨国立大学教授。

208 Masao Kitagawa (1910-1995), Japanese plant taxonomist and authority on plants of northeast China, was born in Dalian, Liaoning Province, China. He graduated from the department of botany, Tokyo Imperial University and served as researcher in Institute of Scientific Research Manchoukuo. He collected specimens and published *Lineamenta Florae Manshuricae* (1939) and *Neo-Lineamenta Florae Manshuricae* (1979). He served in the Agriculture Institute of Technology in 1946. He became professor at Yokohama National University in 1950.

209 江西安福人。

210 祖籍福建漳州，1929 年生于日本京都，1932 年回国定居台湾嘉义，1948 年入台湾大学农学院森林学系，1952 年毕业后任职于林业试验所和台北植物园，1958—1995 年任教于台湾大学森林系（兼职园艺系和植物系），著有拉丁文和树木学等。1990 年曾访问北京、南京、杭州、广州和香港的主要植物标本馆，研究樟科和壳斗科植物；1992 年再访问广州以及昆明的主要植物学机构，特别是研究桑科植物标本。

211 台湾苗栗人，1955 年台湾大学本科毕业、1959 年台湾大学硕士毕业、1965 年美国圣路易斯华盛顿大学博士毕业，1965 年返台后任台湾大学植物系副教授、教授、主任，2001 年退休；曾任台湾植物志第 1 版秘书和第 2 版主编。博士论文：Monograph of *Daphniphyllum* (Daphniphyllaceae) 先后发表于 *Taiwania* 11:57-98, 1965 和 12:137-234,1966；原导师：Robert E. Woodson, Jr. (1904-1963)；论文委员会：Drs. Walter H. Lewis (1930-，主席），John D. Dwyer (1915-2005) 和 Herbert Wayne Nichols (1937-1993).

许建昌（1932—）等台湾第一代植物分类学家。1963 年前往欧美及日本等国研究冷杉属树木的分类与分布，搜集有关图书、标本等资料；于 1970 年获日本东京大学博士学位，1971 年正式发表冷杉属专著。后致力于台湾植物志的编辑工作；1982 年从台大退休后赴洛杉矶定居。

1927, Tang-Shui LIU (Tung-Shui LIU, 1911-1997) went to Japan, entered Tokyo Teachers College in 1929, then Tokyo Imperial University in 1933. He graduated with a bachelor's degree in 1936; returned to the Institute of Natural Science, Shanghai in 1937, then professor, later dean, in the department of biology, Guangdong Science College in 1939. At the same time he was a professor in the Teachers College of National Sun Yat-Sen University. He entered the Taiwan Natural History Museum in 1948 as a research professor and team leader. He became a professor in the forestry department of the Agriculture College, Taiwan University, in 1955, then dean of the forestry department in 1959 and chairman of the Agriculture College in 1972. He established classes for doctoral candidates in dendrology, plant taxonomy, biogeography of trees and local dendrology. His students, the first generation of plant taxonomists in Taiwan, included Hsuan KENG (1923-2009), Jih-Ching LIAO (1929-2013), Tseng-Chieng HUANG (1931-) and Chien-Chang HSU (1932-). He studied *Abies* in Europe, North America and Japan in 1963, then received his Ph. D. degree from the University of Tokyo in 1970, and officially published his A Monograph of the Genus Abies in 1971 after returned to Taiwan. He was also an editor of the *Flora of Taiwan*. He retired from Taiwan University in 1982, and moved to Los Angeles, California.

1927, Ernest H. Wilson, ***Plant Hunting***, 1: 248 p, 2: 276 p; The Stratford Com., Boston, Massachusetts.

1927 年，Ernest H. Wilson，**探寻植物**，第 1 卷：248 页，第 2 卷：276 页；The Stratford Com.，波士顿，麻州。

1927、1928 年，伊藤武夫著，**台湾植物图说**，1 083 页，1927；弘道阁，东京；**续台湾植物图说**，全 1 册，400 页，1928；台湾植物图说发行所，台北。

1927 and 1928，Takeo Ito, '***Illustrated Flora of Formosa***', 1,083 p, 1927; Kodokaku, Tokyo; and '***Illustrated Flora of Formosa Supplement***', 400 p, 1928; Institute of Illustrated Flora of Formosa, Taihoku [Taibei].

1927-1928, Ju-Chiang LIU, Enumeration of plants collected by the late Mr. Nathaniel Harrington Cowdry in Chihli Province (also Chefoo), North China, *Bulletin of the Peking Society of Natural History*, 2(3): 47-194.[212]

1927—1928 年，刘汝强，直隶（附烟台）植物名录，*Bulletin of the Peking Society of Natural History* 2（3）：47-194。

212 Nathaniel H. Cowdry (1849-1925), amateur botanist and banker, was the father of cytologist and anatomist Edmund V. Cowdry (1888-1975) in the US, the first Chairman of the Department of Anatomy, Peking Union Medical College, 1917-1921.

1927-1937, Hsen-Hsu HU and Woon-Young CHUN, ***Icones Plantarum Sinicarum***, vols. 1-5, 1927-1937; 1: 1-50 p, pl. 1-50, 1927; 2: 1-50 p, pl. 51-100, 1929; 3: 1-50 p, pl. 101-150, 1933; 4: 1-50 p, pl. 151-200, 1935; 5: 1-50 p, pl. 201-250, 1937; Commercial Press, Shanghai (vols. 1-2), and the Fan Memorial Institute of Biology, Peking (vols. 3-5).

1927—1937 年，胡先骕、陈焕镛编纂，**中国植物图谱**，5 卷本；第 1 卷：1-50 页，图版：1-50，1927；第 2 卷：1-50 页；图版：51-100，1929；第 3 卷：1-50 页；图版：101-150，1933；第 4 卷：1-50 页；图版：151-200，1935；第 5 卷：1-50 页；图版：201-250，1937；商务印书馆（第 1-2 卷），上海；静生生物调查所（第 3-5 卷），北平。

1927, 1940，Alfred Rehder[213], ***Manual of Cultivated Trees and Shrubs Hardy in North America*** *exclusive of the subtropical and warmer temperate regions*, 930 p, 1927; ed. 2, 996 p, 1940; The MacMillan Co., New York. Many Chinese plants are cited in this book.

1927 年，Alfred Rehder，**北美耐寒栽培乔灌木手册** —— 除亚热带和暖温带以外地区，930 页，1927；第 2 版，996 页，1940；The MacMillan Co.，纽约。本书记载很多有关中国的植物。

1927, 1996, Heinrich R. E. Handel-Mazzetti, ***Naturbilder aus Südwest-China***—*erlebnisse und eindrücke eines österreichischen Forschers während des Weltkrieges, mit einer Karte und 148 Bildern nach aufnahmen des verfassers, darunter 24 Autochromen*, 380 p, 1927; Österreichischer Bundesverlag für Unterricht, Wissenschaft und Kunst, Wien und Leipzig. ***A Botanical Pioneer in South West China***—*experiences and impressions of an Austrian botanist during the First World War*, translated, completed and unabridged with biography of Heinrich R. E. Handel-Mazzetti by David Winstanley, with Introduction by

213 Alfred Rehder (1863-1949), born in Germany to a family of horticulturists, was one of the foremost plant researchers of his day. After attending Gymnasium, he apprenticed to his father and later continued his botanical studies in Berlin. There he worked as a gardener and in 1895 became the associate editor of *Moller's Deutsches Gartner-Zeitung*. In 1898, at age 34, Rehder sailed for the United States to undertake dendrological studies for *Moller's* and to investigate fruit growing and viniculture in the United States for the German government. To supplement his stipend from *Moller's*, Rehder applied to Charles S. Sargent to work on the Arboretum grounds. Sargent, then director of Arnold Arboretum of Harvard University, impressed by his botanical knowledge and work ethic, convinced Rehder to join the Arboretum staff. One of his first assignment was to work on the *Bradley Bibliography*, a massive listing of all the written works on woody plants published before 1900. The research and publication of the bibliography was funded by Abby A. Bradley as a memorial to her father, William L. Bradley (1826-1894) an innovator in the field of chemical fertilizers. By the time it was completed, Rehder had compiled more than 100,000 entries. Ernest H. Wilson's plant collections in China stimulated years of work on the part of Arboretum staff members, not the least being Alfred Rehder. Along with identifying and naming new plants, Rehder collaborated with Wilson to write *Plantae Wilsonianae*, which documented Wilson's Chinese plant collections. In 1913 he was awarded an honorary Master of Arts degree from Harvard University for his work on the *Bradley Bibliography* and in 1918 he became Curator of the Herbarium at the Arnold Arboretum. Over the course of his career, Rehder authored some 1,400 plant names and published more than 1,000 articles in both English and German in botanical and horticultural publications (https://www.arboretum.harvard.edu/library/archive-collection/historical-biographies/alfred-rehder/, accessed 19 September 2017).

Christopher Grey-Wilson, 192 p, 1996; David Winstanley, Brentwood, England.

1927、1996 年，Heinrich R. E. Handel-Mazzetti，**中国西南的自然印象**[214]—— 第一次世界大战期间一位奥地利植物学家的经历与印象，380 页 ,1927 年；Österreichischer bundesverlag für Unterricht, Wissenschaft und Kunst, Wien und Leipzig。**一位植物先驱在中国西南**[215]—— 奥地利植物学家第一次世界大战期间的经历和印象，192 页，1996 年；David Winstanley, Brentwood, 英国。

214 自原书德文版。本书实为作者的旅行记，而后人的英文版翻译更是采用拟人的笔法，以致出现德文原文与英文翻译及本书的中文翻译不尽一致，特此说明。

215 自翻译的英文版。

1928

1928 年春，经胡先骕努力，哈佛大学阿诺德树木园资助中国科学社生物研究所方文培赴四川和西康采集（松潘、灌县、宝兴、天全、康定、峨眉山和南川等地），历时 8 个月，步行数千里，采得标本 4 000 余号。所采标本分别存于美国（A）和中国（NAS）。

1928, Spring, Wen-Pei FANG from the Biological Laboratory of the Science Society of China [Nanjing], explored Sichuan and Sikang [Xikang]. With support from the Arnold Arboretum of Harvard University via Hsen-Hsu HU, he collected 4,000 numbers of specimens in eight months. The specimens were divided between the United States (A) and China (NAS).

1928 年 4 月 23 日，国民政府设立国立中央研究院，蔡元培（1868—1940）被任命为院长。

1928.04.23, Academia Sinica, the first national scientific organization, was established by the Chinese government. Yuan-Pei TSAI (1868-1940) named president.

1928 年 4 月至 12 月，中央研究院组织广西科学调查团，对九万山、苗山、青龙山、八角山、十万大山等地进行植物、动物、人文和地质农林调查；其中秦仁昌负责植物的调查，调查成员有陈长年，8 个月获得标本 3 400 号，3 万余份。

1928.04.12, Guangxi Expedition in

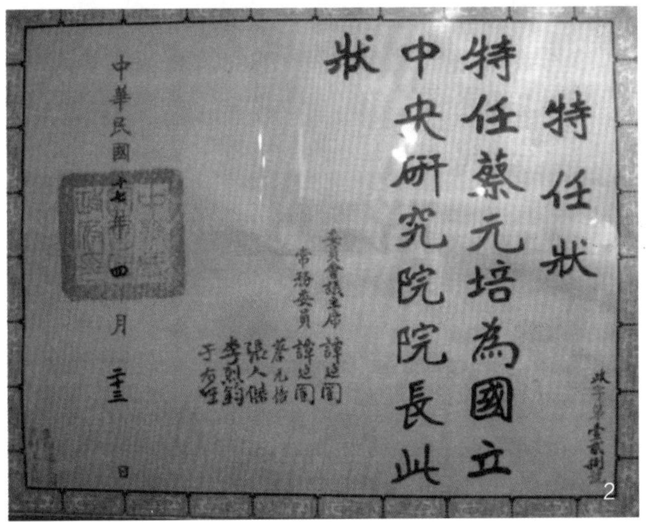

特任狀

特任蔡元培為國立中央研究院院長此狀

中華民國十七年四月廿三日

1. Yuan-Pei TSAI (1868-1940), 2. Certificate of appointment of Yuan-Pei Tsai to president of Academia Sinica (Photo provided by Zong-Gang HU)

1. 蔡元培（1868—1940）；2. 蔡元培任国立中央研究院院长之特任状（相片提供者：胡宗刚）

the fields of botany, zoology, humanity, geology, agriculture and forestry organized by Academia Sinica. Botanical team lead by Ren-Chang CHING, with team member Chang-Nian CHEN; 3,400 numbers and up to 30,000 collections of plant specimens were made in 8 months.

1928 年 10 月 1 日，静生生物调查所于北平举行成立典礼，该所系尚志学会、中华教育文化基金会及范静生（1875—1927）[216] 家人共同创办的私立研究机构，1928—1932 年秉志为所长兼动物部主任，胡先骕为植物部主任；植物标本馆同时诞生（FM，现并入 PE）。

1928.10.01, Fan Memorial Institute of Biology, a private research organization co-founded in Peiping [Beijing] by Sangchi Society, the China Foundation for the Promotion of Education and Culture and Ching-Sheng FAN (1875-1927) family; with Chi PING named director, 1928-1932, and head of department of zoology; Hsen-Hsu HU made head of department of botany. The herbarium FM (now incorporated in PE) was founded at the same time.

The newly constructed building of Fan Memorial Institute of Biology, 1931 (Address: 3 Wen Tsin Chieh, Peiping) (Photo provided by Xiao-Jiang HU)
静生生物调查所 1931 年落成的新楼（地址：北平文津街 3 号）（相片提供者：胡晓江）

216 范源廉，字静生，湖南湘阴人，著名教育家。

Fan Memorial Institute of Biology founded, Peiping [Beijing], 1 October 1928 (front row, first on left, Hsen-Hsu HU) (Photo provided by Zong-Gang HU)

1928 年 10 月 1 日，静生生物调查所于北平成立（前排左一：胡先骕）（相片提供者：胡宗刚）

1928, *Sinojackia*, an endemic genus of Styracaceae from China, was described by Hsen-Hsu HU[217], based on a collections by Ren-Chang CHING and Yi-Li KENG from Nanjing, Jiangsu. It was named for John G. Jack (1861-1949), Hu's teacher at Harvard University. It was the first genus described by a Chinese botanist.

1928 年，秦仁昌和耿以礼在江苏南京采集到新植物，胡先骕定名为安息香科中国特有新属秤锤树属（楗梳木），并以其留学哈佛大学的导师（John G. Jack[218]，1861—1949）命名；此为中国植物学家描述的第一个新属[219]。

1928 年，佐佐木舜一[220]，**台湾植物名录**，I-XXVI，562 页；台湾博物学会，台北。

1928, Syuniti Sasaki, ***List of Plants of Formosa***, I-XXVI, 562 p; The Natural History Society of Formosa, Taihoku [Taibei].

1928, Taiwan Imperial University founded by Japanese scholars in Taipei [Taibei]. Japanese botanists Yushun Kudo (1887-1932), Genkei Masamune (1899-1993) and Yoshimatsu Yamamoto[221] (1893-1947) went to Taiwan successively to collect and study the plants of Taiwan; herbarium (TAI) at Taihoku University founded the following year. Taihoku University was named National Taiwan University after 1945. Some well-known Chinese scholars, such as Hui-Lin LI (1911-2002) and Shun-Ching LEE worked there. TAI was the editorial center for two editions of the *Flora of Taiwan* and both secretaries of the *Flora of Taiwan*, Tseng-Chieng HUANG and Chang-Fu HSIEH (1947-) worked here.

1928 年，日本学者在台湾建立台北帝国大学，先后派工藤祐舜（1887—1932）、正宗严敬（1899—1993）及山本由松（1893—1947）等赴台采集植物并予以研究，第 2 年建立植物标本馆（即今国立台湾大学植物系标本馆，TAI）。1945 年台湾光复后，台北帝国大学更名为国立台湾大学，有诸多著名植物学家在此工作，如李惠林（1911—2002）和李顺卿。国立台湾大学是两版*台湾植物志*的编研中心，两版*台湾植物志*的秘书黄增泉和谢长富（1947—）都在此工作。

1928—1931 年，国立中山大学生物系辛树帜等考察广西大瑶山和大明山，获得植物标本至少 30 000 份（IBSC，SYS）。[222]

---

217 Hsen-Hsu HU, 1928, *Sinojackia*, a new genus of Styracaceae from Southeastern China, *Contributions from the Biological Laboratory of the Science Society of China* 4(1): 1-3, May 1928 and *Journal of Arnold Arboretum* 9(2-3): 130-131, July 1928.

218 Lisa Pearson, 2014, John George Jack: Dendrologist, Educator, Plant Explorer, *Arnoldia* 71(4): 2-11.

219 胡先骕早在 1925 年于哈佛攻读博士时便通过新组合建立兰科风兰属名称 *Neofinetia* 并至今得到承认；详细参见：Hsen-Hsu HU, 1925, Nomenclatorial changes for some Chinese Orchids, *Rhodora* 27(318)：105-107。

220 佐佐木舜一（1888—1961），日本占领台湾时，对台湾植物资源调查重要的植物学者之一，足迹遍及全台及离岛；发表相关文章不下数十篇。

221 Hisakichi Hisazumi, 1948, Yoshimatsu Yamamoto (1893-1947), *Taiwania* 1: 7-12.

222 冯双，2011，*中山大学生命科学院（生物学系）编年史*，36-60 页；中山大学出版社，广州。

Editorial center of two editions of *Flora of Taiwan*, Herbarium (TAI), Taiwan University, 2016 (left-right): Jin-Shuang MA (the editor in chief of this book), Cheng-Fu HSIEH (former curator of TAI, secretary of the second edition and editor in chief of *Taiwania*), Tseng-Chieng HUANG (former curator of TAI, secretary of the first edition and editor in chief of the second edition), Jer-Ming HU (current curator of TAI and director of Institute of Ecology and Evolutionary Biology), and Kuo-Hsiung WANG (standing, managing editor of *Taiwania*) (Photo provided by Jin-Shuang MA)

2016 年于台北的两版台湾植物志的编辑中心、台湾大学植物标本馆留影（从左至右）：马金双（本书主编）、谢长富（前任标本馆馆长、第二版秘书、*Taiwania* 主编）、黄增泉（前任标本馆馆长、第一版秘书、第二版主编）、胡哲明（现任标本馆馆长、生态学与演化生物学研究所所长）、王国雄（站立者，*Taiwania* 执行编辑）（相片提供者：马金双）

1928–1931, Shu-Chih HSIN (S. S. SIN) [223] led exploration team from department of biology, National Sun Yat-Sen University, to Dayao Shan and Daming Shan, Guangxi. More than 30,000 sheets of plant specimens were collected (IBSC, SYS).

223 The derivation of the abbreviation, S. S. Sin, which appears on some specimens and in Chinese literature is unknown, but perhaps from Shu-Shih Sin.

1929

1929 年 1 月，中央研究院自然历史博物馆筹备处在南京成立。1930 年 1 月博物馆正式成立，钱天鹤（1893—1972）任主任。秦仁昌、蒋英[224]（1898—1982）、陈长年、陈少卿[225]（1911—1997）曾在该馆从事植物分类学研究。

1929.01, The Preparatory of the Metropolitan Museum of Natural History was established by Academia Sinica in Nanjing. It was officially founded in January 1930, with Tien-Hao CHIEN (1893-1972) named director. Ren-Chang CHING, Ying TSIANG (1898-1982), Chang-Nian CHEN and Shao-Hsing CHUN (1911-1997) were among the staff member who once worked here.

1929 年 7 月 1 日，总理陵园纪念植物园于南京成立，该园隶属于陵园园林组，傅焕光（1892—1972）[226]任园林组主任。

1929.07.01, The Botanical Garden, Dr. Sun Yat-Sen's Memorial Park, was established in Nanjing with Huan-Kuang FU (1892-1972) as head of the department of horticulture.

1929 年 9 月 9 日，国立北平研究院成立；李煜瀛（1881—1973）任院长。

1929.09.09, National Academy of Peiping [Beijing] was founded in Peiping [Beijing], with Yu-Ying LI (1881-1973) as president.

1929 年 10 月，国立北平研究院设生物部于天然博物馆，下设植物学研究所，1929—1949 年刘慎谔任所长。

1929.10, The Institute of Botany, Department of Biology, National Academy of Peiping [Beijing], was

224 江苏昆山人，1918—1919 年肄业于上海沪江大学文学院，1920 入金陵大学农学院森林系，1925 年毕业，获美国纽约大学林学学士学位，1926 年任安徽安庆农业专门学校教员，1927 年任江苏昆山农民协会主任，1928—1929 年任国立中山大学理学院助教兼植物研究室研究人员，1930—1933 年任南京中央研究院自然历史博物馆技师兼担任植物标本室主任，并带队到贵州、云南、广西、江西、湖南等地采集，1934—1941 年任国立中山大学农学院和研究院副教授、教授、研究员；1942—1945 年任国立中山大学理学院和农学院教授兼农林植物研究所代理主任，1946—1947 年任台湾林业试验所技正兼台北植物园园长，1947—1951 年任国立中山大学农学院教授，兼国立广西大学农学院森林系主任和岭南大学生物系教授，1952—1958 年任华南农学院林学系教授，1958—1962 年任广东林学院教授兼林学系主任，1963—1972 年任中南林学院教授兼科研部部长，1973—1976 年任广东农林学院教授，1977—1982 年任华南农学院教授；研究夹竹桃科和萝藦科以及番荔枝科的著名专家。Ping-Tao LI and Paul Pui-Hay BUT, 1986, In commemoration of Professor Tsiang Ying (1898-1982), *Journal of Wuhan Botanical Research* 4(1): 69-78.

225 贵州桐梓人，中国科学院华南植物研究所植物采集家。

226 一说 1891—1972，字志章，江苏太仓人；早年就读于上海公学，后毕业于菲律宾大学林科。

Members of Department of Biology, National Academy of Peiping, 1931 (second from right: Tchen-Ngo LIOU) (Photo provided by Zong-Gang HU)
国立北平研究院生物部全体职员于 1931 年合影（前排右二：刘慎谔）（相片提供者：胡宗刚）

founded in the Natural History Museum, with Tchen-Ngo LIOU as director, 1929-1949.

1929 年 12 月 4 日，国立中山大学农学院创建农林植物研究所设立植物标本室（今 IBSC），1929—1954 年陈焕镛任主任[227]；致力于广东及海南植物的调查采集，先由蒋英、左景烈[228]奠定基础，后得中华教育文化基金会资助，开展大规模采集，由陈念劬、左景烈、侯宽昭（1908—1959）、高锡朋、黄志、梁向日（梁葵，1908—1975）等担任采集工作，至 1933 年得标本共计 3.1 万号。

1929.12.04, The Botanical Institute and herbarium (today's IBSC), the Agriculture College of National Sun Yat-Sen University, Guangzhou, Guangdong, was founded, with Woon-Young CHUN as director from 1929 to 1954, and focused on plant collecting in Guangdong and Hainan. Ying TSIANG and Ching-Lieh TSO formed the foundation of the herbarium. With support from the China Foundation for the Promotion of Education and Culture, Nien-Kun CHUN, Ching-Lieh TSO, Foon-Chew HOW (1908-1959), Si-Peng KO, Chi WANG and Hsiang-Ri LIANG (Kui LIANG, 1908-1975) increased the collection to 31,000 numbers by 1933.

1929, Hsin-Hsuan CHUNG, The study of botany in Fukien, *Lingnan Science Journal* 7: 121-130.

1929 年，钟心煊，福建植物学研究，岭南科学杂志 7: 121-130。

1929, Friedrich Ludwing E. Diels, ***Pflanzengeographie***, 3rd edition, 159 p; Walter de Gruyter und Co., Leipzig. Translated into Chinese in 1934.

1934 年，董爽秋译，**植物地理学**，222+11 页；国立编译馆出版、商务印书馆发行，上海。

1929 年，中国科学社生物研究所方文培、郑万钧和静生生物调查所唐进（1897—1984）、汪发缵（1899—1985）在西部科学院筹备处的帮助下，组成川康考察团，6 个月获得 5 740 多号标本（CQNM、NAS、PE）。[229]

1929, Wen-Pei FANG and Wan-Chun CHENG from the Biological Laboratory of the Science Society of China, Tsin TANG (1897-1984) and Fa-Tsuan WANG (1899-1985) from the Fan Memorial Institute of Biology, with help from the Preparatory of Science Institute of West China, explored the Sichuan and Kham

227 据冯双考证，1928 年即设立植物研究室。参见：冯双，2011，中山大学生命科学院（生物学系）编年史 1924—2011，第 41 页；中山大学出版社，广州。据中国科学院华南植物研究所大事记记载，1928 年为筹建“植物研究室”，1929 年 12 月“植物研究室”扩展为“植物研究所”。参见：陈忠毅主编，1999，中国科学院华南植物研究所建所七十周年纪念文集，160 页；中国科学院华南植物研究所，广州（内部印制）；魏平主编，2009，根深叶茂竞芳菲：中国科学院华南植物园八十周年纪念文集，211 页；广东科技出版社，广州。

228 字仲伟，湖南长沙人（生卒不详）；1927 年毕业于国立东南大学农科，1929 年主持国立中山大学农林植物研究所采集工作，1932—1933 年分别领导国立中山大学的海南和广西采集，1933 年任国立山东大学生物系讲师并研究兰科植物，同时再次带队赴海南采集。1936 年赴爱丁堡植物园研修，后返国。

229 蒋卓然，1936，中国西部科学院生物所植物部五年来之进展（续），工作月刊 1（4）：71-85。

regions. They collected more than 5,740 numbers of plant specimens within six months (CQNM, NAS, PE).

1929, Ernest H. Wilson, ***China—Mother of Gardens***, 408 p; The Stratford Co., Boston, Massachusetts. Translated into Chinese in 2009, 2015, and 2017.

2009 年、2015 年、2017 年，红音、干文清编译，**威尔逊在阿坝**[230]——100 年前威尔逊在四川西北部汶川、茂县、松潘、小金旅行游记，106 页；四川民族出版社，成都；2015，胡启明译，**中国——园林之母**，305 页；广东科技出版社，广州；[231] 2017，包志毅主译，**中国乃世界花园之母**，580 页；中国青年出版社，北京。[232]

1929-1937, Heinrich R. E. Handel-Mazzetti et al., ***Symbolae Sinicae***, *Botanische Ergebnisse der Expedition der Akademie der Wissenschaften in Wien nach Südwest-China, 1914-1918,* Teil. 7; Teil. 1 (Algae), 1937; Teil. 2 (Fungi), 1937; Teil. 3 (Lichenes), 1930; Teil. 4 (Musci)[233], 1929; Teil. 5 (Hepaticae), 1930; Teil. 6, (Pteridophyta), 1929; Teil. 7, (Anthophyta), lief 1, 1929, lief 2, 1931, lief 3, 1933, lief 4, 1936, lief 5, 1936.

1929—1937 年，Heinrich R. E. Handel-Mazzetti et al.，**中国植物纪要**，1914—1918 年奥地利科学院 Heinrich R. E. Handel-Mazzetti 的中国西南植物考察报告，Teil. 7；第 1 部（藻类），1937；第 2 部（菌类），1937；第 3 部（地衣），1930；第 4 部（藓类），1929；第 5 部（苔类），1930；第 6 部（蕨类），1929；第 7 部（种子），第 1 册，1929，第 2 册，1931，第 3 册，1933，第 4 册，1936，第 5 册，1936。

1929-1948, Fan Memorial Institute of Biology, ***Bulletin of the Fan Memorial Institute of Biology***, vols. 1-4, 1929-1933; ***Bulletin of the Fan Memorial Institute of Biology Botany***, 5-11(1-2), 1934-1941; ***Bulletin of the Fan Memorial Institute of Biology new series***, 1(1), 1943, 1(2 and 3), 1948.

1929—1948 年，静生生物调查所，**静生生物调查所汇报**，第 1-4 卷，1929—1933 年；第 5-11 卷第 1、2 期，1934—1941 年；新系第 1 卷第 1 期，1943；第 1 卷第 2-3 期，1948。

1929-1949, The Metropolitan Museum of Natural History, Academia Sinica, ***Sinensia*** *Contributions from the Metropolitan Museum of Natural History, Academia Sinica*, vols. 1-4, 1929-1934; National Research Institute of Biology, Academia Sinica, ***Sinensia***, *Contributions from the National Research Institute of Biology*, vols. 5-7, 1934-1936; National Institute of Zoology and Botany, Academia Sinica, ***Sinensia***,

230 仅翻译了威尔逊原著中的第 11-15 章，即有关阿坝州的内容；故取名为威尔逊在阿坝。

231 王晨绯，2017 年 7 月 24 日，胡启明 —— 力学如力耕，中国科学报，第 6 版 • 院所。

232 谭文德，2018 年 1 月 3 日，威尔逊 China: Mother of Gardens 两个中译本的比较阅读，中华读书报第 16 版，书评周刊 • 科学。

233 Tong CAO & Timo J. Koponen 2004, Musci in '*Symbolae Sinicae*'; an annotated checklist of mosses collected by H. Handel-Mazzetti in China in 1914-1918, and described by V. F. Brotherus in 1922-1929. *Bryobrothera* 8: 1-34.

Contributions from the National Institute of Zoology and Botany, vols. 8-20, 1937-1949.

1929—1949 年，国立中央研究院自然历史博物馆，**国立中央研究院自然历史博物馆丛刊**，第 1-4 卷，**国立中央研究院动植物研究所丛刊**，第 5-20 卷。[234]

1929—1958 年，黄志在广东（包括香港以及广西）采集植物标本 14 000 号（IBSC）。[235]

1929-1958, Chi WANG collected more than 14,000 numbers of plant specimens in Guangdong (and also in Hong Kong and Guangxi, IBSC).

234 由于多次更改题目与隶属，不但题目与卷册信息不规则，且中英文标题也不是完全相对应。

235 胡启明、曾飞燕，2011，广东植物志，10：321 页；广东科技出版社，广州［附录 2，华南植物研究所（园）植物标本采集简史］。

1930

1930.08.16-08.23, Fifth International Botanical Congress held in Cambridge, England, five Chinese botanists, Woon-Young CHUN, Ren-Chang CHING, Ching-Yueh CHANG, Hsing-Chien SZE (1901-1964) and Chung-Chen LIN, attended. Woon-Young CHUN of National Sun Yat-Sen University, Hsen-Hsu HU of the Fan Memorial Institute of Biology and Albert N. Steward of the University of Nanking [Nanjing] were elected members of the general committee of botanical nomenclature[236]. Woon-Young CHUN visited the Royal Botanic Gardens, Kew and the Royal Botanic Garden, Edinburgh, after the congress.

1930 年 8 月 16 日至 23 日，第五届国际植物学大会于英国剑桥召开，陈焕镛、秦仁昌、张景钺、斯行健（1901—1964）、林崇真出席。国立中山大学陈焕镛、静生生物调查所胡先骕和金陵大学史德蔚[237] 被选为国际植物命名法规委员会的委员。陈焕镛会后还顺访邱园和爱丁堡植物园。[238]

1930 年 9 月，四川成立中国西部科学院，院长卢作孚（1893—1952）；1931 年成立生物研究所（1931—1938 年），并设有植物部。西部科学院植物标本馆成立于 1936 年，1944 年 12 月统计馆藏植物标本有 6.7 万份，该馆为今日重庆自然博物馆标本馆（CQNM）的前身。[239]

1930.09, Science Institute of Western China, founded in Chongqing, Sichuan, with Zuo-Fu LU (1893-1952) as president; department of botany within the Institute of Biology (1931-1938) was established in 1931. The herbarium of the Science Institute of Western China was founded in 1936. By December 1944, the herbarium numbered about 67,000 sheets; today's Chongqing Natural History Museum (CQNM) is its successor.

1930, Ren-Chang CHING, ***The Monograph of Chinese Ferns****—Being a systematic treatment of all ferns known hitherto in China*, 546+32 p; The Metropolitan Museum of Natural History, Academia Sinica, Nanking [Nanjing] (manuscript).

---

236 Woon-Young CHUN, 1931, Recent developments in systematic botany in China, *Report of Proceedings of Fifth International Botanical Congress*, 524-528 p; Cambridge University Press, Cambridge.

237 又译作史迪威。

238 胡宗刚，2013，华南植物研究所早期史，第 74 页；上海交通大学出版社，上海。

239 1943 年 10 月，中国西部科学院与在北碚的中央研究院动植物研究所、气象研究所，经济部中央地质调查所、中央工业试验所、矿冶研究所，农林部中央农业实验所、中央林业试验所、中央畜牧实验所，中国科学社生物研究所、国立江苏医学院、中国地理研究所等单位成立“中国西部博物馆”。中国西部科学院于 1951 年 7 月更名为“西南人民科学馆”，1953 年秋改组为“西南博物院自然馆”。1954 年西南大区撤销后，西南博物院更名为“重庆市博物馆”，“西南博物院自然馆”也相应更名为“重庆市博物馆北碚陈列馆”。1981 年，重庆市博物馆划分为重庆市博物馆和重庆市自然博物馆两个机构后，原北碚陈列馆再次更名为“重庆市自然博物馆北碚陈列所”，至今仍发挥着“辅助教育及促进学术研究”的作用。赵宇晓、陈益升，1991，中国西部科学院，中国科技史料 12（2）：72-83。

Taxonomy Group, 5th International Botanical Congress, Cambridge, England, 16–23 August 1930: Top Row: 7, Ren-Chang CHING, and 8, Woon-Young CHUN (Photo provided by Harvard University Herbaria)

FIFTH INTERNATIONAL BOTANICAL CONGRESS

CAMBRIDGE, 1930

TAXONOMY 6

Front Row—

1. Moreau, F.
2. Du Rietz, G. E.
3. Pampanini, R.
4. Alston, A. H. G.
5. Fogg, J. M.
6. Black, J. M.
7. Allan, H. H.
8. Lalonde, L.
9. Kern, F. D.
10. Degen, A. de
11. Mattfeld, J.
12. Gagnepain, H.
13. Maillefer, A.
14. Hauman, L.
15. Clemens,
16. Clemens, Mrs.
17. Van Eseltine, G.P.
18. Cotton, A. D.
19. Dandy, J. E.
20. Taylor, G.
21. Guillaumin, A.
22. Huber, B.
23. Hoffman, W. E.

Second Row—

1. Taylor, F. B.
2. Schellenberg, G.
3.
4. Hitchcock, A. E.
5. Stapf, O.
6. Rendle, A. B.
7. Maire, R.
8. Harms, H.
9. Fries, R. E.
10. Hill, A. W.
11. Merrill, E. D.
12. Diels, L.
13. Briquet, J. I.
14. Mangin, L.
15. Arthur, J. C.
16. Barnhart, J. H.
17. Hofker, H.
18. Sprague, T. A.
19. Rehder, A.
20. Nelson,
21. Cratty, R. I.
22. Pfeiffer, H.
23. Grout, A. J.

Third Row—

1. Cowan, J.
2. Patton, R. T.
3. Ramsbottom, J.
4. Smith, W. W.
5. Knowlton, C. H.
6. Knowlton, Mrs.
7.
8. Lawrence, J. R.
9. Ottley, A. M.
10. Juel, H. O.
11. Briquet, N.
12. Grant, A. L.
13.
14. Catter, N.
15. Bayliss Elliot, J. S.
16. Noel, E. F.
17. Steiger, J. L.
18. Pool, R.
19.
20. Jepson, W. L.
21. Buchanan, R. E.
22. Davidson, J.
23. Prain, D.
24. Loder, G. W. E.
25. Parker, R. N.
26. Regel, C.
27. Domin, K.
28. Clinton, G. P.
29. Borza, A.
30. Handel Mazzetti, H.
31. Marquand, C. V. B.
32.
33. Rimo Bacigalupi

Fourth Row—

1. Pia, J.
2. Donk, M. A.
3. Davy, J. B.
4. Knoche, L. H.
5. Fries, Mrs. R. E.
6. Robyns, W.
7. Christensen, Mrs. C.
8. Eastwood, A.
9. Poulton, E. M.
10. Williams, R. O.
11. Henrard
12. Pulle, A.
13. Greene, M. L.
14. Gleason, H. A.
15. Baxter, D. V.
16.
17. Maxon, W. R.
18. Tandy, G.
19. Lakshnakara, M. C.
20. Adamson, R. S.
21. Pugsley, H. W.
22. Fernald,
23. Malte, M. O.
24. Long, B.
25. Gundersen, A.

Fifth Row—

1. Shear, C. L.
2. Johnson, F. W.
3. Christensen, C.
4. Janchen, E.
5. Thompson, S.
6. Fuentes, F.
7. Pillans, N. S.
8. Markgraf, F.
9. Mansfeld, R.
10. Francis, W. D.
11. Summerhayes, V. S.
12. Ballard, F.
13. Howes, F. N.
14. Holttum, R. E.
15. Bjorklund, E.
16. Preston,
17. Jenkin, T. G.
18. Podpera, J.
19. Gager, C. S.
20. Sherrin, W. R.
21.
22. Naveau,
23. Dixon, H. N.

Top Row—

1.
2. Hill, A. F.
3. Miller, J.
4. Pennell, F. W.
5. Schaffner, J. H.
6. Smith, C. A.
7. Ching, R. C.
8. Chun, W. Y.
9. Carr, C. E.
10. Epling, C.
11. Crook, A. H.
12. Burkill, I. H.
13. Jeswiet, J.
14.
15. Wilmott, A. J.
16. Hyde, H. A.

Taxonomy Group list of participants, 5th International Botanical Congress, Cambridge, England, 16–23 August 1930, Top Row: 7, Ren-Chang CHING, and 8, Woon-Young CHUN (Photo provided by Harvard University Herbaria)

1930 年 8 月 16 日至 23 日，第五届国际植物学大会植物分类学组于英国剑桥合影名单；上排：7. 秦仁昌，8. 陈焕镛（相片提供者：哈佛大学标本馆）

1930 年，秦仁昌，**中国蕨类专著——目前已知的中国蕨类植物系统学处理**，546 页 +32 页；中央研究院自然历史博物馆，南京（手稿）。[240]

1930, Euan H. M. Cox, ***The Plant Introductions of Reginald Farrer***, 113 p; New Flora and Silva, London.

1930 年，Euan H. M. Cox，**Reginald Farrer 的植物引种**，113 页；New Flora and Silva，伦敦。

1930 年，耿以礼[241]（1897—1975）赴美国留学，在史密森学会[242]研究禾本科，1932 年获乔治·华盛顿大学硕士学位，1933 年获博士学位，然后转赴欧洲各大主要标本馆修订中国禾本科；1934 年回国后任国立中央大学教授，后一直在此从事禾本科研究；并于 1959 年主编中国主要植物图说——禾本科等专著。

1930, Yi-Li KENG (1897-1975) traveled to the United States to study Poaceae at the Smithsonian Institution; received a master's degree in 1932, and a Ph. D. degree in 1933 from George Washington University[243]; traveled to various major herbaria in Europe to revise Chinese grasses; returned to China in 1934 to become professor at National Central University, Nanjing, and to continue his studies of grasses there. Chief editor of monograph *Flora Illustralis Plantarum Primarum Sinicarum—Gramineae* in 1959.

Yi-Li KENG (1897-1975) (Photo provided by Zong-Gang HU)
耿以礼（1897—1975）（相片提供者：胡宗刚）

1930 年，满铁中等教育研究会博物分科会编纂，**满洲植物目录**，292 页；满铁中等教育

240 张宪春，2004，中国蕨类植物系统分类学研究百年之回顾与前瞻，中国花卉协会蕨类植物分会简讯 10：1-9。

241 江苏江宁人，1918—1921 年就读于南京高等师范学堂农科，1921—1926 年任教于金华大学、集美大学以及厦门大学等，1926—1927 年入中国科学社生物研究所，同时在国立东南大学生物学补读学分，1927 年 3 月获学士学位，后留校任教。

242 Smithsonian Institution 一词的中文表述还有史密森研究院、史密森博物馆以及史密森尼学会、史密森尼研究院等。在此依据辞海作史密森学会记载。**史密森学会**是一个庞大的机构；凡是在史密森学会从事植物分类的，其实都是在史密森学会的国家自然历史博物馆的植物学部（Department of Botany, National Museum of Natural History），即美国国家植物标本馆（United States National Herbarium，US）。

243 Ph.D. dissertation: The Grass of China, Advisor: Albert S. Hitchcock (1865-1935).

研究会博物分科会出版，奉天[244]。

1930, The Nature-Study Society of the Secondary School of South Manchuria Railway Co., ***A List of the Manchurian Plants***, 292 p; The Nature-Study Society of Secondary School of South Manchuria Railway Co., Hoten.

1930 年，佐佐木舜一，**林业部腊叶标本馆目录**[245]，592 页；台湾总督府中央研究所，台北，台湾。

1930, Syuniti Sasaki, ***A Catalogue of the Government Herbarium***, 592 p; Government Research Institute, Taihoku [Taibei], Formosa [Taiwan].

1930, Francis Kingdon Ward, ***Plant Hunting on the Edge of the World***, 383 p; Victor Gollan, Co. Ltd., London.

1930 年，Francis Kingdon Ward，**在世界边缘探寻植物**，383 页；Victor Gollan, Co. Ltd.，伦敦。

1930 年，中央研究院自然历史博物馆制定"限制外人在华采集动植物标本条件"，代表中国政府与来华作动植物标本采集的国外学术团体和个人签订具体协议，并对协议执行予以监管。

1930, Metropolitan Museum of Natural History, Academia Sinica, limited collecting of animals and plants in China by foreigners, represented Chinese government in signing agreements with foreign academic institutions and individuals and supervised execution of permits.

1930 年 4 月至 1931 年 3 月，中央研究院自然历史博物馆组织贵州自然科学调查团，其中植物部分由蒋英负责，采得标本 7 000 余号，10 万余份。

1930.04-1931.03, The Metropolitan Museum of Natural History, Academia Sinica, organized an expedition to Guizhou with Ying TSIANG as botanical leader; about 7,000 numbers and up to 100,000 plant specimens were collected.

1930 年 6 月至 1932 年 9 月，中央研究院自然历史博物馆秦仁昌赴丹麦研究蕨类植物，在欧洲期间经胡先骕努力，中华教育文化基金会资助秦仁昌 3 000 美元，用于拍摄欧洲主要植物标本馆（特别是邱园和自然历史博物馆）收藏的中国植物标本；照片共有 1.8 万多张（详细参见 1935 年条目），为日后中国植物学研究奠定了重要基础。秦仁昌回国后任北平静生生物调查所植物标本馆（FM）[246]馆长。

1930.06-1932.09, Ren-Chang CHING of Metropolitan Museum of Natural History, Academia Sinica, went to Denmark to study ferns, returned to the Fan Memorial Institute of Biology in Peiping [Beijing] to become curator of the Herbarium (FM). With support from the China Foundation for the Promotion of

244 今沈阳的旧称。

245 台湾总督府中央研究所林业部报告第九号；昭和五年（1930）。

246 即今日 PE。

Education and Culture via Hsen-Hsu HU of the Fan Memorial Institute of Biology, Ren-Chang CHING received $3,000 to take more than 18,000 photos of Chinese specimens in major herbaria in Europe (especially in K and BM). The photos played a key role in plant taxonomic research in China (see 1935 for details).

Ren-Chang CHING (1898-1986) (Photo provided by Zong-Gang HU)
秦仁昌（1898—1986）（相片提供者：胡宗刚）

1930—1936 年，国立北平研究院，**国立北平研究院院务汇报**，1-7 卷。

1930-1936, National Academy of Peiping [Beijing], ***Bulletin of the National Academy of Peiping***, vols. 1-7.

1930-1941, ***The Hong Kong Naturalist***, vols. 1-10 and *Supplements* 1-6; The Newspaper Enterprise Ltd., Hong Kong. This is the earliest journal on natural history from Hong Kong regarding botany. [247]

1930—1941 年，**香港博物学家**，第 1-10 卷及增刊 1-6；The Newspaper Enterprise Ltd.，香港。该刊是香港最早涉及植物学的博物学杂志。

1930-1941, 1946, 1948, ***Sunyatsenia***, *Journal of the Botanical Institute, College of Agriculture and College of Science, National Sun Yat-Sen University*, vols. 1-6(2), 6(3-4), 7(1-2). [248]

1930—1941 年、1946 年、1948 年，**国立中山大学农林植物研究所专刊**，vols. 1-6(2)、6(3-4), **国立中山大学理学院植物研究所专刊**，7(1)、7(2)。

1930-1958, 2011, Hsen-Hsu HU and Ren-Chang CHING, ***Icones Filicum Sinicarum***, vol. 1; Ren-Chang CHING, ***Icones Filicum Sinicarum***, vols. 2-5; 1, pl. 1-50, 1930; 2, pl. 51-101, 1934; 3, pl. 101-150, 1935; 4, pl. 151-200, 1937; 5, pl. 201-250, 1958; The Metropolitan Museum of Natural History, Academia Sinica, Nanking [Nanjing] and the Fan Memorial Institute of Biology, Peiping [Beijing] (vol. 1), the Fan

247 John Hodgkiss, 1988, A guide to articles in the Hong Kong Naturalist (1930-1941) and in the Memoirs of the Hong Kong Natural History Society (1953-1988), *Memoirs of the Hong Kong Natural History Society* 18: 67-71.

248 Wan-Yi ZHAO, Ke-Wang XU, Qiang FAN and Wen-Bo LIAO, 2016, Contributions to the botanical journal *Sunyatsenia* from 1930 to 1948, *Phytotaxa* 269(4): 237-270.

Memorial Institute of Biology, Peping [Beijing] (vols. 2-4) and Science Press, Beijing (vol. 5). Reprinted Ren-Chang CHING, ***Icones Filicum Sinicarum***, 522 p, 2011; Peking University Press, Beijing.[249]

1930—1958 年，2011 年，胡先骕、秦仁昌，**中国蕨类植物图谱**，第 1 册，图版 1-50，1930；秦仁昌，**中国蕨类植物图谱**，第 2 册，图版 51-101，1934；第 3 册，图版 101-150，1935；第 4 册，图版 151-200，1937；第 5 册，图版 201-250，1958；国立中央研究院自然历史博物馆，南京、静生生物调查所，北平（第 1 册），静生生物调查所，北平（第 2-4 册），科学出版社，北京（第 5 册）。秦仁昌编著，**中国蕨类植物图谱**，522 页，图版 1-250；2011[250]；北京大学出版社，北京。

249 Reprinted as one volume.

250 重印合订本。参见：陈斌惠，2011，秦仁昌先生的《中国蕨类植物图谱》（*Icones Filicum Sinicarum*）已再版，*Sinopteris* 中国蕨 17：26-27。

1931

1931 年 4 月，秦仁昌，**今后发展我国植物分类学应取之途径**，37 页；国立中山大学农林植物研究所，广州。[251]

1931.04, Ren-Chang CHING, '***The Way of the Development of Our Plant Taxonomy from Now***', 37 p; Botanical Institute, College of Agriculture and College of Science, National Sun Yat-Sen University, Canton [Guangzhou].

1931 年 9 月，刘毅然[252]再度留美，1933 年 12 月获威斯康星大学麦迪逊分校植物学与植物生理学博士学位，1934 年 1 月至 1934 年 7 月任燕京大学教授，1934 年 8 月至 1936 年 8 月任国立河北农学院教授，1936 年 8 月至 1937 年 7 月任国立北平师范大学教授，1937 年 7 月至 1950 年 8 月任国立西北大学教授兼系主任，1950 年 9 月至 1986 年 8 月任南开大学教授，1986 年 90 岁高龄从南开大学生物系退休[253]。另参见 1918 年条目。

1931.09, Yi-Ran LIU returned to the United States of America to study botany; received Ph. D. degree in botany and plant physiology from the University of Wisconsin, Madison, Wisconsin, in December 1933[254]; returned to China to become professor at Yenching University, January 1934 to July 1934, then professor of National Hebei Agriculture College, August 1934 to August 1936, professor of National Peking Normal University, August 1936 to July 1937, professor and dean of National Northwest University, July 1937 to August 1950, professor of Nankai University in September 1950 until retirement in August 1986 at 90 years of age. See also 1918.

1931, Edward I. Farrington, ***Ernest H. Wilson Plant Hunter****—with a list of his most important introductions and where to get them*, 197 p, 34 plates; The Stratford Co., Boston, Massachusetts.

1931 年，Edward I. Farrington，**植物猎人威尔逊**——附他最重要的引种植物名录及由何处能得到它们，197 页，34 图版；The Stratford Co.，波士顿，麻州。

251 文前有左景烈所撰写的小引，文后有秦仁昌参加 1930 年第五届国际植物学会详记。

252 1925 年 1 月至 1930 年 7 月任北平协和医学院药学助教，1930 年 8 月至 1931 年 7 月任燕京大学教授，期间主要从事植物分类学和药用植物学研究，并参与本草纲目的英文翻译，同时还写过植物分类学专著、出版过植物学教科书。

253 本书所有记载的刘毅然先生的简历，均由刘毅然先生生前的最后工作单位、南开大学的档案馆管理服务科科长张兰普先生根据刘毅然先生自己填写的简历而提供（2015 年 10 月 15 日星期四）；特此致谢。

254 Ph.D. dissertation: Investigation on the sexual behavior of the apple rust fungus.

1931, Chi PING and Hsen-Hsu HU, ***Biological Science***[255], Sophia H. CHEN ZEN (ed.), ***Symposium on Chinese Culture***, 194-205 p; China Institute of Pacific Relations, Shanghai. The Chinese edition published in 2009.

2009 年，王宪明、高继美译，**生物科学**，陈衡哲编，**中国文化论集**，154-165；福建教育出版社，福州。

1931, William W. Smith, The contribution of China to European gardens, *Notes from the Royal Botanic Garden Edinburgh* 79: 215-221.

1931 年，William W. Smith，中国对欧洲园林的贡献，*Notes from the Royal Botanic Garden Edinburgh* 79：215-221。

1931, Francis Kingdon Ward, ***Plant Hunting in the Wilds***, 78 p; Figurehead, London.

1931 年，Francis Kingdon Ward，**野外植物探寻**，78 页；Figurehead，伦敦。

1931 年 5 月至 1933 年 2 月，北平研究院植物学研究所刘慎谔参加中法西北科学考察队，后经新疆和西藏至印度，再经印度加尔各答返回[256]；先后采集植物标本约 4 500 号，现存于中国科学院植物研究所（PE）。

1931.05-1933.02, Tchen-Ngo LIOU, the Institute of Botany, National Academy of Peiping [Beijing], joined Chinese-French Expedition to northwest China, from Xinjiang and Xizang to India, returned to China via Calcutta; over 4,500 numbers of plant specimens (PE) were collected.

1931.06.20-1932.01.10, Albert N. Steward[257](1897-1959) of the University of Nanking [Nanjing] collaborated with the Arnold Arboretum and Farlow Herbarium of Harvard University, and New York

255 原英文目录为：*Recent Progress of Biological Science*，但是正文却是 *Biological Science*。

256 姜玉平，2003，北平研究院植物学研究所的二十年，中国科技史料 24（1）：34-46。

257 Steward attended Oregon Agricultural College from 1917 to 1920 and earned a bachelor degree of science in Agriculture in 1921. In the same year, he became a faculty member in botany at the University of Nanking in Nanking [Nanjing], China. He and his wife, Celia Belle Speak Steward, were appointed educational missionaries by the Methodist Board of Missions. He returned to the United States for several years in the late 1920s to complete master of arts and Ph.D. degrees in biology at Harvard University. Steward spent most of the 1930s and 1940s in China. He was interned at Chapei Camp in Shanghai from 1943 to 1945. He returned to the United States permanently in 1950. Albert N. Steward was appointed associate professor of botany, herbarium curator and associate botanist for the Agricultural Experiment Station, Oregon State College, in 1951. He remained there until his death in 1959. He specialized in systematic, economic and geographic botany, the flora of central China and aquatic plants of northwestern North America. See Helen M. Gilkey, 1959, Albert Newton Steward, *Bulletin of the Torrey Botanical Club* 86(5): 342-344; Rebecca Huot, 2003, Albert N. Steward (1897-1959): Twenty-Six Years in China and Curator of OSU, *Oregon Flora Newsletter* 9(1):1,4-5.

Botanical Gardens, and led a team with Chi-Yuen CHIAO (Chi-Yuan CHIAO, 1901–1968) et al. to collect plants in Kweichow [Guizhou] (via Hubei, Sichuan as well as Hunan), collected 956 vascular numbers (about 15,000 sheets, A, K, N, NY, US), 893 numbers of cellular cryptogams (FH), 115 numbers of seeds and fruits (USDA), and 73 woods numbers (Forest Service, USDA)[258].

1931年6月20日至1932年1月10日，南京金陵大学 Albert N. Steward（史德蔚，1897—1959）与哈佛大学阿诺德树木园和隐花植物标本馆以及纽约植物园合作，带领焦启源（1901—1968）等赴贵州（途经湖北、四川以及湖南）采集维管束植物标本 956 号（约 1.5 万份，A、K、N、NY、US），隐花植物 893 号（FH），另有种子及果实标本 115 号（美国农业部），木材标本 73 号（美国农业部林务局）。

Albert N. Steward (1897–1959) (Photo provided by Oregon State University)
Albert N. Steward（1897—1959）（相片提供者：俄勒冈州立大学）

1931, 1934, Ju-Chiang LIU, ***Systematic Botany of the Flowering Families in North China***, 212 p, 1931; Henri Vetch the French Bookstore, Peiping [Beijing]; ***Systematic Botany of the Flowering Families in North China***, ed. 2, 218 p, 1934; Henri Vetch the French Bookstore, Peiping [Beijing].

1931 年、1934 年，刘毅然[259]，**植物分类学**，212 页，1931；法文图书馆，北平 [北京]；**植物分类学**，第 2 版，218 页，1934；法文图书馆，北平 [北京]。

1931–1936，Tchen-Ngo LIOU, Yong LING, Ching-Sheng HAO and Hsien-Wu KONG, ***Flore Illustrée du Nord de la Chine Hopei (Chihli) et ses Provinces Voisines***, Fascicules 1–5; Academie Nationale de Peiping, Peiping [Beijing].

1931—1936 年，刘慎谔、林镕、郝景盛、孔宪武，**中国北部植物图志：河北及其邻省**，1–5 册；国立北平研究院，北平。

258 Albert N. Steward & Chi-Yuen CHIAO, 1933, Recent botanical explorations in Kweichow, *The China Journal* 18(1, 2, & 3): 19–27, 79–87, & 132–143.

259 原书作者英文名字为 J. C. Liu，但是中文使用的却是刘毅然。同样，该书的英文与中文名称原标题也不是完全相同。

1931—1940 年，曾怀德在广东和广西（以及湖南）采集约 31 000 号（IBSC，PE，SYS）。[260]

1931-1940, Wai-Tak TSANG collected over 31,000 numbers of plant specimens (IBSC, PE, SYS) in Guangdong and Guangxi (and Hunan).

1931-1949, National Academy of Peiping [Beijing], ***Contributions from the Institute of Botany, National Academy of Peiping***, vols. 1-6(4); as ***Contributions from the Laboratory of Botany, National Academy of Peiping*** **[Beijing]**, title of volume 1(1).

1931—1949 年，国立北平研究院，**国立北平研究院植物学研究所丛刊**，第 1–6 卷第 4 期。其中，第 1 卷英文标题与其他不同。

1931-1950, Musée Heude, ***Notes de Botanique Chinoise***, volumes 1-10[261], Chang Hai [Shanghai].

1931—1950 年，震旦博物馆，**中国植物学记录**，1–10 卷，上海。

260 胡启明、曾飞燕，2011，广东植物志，10：328–329 页；广东科技出版社，广州（附录 2，华南植物研究所（园）植物标本采集简史）。

261 Volumes 1, 1931; 2, 1933; 3, 1942; 4, 1943; 5, 1943; 6, 1943; 7, 1945; 8, 1946; 9, 1949; and 10, 1950.

1932

1932 年 1 月 1 日，秉志辞去静生生物调查所所长职务，胡先骕接任（1932—1949）。

1932.01.01, Chi PING resigned as director of the Fan Memorial Institute of Biology, succeeded by Hsen-Hsu HU (1932-1949).

1932 年，方文培，中国植物学发达史略，科学世界 1(2): 125-130。

1932, Wen-Pei FANG, 'Development history of botany in China', *Scientific World* 1(2): 125-130.

1932，Yushun Kudo, Genera Plantarum Formosanarum, or a description of all the genera of the vascular plants indigenous to Formosa and an enumeration of all the species, varieties and forms hitherto known in Formosa. I. Sauruaceae-Rosaceae, *Annual Reports of the Taihoku Botanic Garden, Faculty of Science and Agriculture, Taihoku Imperial University, Formosa, Japan* 2: 1-141.

1932 年，工藤祐舜，台湾植物属志，I. 三白草科 – 蔷薇科，台北帝国大学理农学部附属植物园年报 2: 1-141 页。

1932 年，林镕，西藏植物采集意见报告书[262]，国立北平研究院院务汇报 3(3): 1-3。

1932, Yong LING, 'Report on plant collection in Tibet', *Bulletin of National Academy of Peiping* 3(3): 1-3.

1932 年，钟观光，论植物邦名之重要及其整理法，国立中央研究院自然历史博物馆丛刊 3(1): 1-8。

1932, Kuan-Kuang TSOONG, On the importance of Chinese names for plants with suggestions for a proper system of nomenclature, *Sinensia* 3(1): 1-8.

1932 年，吴韫珍，河北省植物发见史概略，清华周刊 38(10-11): 97-110。

1932, Yun-Chen WU, Personnels having contribution to our Chili Flora, *Tsing Hwa Weekly* 38(10-11): 97-110.

1932 年，钟观光，科学名词审查会植物学名词审查本植物属名之校订，国立中央研究院自然历史博物馆丛刊 3(1): 9-52。本文更正并修订的很多中文名至今依然在使用。

262 该文报告刘慎谔在西藏考察情况。

1932, Kuan-Kuang TSOONG, Criticisms and corrections of the botanic terms, *Sinensia* 3(1): 9-52. Many Chinese names corrected and revised in this work are still in use today.

1932—1934年，静生生物调查所蔡希陶[263]（1911—1981）赴云南采集，获得万余号标本（A、HITBC、KUN、PE）。

1932-1934, Hse-Tao TSAI (Hsi-Tao TSAI, 1911-1981), the Fan Memorial Institute of Biology, collected over 10,000 numbers of plant specimens in Yunnan (A, HITBC, KUN, PE).

Hse-Tao TSAI (1911-1981), and Hse-Tao TSAI (left) with his local guide (Photo provided by Zong-Gang HU)
蔡希陶（1911—1981）、蔡希陶（左）与当地向导（相片提供者：胡宗刚）

1932—1935年，西部科学院生物研究所俞德浚（1908—1986）受哈佛大学阿诺德树木园与静生生物调查所的合作项目资助于四川采集，获得大量珍贵植物标本（A, GH, PE）。

1932-1935, Te-Tsun YU (1908-1986), Department of Biology, Science Institute of Western China, with support from a cooperative program between the Arnold Arboretum of Harvard University and the Fan Memorial Institute of Biology, collected large sets of plant specimens in Sichuan (A, GH, PE).

263 字侃如，浙江东阳人；上海光华大学肄业，1930年进入静生生物调查所。1932年进入云南考察，1938年于昆明黑龙潭协助胡先骕创办云南农林植物研究所（今中国科学院昆明植物研究所）；1958年在西双版纳的葫芦岛筹建热带植物园（今中国科学院西双版纳热带植物园），生前一直任主任。

1933

1933 年 8 月 20 日，胡先骕、辛树帜、李继侗[264]（1897—1961）、张景钺、裴鉴、李良庆[265]（1904—1952）、严楚江[266]（1900—1978）、钱天鹤、董爽秋（董桂阳）、叶雅各[267]（1894—1967）、秦仁昌、钱崇澍、陈焕镛、钟心煊、刘慎谔、吴韫珍、陈嵘、张珽、林镕共 19 人发起，于四川重庆北碚西部科学院成立中国植物学会[268]；钱崇澍任第一任会长，副会长陈焕镛，书记张景钺，会计秦仁昌；会员 105 人，其中与植物分类学相关者有方文培、王启无（1913—1987）、左景烈、史久庄[269]、朱王家玮[270]（1903—？）、李良庆、李荫祯（李构堂，1903—1992）、李顺卿、汪发缵、吴印禅[271]（1902—1959）、

264 江苏兴化人，1918—1921 年南京金陵大学林科毕业，1921 年考取清华学校公费留美，入美国耶鲁大学林学院，1923 年获硕士学位，1925 年获耶鲁大学博士学位；同年回国在金陵大学任教；先后任教于南开大学、国立清华大学、国立西南联合大学和北京大学，1955 年任中国科学院生物学部委员、1957—1961 年任内蒙古大学教授、第一任副校长，并主持工作。1961 年 12 月 12 日在呼和浩特市逝世。参见：于北辰主编，1986，李继侗文集，422 页；科学出版社，北京。

265 河北清丰人，1921 年入北平协和医学院预科，1927 年毕业于南开大学，1927—1929 年任教于南开中学，1929 年赴美，1930 年和 1932 年分别获美国俄亥俄州立大学硕士和博士学位；归国后历任静生生物调查所研究员兼植物标本室主任，并先后在北京大学、国立清华大学、国立北平师范大学、辅仁大学等兼任教授；1947 年赴青岛任国立山东大学植物系教授，1951 年起兼任系主任；1952 年 12 月 8 日因心脏病逝世；著名藻类学家。

266 江苏（上海）崇明人，1926 年毕业于国立东南大学园艺系，1929 年留学美国芝加哥大学，1932 年获博士学位；同年回国，先后任教于国立中央大学、国立北平师范大学、云南大学、厦门大学；著名植物形态学家。

267 又名雅谷，广东番禺人，早年就读于广州岭南学堂，1916 年赴菲律宾大学学习，1917 年赴美留学，1918 年获宾夕法尼亚州立大学学士学位、1921 年获耶鲁大学林学硕士学位，同年经欧洲回国；先任职于南京金陵大学，1928 年受国民政府大学院委派，赴武汉筹建国立武汉大学新校址，并长期担任国立武汉大学生物系教授、农学院院长等职。详细参见：武汉大学生命科学院院史编纂委员会，2012，奋进岁月 铸就辉煌——武汉大学生命科学学院院史，481 页；武汉大学出版社，武汉。

268 植边，1983，中国植物学会五十周年介绍，植物杂志 3：1-2。

269 1934 年曾经担任国立中央大学生物系助教，并从事悬钩子属研究；参见：胡先骕，1934，中国近年植物学进步之概况，中国植物学杂志 1（1）：3-10。

270 1927 年毕业于金陵女子大学（女）；1934 年从事植物分类学；参见：胡先骕，1934，中国近年植物学进步之概况，中国植物学杂志 1（1）：3-10。

271 字韬甫，江苏沭阳人，1916 年入江苏省立第八师范学校，1920 年毕业留校工作；次年回沭阳县第一小学教书；1925 年入武昌高等师范学校生物系，1928 年毕业任教于国立中山大学；1934 年 5 月赴德国柏林大学学习，1940 年回国任国立中山大学生物系教授；1941—1945 年任教于同济大学并兼任系主任及代理理学院院长，1946 年以后任教于国立中山大学并兼任农林植物研究所所长以及中国科学院华南植物研究所副所长等职。

吴韫珍、辛树帜、林镕、周汉藩[272]（1877—1944）、金维坚[273]（1902—1986）、胡先骕、马心仪[274]、俞德浚、郝景盛[275]（1903—1955）、陆文郁[276]（1887—1974）、陈焕镛、陈封怀（1900—1993）、陈嵘、陈邦杰、孙祥钟[277]（1907—1994）、孙雄才（1898—1964）、耿以礼、秦仁昌、唐进、黄以仁、董爽秋（董桂阳）、裴鉴、邓叔群、蒋英、郑万钧、刘慎谔、刘厚（刘大悲）、刘振书（刘式民）、蔡𬞟、钱崇澍、钟观光、钟补求[278]（1906—1981）、戴芳澜[279]（1893—1973）等。

1933.08.20, Botanical Society of China founded by 19 Chinese botanists: Hsen-Hsu HU, Shu-Chih HSIN, Chi-Tung LI (1897-1961), Ching-Yueh CHANG, Chien PEI, Liang-Ching LI (1904-1952), Chu-Chiang YAN (1900-1978), Tien-Hao CHIEN, Shuang-Chiu TUNG (Kwei-Yang TUNG), Ya-Ko YEH (Nga-Kok IP, 1894-1967), Ren-Chang CHING, Sung-Shu CHIEN, Woon-Young CHUN, Hsin-Hsuan CHUNG, Tchen-Ngo LIOU, Yun-Chen WU, Yung CHEN, Ting CHANG and Yong LING. First meeting was held at the Science Institute of Western China, Beibei, Chongqing, Sichuan; Sung-Shu CHIEN was chosen to be the first president of the Botanical Society of China; 105 members from throughout China. Plant taxonomists (alphabetically) included Feng-Hwai CHEN (1900-1993), Pan-Chieh CHEN, Yung CHEN, Wan-Chun CHENG, Sung-Shu CHIEN, Ren-Chang CHING, Hang-Fan CHOW (Han-Fan CHOU, 1877-1944), Woon-Young CHUN, Wen-Pei FANG, Ching-Sheng HAO (Kin-Shen HAO, 1903-1955), Shu-Chih HSIN, Hsen-Hsu HU, Yi-Li KENG, Wei-Jean KING (Wei-Chien CHEN, 1902-1986), Shun-Ching LEE, Liang-Ching LI, Yin-Chen LI (Ko-Tang LI, 1903-1992), Yong LING, Tchen-Ngo LIOU, Chen-Shu LIU, Hou LIU, Wen-Yu LU (1887-1974), Hsin-Yih MA, Chien PEI, Chiu-Chuang SHIH, Hsiang-Chung SUN (1907-1994), Yon-Zai SUN (Hsiung-Tsai SUN, 1898-1964), Fang-Lan TAI (1893-1973), Tsin TANG, Shu-Chun TENG, Tan TSAI, Ying TSIANG, Ching-Lieh TSO, Kwan-Kwang TSOONG, Pu-Chiu TSOONG (1906-1981), Shuang-

272 原名周勺泉，湖南平江人，早年就读于长沙湖南优级师范博物科，1924 年任湖南省教育会博物馆馆长；后又为蔡元培所聘就教于北京大学，兼任北平博物馆馆长，后任职于静生生物调查所。

273 又名鸿豪，字叔闻，号万绿庐主，浙江金华人；毕业于南京国立东南大学，曾任国立浙江大学副教授，浙江省立西湖博物馆馆长；退休前任山东农学院植物系教授；植物学家，业余爱好集邮、书法以及收藏等。

274 山东青岛人（女，生卒不详），1930 年德州大学植物生理学博士，1934 年曾担任广西大学生物系教授，从事土壤菌研究；参见：胡先骕，1934，中国近年植物学进步之概况，中国植物学杂志 1（1）：3-10。

275 河北正定人，字键君，1924 年入旅顺工科大学，1925 年考入北京大学预科，1927—1931 年北京大学生物系学习；在此期间曾到北平研究院植物学研究所工作，并兼任国立女子大学生物学讲师；1931—1933 年任北平研究院植物学研究所助理研究员；1931 年曾参加中法西北考察团。另参见 1934 年条目。

276 字莘农，别署老辛、百蜨庵主；天津现代著名花卉画家，也是著名的博物馆学家、植物学家和地方史学家。

277 安徽桐城人，1929 年考入国立武汉大学生物系，1936 年赴爱丁堡皇家植物园留学深造；长期在武汉大学任教，中国水生植物学的奠基人。

278 浙江镇海（今宁波）人，钟观光幼子；1927 年毕业于上海新华艺术大学，先后任职于西湖博物馆、中央研究院自然历史博物馆、北平研究院植物学研究所、中国西北植物调查所等；1947 年赴英法研修，1950 年回国后任中国科学院植物分类研究所研究员；专长桔梗科、玄参科以及槐树属；培养的著名弟子包括洪德元。

279 字观亭，湖北江陵人，1912—1913 年入清华学校留美预备班，1914—1919 年先后获美国康奈尔大学农学院学士学位和哥伦比亚大学研究生院硕士学位；1920 年回国后先后任教于广东省农业专门学校、国立东南大学、南京金陵大学，1934—1935 年美国纽约植物园及康奈尔大学研究院专攻真菌遗传学，回国后任国立清华大学教授兼农业研究所植物病理研究室、农学院植物病理系主任、北京农业大学教授，兼任中国科学院真菌植病室主任、中国科学院应用真菌学研究所所长、中国科学院微生物研究所所长等职。

Chiu TUNG (Kwei-Yang TUNG), Chi-Wu WANG (1913-1987), Chia-Wei CHU-WANG (1903-?), Fa-Tsuan WANG, I-Jen WANG, Yin-Chan WU (1902-1959), Yun-Chen WU and Te-Tsun YU.

1933 年，焦启源[280]（1901—1968）赴美留学，1936 年获威斯康星大学麦迪逊分校博士学位，回国后任教于金陵大学（任系主任）、国立四川大学、国立武汉大学、南京农学院等院校；主要从事植物生理学研究，1953 年调入复旦大学，任植物生理教研室主任。

1933, Chi-Yuen CHIAO (Chi-Yuan CHIAO, 1901-1968) traveled to the United States of America to study at the University of Wisconsin, Madison, Wisconsin; received Ph. D. degree in 1936[281]; served as professor, University of Nanking (dean of department), National Sichuan University, National Wuhan University and Nanjing Agriculture College, with research focus on plant physiology, then to Fudan University in 1953, as leader of plant physiology program.

Chi-Yuan CHIAO (1901-1968) (Photo provided by Zong-Gang HU)
焦启源（1901—1968）（相片提供者：胡宗刚）

1933 年，杜亚泉编，**高等植物分类学**，243 页；商务印书馆，上海。[282]

1933, Ya-Tsuan TU, '***Taxonomy of Higher Plants***', 243 p; Commercial Press, Shanghai.

1933—1935 年，国立中央大学生物系张肇骞[283]（1900—1972）赴英伦留学，先后在英国皇家植物园邱园和爱丁堡植物园研修菊科；回国后任教于广西大学、国立浙江大学、国立中正大学（兼生物系系主任）等；1946 年赴北平静生生物调查所工作并兼北京大学生物系教授；1949 年后任中国科学院植物分类研究所（后改名为植物研究所）研究员，1953 年起兼任该所副所长，1955 年被选聘为中国科学院学部委员，同年被调到广州协助陈焕镛组建中国科学院华南植物研究所，历任该所副所长、

280 江苏镇江人，1923 年毕业于金陵大学，并留校任教，期间曾在山东和贵州等地大规模采集植物，并从事植物分类学研究。

281 Ph.D. dissertation: The influence of environmental factors on the development of anthocyanin and the physiological significance of this pigment in *Amarnthus cordatus*.

282 此为百科小丛书之一，1934 年作为万有文库再次出版发行。

283 字冠超，浙江永嘉人，1920—1925 年在金陵大学农学院学习，1925—1926 年在国立东南大学生物系学习；1926—1933 年任国立中央大学生物系助教、讲师。

代所长等职。

1933-1935, Chao-Chien CHANG (1900-1972), department of biology, National Central University, traveled to the United Kingdom to study Asteraceae at the Royal Botanic Gardens, Kew, and at the Royal Botanic Garden, Edinburgh; returned to teach at Kwangsi [Guangxi] University, National Chekiang University and National Chiang Kai Shek University (as dean of department of biology), research professor in the Fan Memorial Institute of Biology in Peiping [Beijing] in 1946, at the same time professor in department of biology, Peking University; joined the Institute of Plant Taxonomy (later renamed Institute of Botany), Chinese Academy of Sciences, in 1949, vice director of Institute of Botany, Chinese Academy of Sciences, in 1953, then vice director of South China Institute of Botany, Chinese Academy of Sciences, in 1955; elected academician in Chinese Academy of Sciences in the same year.

Chao-Chien CHANG (1900-1972) (Photo provided by Zong-Gang HU)
张肇骞（1900—1972）（相片提供者：胡宗刚）

1933—1937 年，叶培忠[284]（1899—1978）任总理陵园纪念植物园植物科主任。

1933-1937, Pei-Chung YIEH (1899-1978) named head of botany, Botanical Garden, Dr. Sun Yat-Sens Memorial Park.

1933 and 1943，Liberty H. Bailey, ***How Plants Get Their Names***, 209 p, 1933; The MacMillan Co., New York; George F. Zimmer, ***A Popular Dictionary of Botanical Names and Terms with Their English Equivalents****: for the use of botanists and horticulturists, as well as for lovers of the flowers of garden, field and wood*, 122 p, 1943; Routledge and Kegan Paul Ltd., London. Translated into Chinese in 1957.

1957 年，丁广奇编译[285]，侯宽昭校订，**植物种名释**，114 页；科学出版社，北京。

284 江苏江阴人，原名沈培忠；1921 年考入南京金陵大学，主修森林学，辅修园艺学，1927 年毕业，先后任职母校助教和广西柳州沙塘城林场场长，1929 年入南京总理陵园纪念植物园任筹备助理员，1930 年赴爱丁堡植物园研修园艺，1932 年返国。抗战期间先后于长沙、重庆、峨眉、天水等地林业部门任职，抗战胜利初期任职国民政府农林部，1948 年任国立武汉大学森林系教授，1952 年任华中农学院教授，1956 年任南京林业大学教授同时兼任南京植物园研究员。参见：叶和平，2009，叶培忠，204 页；中国林业出版社，北京；《遍三点水加西绿荫》—— 叶培忠纪念文集编委会，2010，遍三点水加西绿荫 —— 叶培忠纪念文集，308 页；中国林业出版社，北京。

285 基于上述两本著作编译而成。

1933-1950, Frank Ludlow (1885-1972) and George Sherriff (1898-1967) explored Bhutan and southeastern Xizang (8 trips to Xizang); collected more than 21,000 numbers of plant specimens (about half from Xizang). Those collections are mainly in England (BM)[286] with duplicates elsewhere.

1933—1950 年, Frank Ludlow (1885—1972) 和 George Sherriff (1898—1967) 考察不丹和藏东南(其中 8 次进入西藏), 采集约 21 000 号植物标本(大约一半采自西藏); 所采标本主要存于英国(BM), 其他标本馆也有复份标本。

Frank Ludlow (1885-1972) (Photo provided by Royal Botanic Gardens, Kew)

Frank Ludlow (1885—1972)(相片提供者:邱园)

286 William T. Stearn, 1976, Frank Ludlow (1885-1972) and the Ludlow-Sherriff expeditions to Bhutan and south-eastern Tibet of 1933-1950, *Bulletin of the British Museum (Natural History) Botany* 5(5): 243-268.

1934

1934 年 8 月 20 日，静生生物调查所与江西省农业院合办庐山森林植物园于江西九江庐山成立；1934—1946 年秦仁昌任第一任主任兼标本馆（LBG）馆长。

1934.08.20, Lushan Forestry Botanical Garden, organized by the Fan Memorial Institute of Biology and Jiangxi Academy of Agriculture, founded in Lushan, Jiujiang, Jiangxi, with Ren-Chang CHING as first director, 1934-1946, and director of Herbarium (LBG).

Lushan Forestry Botanical Garden founded, Lushan, 20, August 1934, front row (left-right): Hsen-Hsu HU, Chi PING, Ren-Chang CHING, back row: Chong-Lun ZENG, Yu-Shi LIU, Zao TU, Kuo-Mei FENG, Zhen LEI, Ju-Yuan WANG) (Photo provided by Zong-Gang HU)

1934 年 8 月 20 日，庐山森林植物园于庐山建立，前排（从左至右）：胡先骕、秉志、秦仁昌，后排：曾仲伦、刘雨时、涂藻、冯国楣、雷震、汪菊渊）（相片提供者：胡宗刚）

1934 年 8 月 21 日，中国植物学会第一届年会（第二届全国代表大会）在江西九江庐山举行；选举胡先骕为会长、陈焕镛为副会长、张景钺为书记、秦仁昌为会计。胡先骕提案编写中国植物志。

1934.08.21, First anniversary meeting (Second National Congress) of the Botanical Society of China

View of Lushan Forestry Botanical Garden, 1936 (Photo provided by Zong-Gang HU)
1936 年庐山森林植物园远景（相片提供者：胡宗刚）

Gate of Lushan Botanical Garden, 1954 (Photo provided by Zong-Gang HU)
1954 年庐山植物园大门（相片提供者：胡宗刚）

Gate of Lushan Botanical Garden, 1962 (Photo provided by Zong-Gang HU)
1962 年庐山植物园大门（相片提供者：胡宗刚）

held in Lushan, Jiujiang, Jiangxi, Hsen-Hsu HU was elected president. Hsen-Hsu HU proposed compiling a *flora of China*.

1934年，郝景盛[287]（1903—1955）赴德入柏林大学理学院和爱北瓦林业专科大学，攻读植物地理和植物生理，1937年获柏林大学自然科学博士学位；1938年6月在爱北瓦林业专科大学获林学博士学位。1939年初经香港、河内到达昆明，在云南省建设厅林务处任技正，并兼国立中山大学林学教授。1940—1945年任位于重庆的国立中央大学森林系教授，其中1943年兼任昆明北平研究院植物学研究所研究员，1946—1947年任东北大学农学院森林系教授，兼农学院院长，1947—1949年任北平研究院植物学研究所研究员，1950—1954年任中国科学院植物分类研究所研究员，1954—1955年任林业部总工程师兼林业部技术委员会主任。

Ching-Sheng HAO (1903-1955) (Photo provided by Zong-Gang HU)
郝景盛（1903—1955）（相片提供者：胡宗刚）

1934, Ching-Sheng HAO (Kin-Shen HAO, 1903-1955) attended College of Arts and Science, Universität zu Berlin and Forestry College Eberswalde, Germany, to study phytogeography and physiology, received Ph. D. degree in science from Universität zu Berlin in 1937[288] and Ph. D. degree in forestry from Forestry College, Eberswalde, in 1938[289]. Hao returned to China via Hong Kong and Hanoi and arrived in Kunming in 1939; served in the Department of Forestry, Bureau of Construction of Yunnan, at the same time as professor in the Department of Biology, National Sun Yat-Sen University; later professor in Department of Forestry, National Central University, Chungching [Chongqing], 1940-1945, while also research professor of the Institute of Botany, National Academy of Peiping [Beijing], in 1943; professor in department of forestry and dean of School of Agriculture, Northeast Agriculture College, Shenyang, 1946-1947; research professor in the Institute of Botany, National Academy of Peiping [Beijing], between 1947-1949; research professor, Institute of Plant Taxonomy, Chinese Academy of Sciences, 1950-1954; General Engineer of Administrative Bureau of Forestry and director of Scientific Committee of Administrative

287 详细参见：王宗训，1955，林学家、植物学家郝景盛先生逝世，科学通报 6：75-76。

288 Berlin, Math.-Naturwiss. Diss: Kin-Shen HAO, 1938, Pflanzengeographische Studienüber den Kokonor-See und über das angrenzende Gebiet, *Botanische Jahrbücher fur Systematik, Pflanzengeschichte und Pflanzengeographie* 68: 515-668; Advisors: Friedrich Ludwing E. Diels (1874-1945) and Robert K. F. Pilger (1876-1953).

289 Fachhochschule Dissertation: Ching-Sheng HAO, 1939, Über Saatgutprüfung auf biochemischem Wege, *Zeitschrift für Forst- und Jagdwesen* 71 (3, 4 und 5): 141-156, 187-204, 249-269.

Bureau of Forestry, 1954-1955.

1934, Hsen-Hsu HU, Recent progress of the botanical sciences in China, *Peking Natural History Bulletin* 9(2): 71-76.

1934 年，胡先骕，中国近年植物科学的进展，北平博物杂志 9（2）：71-76。

1934 年，胡先骕，中国近年植物学进步之概况，中国植物学杂志 1(1): 3-10。

1934, Hsen-Hsu HU, 'Overview of progress of Chinese botany in recent years', *The Journal of the Botanical Society of China* 1(1): 3-10.

1934 年，胡先骕，植物分类学研究之方法，中国植物学杂志 1(3): 306-317。

1934, Hsen-Hsu HU, 'Methods in the study of plant taxonomy', *The Journal of the Botanical Society of China* 1(3): 306-317.

1934, Hsen-Hsu HU, Notulae Systematicae ad Floram Sinensem V, *Bulletin of the Fan Memorial Institute of Biology, Botany* 5: 305-318, including Torricelliaceae, a new family. This is the first plant family described by a Chinese botanist.

1934 年，胡先骕，中国植物分类小志五，静生生物调查所汇报，5: 305-318；包括新科——鞘柄木科。此为中国植物学者发表的第一个植物科。

1934, Francis Kingdon Ward, ***A Plant Hunter in Tibet***, 317 p; Jonathan Cape Ltd., London.

1934 年，Francis Kingdon Ward，**一个植物猎人在西藏**，317 页;Jonathan Cape Ltd.，伦敦。

1934 年 5 月至 1935 年 6 月，中国科学社生物研究所吴中伦和国立中央大学农学院森林系陈谋[290]

Mou CHEN (1903-1935) (Photo provided by Ning-Sheng CHEN)
陈谋（1903—1935）（相片提供者：陈宁生）

290 字尊三，浙江诸暨人；1924 年入国立北平农科大学，两年后因家境退学，1926 年入国立浙江大学农学院生物教研室任技术员，1932 年 9 月入国立中央大学农学院森林系任树木学助教。参见：陈谋君为学术而牺牲，国立中央大学日刊，1935 年 5 月 4 日；李寅恭，陈谋先生事略，国立中央大学日刊，1935 年 5 月 28 日；陈谋先生殉学纪念专刊，国立中央大学日刊，1935 年 6 月 3 日；李寅恭，1946，记陈尊三君子殉学事，科学 28（3）：163-164。

（1903—1935；1935 年 4 月 27 日病故于云南墨江）赴云南考察采集，采集植物标本数千号。

1934.05-1935.06, Chung-Lwen WU (Chung-Lun WU), the Biological Laboratory of the Science Society of China and Mou CHEN (1903-1935, died of illness in Mojiang, Yunnan, April 27, 1935), Department of Forestry, College of Agriculture, National Central Universithety, explored in Yunnan; collected several thousand numbers of plant specimens.

1934—1935 年，汪振儒，广西种子植物名录初稿（I-XXVII），广西大学周刊 6(2-9), 6(12-17), 7(1-8，10-12)，7（13），1934; 8(1-7，9-11), 1935。

1934-1935, Y.C. Hawthorne WANG[291], A primary checklist of flowering plants known from Kwangsi [Guangxi], *Kwangsi University Weekly* 6(2-9), 6(12-17), 7(1-8, 10-12), 7(13), 1934; 8(1-7, 9-11), 1935.

1934—1936 年，静生生物调查所陈封怀[292]（1900—1993）赴苏格兰爱丁堡植物园学习。回国后，1936—1938 年任庐山森林植物园技师，1938—1942 年在昆明任静生生物调查所研究员，1943—1945 年任江西泰和国立中正大学生物系教授，1946—1948 年任庐山森林植物园主任兼国立中正大学教

Feng-Hwai CHEN (1900-1993) (Photo provided by Zong-Gang HU)
陈封怀（1900—1993）（相片提供者：胡宗刚）

291 Yen-Chieh WANG.

292 江西修水人，1922—1924 年在金陵大学农学系学习；1925—1926 年转入国立东南大学农科学习；1927—1929 年任吴淞中国公学、沈阳文华中学教员；1929—1930 年任国立清华大学助教；1930—1934 年在北平静生生物调查所工作。

授，1949—1950 年任江西省农业科学研究所副所长，1950—1953 年任中国科学院庐山植物园主任，1954—1957 年任中国科学院南京中山植物园副主任，1958—1962 年任中国科学院武汉植物园主任，1962—1978 年任中国科学院华南植物研究所副所长兼华南植物园主任，1979—1983 年任中国科学院华南植物研究所所长、华南植物园主任，1983—1993 年任中国科学院华南植物研究所名誉所长。植物园与报春花科专家。

1934-1936, Feng-Hwai CHEN (1900-1993), the Fan Memorial Institute of Biology, studied at Royal Botanic Garden, Edinburgh, Scotland, returned to China to become horticultural engineer at Lushan Forestry Botanical Garden, 1936-1938; research professor at the Fan Memorial Institute of Biology, Kunming, 1938-1942; professor of National Chiang Kai Shek University, Taihe, Jiangxi, 1943-1945; director of Lushan Forestry Botanical Garden and professor of National Chiang Kai Shek University; vice director of Jiangxi Agriculture Academy, 1949-1950; director of Lushan Botanical Garden, 1950-1953; vice director of Nanjing Botanical Garden Memorial Sun Yat-Sen, Chinese Academy of Sciences, 1954-1957; director of Wuhan Botanical Garden, Chinese Academy of Sciences, 1958-1962; vice director of South China Institute of Botany, Chinese Academy of Sciences; director of South China Botanical Garden, Chinese Academy of Sciences, 1962-1978; director of South China Institute of Botany and director of South China Botanical Garden, Chinese Academy of Sciences, 1979-1983; emeritus director, South China Institute of Botany, Chinese Academy of Sciences, 1983-1993. Specialist in botanical gardens and Primulaceae.

1934-1936, Shigeyasu Tokunaga, ***Report of the First Scientific Expedition to Manchoukuo***, IV Botany, Takenoshin Nakai and Masao Kitagawa, Part I, 71 p, plates 20, 1934; Takenoshin Nakai, Masaji Honda, Masao Kitagawa, Part II, 187 p, plates 19, 1935; Yoshikadzu Emoto and Motoo Takahasi, Part III, 3+55 p, plates 1+85, 1936; Takenoshin Nakai, Masaji Honda, Yoshisuke Satake, Masao Kitagawa, Part IV, 108 p, plates 3, 1936; Office of the Scientific Expedition to Manchoukuo, Faculty of Science and Engineering, Waseda University, Tokyo.

1934—1936 年，德永重康，**第一次满蒙学术调查研究团报告**[293]，第四部：植物学，中井猛之进[294]、北川政夫，第一篇，71 页，图版 20，1934；中井猛之进、本田正次、北川政夫，第二篇，187 页，图版 19，1935；江本义数、高桥基生，第三篇，3 页 +55 页，图版 1+85，1936；中井猛之进、本田正次、佐竹义辅、北川政夫，第四篇，108 页，图版 3，1936；早稻田大学理工学部内满蒙学术调查研究团事务所，东京。

293 1933 年 6 月至 10 月，由日本早稻田大学德永重康（1874—1940）领衔的满蒙学术调查研究团对日本占领下的满洲国等地进行综合考察；其中植物学调查范围大致在今承德、赤峰、朝阳一带；参加人员包括东京帝国大学教授中井猛之进（1882—1952）、本田正次（1897—1984）和北川政夫。全书共 6 部计 25 册，3 937 页，图版 820；其中第四部植物学共 4 册，记载 81 个新种和 43 个新亚种。详细参见：德永重康与中井猛之进，1940，第一次满蒙学术调查研究团报告，结文：1-4 页，追记：5-6 页，总目录：ⅰ - ⅸ页。

294 Takenoshin Nakai (1882-1952), Japanese plant taxonomist and authority on the flora of Korea; professor at University of Tokyo and director of Koishikawa Botanical Garden, 1930; director of Bogor Botanical Garden, 1943-1945; director of National Science Museum, Japan, 1949.

1934—1937 年，方文培[295]（1899—1983）获中华教育文化基金会资助，赴英伦学习槭树科和杜鹃花科分类；1937 年 6 月获爱丁堡大学博士学位；1937—1947 年任国立四川大学生物系教授；1948—1949 年赴美国研修[296]；1950—1983 年任四川大学生物系教授；著名的槭属和杜鹃花属专家。

1934-1937, Wen-Pei FANG (1899-1983) studied Aceraceae and Rhododendron at University of Edinburgh and completed Ph. D. degree by June 1937[297]; returned to China to become professor at National Sichuan University, Chengdu, 1937-1947; studied、in the United States, 1948-1949; professor at Sichuan University, Chengdu, 1950-1983; specialist in *Acer* and *Rhododendron*.

1934—1944 年，中央研究院自然历史博物馆改组为动植物研究所，所长王家楫（1898—1976）[298]，裴鉴、耿以礼为该所兼任研究员，单人骅（1909—1986）为助理研究员。

1934-1944, The Metropolitan Museum of Natural History, Academia Sinica, renamed Institute of Zoology and Botany, Academia Sinica; Chia-Chi WANG (1898-1976) as director with Chien PEI and Yi-Li KENG as research professor and Ren-Hwa SHAN (1909-1986) as an assistant research professor.

1934—1952 年，中国植物学会，**中国植物学杂志**，1–6 卷，其中 1934—1937 年为 1–4 卷[299]，而 1950—1952 年为 5–6 卷。

1934-1952, Chinese Botanical Society[300], ***The Journal of the Botanical Society of China***, vols. 1-6, 1934-1952; between 1934-1937 as vols. 1-4 and 1950-1952 as vols. 5-6.

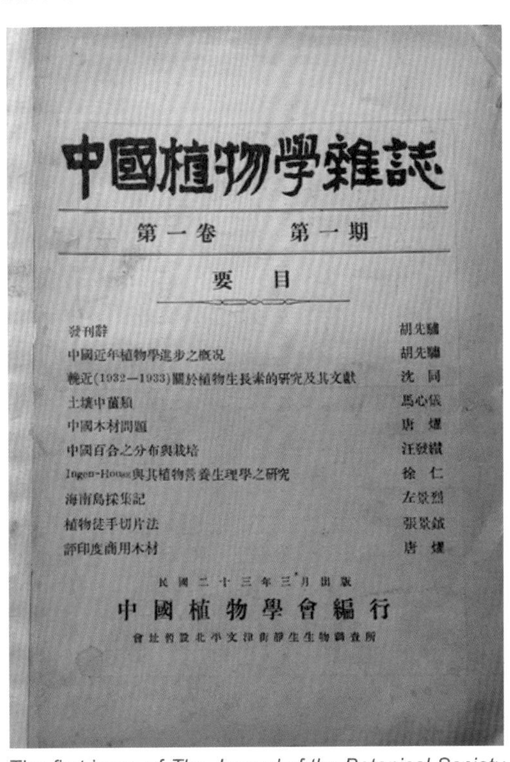

中國植物學雜誌

第一卷 第一期

要目

發刊辭	胡先驌
中國近年植物學進步之概況	胡先驌
晚近(1932—1933)關於植物生長素的研究及其文獻	沈同
土壤中菌類	馬心儀
中國木材問題	唐燿
中國百合之分布與栽培	汪發纘
Ingen-Housz與其植物營養生理學之研究	徐仁
海南島採集記	左景烈
植物徒手切片法	張景鉞
評印度商用木材	唐燿

民國二十三年三月出版

中國植物學會編行

會址暫設北平文津街靜生生物調查所

The first issue of *The Journal of the Botanical Society of China*, 1934 (Photo provided by Xiao-Jiang HU)
1934 年中国植物学杂志创刊号（相片提供者：胡晓江）

295 四川忠县人，1921 年（22 岁）考入国立东南大学首次创办的生物系，毕业后曾在河南中州大学短期任教；1927 年 28 岁时考入南京中国科学社生物研究所读研究生，在钱崇澍教授直接指导下专攻植物分类学，1928—1932 年间多次受派遣前往植物丰富的四川省考察与采集。

296 其间拍摄“东亚植物原种照片”4 934 张，署名为“方文培、王启无、戴清旭摄”，时间为 1949 年；其中全套照片目前至少在云南大学植物标本馆（YUKU 与 PYU 合并）有收藏。

297 Ph.D. dissertation: Wen-Pei FANG, 1939, A Monograph of Chinese Aceraceae, *Contributions from the Biological Laboratory of the Science Society of China* 11: 1-346; Advisor: John M. Cowan (1892-1960).

298 王家楫（1898—1976），江苏（上海）奉贤人，1920 年毕业于南京高等师范学校，1924 年获国立东南大学农学学士学位，1928 年获美国宾夕法尼亚大学哲学博士学位，1948 年选聘为中央研究院院士；中国科学院水生生物研究所研究员、所长，中国原生动物学的奠基人。

299 全国图书馆文献缩编复制中心，2009，中国近代植物与昆虫文献史料汇编，第 1–3 册（全书 12 册）曾将此刊的第 1–4 卷重新刊印。

300 i.e. Botanical Society of China.

1935

1935 年 3 月，广西大学校长马君武[301]（1881—1940）邀请国立中山大学农林植物研究所所长陈焕镛，赴广西梧州合办广西大学植物研究所，陈焕镛兼任该所所长（1935—1938）；同年建立植物标本室（IBK）。[302]

1935.03, Jun-Wu MA (1881-1940), president of Kwangsi [Guangxi] University, invited Woon-Young CHUN, director of the Botanical Institute, National Sun Yat-Sen University, to establish joined Kwangsi Institute of Botany, Kwangsi University, in Wuzhou, Guangxi, with Woon-Young CHUN as director, 1935-1938. Herbarium (IBK) was established in the same year.

1935.05, The Fan Memorial Institute of Biology, ***The Photographs of Chinese Plants***[303], nos. 2-17737 + 17738-18337; ***Catalogue of the Photographs of Chinese Plants***, 750 + 31 p; the Fan Memorial Institute of Biology, Peiping [Beijing]. The online database is at http://q.plantphoto.cn/ (accessed 28 August 2017).

1935 年 5 月，静生生物调查所，**中国植物照片集**[304]，编号 2-17737、17738-18337；**中国植物照片集目录**，750 + 31 页；静生生物调查所，北平。数据库网络地址：http://q.plantphoto.cn/（2017 年 8 月 28 日进入）。

1935 年 8 月 12 日至 15 日，中国植物学会第二届年会（第三届全国代表大会）在广西南宁举行[305]；选举陈焕镛为会长，戴芳澜为副会长，张景钺为书记，秦仁昌为会计。

---

301 字厚山，号君武，祖籍湖北蒲圻，生于广西桂林；1901 年留学日本，1907 年和 1913 年先后两次留学德国并获得博士学位，为大夏大学（今华东师范大学前身）、广西大学的创建人和首任校长；曾翻译物种起源。

302 刘演，2005，广西植物标本馆 70 年的回顾与展望，中国科学院广西植物研究所发展七十年（1935—2005），第 28–30 页；广西壮族自治区、中国科学院广西植物研究所，桂林（内部印制）。

303 As *Album of Chinese Plants* by the Institute of Botany, National Academy of Peiping [Beijing] (1936).

304 此为秦仁昌 1930 年 6 月至 1932 年 9 月于丹麦哥本哈根大学植物博物馆研究蕨类期间，从欧洲主要植物标本馆（特别是英国伦敦的邱园和自然历史博物馆，以及丹麦哥本哈根、瑞典斯德哥尔摩和乌普萨拉、德国柏林和莱比锡、捷克布拉格、奥地利维也纳、法国巴黎等）获得的中国植物标本照片。当时全国至少有数个单位分别装订成册，其中静生生物调查所装订的时间为 1935 年，而北平研究院植物学研究所装订的时间为 1936 年。此外，国立清华大学、岭南大学、总理陵园、国立武汉大学和中国科学院华南植物园（原国立中山大学农林植物研究所）的具体装订时间不详。如今北京大学生命科学学院（原生物系）、北京师范大学生命科学学院（原生物系）、北京大学药学院（原北京医科大学药学院）、江苏省中国科学院植物研究所 南京中山植物园（原中央研究院自然历史博物馆）、西北农林科技大学（原中国科学院西北植物研究所）、云南大学等标本馆有收藏。这些标本的底片现存于中国科学院植物研究所植物标本馆（李敏办公室）。

305 范柏樟、黄启文，1990，三十年代的一次科学盛会，中国科技史料 11（4）：55-63。

Photographs of Chinese Plant Specimens of Northwest Institute of Biology and Pedology, Chinese Academy of Sciences (Photo provided by Herbarium, Northwest A&F University)
中国科学院西北生物土壤研究所的中国植物标本照片集（相片提供者：西北农林科技大学植物标本馆）

Negatives of Chinese Plant Specimens at Institute of Botany (PE, Min LI's office) (Photo provided by Min LI)
植物研究所的中国植物标本照片底片（李敏办公室）（相片提供者：李敏）

Part of Photographs of Chinese Plant Specimens in Peking University (Photo provided by Shi-Yong MENG)
北京大学的部分中国植物标本照片集（相片提供者：孟世勇）

1935.08.12-08.15, Second anniversary meeting (Third National Congress) of the Botanical Society of China, with Woon-Young CHUN as president, held in Nanning, Guangxi.

1935.09.07, Sixth International Botanical Congress held in Amsterdam, The Netherlands. Woon-Young CHUN and Chi-Tung LI were among the Chinese delegates; Woon-Young CHUN was elected vice president of the plant taxonomy and nomenclature session. Woon-Young CHUN visited England, the United States of America and Hong Kong after the congress.

1935 年 9 月 7 日，第六届国际植物学大会在荷兰阿姆斯特丹举行，陈焕镛、李继侗等中国代表出席；陈焕镛被推选为植物分类与命名组副主席。会后，陈焕镛顺访英国、美国和中国香港。[306]

1935 年 9 月 7 日，中央研究院成立评议会，首届评议员 29 人，包括植物分类学家胡先骕和陈焕镛。

1935.09.07, Council of Academia Sinica founded, with first 29 councilors, including plant taxonomists Hsen-Hsu HU and Woon-Young CHUN.

Chen-Ju WANG (Yen-Chieh WANG, 1908-2008) (Photo provided by Beijing Forestry University)
汪振儒（汪燕杰，1908—2008）（相片提供者：北京林业大学）

1935 年 9 月，汪振儒（1908—2008）[307] 入美国康奈尔大学学习林业，1936 年获得了理科硕士学位，同时转入杜克大学，1939 年获杜克大学林学院哲学博士学位；1939 年回国后任国立广西大学森林系教授兼系主任和植物研究所主任，1943 年任国立广西大学农学院院长；1946 年任北京大学农学院森林系教授、系主任；1952 年任北京林学院（1985 年改称北京林业大学）教授，兼林业系主任等职务直至 1989 年退休。

1935, Yen-Chieh WANG (as Chen-Ju WANG, 1908-2008) attended graduate school at Cornell University, Ithaca, New York, United States of America, to study forestry; received master's degree in 1936, then to Duke University and

306 胡宗刚，2013，华南植物研究所早期史，第 75-76 页；上海交通大学出版社，上海。

307 1925 年中学毕业后考取国立清华大学生物系；1929 年 8 月毕业，获理学学士学位。毕业后先后在南京中国科学社、国立清华大学生物系和国立广西大学任职。著名林学家与林木生理学家，早年曾从事植物分类学工作。

received Ph.D. degree[308] in 1939; returned to National Kwangsi University as professor in 1939, then dean of department of forestry and director of Institute of Botany, and chairman of School of Agriculture, National Kwangsi University in 1943; professor and dean of department of forestry, School of Agriculture, Peking University in 1946; professor and dean of department of forestry, Beijing Forestry College in 1952 (Beijing Forestry University since 1985) until retirement in 1989.

1935, John I. Briquet, ***International Rules of Botanical Nomenclature***, revised by the International Botanical Congress, Cambridge, 1930, 152 p; G. Fischer, Jena. Translated into Chinese in 1936-1937.

1936—1937 年，俞德浚译，国际植物学命名法规，中国植物学杂志 3（1）：873-893，3（2）：957-976，3（3）：1 109-1 136，4（1）：79-103。

1935 年，华汝成编，**植物学纲要**，304 页；中华书局，上海。

1935, Ru-Cheng HUA, '***Outline of Botany***', 304 p; Chinese Books, Shanghai.

1935, Scottish Rock Garden Club, Rowland E. Cooper, A. O. Curle and W. S. Fair (eds.), ***George Forrest V.M.H. Explorer and Botanist***[309] *who by his discoveries and plants successfully introduced has greatly enriched our gardens*, 89 p; Stoddart and Malcolm Ltd., Edinburgh.

1935 年，Scottish Rock Garden Club, Rowland E. Cooper, A. O. Curle 和 W. S. Fair 编辑，**考察者与植物学家、维多利亚荣誉勋章获得者 George Forrest** 发现并成功引入的植物极大地丰富了我们的植物园；89 页；Stoddart and Malcolm Ltd，爱丁堡。

1935, Francis Kingdon Ward, ***The Romance of Gardening***, 271 p; Jonathan Cape, London.

1935 年，Francis Kingdon Ward，**园艺传奇**，271 页；Jonathan Cape，伦敦。

1935，Francis Kingdon Ward, A sketch of the geography and botany of Tibet being materials for a flora of that country, *Journal of the Linnean Society Botany* 50：239-265.

1935 年，Francis Kingdon Ward，作为西藏植物志资料的西藏地理与植物概况，*Journal of the Linnean Society Botany* 50：239-265。

1935 年，王启无，关于中国苔藓类植物之研究及其文献，中国植物学会杂志 1（4）：415-427。

1935, Chi-Wu WANG, 'The study and literature of Chinese Bryophytes', *The Journal of Botanical*

308 Yen-Chieh WANG, 1939, Studies of interactions of certain site factors and young loblolly pine (*Pinus taeda* Linn.) plantations, 220 p; Advisor: Clarence F. Korstian (1889-1968).

309 Victoria Medal of Honour：维多利亚荣誉勋章。

Society of China 1(4): 415-427.

1935 年，国立中山大学农林植物研究所陈焕镛招收研究生李日光和王孝，两人分别于 1937 和 1938 年毕业。

1935, Ri-Guang LI and Xiao WANG admitted as graduated students to the Botanical Institute, National Sun Yat-Sen University, with Woon-Young CHUN as their tutor. They graduated in 1937 and 1938 respectively.

1935-1937, Botanical Society of China, ***Bulletin of the Chinese Botanical Society*** vols. 1-3.

1935—1937 年，中国植物学会，**中国植物学会汇报**，1-3 卷。

1935—1937 年，静生生物调查所王启无赴云南采集，获得植物标本 22 000 余号（A, KUN, PE）。[310]

1935-1937, Chi-Wu WANG, Fan Memorial Institute of Biology, explored in Yunnan, collected around 22,000 numbers of plant specimens (A, KUN, PE).

1935 年 10 月至 1938 年 10 月，汪发缵[311]、

Fa-Tsuan WANG (1899-1985) (Photo provided by Zong-Gang HU)
汪发缵（1899—1985）（相片提供者：胡宗刚）

310 参见：汪发缵，1941，本所之回顾与前瞻，云南农林植物研究所丛刊 1（1）：3-5；王印政、覃海宁、傅德志，2004，中国植物采集简史，中国植物志 第 1 卷：699 页；科学出版社，北京。

311 安徽祁门人；1922 年入国立东南大学，1927 年生物系毕业；1927—1929 年在安徽休宁二中任教，1929—1935 年在北平静生生物调查所任助理研究员；1938—1946 年先后在昆明云南大学任教授，云南农林植物研究所任副所长、研究员，四川北碚复旦大学农学院任主任、教授，四川歌乐山中央林业试验所任技正；1946—1985 年在北平研究院植物学研究所、中国科学院植物研究所任研究员；为中国著名的单子叶植物学家，除禾本科外几乎所有科属均深入研究过，尤其是百合科、兰科和莎草科。

唐进[312]受中华教育文化基金会资助赴英国、法国、德国、瑞士、意大利、奥地利等欧洲主要植物标本馆（包括B、G、K、P、W、WU）从事兰科和百合科研究。[313]

1935.10-1938.10, Fa-Tsuan WANG and Tsin TANG, supported by the China Foundation for the Promotion of Education and Culture, went to the United Kingdom, France, Germany, Switzland, Italy and Austria, to visit major herbaria (including B, G, K, P, W, WU) to study Orchidaceae and Liliaceae.

Tsin TANG (1897-1984) (Photo provided by Zong-Gang HU)
唐进（1897−1984）（相片提供者：胡宗刚）

1935 and 1973, Shun-Ching LEE, ***Forest Botany of China***, 991 p, 1935[314]; Commercial Press, Shanghai; and ***Forest Botany of China Supplement***, 477 p, 1973; Chinese Forestry Association, Taipei [Taibei].

1935、1973年，李顺卿编著，**中国森林植物学**，991页，1935；商务印书馆，上海；**中国森林植物学续篇**，477页，1973；中华林学会，台北。

312 江苏吴江人，上海南洋中学毕业，1921年赴菲律宾大学农学院学习，1922年转入岭南大学农学院，1926年毕业于北平农业大学；1926—1927年在浙江、江苏和河北采集，1928年初入中国科学社生物研究所，后转入静生生物调查所，1944—1947年任职于江苏宜兴美术专门学校，并担任过江苏宜兴高级农业学校校长，1947年再次返回静生生物调查所；新中国成立后任中国科学院植物分类研究所、植物研究所研究员。长期从事植物分类研究，是中国单子叶植物，特别是兰科、百合科、莎草科等研究的创始人之一。

313 中国科学院植物研究所志编纂委员会编，2008，中国科学院植物研究所所志，713页，734–735页；高等教育出版社，北京。

314 Preface (pages iii–iv) by Alfred Rehder, Arnold Arboretum, Harvard University.

1936

1936 年 8 月，中国植物学会第三届年会（第四届全国代表大会）在北平国立清华大学举行；选举戴芳澜为会长。

1936.08, Third anniversary meeting (Fourth National Congress) of Botanical Society of China, with Fang-Lan TAI as president, held in National Tsinghua University, Peiping [Beijing].

1936 年 10 月 13 日至 17 日，国立中山大学农林植物研究所和广西植物研究所派往贵州采集的邓世纬[315]等数人先后患疟疾病故。

1936.10.13–10.17, Botanical Institute, National Sun Yat-Sen University and Guangxi Institute of Botany, sent a team to Guizhou to collect specimens; Shi-Wei TENG and other staff members died there of malaria.

1936年11月18日，北平研究院植物学研究所刘慎谔和国立西北农林专科学校辛树帜合组中国西北植物调查所在陕西武功成立，同时建立植物标本室（WUK），刘慎谔主持；北平植物研究所则由林镕主持。[316]

Botanical Survey of Northwestern China established in Wukong (second left: Tchen-Ngo LIOU) (Photo provided by Zong-Gang HU)

中国西北植物调查所于武功成立（左二：刘慎谔）（相片提供者：胡宗刚）

1936.11.18, Botanical Survey of Northwestern China established in Wukong [Wugong], Shaanxi, co-founded jointly by Tchen-Ngo LIOU,

315 贵州清镇人；参见：中国植物学杂志第 3 卷第 4 期第 1 284–1 285 页，1936。

316 王宗训，1985，回忆北平研究院植物研究所，中国科技史料 6（2）：16–20。姜玉平，2003，北平研究院植物学研究所的二十年，中国科技史料 24（1）：34–46。

National Academy of Peiping [Beijing] and Shu-Chih HSIN, National Northwest School of Agriculture and Forestry, with Tchen-Ngo LIOU as head; herbarium (WUK) also established; Yong LING became head of the Institute of Botany, National Academy of Peiping [Beijing].

1936 年，陈秀英[317]（1910—1949）留学美国俄亥俄州的 Oberlin 学院研修植物学，于 1938 年获硕士学位；同年入哈佛大学 Radcliffe College，于 1942 年获得博士学位；并成为哈佛大学植物分类学第一个华裔女博士。陈秀英于 1935—1936 年发表有关学术文章，为中国首位发表植物分类学研究成果的女性[318]。第二次世界大战时曾服务于中国战区医务部门，第二次世界大战后与 Franklin P. Metcalf 结婚并于 1947 年赴 Oberlin 学院任教。1949 年 6 月 7 日病逝于俄亥俄州 Oberlin。

1936, Luetta Hsiu-Ying CHEN[319](1910-1949), studied botany at Oberlin College, Ohio, United States of America, and received her master's degree in 1938. She entered Radcliffe College[320] in 1938 for further study and received her Ph. D. degree in 1942[321]. She was the first Chinese woman to earn a Ph. D. degree at Harvard University for research in plant taxonomy. She was the first Chinese woman to conduct taxonomic research on plants and to publish academic papers, 1935-1936. During World War II, she served as a commissioned officer in the Chinese Army medical service. She was married to Franklin P. Metcalf at the close of the war and they went to Oberlin College in 1947. She died by long illness in Oberlin in June 7, 1949.

1936 年，陈邦杰[322]（1907—1970）赴德国柏林大学（柏林达莱植物园）攻读植物学，1940 年获博

317 福建福州南台（今仓山区）人（女），1930 年毕业于华南女子大学（即位于福州的华南女子文理学院，英文 Hwa Nan College），后在广州岭南大学理学院生物系任教。故去时当地新闻记载其出生时间为 1908 年 4 月 7 日。

318 Luetta (Hsiu-Ying) Chen, 1935a, *Eucalyptus* on Lingnan campus, *Lingnan Agricultural Journal* 1(4): 139-154, fig. 1-11; 1935b, *Diospyros* in southeast China, *Lingnan Science Journal* 14: 665-685; 1936, The flowers of *Diospyros foochowensis* and *D. susarticulata*, *Lingnan Science Journal* 15: 119-120, fig. 1-4.

319 Shiu-Ying (also as Siu-Ying) CHEN, from Nantai (today's Cangshan), Fuzhou, Fujian, graduated from Hwa Nan College of Foochow in 1930, worked at Lingnan University before traveling to the United States. Her doctoral dissertation was entitled "A revision of the genus *Sabia* Colebrooke", *Sargentia*, a continuation of the Contributions from the Arnold Arboretum of Harvard University 3: 1-75, 1943. Her obituary, as reported in the local newspaper, indicated that she was born on April 7, 1908.

320 Radcliffe College was a women's liberal arts college in Cambridge, Massachusetts, and functioned as a female coordinate institution for the all-male Harvard College. Radcliffe conferred Radcliffe College diplomas to undergraduates and graduate students for the first 70 or so years of its history and then joint Harvard-Radcliffe diplomas to undergraduates beginning in 1963. The degrees were signed by the presidents of both Radcliffe College and Harvard University. Radcliffe stopped awarding degrees and admitting women in 1962. A formal "non-merger merger" agreement with Harvard was signed in 1977, with full integration with Harvard completed in 1999. For details, see https://en.wikipedia.org/wiki/Radcliffe_College (Accessed 18 June 2018).

321 Ph.D. dissertation: Luetta CHEN, 1943, A revision of the genus *Sabia* Colebrooke, *Sargentia*, a continuation of the Contributions from the Arnold Arboretum of Harvard University 3: 1-75; advisor: Elmer D. Merrill.

322 字逸尘，江苏丹徒人；1921 年就读于江苏省立第五师范学校，后考入国立中央大学生物系，1931 年毕业，1932 年任四川重庆乡村建设学院教师。中国苔藓植物学奠基人。

Luetta Hsia-Ying CHEN at Harvard University: 1. Working; 2. Standing; 3. Library of Arnold Arboretum of Harvard University (second from right) (Photo provided by Arnold Arboretum of Harvard University)

陈秀英于哈佛大学：1. 工作；2. 站立；3. 哈佛大学阿诺德树木园图书馆（右二）(相片提供者：哈佛大学阿诺德树木园)

士学位并回国，历任国立中央大学、同济大学、国立南京大学教授；1952年后任南京师范学院教授兼生物系主任，1954年兼任中国科学院植物研究所研究员；1955年以后在南京举办苔藓学进修班并为国内主要植物学单位代培在职研究人员，1957年后又招收研究生。详细参见1954年条目和1955年条目。

1936, Pan-Chieh CHEN[323] (1907-1970) went to Universität zu Berlin (Botanischer Garten und Botanisches Museum Berlin-Dahlem), Germany, to study botany. He received a Ph. D. degree in 1940[324] and returned to China, in 1940 as professor at National Central University, Tongji University and National Nanking University. He served as professor and chairman of the department of biology, Nanking Normal College, in 1952, concurrently research professor at the Institute of Botany, Chinese Academy of Sciences, in 1954; organized a training class in bryology in Nanjing for the national institutions in 1955 and admitted graduate students in bryology in 1957. See also 1954 and 1955.

Pan-Chieh CHEN (1907-1970) (Photo provided by Pan-Cheng WU)
陈邦杰（1907—1970）（相片提供者：吴鹏程）

1936, The Chinese Society of Biological Sciences, ***The Chinese Journal of Botany***, vol. 1(1).

1936年，中国生物科学学会，**中国植物学杂志**[325]，vol. 1（1）。

1936年，胡先骕，二十年来中国植物学之进步，科学19（10）：1 555-1 559；1937，刘咸（编），中国科学二十年，192-200页；中国科学社，上海。

1936, Hsen-Hsu HU, ‘Progress of Chinese botany in the last twenty years’, *Science (China)* 19(10): 1,555-1,559; 1937, Hsien Liu (ed.), *Twenty Years of Science in China*, 192-200 p; Science Society of China, Shanghai.

323 C. L. Wan (As Zon-Ling WAN but no details is given in the original report), 1980. P. C. Chen (1907-1970), founder of Chinese bryology, *Taxon* 29(5-6): 671-672; Hong GUI, 2007, Mr. Pan-Chieh Chen in my heart, *Chenia* 9: 13-15; Tong CAO & Cai-Hua GAO, 1991 “1990”, Memory of the late Professor Chen PC, *Acta Bryolichenologica Asiatica* 2: 81-82.

324 Ph.D. dissertation: Pang-Chieh CHEN, 1941, Studien über die ostasiatischen Arten der Pottiaceae (I & II), *Hedwigia,* 80 (I & II): 1-76 and 141-322; advisor: Hermann Reimers (1893-1961).

325 此刊与中国植物学会1934年创办的中国植物学杂志中文同名，但英文不同。

1936, Hsen-Hsu HU, The characteristics and affinities of Chinese flora, *Bulletin of the Chinese Botanical Society* 2(2): 67-84.

1937 年，胡先骕[326]，中国植物之性质与关系，中国植物学会杂志 4（1）：7-25。

1936 年，卢开运编，**高等植物分类学**，405 页；中华书局，上海。

1936, Kai-Yun LU, '***Taxonomy of Higher Plants***', 405 p; Chinese Books, Shanghai.

1936 年，正宗严敬，**最新台湾植物总目录**，410 页；Kudoa，台北。

1936, Genkei Masamune, ***Short Flora of Formosa*** *or an enumeration of the higher cryptogamic and phanerogamic plants hitherto known from the Island of Formosa and its adjacent islands*, 410 p; Kudoa, Taihoku [Taibei].

1936 年 6 月至 1943 年 12 月，满洲帝国国务院大陆科学院[327]，**大陆科学院研究报告**，1（1）-7（10）；大同大街，新京[328]。

1936.06-1943.12, Institute of Scientific Research Manchoukuo, ***Report of the Institute of Scientific Research Manchoukuo***, vols. 1(1)-7(10); Tatsung Tachie, Hshinching [Changchun].

1936—1974 年，邓良在广东和海南采集 11 000 号（IBSC）。[329]

1936-1974, Liang TENG collected more than 11,000 numbers of plants in Guangdong and Hainan (IBSC).

326 中文为张英伯翻译，且于不同时间刊登于不同刊物，故中英文期刊页码不一致。

327 日本侵华期间在长春设立的研究机构。

328 伪满时期吉林省长春市的旧称。

329 胡启明、曾飞燕，2011，广东植物志，10：323 页；广东科技出版社，广州［附录 2，华南植物研究所（园）植物标本采集简史］。

1937

1937 年 5 月，1955 年，贾祖璋、贾祖珊编，**中国植物图鉴**，1 460 页；开明书店，上海；胡先骕、陈焕镛、钱崇澍、董爽秋作序。1955 年，重印版，1 529+62 页；中华书局，上海。

1937.05, 1955, Tchou-Tsang KIA (Tsu-Chang CHIA) and Tchou-Shan KIA (Tsu-San CHIA), ***Plantae Sinicae cum Illustrationibus***, 1,460 p, 1937; Kai Ming, Shanghai. Reprint Edition, 1,529+62 p, 1955; Chinese Books, Shanghai.

1937 年 8 月，中央研究院动植物研究所迁往湖南衡阳，12 月又迁广西阳朔。1940 年，中央研究院动植物研究所再迁四川重庆北碚，所长王家楫。

1937.08, The Institute of Zoology and Botany, Academia Sinica, was moved to Hengyang, Hunan; then to Yangshuo, Guangxi, in December; The Institutes of Zoology and Botany, Academia Sinica, were moved to Beibei, Chongqing, Sichuan, in 1940 under the directorship of Chia-Chi WANG.

1937 年，侯宽昭入国立中山大学农林植物研究所读研究生，导师为陈焕镛。

1937, Foon-Chew HOW was admitted as a graduate student to the Botanical Institute, National Sun Yat-Sen University; Woon-Young CHUN was his tutor.

1937, Hsen-Hsu HU, Chronicle of the biological sciences in China, *T'ien Hsia Monthly* 4(5): 484–497.

1937 年，胡先骕，中国生物科学史，天下 4(5): 484–497。

1937, Masao Kitagawa, Brief history of botanical investigations in Manchuria, *Report of Institute of Scientific Research Manchoukuo* 1: 91–106.

1937 年，北川政夫，满洲国植物调查研究史，大陆科学院研究报告 1: 91–106。

1937 年，北平研究院植物学研究所迁往陕西武功，1938 年 3 月林镕抵达武功并主持工作。

1937, The Institute of Botany, National Academy of Peiping [Beijing] moved to Wukong [Wugong], Shaanxi, Yong LING arrived in Wugong in March 1938 to oversee the daily work.

1937, Francis Kingdon Ward, ***Plant Hunter's Paradise***, 347 p; J. Cope, London.

1937 年，Francis Kingdon Ward，**植物猎人的天堂**，347 页；J. Cope，伦敦。

Te-Tsun YU (far left) and field assistants in Dulongjiang, 1937-1938 (Photo provided by Zong-Gang HU)
1937—1938 年，俞德浚（左）及采集队在独龙江（相片提供者：胡宗刚）

1937—1938 年，静生生物调查所与英国皇家植物园和英国皇家园艺学会合作，派俞德浚和刘瑛赴云南采集，获得植物标本 2 万余号（A、E、K、KUN、PE）。

1937-1938, The Fan Memorial Institute of Biology, in cooperation with the Royal Botanic Gardens and Royal Horticultural Society, sent Te-Tsun YU and Ying LIU to explore Yunnan; they collected over 20,000 numbers of plant specimens (A, E, K, KUN, PE).

1937、1950 年，钱崇澍（第 1 册），钱崇澍、杨衔晋（第 2 册），**中国森林植物志**，第 1 册：1–50 图版，1937；实业部林垦署、中国科学社生物研究所合编，北碚；第 2 册：51–100 图版，1950；中国科学社生物研究所，上海。

1937 and 1950, Sung-Shu CHIEN (Part 1), Sung-Shu CHIEN and Yen-Chin YANG (Hsien-Chin YANG, Part 2), ***Icones of Chinese Forest Trees***, Parts. **1:** 1–50 plates, 1937; The Forestry and Agriculture

Department of the Ministry of Industry and the Biological Laboratoy of the Science Society of China, Beibei; **2:** 51-100 plates, 1950; The Biological Laboratories, Chinese Association for the Advancement of Science, Shanghai.

1937、1953 年，陈嵘著，**中国树木分类学**，1 086 页，1937；中华农学会，南京；陈嵘著，**中国树木分类学**，修订版，1 191 页，1953；中国图书，南京。

1937 and 1953, Yung CHEN, ***Illustrated Manual of Chinese Trees and Shrubs***, 1,086 p, 1937; Agricultural Association of China, Nanjing; Yung CHEN, ***Illustrated Manual of Chinese Trees and Shrubs***, Revised and Enlarged, 1,191 p, 1953; Chinese Books, Nanjing.

1938

1938 年 10 月，静生生物调查所和云南教育厅在昆明合办云南农林植物研究所[330]；胡先骕任所长（1938—1949），北平部分人员来此工作；刘瑛、王启无、张英伯[331]（1913—1984）、郑万钧、汪发缵、邓祥坤、梁国贤等人大规模采集，共计有 2.5 万号以上。

1938.10, Yunnan Botanical Institute[332] co-founded by the Fan Memorial Institute of Biology and the Education Bureau of Yunnan Province, with Hsen-Hsu HU as director (1938-1949); some staff members from the Fan Memorial Institute of Biology joined the new institute. Collections made by Ying LIU, Chi-Wu WANG, Ying-Po CHANG (1913-1984), Wan-Chun CHENG, Fa-Tsuan WANG, Hsiang-Kun TENG and Kuo-Hsian LIANG totaled nearly 25,000 numbers of plant specimens.

1938 年 10 月，庐山被日本占领前夕，秦仁昌等离开庐山，经湖南长沙到达云南昆明。1939 年 1 月秦仁昌等离开昆明赴云南丽江，建立庐山森林植物园丽江工作站。

1938.10, Prior to occupation of Lushan by the Japanese, Ren-Chang CHING and staff members traveled via Changsha, Hunan, to Kunming, Yunnan. In January 1939, Ren-Chang CHING and others left Kunming for Lijiang, Yunnan, where the Lijiang Station of Lushan Forestry Botanical Garden was established.

1938, Hsen-Hsu HU, Recent progress in botanical exploration in China, *Journal of the Royal Horticultural Society of London* 63: 381-389, figures 99-106.

1938 年，胡先骕，中国植物学考察的新进展，*Journal of the Royal Horticultural Society of London* 63：381-389，figures 99-106。

1938 and 1960, Elmer D. Merrill and Egbert H. Walker, ***A Bibliography of Eastern Asiatic Botany***, 719 p, 1938, The Arnold Arboretum of Harvard University, Jamaica Plain, Massachusetts; Egbert H. Walker,

---

330 胡宗刚，2001，云南农林植物研究所创办缘起，中国科技史料，22（3）：238-248。

331 河北（天津）武清人，1932—1937 年就读于国立北平师范大学，1938—1940 年任北平静生生物调查所助理研究员，1940—1946 年任云南农林植物研究所和中央研究院工学研究所副研究员，1946—1947 年在美国耶鲁大学林学及理工研究院攻读研究生，获硕士学位，1947—1951 年在密执安大学资源学院及理学院攻读学位，获硕士和博士学位；1951—1955 年任美国威斯康星大学林产研究所研究员；1956—1984 年任中国林业科学研究院林业研究所研究员，1957—1983 年兼任中国科学院植物研究所研究员。

332 The English title for this institute should be 'Yunnan Institute of Agricultural and Forestry Botany' which was used by Li-Kuo FU (1993), but here follows the first journal of the institute, *Bulletin of the Yunnan Botanical Institute*, by Yunnan Botanical Institute. See 1941 for details.

Early days of Yunnan Botanical Institute: 1. Institute, 2. Office (Photo provided by Kunming Institute of Botany)
初期的云南农林植物研究所：1. 研究所；2. 办公室（相片提供者：昆明植物所）

A Bibliography of Eastern Asiatic Botany, Supplement I[333], 552 p, 1960, American Institute of Biological Sciences, Washington, D.C. These two authoritative references on systematics botany in eastern Asia are still important for the study of plants in China.

1938、1960 年，Elmer D. Merrill、Egbert H. Walker，**东亚植物文献目录**，719 页；哈佛大学阿诺德树木园，Jamaica Plain，麻州，1938；Egbert H. Walker，**东亚植物文献目录 增补 I**，552 页；American Institute of Biological Sciences，华盛顿特区，1960。这两部东亚植物系统学权威性文献[334]，直至今日依然是中国植物学研究的重要参考书。

333 正篇收载截至 1936 年，补编收载 1937—1958 年。

334 其中古代文献部分参见：蒋英，1977，对《东亚植物学文献》附录中“中国古代文献”部分的订正，植物分类学报 15（1）：86-96。该书 1938 年的正篇没有中文书名，此处的东亚植物文献目录记载是根据 1960 年补编的中文书名记载；特此说明。另外，蒋英（1977）在订正该书的中国古代文献时记载该书的中文书名为东亚植物学文献。

1939

1939 年 4 月，福建省研究所于厦门大学在抗日战争期间的所在地福建长汀成立；1940 年 11 月，福建省研究所改名为福建省研究院并移至永安；1945 年，抗日战争胜利后迁至福州；1951 年，福建省研究院动植物研究所和工业研究所合并为自然科学研究所，归属于福州大学；1953 年 9 月福州大学更名为福建师范学院；1972 年福建师范学院更名为福建师范大学。林镕于 1942—1944 年任动植物研究所所长。1945 和 1947 年先后发行**福建省研究院研究汇报** 2 期。

1939.04, Fukien [Fujian] Institute was established within the campus of Amoy University then moved to Changting, Fukien, during the anti-Japanese war; it was renamed Fukien Academy in November 1940 when it moved to Yongan. It was moved to Fuzhou after the war in 1945. The Institute of Zoology and Botany, Fukien Academy merged with the Institute of Industry to form the Institute of Natural Sciences under the former Fuzhou University in 1951. Fuzhou University was renamed Fukien Normal College in September 1953, then Fujian Normal University in 1972. Yong LING was director of the Institute of Zoology and Botany, Fukien Academy, from 1942 to 1944. ***Research Bulletin, The Fukien Academy***, nos. 1 and 2 was issued in 1945 and 1947.

Wan-Chun CHENG (1904-1983) (Photo provided by Zong-Gang HU)
郑万钧（1904—1983）（相片提供者：胡宗刚）

1939 年 4 月至 11 月，郑万钧[335]（1904—1983）赴法国图卢兹大学林学院学习，获博士学位；回国后 1939 年至 1944 年任教于云南大学并兼任云南农林植物研究所副所长，1944—1949 年任教于国立中央大学森林系教授兼系主任，1950—1952 年任教于南京大学林学系教授兼系主任，1952—1961 年任职于南京林学院（自南京大学独立）教授、副院

335 江苏徐州人，1924 年毕业于江苏省立第一农业学校林科，1924 年 8 月至 1925 年 7 月任教于国立东南大学，1929 年 8 月至 1939 年 12 月任中国科学社生物研究所植物研究员。

长、院长，1962—1983 年任中国林业科学研究院研究员、副院长、院长。1955 年当选中国科学院学部委员。

1939.04-1939.11, Wan-Chun CHENG (1904-1983) studied forestry at Toulouse University, France, where he received his Ph. D. degree in 1939[336]; returned to China to become professor at Yunnan University, 1939-1944, concurrently vice director of Yunnan Botanical Institute, then professor and dean of the Department of Forestry, National Central University, 1944-1949 and Nanjing University, 1950-1952, then professor, vice president and president of Nanjing Forestry College (independent from Nanjing University), 1952-1961, then research professor, vice president and president of the Chinese Academy of Forestry, Beijing, 1962-1983. He was elected an academician of the Chinese Academy of Sciences in 1955.

1939 年 8 月，汪振儒任国立广西大学植物研究所主任（1939—1947）。

1939.08, Chen-Ju WANG, director, Guangxi Institute of Botany, National Kwangsi University (1939-1947).

1939 年 10 月，中国西北植物调查所，**中国西北植物调查所丛刊**（仅出版 1 卷 2 期）。

1939.10, Botanical Survey of North-Western China, ***Contributions from the Botanical Survey of North-Western China*** (1 volume with 2 parts only).

1939—1940 年，刘慎谔赴云南昆明设立工作站，恢复北平研究院植物学研究所。

1939-1940, Tchen-Ngo LIOU went to Kunming, Yunnan and established the Kunming Station, to restore the Institute of Botany, National Academy of Peiping [Beijing].

Kuo-Mei FENG (1917-2007) (Photo provided by Zong-Gang HU)
冯国楣（1917—2007）（相片提供者：胡宗刚）

336 Ph.D. dissertation: Wan-Chun CHENG, 1939, Les forêts du Se-Tchouan et du Si-kang oriental, Toulouse: Faculté des Sciences, *Toulouse Université, Laboratoire Forestier, Traväus* t. 5, v. 1, art. 2, 1-237 p; Advisor: Henri M. Gaussen (1891-1981). 其中第 19-87 页关于四川植物的采集历史部分被翻译成中文并先后发表两次，且第二次有更正与补充。参见管中天，1981，四川植物与森林考察简史（1869-1938），四川林业科技 3:66-76；管中天，2005，中华人民共和国成立前中外学者在四川考察森林植物简史注释（1869—1942），管中天，森林生态研究与应用，364-372 页；成都：四川科学技术出版社。

1939—1941 年，冯国楣[337]（1917—2007）于云南西北部大规模采集，收获标本约 1.5 万号（A、KUN、PE）。

1939-1941, Kuo-Mei FENG (1917-2007) collected more than 15,000 numbers of plant specimens in northwestern Yunnan (A, KUN, PE).

1939、1979 年，北川政夫，**'满洲国'植物考**，487 页，1939；满洲大陆科学院，新京（长春）；北川政夫，**满洲植物新考**，修订版，715 页，1979；J. Cramer，Lehre。

1939 and 1979, Masao Kitagawa, ***Lineamenta Florae Manshuricae*** *or an enumeration of all of the indigenous vascular plants hitherto known from Manchurian Empire together with their synonymy, distribution and utility*, 487 p, 1939; Institute of Scientific Research Manchoukuo, Hshinching [Changchun]; Masao Kitagawa, ***Neo-Lineamenta Florae Manshuricae*** or *enumeration of the spontaneous vascular plants hitherto known from Manchuria (Northeastern China) together with their synonymy and distribution*, revised ed., 715 p, 1979; J. Cramer, Lehre.

337 江苏宜兴人，1934—1938 年任庐山森林植物园练习生、技佐，1939—1944 年任职于庐山森林植物园丽江工作站，1944—1946 年任云南省农林部金沙江森林管理处技士，1946—1949 年任云南农林植物研究所助理研究员，1950—1957 年任中国科学院植物（分类）研究所昆明工作站助理研究员，1958—1990 年任中国科学院昆明植物研究所植物园主任、副研究员、研究员；1990 年退休。

1940

1940 年 10 月 31 日，胡先骕任江西泰和新成立的国立中正大学校长（1940—1944）。

1940.10.31, Hsen-Hsu HU named president (1940-1944) of newly founded National Chiang Kai Shek University, Taihe, Jiangxi.

1940—1948 年，王云章[338]（1906—2012）代理陕西武功中国西北植物调查所所长。

1940-1948, Yun-Chang WANG (1906-2012) became acting director of the Botanical Survey of North-Western China, Wukong [Wugong], Shaanxi.

1940 年，中央研究院遴选出第二届评议员 30 人，其中植物分类学家有戴芳澜、胡先骕、陈焕镛[339]、钱崇澍[340]。

1940, second set of 30 councilors of Academia Sinica were elected, including plant taxonomists Sung-Shu CHIEN, Woon-Young CHUN, Hsen-Hsu HU and Fang-Lan TAI.

1940, Ren-Chang CHING, on natural classification of the family Polypodiaceae, *Sunyatsenia*, 5(4): 201-268. More than 30 new families of ferns were described, combined or confirmed by Ren-Chang CHING

---

338 字蔚青，河南内黄人；1927 年入北平大学农学院，毕业后 1931—1936 年任北平研究院植物学研究所助理员，1936—1938 年在比利时卢万大学攻读研究生，获理学博士学位，1938—1939 年在加拿大温尼伯自治领锈菌研究室从事博士后研究，1939—1941 年任国立浙江大学教授，1941—1948 年任国立西北农学院兼中国西北植物调查所所长，1948—1949 年任北平研究院植物学研究所研究员，1950 年后分别任职于中国科学院植物研究所真菌植物病理研究室，中国科学院应用真菌学研究所和中国科学院微生物研究所。

339 1940—1942 年。

340 1943 年之后。

between 1940 and 1978.[341]

1940 年，秦仁昌，水龙骨科的自然分类，国立中山大学农林植物研究所专刊，5(4): 201-268。自 1940—1978 年，秦仁昌发表、组合或确认了 30 多个蕨类新科。

1940 年，李惠林[342]（1911—2002）赴美求学，1942 年获哈佛大学博士学位，1943—1945 年于宾夕法尼亚大学及费城科学院从事马先蒿属及玄参科研究，1946 年返回母校东吴大学任教，1947 年受聘于国立台湾大学植物系主任，1948 年创刊 *Taiwania*；1950 年赴美在弗吉尼亚大学 Boyce 分校从事玄参科细胞学研究，1951 年在史密森学会研究东亚和北美的植物区系关系，1952 年赴宾夕法尼亚大学及附属的莫里斯树木园从事杜鹃花的细胞学研究，1958 年晋升为宾夕法尼亚大学副教授，1963 年晋升为教授，同年出版**台湾木本植物志**[343]，1964 年当选台湾中央研究院院士，1971 年任莫里斯树木园执行园长，1972 年转为正式园长，1974 年任宾夕法尼亚大学巴群植物学及园艺学讲座教授，1975—1979 年主编台湾植物志，1979 年退休。[344]

1940, Hui-Lin LI (1911-2002) studied in the United States of America and earned his Ph. D. degree from Harvard University in 1942[345]; further studied *Pedicularis* and Scrophulariaceae at the University of

341 Adiantaceae (C. Presl) Ching, *Sunyatsenia* 5: 221. 1940; Antrophyaceae (Link) Ching, *Sunyatsenia* 5: 231. 1940; Blechnaceae Ching, *Sunyatsenia* 5: 241. 1940; Bolbitidaceae (Pic. Serm.) Ching, *Acta Phytotaxonomica Sinica* 16(4): 16. 1978; Christenseniaceae Ching, *Bulletin of the Fan Memorial Institute of Biology, Botany* 10: 227. 1941; Culcitaceae Ching, *Sunyatsenia* 5: 214. 1940; Dennstaedtiaceae Ching, *Sunyatsenia* 5: 214. 1940; Dictyoxiphiaceae Ching, *Sunyatsenia* 5: 218. 1940; Didymochlaenaceae Ching, *Sunyatsenia* 5: 253. 1940; Drynariaceae Ching, *Acta Phytotaxonomica Sinica* 16(4): 19. 1978; Dryopteridaceae Ching, *Acta Phytotaxonomica Sinica* 10: 1. 1965; Elaphoglossaceae Ching, *Sunyatsenia* 5: 265. 1940; Grammitidaceae (C. Presl) Ching, *Sunyatsenia* 5: 264. 1940 ('Grammitaceae'); Gymnogrammaceae Ching, *Sunyatsenia* 5: 227. 1940; Gymnogrammitidaceae Ching, *Acta Phytotaxonomica Sinica* 11: 12, pl. 3. 1966; Helminthostachyaceae Ching, *Bulletin of the Fan Memorial Institute of Biology, Botany* 10: 235. 1941; Hypoderriaceae Ching, *Sunyatsenia* 5: 221. 1940; Hypodematiaceae Ching, *Acta Phytotaxonomica Sinica* 13: 96. 1975; Hypolepidaceae Ching, *Sunyatsenia* 5: 221. 1940; Lindsaeaceae Ching, *Sunyatsenia* 5: 212. 1940; Loxogrammaceae Ching, *Sunyatsenia* 5: 233. 1940; Monachosoraceae Ching, *Sunyatsenia* 5: 241. 1940; Oleandraceae Ching, *Sunyatsenia* 5: 221. 1940; Onocleaceae Ching, *Sunyatsenia* 5: 244. 1940; Peranemataceae (C. Presl) Ching, *Sunyatsenia* 5: 246. 1940; Platyceriaceae (Nayar) Ching, *Acta Phytotaxonomica Sinica* 16: 18. 1978; Pleurosoriopsidaceae Kurita et Ikebe, *Journal of Japanese Botany* 52: 39-49. 1977, nom. nud., Ching; *Acta Phytotaxonomica Sinica* 16(4): 17. 1978; Pteridaceae Ching, *Sunyatsenia* 5: 222, 1940; Pteridiaceae Ching, *Acta Phytotaxonomica Sinica* 13(1): 96. 1975; Sphaerostephanaceae Ching, *Sunyatsenia* 5: 240. 1940; Stenochlaenaceae Ching, *Acta Phytotaxonomica Sinica* 16(4): 18. 1978; Thelypteridaceae, *Sunyatsenia* 5: 237. 1940; Vittariaceae (C. Presl) Ching, *Sunyatsenia* 5: 232. 1940; Woodsiaceae Ching, *Sunyatsenia* 5: 245. 1940. Details provided by Xian-Chun ZHANG (PE), 15 December 2017.

342 江苏苏州人，1930 年本科毕业于东吴大学生物系，1932 年硕士毕业于北平燕京大学生物系。1979 年和 1983 年分别回国访问，并在北京和昆明做学术报告。

343 Hui-Lin LI, 1963, ***Woody Flora of Taiwan***, A Morris Arboretum Monograph, 974 p, The Morris Arboretum, Philadelphia, Pennsylvania, and the Livingston Publ. Co., Narbeth, Pennsylvania.

344 李惠林，1982，植物学论丛，527 页；时代出版社，台北。

345 Ph. D. dissertation, Hui-Lin LI, 1942, The Araliaceae of China, *Sargentia* 2: 1-134; Advisor: Elmer D. Merrill.

Pennsylvania and Philadelphia Academy of Natural Sciences, 1943-1945; returned to his home school, Soochow [Suzhou] University in 1946, to National Taiwan University as dean of the Department of Botany in 1947, and issued *Taiwania* in 1948; returned to the United States of America for further work on cytology of Scrophulariaceae at the University of Virginia, Boyce, Virginia in 1950; studied floristic relationship between eastern North America and eastern Asia at the Smithsonian Institution in 1951; studied cytology of *Rhododendron* at the University of Pennsylvania and Morris Arboretum in 1952; promoted to associate professor in 1958, then full professor in 1963, elected academician of Academia Sinica in Taiwan in 1964; acting director of Morris Arboretum, University of Pennsylvania in 1971, then director in 1972; John Bartram Professor of Botany and Horticulture, University of Pennsylvania, in 1974; Editor in Chief for *Flora of Taiwan* from 1975-1979; retired in 1979. [346]

Hui-Lin LI (1911-2002) in early 1980s (Photo provided by Zong-Gang HU)
20 世纪 80 年代初的李惠林（1911—2002）（相片提供者：胡宗刚）

1940, Yin-Chan WU, Beiträge zur Kenntnis der Flora Süd-Chinas, *Botanische Jahrbücher für Systematik, Pflanzengeschichte und Pflanzengeographie*, 71(2): 169-199.

1940 年，吴印禅，今日华南植物学的成就，*Botanische Jahrbücher für Systematik*，*Pflanzengeschichte und Pflanzengeographie,* 71(2): 169-199。

1940—1942 年，汪振儒，广西种子植物名录 I-Ⅲ，广西农业 1（2）：68-77，1（6）：403-415，2（2）：134-141，1940；汪振儒、钟济新、陈立卿，广西种子植物名录 Ⅳ-Ⅸ，2（3）：223-229，2（4）：285-294，2（5）：371-384，2（6）：468-472，1941；3（1）：57-60，3（2）：121-124，1942。

1940-1942, Chen-Ju WANG, An enumeration of seed plants collected in Kwangsi [Guangxi], I-III, *The Kwangsi Agriculture* 1(2): 68-77, 1(6): 403-405, 2(2): 134-141, 1940; Chen-Ju WANG, Ghi-Hsing CHUNG[347] and Li-Ching CHEN, An enumeration of seed plants collected in Kwangsi [Guangxi], IV-IX, *The Kwangsi Agriculture* 2(3): 223-229, 2(4): 285-294, 2(5): 371-385, 2(6): 468-472, 1941; 3(1): 57-60, 3(2): 121-124, 1942.

---

346 Hui-Lin LI, 1982, *Contributions to Botany Studies in Plant Geography, Phylogeny and Evolution, Ethnobotany and Dendrological and Horticultural Botany*, 527 p; Epoch Publishing Co., Ltd., Taiwan.

347 As Chi-Hsing CHUNG.

1941

1941 年 6 月，云南农林植物研究所，**云南农林植物研究所丛刊**（仅出版 1 卷 1 期，88 页）。[348]

1941.06, Yunnan Botanical Institute, ***Bulletin of the Yunnan Botanical Institute,*** vol. 1 (1 part only, 88 pp).

1941 年 6 月，汪发缵，本所之回顾与前瞻[349]，云南农林植物研究所丛刊 1(1): 3–5。

1941.06, Fa-Tsuan WANG, An historical account on the establishment of the Yunnan Botanical Institute, *Bulletin of the Yunnan Botanical Institute* 1(1): 3–5.

1941 年 6 月，云南农林植物研究所编印，**云南农林植物研究所概况**，14 页；云南农林植物研究所，昆明。

1941.06, Yunnan Botanical Institute, '***Outline of the Yunnan Botanical Institute***', 14 p; Yunnan Botanical Institute, Kunming.

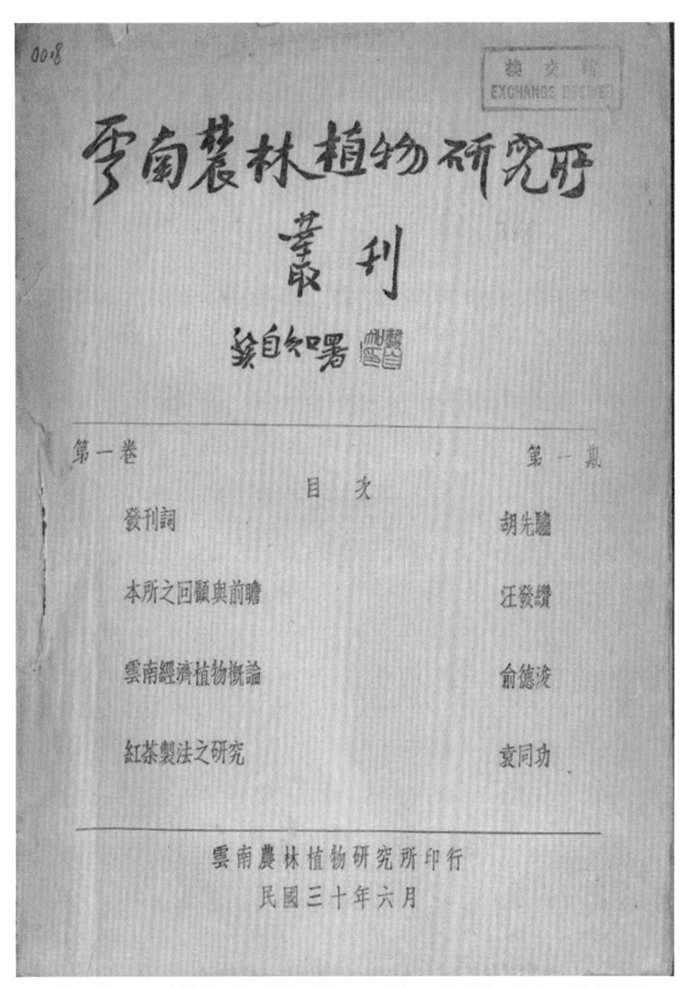

雲南農林植物研究所

叢刊

第一卷 第一期

目次

發刊詞	胡先驌
本所之回顧與前瞻	汪發纘
雲南經濟植物概論	俞德浚
紅茶製法之研究	袁同功

雲南農林植物研究所印行

民國三十年六月

The first issue of *Bulletin of the Yunnan Botanical Institute*, 1941 (Photo provided by Xiao-Jiang HU)

1941 年云南农林植物研究所丛刊创刊号（相片提供者：胡晓江）

348 第 1 期中预告第 2 期的 4 篇文章均未查获（尽管中国植物学文献目录也有记载，但均无具体页码）。

349 即云南农林植物研究所之回顾与前瞻。

1942

1942, Luetta Hsiu-Ying CHEN (1910-1949) became the first Chinese woman to earn a Ph. D. degree at Harvard University for research in plant taxonomy (for details, see 1936).

1942 年，陈秀英（1910—1949）成为第一位在哈佛大学获得植物分类学研究博士学位的中国女性（参见 1936 年条目）。

1942, Franklin P. Metcalf, ***Flora of Fukien and Floristic Notes on Southeastern China***, 82 p; Lingnan University, Canton [Guangzhou].

1942 年，Franklin P. Metcalf，**福建植物志和中国东南部植物区系注释**[350]，82 页；岭南大学，广州。

1942—1946，方文培主编，**峨眉植物图志**，两卷本（各 2 期）；国立四川大学，成都。

1942-1946, Wen-Pei FANG (editor in chief), ***Icones Plantarum Omeiensium***, vols. 1-2 (2 issues each); National Szechuan [Sichuan] University, Chengtu [Chengdu].

350 仅包括裸子植物和被子植物的木麻黄科至壳斗科。

1943

1943 年，徐祥浩（1920—2017）、谭景燊（1918—2011）入国立中山大学农林植物研究所读研究生，师从蒋英。其时，国立中山大学农学院迁至粤北石坪。

1943, Hsiang-Hao HSUE (1920-2017) and Ching-Shen TAN (1918-2011) were admitted as graduate students under the supervision of Ying TSIANG to the Botanical Institute, National Sun Yat-Sen University, which was moved to Shiping, northern Guangdong.

1943 年，胡先骕，中国生物学研究之回顾与前瞻，科学 26(1): 5-8。

1943, Hsen-Hsu HU, 'Review and prospect of Chinese biological research', *Science (China)* 26(1): 5-8.

1943 年，正宗严敬，**海南岛植物志**，443 页；台湾总督府外事部，台北。

1943，Genkei Masamune, ***Flora Kainantensis*** *sive enumeration plantarum spone crescent hucusque rite cognitarum*, 443 p; Taiwan Sotokufu Gaijabu (Taiwan Ministry of Foreign Affairs), Taihoku [Taibei].

1943—1963 年，陈少卿在广东、广西、海南（以及湖南）采集 1.8 万号（IBK、IBSC、PE）。[351]

1943-1963, Shao-Hsing CHUN collected over 18,000 numbers of plant specimens (IBK, IBSC, PE) in Guangdong, Guangxi, Hainan and Hunan.

351 胡启明、曾飞燕，2011，广东植物志，10：322 页；广东科技出版社，广州［附录 2，华南植物研究所（园）植物标本采集简史］。

1944

1944年，由于战争国立广西大学植物研究所，迁往贵州榕江；迁移过程中，植物标本先后遭遇火灾、水灾，损失殆尽。

1944, Kwangsi Institute of Botany, National Kwangsi University, moved to Rongjiang, Guizhou, during the anti-Japanese war. Since the move, all plant specimens were lost to either fire or flood.

1944, Hui-Lin LI, Botanical exploration in China during the last twenty-five years, *Proceedings of the Linnean Society London* 156(1): 25-44.

1944年，李惠林，过去二十五年中国植物的考察，*Proceedings of the Linnean Society London* 156(1): 25-44。

1944—1949年，中央研究院动植物研究所分为两个所，即植物研究所和动物研究所。罗宗洛（1898—1978）任植物研究所所长兼植物生理室主任，饶钦止[352]（1900—1998）任藻类室主任，裴鉴任高等植物分类室主任，其他人员包括单人骅、刘玉壶[353]（1917—2004）等。

1944-1949, Institute of Zoology and Botany, Academia Sinica, was divided into two institutes; the Botany Institute, directed by Tsung-Le LOO (1898-1978), was composed of the Lower Plants Department headed by Chin-Chih JAO (1900-1998) and the Higher Plants Department headed by Chien PEI, with Ren-Hwa SHAN and Yu-Wu LAW (1917-2004) as staff members.

352 四川重庆人，1920年毕业于国立成都高等师范学校博物学系，1922年毕业于北京高等师范学校研究生科，留校任助教，兼任讲师，1931年任副教授，1932—1935年在美国密执安大学研究生院，先后获文学硕士和哲学博士学位；1935年先后在美国西海岸的华盛顿大学和夏威夷大学从事藻类研究；1936—1941年任中央研究院动植物研究所研究员，1941—1950年任“中央”研究院植物研究所研究员，1950—1989年任中国科学院水生生物研究所研究员，1961年起兼任副所长；1989年退休；著名藻类学家。

353 广东中山人，1942年毕业于位于重庆的国立中央大学，1943—1944年留校任教；1945—1961年先后在“中央”研究院植物研究所、中国科学院华东工作站、南京中山植物园、中国科学院武汉植物园等单位从事植物分类学研究；1962年调入中国科学院华南植物研究所，历任助理研究员、副研究员、研究员；著名木兰科专家。

1945

1945 年 10 月 17 日，夏纬琨[354] 负责静生生物调查所战后复员。[355]

1945.10.17, Wei-Kun XIA in charge of recovery of the Fan Memorial Institute of Biology after the war.

1945 年，郝景盛著，**中国木本植物属志**，上卷，244 页；中华书局，重庆。

1945, Ching-Sheng HAO, ***The Genera of Chinese Woody Plants***, part 1, 244 p; Chung Hwa Book Co., Chongqing.

1945 年，郝景盛著，**中国裸子植物志**，152 页；正中书局，重庆；正中书局，上海，1947；人民出版社，北京，1951。

1945, Ching-Sheng HAO, ***Gymnospermae Sinicae***, 152 p; Chongcheng Books, Chongqing; 1947, Chongcheng Books, Shanghai; 1951, People's Press, Beijing.

1945 and 1986, Euan H. M. Cox[356], ***Plant Hunting in China****—A History of Botanical Exploration in China and the Tibetan Marches*, 230 p, 1945; William Collins Sons and Co. Ltd., London; Reprinted by Oxford University Press, 230 + 24 p, 1986, with an introduction by Peter A. Cox (1934–).

1945、1986 年，Euan H. M. Cox，**中国植物探寻** —— 在中国和西藏边区的植物学考察历史，230 页，1945；William Collins Sons and Co. Ltd.，伦敦；重新印刷，230+24 页，1986；牛津大学出版社，且增加了 Peter A. Cox（1934— ）的序言。

---

354 字玉峰，河北柏乡人（生卒不详；与北平研究院植物学研究所夏纬瑛为堂兄弟），静生生物调查所植物标本馆管理人员，1949 年中国科学院合组植物分类研究所时被留用，1957 年被中国科学院植物研究所辞退回乡。

355 胡宗刚，2005，静生生物调查所史稿，185–191 页；山东教育出版社，济南。

356 Euan H. M. Cox (1893–1977), Scottish plant collector, botanist and horticulturist, accompanied Reginald Farrer on his last botanical expedition (1919–1920) to Burma [Myanmar] and its border with China. He was a successful propagator of rhododendrons and had an extensive collection of them in his garden at Glendoick, Perthshire, Scotland, which formed the basis of his commercial nursery, later run by his son, Peter A. Cox and grandson, Kenneth N. E. Cox. Kenneth N. E. Cox also led several expeditions to China: 1992, Sino-Scottish Expedition to N.W. Yunnan (Zhongdian, Baima Shan); 1993, N.W. Yunnan; 1994, N.W. Yunnan (Weixi); 1995, S.E. Tibet, Namjiabawa region; 1996, S.E. Tibet, Yarlongtsangpo Gorges, Pemako, Namjiabawa; 1997, S.E. Tibet, Zayul, Bomi; 1998, Tsari, S.E. Tibet; 1999, Tsari, S.E. Tibet; 2001, S. Tibet (Arunachal Pradesh), upper Siang; 2002, S. Tibet (Arunachal Pradesh), Subansiri-Siyom divide; 2003, S. Tibet (Arunachal Pradesh), Dibang & Tawang / Kameng; 2005, S. Tibet (Arunachal Pradesh), Upper Siang, Riutala pass and Kora; 2011, N. Vietnam; 2012, Sichuan, (https://www.glendoick.com/Kenneth-Cox-meet-the-grower, accessed 19 August 2017).

1946

1946年1月，吴中伦[357]（1913—1995）赴美留学，1947年取得耶鲁大学林学硕士学位，1951年获杜克大学林学博士学位；1951—1956年任林垦部（后为林业部）工程师、总工程师，1956年任中国林业科学研究院林业研究所研究员、1959年任副所长、1978年任中国林业科学院副院长。1980年当选为中国科学院学部委员。

1946.01, Chung-Lwen WU (Chung-Lun WU, 1913-1995) went to United States of America to study forest in Yale University, received master's degree in 1947, then study forest in Duke University, received Ph.D. degree in 1951[358], returned as engineer and chief engineer in China Central Forestry Administration in Beijing from 1951 to 1956, research professor since 1956, and vice director since 1959, Institute of Forestry, Chinese Academy of Forestry, and vice president of Chinese Academy of Forestry since 1978. He was elected an academician of the Chinese Academy of Sciences in 1980.

1946年9月，北平研究院植物学研究所战后复员北平，林镕负责。

1946.09, The Institute of Botany, National Academy of Peiping [Beijing], was restored in Peiping [Beijing], Yong LING was in charge of the daily work.

1946年9月15日，胡秀英[359]（1908—2012）赴美国哈佛大学Radcliffe College学习，1949年获博士学位，之后任职于哈佛大学

Shiu-Ying HU (1908-2012), 2008 (Photo provided by Zong-Gang HU)
2008年的胡秀英（1908—2012）（相片提供者：胡宗刚）

357 浙江诸暨人，1933—1936年中国科学社采集员；1936—1941年于金陵大学农学院林学系学习；1941—1942年国立中央大学助教，1942—1945年农林部技术员，1945—1946年云南大学讲师。

358 Chung-Lwen WU, 1950, Forest regions in China with special reference to the natural distribution of pines, 151 p; Advisor: Clarence F. Korstian (1889-1968). 吴中伦，1956，中国松属的分类与分布，植物分类学报5（3）：131-163。

359 江苏徐州人（女），1929—1933年于南京金陵大学获学士学位，1934—1937年于广州岭南大学获硕士学位，1938—1946年任教于成都华西联合大学。

阿诺德树木园直到 1976 年退休[360]。1968 年起兼任香港中文大学教授，并致力于香港植物、中国药用植物以及食用植物的研究，从香港采集很多标本（A，CUHK 等）；晚年将自己私人的部分书刊及通信等资料，捐给香港中文大学和仙湖植物园[361]。其他资料包括信件、笔记、出版物等，则主要入哈佛大学阿诺德树木园档案和哈佛大学植物图书馆。2012 年 10 月 14 日，香港中文大学的植物标本馆（CUHK）正式更名为胡秀英植物标本馆。

1946.09.15, Shiu-Ying HU[362](1908-2012) attended Radcliffe College to study systematic botany, received Ph. D. in 1949[363], then worked for the Arnold Arboretum of Harvard University until retirement in 1976. She served as professor at the Chinese University of Hong Kong, starting in 1968; studied plants of Hong Kong, Chinese medicinal plants and edible plants; collected many plant specimens in Hong Kong (A, CUHK). She gave part of her private collection of books, journals and personal letters to the Chinese University of Hong Kong and to Fairylake Botanical Garden; additional items, including her private letters, notebooks, publications are in the archives of the Arnold Arborctum of Harvard University and in the Botany Library of Harvard University. The herbarium (CUHK) was officially named the Shiu-Ying Hu Herbarium, October 14, 2012.

1946 年 9 月 15 日，单人骅[364]（1909—1986）赴美国加州大学伯克利分校植物系学习，1947 年获硕士学位，1949 年 5 月获博士学位，1949 年 4 月回国任“中央”研究院植物研究所研究员。然而，由于当时急于搭乘最后一班船回国而没有及时拿到博士学位证书，直到 30 年后（1979 年）中国改革开放时，伯克利的导师 Lincoln

Ren-Hwa SHAN (Jen-Hua SHAN, 1909-1986) (Photo provided by Zong-Gang HU)
单人骅（1909—1986）（相片提供者：胡宗刚）

360 The Directors Report, Arnold Arboretum of Harvard University, p. 238, 1976; 胡秀英等，2003，秀苑撷英 —— 胡秀英教授论文集，349 页；商务印书馆，香港。

361 前者主要是出版物以及单行本等，后者则包括部分出版物和单行本，以及部分私人通信。

362 Paul Pui-Hay BUT, 2013, Professor Shiu-Ying Hu (1908-2012), *Journal of Systematics and Evolution* 51(2): 235-239.

363 Ph.D. dissertation: Shiu-Ying HU, 1949-1950, The Genus *Ilex* in China, *Journal of Arnold Arboretum* 30: 233-344, & 348-387, 1949; 31: 39 - 80, 214-240, & 242-263, 1950; advisor: Elmer D. Merrill.

364 江西高安人，1930 年入南京金陵大学森林系，1931 年转入国立中央大学生物系，1934 年毕业，1934—1945 年任职于中央研究院动植物研究所。

Constance（1909—2001）才将学位证书连同书籍和银行存款等寄来。[365]

1946.09.15, Ren-Hwa SHAN (Jen-Hua SHAN, 1909-1986) studied at the University of California, Berkeley, United States of America; received master's degree in 1947 and Ph. D. degree in May 1949[366]; returned to the Institute of Botany, Academia Sinica, as a research professor in April 1949. To catch the last ship to China, he did not officially receive his Ph. D. certificate until 30 years later when China began to reopen to the West. Lincoln Constance (1909-2001), his tutor in Berkeley, sent it to him along with his books and the money remaining in his bank account.

1946年，王启无（1913—1987）[367]赴美留学，1947年于耶鲁大学获得硕士学位，1953年于哈佛大学获得博士学位，并参加哈佛大学阿诺德树木园胡秀英领衔的中国植物志项目，于1954年受聘于佛罗里达大学，1959年受聘于明尼苏达大学，1960年之后又受聘于爱达荷大学并研究林木遗传育种。

1946, Chi-Wu WANG (1913-1987) went to United States of America to study forest in Yale University, received master's degree in 1947, then study forest in Harvard University, received Ph.D. degree in 1953[368], joined Flora of China project led by Shiu-Ying HU in Arnold Arboretum. He studied forest genetics and breeding in University of Florida since 1954, University of Mininsota since 1959, and University of Idaho since 1960.

Chi-Wu WANG (1913-1987) (Photo provided by Zong-Gang HU)
王启无（1913—1987）（相片提供者：胡宗刚）

365 刘启新，2010，育才尽瘁 事业流芳 —— 纪念单人骅教授百年诞辰，72页；江苏省中国科学院植物研究所伞形科项目组，南京（内部印制）。

366 Ph.D. dissertation: Ren-Hwa SHAN, 1950, The Old World species of *Sanicula* with reference to the New World, *Science (China)* 32: 25-26; Ren-Hwa SHAN and Lincoln Costance, 1951, The genus *Sanicula* (Umbelliferae) in the old world and the new world, *University of California Publications in Botany* 25: 1-78; Advisor: Lincoln Costance.

367 天津人；1933年毕业于国立清华大学，1933—1943年在静生生物调查所，特别是在云南等地采集而闻名中外；1943—1946年在国立广西大学任副教授。1979年3月5日至4月20日在南京林产工业学院系统讲林木遗传育种学，并于5月随同国家林业部邀请的爱达荷大学林业代表团访问昆明和北京等地。

368 Chi-Wu WANG, 1953, The forest vegetation of continental eastern Asia and its development; Advisor: Hugh M. Raup (1901-1995). Chi-Wu WANG, 1961, The forest of China with a survey of grassland and desert vegetation, 313 p; Harvard University, Cambridge.

1946年，中央研究院接收日本人1931年在上海建立的自然科学研究所（所址在徐汇区祁齐路320号，今岳阳路320号）[369]，同时“中央”研究院植物研究所从重庆迁至上海。

1946, Academia Sinica accepted the Shanghai Science Institute in Shanghai (320 Yueyang Road, Xuhui District, formerly 320 Qiqi Road), which had been founded in 1931 by the Japanese. The Institute of Botany, Academia Sinica, returned to Shanghai from its wartime location in Chongqing.

1946年，何景，**兰州植物志**，176页；国立甘肃科学教育馆专刊第五号，兰州。此书记载兰州所产维管植物66科209属354种，为中国植物分类学家所编写的第一部地方植物志。

1946, Ching HO, ***The Studies of Plants in the Vicinity of Lanchow***, 176 p; National Gansu Science Education Museum, special issues No. 5, Lanzhou. Sixty-six families 209 genera, and 354 species of vascular plants were recorded. This is the first local flora prepared by a Chinese taxonomist.

1946, Hsen-Hsu HU, Notes on a Palaeogene species of *Metasequoia* in China, *Bulletin of the Geological Society of China* 26: 105-107. This is the first report of the discovery of living *Metasequoia* in China.

1946年，胡先骕，记古新期之一种水杉，中国地质学会志26: 105-107。此为水杉活植物在中国发现后的首次报道。

1946年，国立广西大学植物研究所在广西桂林雁山复员。

1946, Kwangsi Institute of Botany, National Kwangsi University, was retored in Yanshan, Guilin, Guangxi.

1946年，俞德浚，八年来云南之植物学研究（民国二十七年至三十四年），教育与科学2(2): 12-16。

1946, Te-Tsun YU, ‘Botanical study of Yunnan in the past eight years (1938-1945)’, *Education and Science* 2(2): 12-16.

1946年11月至1947年6月，国立中山大学派蒋英赴台湾，在台湾林业试验所、国立台湾大学、台北植物园等地鉴定标本与活植物，并摄制台湾和海南植物照片2 000余种[370]，购买图书1 000余册，

369 梁波、翟文豹，2002，日本在中国的殖民科研机构 —— 上海自然科学研究所，中国科技史料23（3）：189-198。

370 云南大学植物标本馆（PYU与YUKU合并）收藏A Catalogue of Photographs of Type—historical collections deposited in the herbarium Taiwan Forest Research Institute, annotated by Ying TSIANG, March-May 1947，1 421张。

带回约 3 000 份植物标本。[371]

1946.11-1947.06, Ying TSIANG, National Sun Yat-Sen University, was sent to Taiwan to work at Taiwan Forestry Research Institute, National Taiwan University and at the Taipei [Taibei] Botanical Garden. He returned to the mainland with photos of nearly 2,000 species from Taiwan and Hainan, 1,000 books purchased and about 3,000 plant specimens.

1946—1947 年，盛诚桂[372]（1912—2002）任南京总理陵园植物园主任。

1946-1947, Cheng-Kuei SHENG (1912-2002) named director, Nanjing Botanical Garden Memorial Sun Yat-Sen, Nanjing.

1946—1949 年，陈封怀任江西省农业院、静生生物调查所庐山森林植物园主任，并负责战后复员。

1946-1949, Feng-Hwai CHEN named director of Lushan Forestry Botanical Garden, Jiangxi Agriculture Institute and the Fan Memorial Institute of Biology and was placed in charge of recovery after the war.

Cheng-Kuei SHENG (1912-2002) (Photo provided by Zong-Gang HU)
盛诚桂（1912—2002）（相片提供者：胡宗刚）

371 国立中山大学，1947，农学院蒋英教授赴台湾考察植物公毕返校，国立中山大学校刊，第 6 期第 3 版；蒋英，1947，考察台湾植物简报，科学 29（11）：344 – 345；蒋英，1947，考察台湾植物简报，国立中山大学校刊，第 6 期第 4 版；蒋英，1948，考察台湾植物简报（续一、续二和续三），国立中山大学校刊，第 7 期第 4 版（续一）、第 8 期第 4 版（续二）和第 9 期第 4 版（续三）。

372 江苏（上海）松江人。1931—1936 年南京金陵大学农学院园艺系学习，1936—1937 年山东济南溥益甜菜制糖厂农业技师，1937—1941 年南京国民革命军遗族学校农业技师，1941—1945 年重庆中国乡村建设育才学院副教授，1945—1946 年美国康奈尔大学和马里兰大学学习，1946—1947 年南京总理纪念植物园主任，1948—1949 年国立山东大学园艺系主任、教授，1949—1957 年南京金陵大学、南京农学院教授，1957—1983 年江苏省植物研究所副所长，南京中山植物园研究员、主任；1983 年后任南京中山植物园名誉主任；著名园艺学家。

1946—1950 年，国立北平研究院钟补求赴英国皇家植物园邱园研修，1950 年 10 月回国后任职于中国科学院植物分类研究所。

1946-1950, Pu-Chiu TSOONG, National Academy of Peiping [Beijing], studied at the Royal Botanic Gardens, Kew; returned to the Institute of Plant Taxonomy, Academia Sinica in October 1950.

Pu-Chiu TSOONG at Royal Botanic Gardens, Kew (Photo provided by Royal Botanic Gardens, Kew)
钟补求于邱园（相片提供者：邱园）

1947

1947 年 7 月，中美拟合作编纂中国植物志[373]，中方编辑会议于上海举行，北平研究院植物学研究所所长刘慎谔、静生生物调查所所长胡先骕的代表唐进、中央研究院植物研究所所长罗宗洛的代表裴鉴，讨论形成中美合作编纂中国植物志事宜。此项计划后未曾实施。

1947.07, Botanists from China and the United States propose collaboration to produce a flora of China; Chinese editorial committee meeting held in Shanghai, attended by Tchen-Ngo LIOU of the Institute of Botany, National Academy of Peiping [Beijing], Tsin TANG, on behalf of Hsen-Hsu HU of the Fan Memorial Institute of Biology and Pei CHIEN, on behalf of Chung-Luo LUO of the Institute of Botany, Academia Sinica. The effort ended without results.

1947 年 8 月 3 日，刘慎谔抵达北平，主持北平研究院植物学研究所复员。

1947.08.03, Tchen-Ngo LIOU returned to Peiping [Beijing] and placed in charge of restoration of the Institute of Botany, National Academy of Peiping [Beijing].

1947 年，张肇骞，中国三十年来之植物学，科学 29(5): 131-137。

1947，Chao-Chien CHANG, Botanical Sciences in China during the last three decades, *Science (China)* 29(5): 131-137.

1947—1948 年，黄成就[374]（1922—2002）和贾良智[375]（1921—2004），考取国立中山大学农林植物研究所陈焕镛的植物分类学研究生。

1947-1948, Ching-Chieu HUANG (1922-2002) and Liang-Chih CHIA (1921-2004) were admitted to the Botanical Institute of National Sun Yat-Sen University, Guangzhou, as graduate students in plant

---

373 美方合作者是史密森学会的 Egbert H. Walker。

374 广东新会人。抗战时在国立西南联合大学读书两年，抗战胜利后于 1947 年在北京大学毕业，同年考取国立中山大学研究生，1950 年获国立中山大学硕士学位；1951 年到中国科学院植物研究所工作，后被打成右派，1959 年调往云南昆明中国科学院昆明植物研究所，1963 年又调往广东广州中国科学院华南植物研究所；“文化大革命”中倍受摧残，经历诸多磨难。专长芸香科和壳斗科，以及牻牛儿苗科。详细参见：张永田，2004，怀念恩师黄成就教授，广东省植物学会会刊 13-14：103-105。

375 四川成都人。1942—1946 年就读于华西大学生物系，毕业留校任教两年；1948—1951 年国立中山大学植物研究所研究生，1951—1954 年任国立中山大学植物研究所助理研究员，1954—1964 年任中国科学院华南植物研究所助理研究员，1964—1984 年任副研究员，1986 年任研究员；1975 年组建中国科学院华南植物研究所资源室并任主任；著名竹子专家。详细参见：朱亮锋、李用华、周文华、何其敏，2009，忆植物资源室的创始人——贾良智教授；魏平主编，根深叶茂竞芳菲——中国科学院华南植物园八十周年纪念文集，166-167 页；广东科技出版社，广州。

taxonomy under the supervision of Woon-Young CHUN.

1947—1948 年，郑万钧任南京总理陵园植物园主任。

1947-1948, Wan-Chun CHENG named director, Nanjing Botanical Garden Memorial Sun Yat-Sen, Nanjing.

1947-1949, ***Botanical Bulletin, Academia Sinica***, issued in Shanghai, vols. 1-3.

1947—1949 年，**国立中央研究院植物学汇报**发刊于上海，vols. 1-3。

1947—1950 年，云南农林植物研究所俞德浚[376]（1908—1986）获皇家植物园资助，赴苏格兰爱丁堡植物园工作两年，英格兰邱园工作一年，返回后任中国科学院植物分类研究所研究员，曾任中国植物志第四任主编（1977—1986 年）。俞德浚 1980 年入选中国科学院学部委员；著名蔷薇科及植物园专家。

1947-1950, Te-Tsun YU (1908-1986), Yunnan Botanical Institute, worked at the Royal Botanic Garden, Edinburgh, Scotland, for two years and at the Royal Botanic Gardens, Kew, England, for one year; supported by Royal Botanic Gardens; returned to China to become a research professor in the Institute of Plant Taxonomy, Academia Sinica. He served as the fourth editor in chief (1977-1986) of *Flora Reipublicae Popularis Sinicae*. He was elected an academician of the Chinese Academy of Sciences in 1980. His lifelong research focused on the Rosaceae and botanical gardens.

1947, 1953, 1965, 1974, Ronald Good, ***The Geography of the Flowering Plants***, 403 p, 1947; ed. 2, 452 p, 1953; ed. 3, 518 p, 1964; ed. 4, 557 p, 1974; Longman, London. Translated into Chinese in 1987.

1987 年，李锡文译，**显花植物地理**[377]，387 页；中国科学院昆明植物研究所，云南省植物学会编印，昆明（内部印制）。

1947—1954 年，广西大学植物研究所更名为广西大学经济植物研究所，陈焕镛兼任所长。

1947-1954, Kwangsi Institute of Botany, Kwangsi University, renamed Economic Institute of Botany, Kwangsi University, with Woon-Young CHUN concurrently as director.

---

376 北京人。1928—1931 年就读于国立北平师范大学生物系，1931 年毕业后被胡先骕引入静生生物调查所；1932—1934 年静生生物调查所接受美国哈佛大学阿诺德树木园的资助，合作进行四川植物采集工作，俞德浚被派任中国西部科学院植物部主任，具体负责与哈佛大学的合作任务，率队在四川西部进行了 3 年的调查采集工作；1937—1939 年，静生生物调查所又接受英国皇家植物园和英国皇家园艺学会的委托，采集云南高山植物种子，俞德浚又接受任务赴云南考察采集。1939—1947 年俞德浚任云南农林植物研究所研究员，1945 年起任副所长两年，并先后兼任云南大学生物系和农学院讲师、副教授职务。

377 据李锡文研究员，译自第 4 版。

1948

1948 年 4 月 1 日，中央研究院于南京正式选举产生了第一届 81 名院士，生物组 25 人，包括植物分类学家胡先骕、钱崇澍、戴芳澜和邓叔群等。

1948.04.01, Eighty one academicians were elected during the first council meeting of Academia Sinica, Nanjing; among them, 25 from biology, including taxonomists Hsen-Hsu HU, Sung-Shu CHIEN, Fang-Lan TAI and Shu-Chun TENG.

First academicians of Academia Sinica, Nanking [Nanjing], 23 September 1948, including plant taxonomists Sung-Shu CHIEN (rear row, 6th from left), Hsen-Hsu HU (row 5, 6th from left), Fang-Lan TAI (row 4, 8 from left) and Shu-Chun TENG (row 5, 1st on left) (Photo provided by Zong-Gang HU)

1948 年 9 月 23 日，中央研究院首届院士于南京，包括植物分类学家钱崇澍（后排左六）、胡先骕（五排左六）、戴芳澜（四排左八）和邓书群（五排左一）（相片提供者：胡宗刚）

1948 年 9 月，王振华[378]（1908—1976）代理陕西武功中国西北植物调查所所长。

1948.09, Chen-Hwa WANG (1908-1976) named acting director of the Botanical Survey of Northwest Institute, Wukong [Wugong], Shaanxi.

1948 年 9 月 24 日，第三届中央研究院院士会议于南京选举产生 32 位评议员，包括植物分类学胡先骕和钱崇澍等。

1948.09.24, Thirty two council members elected at the third meeting of Academia Sinica, Nanjing, including taxonomists Hsen-Hsu HU and Sung-Shu CHIEN.

1948, Hsiu-Chien CHAO, Discovery of Podostemaceae in China, *Contributions from the Institute of Botany, National Academy of Peiping* 6(1): 1-16; Remarks on Chinese plants of family Podostemaceae, *Science (China)* 30(9): 275-276.

1948 年，赵修谦[379]，中国川草科植物之发现，国立北平研究院植物学研究所丛刊 6(1): 1-16；中国川草科植物之发现，科学 30(9): 275-276。

1948, Hsen-Hsu HU and Wan-Chun CHENG, On the new family Metasequoiaceae and on *Metasequoia glyptostroboides,* a living species of the genus *Metasequoia* found in Szechuan [Sichuan] and Hupeh [Hubei], *Bulletin of the Fan Memorial Institute of Biology new series* 1: 153-161. The original description of living *Metasequoia glyptostroboides* was published in this paper.

1948 年，胡先骕、郑万钧，水杉新科及生存之水杉之新种，静生生物调查所汇报新系 1: 153-161。这是中国水杉新种的正式发表。

1948，Hsen-Hsu HU, ***The Silva of China****—a description of the trees which grow naturally in China*, Volume 2, Betulaceae-Corylaceae, 209 p; the Fan Memorial Institute of Biology, Peking [Beijing] and National Forestry Research Bureau Ministry of Agriculture and Forestry, Nanking [Nanjing].

1948 年，胡先骕，**中国森林树木图志**，第 2 册[380]，桦木科 — 榛科，209 页；静生生物调查所，北平；农林部中央林业试验所，南京。

---

378 字健公，安徽太平（今黄山）人；1935 年毕业于国立中央大学，1935 年 10 月 至 1936 年 8 月曾在北平研究院植物学研究所工作，师从著名植物学家刘慎谔教授，任助理研究员；1936 年 8 月至 1950 年 9 月在中国西北植物调查所工作，先后任助理员、副研究员、代理所长、所长。期间从 1941 年 8 月至 1950 年 1 月还兼任国立西北农学院副教授、教授、院务委员会委员及植物保护系系主任、国立西北大学植物学教授；新中国成立后曾担任西北农学院教务长，1956 年任西北农学院副院长。

379 福建福州人（1911—2001），1935 年毕业于国立清华大学后，一直在厦门大学任教。

380 本书第一册未出版。

1948、1951、1958 年，耿以礼编，**中国种子植物分科检索表**，74 页，1948；中国科学图书仪器公司，上海；耿以礼、耿伯介合编，**中国种子植物分科检索表**，修订版，103 页，1951；中国科学图书仪器公司，上海；耿以礼、耿伯介合编，**中国种子植物分科检索表**，增订版，108 页，1958；科学出版社，北京。

1948, 1951, 1958, Yi-Li KENG, '***A Key to the Families of Phanerogams in China***', 74 p, 1948; Chinese Scientific Books, Shanghai; Yi-Li KENG and Pai-Chieh KENG, '***A Key to the Families of Phanerogams in China***', revised ed., 103 p, 1951, Chinese Scientific Books, Shanghai; Yi-Li KENG and Pai-Chieh KENG, '***A Key to the Families of Phanerogams in China***', enlarged ed., 108 p, 1958; Science Press, Beijing.

1948+，**国立台湾博物馆学刊**，vols. 1-35，1948—1982；原名为**台湾省立博物馆季刊**，季刊，英中文版；**台湾省立博物馆半年刊**，36-52(1)，1983—1999；**"国立"台湾博物馆半年刊**[381]，52(2)-57(2)，1999—2004；2005 年与中文的"国立"台湾博物馆年刊合并改为现名同时接受中英文稿件，2007 年至今季刊。

1948+, ***Quarterly Journal of the Taiwan Museum***, vols. 1-35, 1948-1982, quarterly; ***Journal of Taiwan Museum***, vols. 36-52(1), 1983-1999, biennially; combined with ***Annals of the National Taiwan Museum*** 52(2)-57(2), 1999-2004; in both English and Chinese since 2005; quarterly since 2007.

1948+, The Laboratory of Systematic Botany, Department of Botany, College of Science, National Taiwan University, ***Taiwania***[382], 1948+; vols. 1-34, annually or biannually, 1948-1989; quarterly since 1990.

1948+，**国立台湾大学理学院植物学系研究报告**[383]，1954 年第 5 号更名为**台湾植物分类地理丛刊**，vols. 1-34，年刊或半年刊；1990 年 35 卷改为季刊。

1948—1949 年，焦启源任南京总理陵园植物园主任。

1948-1949, Chi-Yuen CHIAO named director, Nanjing Botanical Garden Memorial Sun Yat-Sen, Nanjing.

381 李子宁主编，1999，台湾省立博物馆创立九十年专刊，325 页；台湾省立博物馆，台北。

382 College of Arts and Sciences, National Taiwan University, ***Acta Botanica Taiwanica***, 1947, only one issue published; replaced by *Taiwania* in 1948 with new series number.

383 *Acta Botanica Taiwanica* 创刊于 1947，只出版一期；1948 年由 *Taiwania* 取代，且刊号另起；中文名称及隶属单位更换频繁，年刊、半年刊也不是很规则；详细参见：http://tai2.ntu.edu.tw/taiwania/about.php（2017 年 12 月进入）。

1949

1949 年 7 月 14 日，中国植物学会在北京大学举行（第五届全国代表大会），张景钺任会长。

1949.07.14, Meeting of Botanical Society of China (Fifth National Congress) held at Peking University, with Ching-Yueh CHANG as president.

1949 年 11 月 1 日，中国科学院于北京成立（院址：文津街 3 号，原静生生物调查所大楼）。静生生物调查所与北平研究院植物学研究所合并，组建中国科学院植物分类研究所（所址：动物园，原北平研究院植物学研究所），1949—1965 年钱崇澍任所长。[384]

1949.11.01, Chinese Academy of Sciences established in Beijing (Address: 3 Wen Tsin Chieh, original building of the Fan Memorial Institute of Biology); the Fan Memorial Institute of Biology and the Institute of Botany, National Academy of Peiping [Beijing], were merged into a new Institute of Plant Taxonomy, Chinese Academy of Sciences (Address: Beijing Zoo, original building of Institute of Botany, National Academy of Peiping), with Sung-Shu CHIEN as director, 1949-1965.

1949, Helen M. Fox, ***Abbé David's Diary****—being an account of the French naturalist's journeys and observations in China in the years 1866 to 1869*, 302 p; Harvard University Press, Cambridge, Massachusetts.[385]

1949 年，Helen M. Fox，**Abbé David 日记** ——1866—1869 年法国博物学家在中国的旅行与观测之记载，302 页；哈佛大学出版社，剑桥，麻州。

1949, H. Zanyin GAW, The status of biology in China, *American Scientist* 37(2): 263-265 and 315-317.

1949 年，高尚荫[386]，中国生物学的现状，*American Scientist* 37（2）：263-265、315-317。

1949 年，林镕编，**河北植物名录**，43+11+1 页；（北京）师范大学生物学系，北京（内部印制）。

---

384 胡宗刚，2007，1949 年前后的中国植物分类学，科学文化评论 4（3）：107-110。

385 An abridged translation of the author's Journal d'un voyage en Mongolie fait en 1866 and Journal d'un voyage dans le centre de la Chine et dans le Thibet oriental, which appeared in *the Bulletin des nouvelles archives du muséum, Paris*, 1867-1874.

386 浙江嘉善人（1909—1989），1926 年入苏州东吴大学，1930 年获学士学位；1930—1935 年留学美国，获耶鲁大学博士学位；1935 年 2 月赴英国从事研究后，并于同年 8 月回国，受聘为国立武汉大学教授；1943 年高尚荫再次赴美从事病毒学研究；1947 年回到国立武汉大学，继续教学和病毒学研究工作；1949 年任国立武汉大学生物系主任、教务长、理学院副院长、副校长等；1956 年后兼任中国科学院武汉分院副院长、中南微生物研究所所长、武汉病毒研究所所长和名誉所长；1980 年当选为中国科学院学部委员。

1949, Yong LING, '***Checklist of Hebei Plants***', 43+11+1 p; Department of Biology, Beijing Normal University, Beijing (internal publication).

1949, Alfred Rehder, ***Bibliography of Cultivated Trees and Shrubs*** *hardy in the cooler temperate regions of the northern hemisphere*, 825 p; Arnold Arboretum of Harvard University, Jamaica Plain, Massachusetts. This work recorded many plants that were originally from China. The work is still useful today.

1949 年 , Alfred Rehder, **栽培乔灌木文献目录**——北半球寒温带，825 页；哈佛大学阿诺德树木园，Jamaica Plain，麻州。本书记载很多来自于中国的植物，即使今日仍有参考价值。

Alfred Rehder (1863-1949) (Photo provided by Arnold Arboretum of Harvard University)

Alfred Rehder（1863–1949）（相片提供者：哈佛大学阿诺德树木园）

1950

1950年8月26日至9月1日，中国科学院在北京召开全国第一届植物分类专门会议，38人参加（包括邀请部分京外专家）[387]，讨论以中国科学院植物分类研究所和国立中山大学农林植物研究所为中心，开展全国范围内植物标本采集及植物学书刊订购补充，发行植物分类学报，建立植物园，为编写中国植物志做好准备；并成立了中国植物志工作筹备委员会，推荐了以钱崇澍、陈焕镛为首的，28人组成委员会，包括东北的刘慎谔、杨衔晋[388]（1913—1984），华北的俞德浚、汪振儒、吴征镒、林镕、胡先骕、唐进、汪发缵、郝景盛、张肇骞、钟补求，华东的郑万钧、陈嵘、耿以礼、裴鉴、单人骅，华中的孙祥钟、钟心煊、陈封怀，华南的蒋英、何景[389]（1912—1978）、钟济新[390]（1909—1993），西南的方文培、蔡希陶、秦仁昌，西北的王振华、孔宪武[391]（1897—1984）。会议同时决定加印中国植物照片集，分给各机构使用。

1950.08.26-09.01, First national special meeting on plant taxonomy, with 38 attendees (including participants from outside of Beijing), held in Beijing. It was decided that the Institute of Plant Taxonomy, Chinese Academy of Sciences, and the Botanical Institute of National Sun Yat-Sen University were to be the centers to explore Chinese plants nationwide; the collections of the botanical libraries would be enlarged, *Acta Phytotaxonomica* would be published, botanical gardens would be established and a committee to produce *Flora Reipublicae Popularis Sinicae* would be organized. Sung-Shu CHIEN and Woon-Young CHUN were proposed to serve as lead editors; twenty eight additional taxonomists would participate, including Tchen-Ngo LIOU, Yen-Chin YANG (Hsien-Chin YANG, 1913-1984) from the northeast, Te-Tsun YU, Chen-Ju WANG, Cheng-Yih WU, Yong LING, Hsen-Hsu HU, Tsin TANG, Fa-Tsuan WANG, Ching-Sheng HAO, Chao-Chien CHANG, Pu-Chiu TSOONG from the north, Wan-Chun CHENG, Yung CHEN, Yi-Li KENG, Chien PEI, Ren-Hwa SHAN from the east, Hsiang-Chung SUN, Hsin-Hsuan CHUNG, Feng-

387 胡宗刚、夏振岱，2016，中国植物志编纂史，34-39；上海交通大学出版社，上海。

388 浙江嘉兴人。1931年考入国立中央大学农学院森林系，1935年毕业留校任助教；1937年任中国科学社生物研究所编辑和研究员；1942年在四川北碚任复旦大学农学院教授；1945年赴美国耶鲁大学林学院进修，1946年回国后任复旦大学农学院教授，兼任河南大学农学院、上海同济大学理学院教授；1950年任东北农学院森林系教授兼系主任；1952年任东北林学院教授兼林工系主任；1956年任教务处主任；1962年任东北林学院副院长；1979年任东北林学院院长。著名森林植物学家，专长是樟科、豆科等。

389 字星叔，江苏泰兴人。1935年毕业于国立中央大学并留校任教，1940年后在西北药科学校、西北师范学院、福建研究院和福建协和学院等地任教，1951年任厦门大学教授兼任中国科学院亚热带植物研究所所长；专长五加科。

390 广西苍梧人。1935年毕业于广西大学并留校任教，1954年后历任中国科学院广西植物研究所副所长、所长，广西科学院副院长等，1955年带领科考队在广西花坪发现著名活化石银杉。参见钟树林编辑，2011，钟济新传记，83页（自印）。

391 河北高邑人。1917年入北京高等师范学校博物系；1921年毕业后在山东七中和山西第一师范任教；1923—1925年在（北平）师范大学博物系研究科学习；1925年后任（北平）师范大学和河北大学教授；1929年至北平研究院植物学研究所先后任练习生、助理研究员和副研究员；1936年后赴中国西北植物调查所并兼任国立西北农学院教授；1939年奔赴兰州，先后任西北技艺学校和西北师范学院教授、系主任，植物研究所所长、甘肃师范学院副院长等职。

First national special meeting on plant taxonomy, Beijing, 1950. Standing (left-right, back two rows): Wan-Chun CHENG, Chao-Chian CHANG, Wei-Ying HSIA, Wei-Kun HSIA, Yi-Li KENG, Chen-Ju WANG, Ying TSIANG, Tsin TANG, Shu-Hsia FU, Hsen-Hsu HU, Ke-Zen KUANG, Chen-Hwa WANG, Cheng-Yih WU, Wen-Pei FANG, Tchen-Ngo LIOU, Fa-Tsuan WANG, Yong LING, Ching-Sheng HAO; Sitting (left-right, front two rows): Sheng-Chen LI, Lie-Ying LÜ, Yen-Cheng TANG, Jia-Wen FENG, Ke-Chien KUAN, Ji-Ding ZHAO, Zuo-Min YANG, Cho-Po TSIEN, Wen-Tsai WANG, Yu-Chuan MA, Shu-Chih HAN, Fu-Quan WANG, Lian-Wang XU, Zong-Xun WANG (Photo provided by Zong-Gang HU)

1950 年，全国第一届植物分类学特别会议于北京。站席（从左至右，后两排）：郑万钧、张肇骞、夏纬瑛、夏纬琨、耿以礼、汪振儒、蒋英、唐进、傅书遐、胡先骕、匡可任、王振华、吴征镒、方文培、刘慎谔、汪发缵、林镕、郝景盛；座席（从左至右，前两排）：黎盛臣、吕列英、汤彦承、冯家文、关克俭、赵继鼎、杨作民、简焯坡、王文采、马毓泉、韩树金、王富全、徐连旺、王宗训（相片提供者：胡宗刚）

Hwai CHEN from the central, Ying TSIANG, Ching HO (1912-1978), Chi-Hsing CHUNG (1909-1993) from the south, Wen-Pei FANG, Hse-Tao TSAI, Ren-Chang CHING from the southwest, and Chen-Hwa WANG, Hsien-Wu KUNG (1897-1984) from the northwest. It was decided at the meeting that additional copies of *the Photographs of Chinese Plants* would be printed for the institutions that lacked them.

1950 年 8 月，胡先骕，被子植物的一个多元的新分类系统，中国科学 1（1）：243-254。[392]

1950.12, Hsen-Hsu HU, A polyphyletic system of classification of angiosperms, *Science Record (Peking)* 3(2-4): 221-230.[393]

1950 年，东北农学院成立东北植物调查所。刘慎谔任所长，并对东北植物资源进行了调查采集。[394]

1950, Northeast Institute of Botanical Survey established in Northeast Agriculture College, with Tchen-Ngo LIOU as director; and field exploration for plants in northeast China was also initiated.

1950，Ying TSIANG, The development of plant taxonomy in China during the past thirty years, *Quarterly Journal of the Taiwan Museum* 3: 1-13.

1950 年，蒋英，中国近三十年来植物分类学的发展，台湾省立博物馆季刊 3: 1-13。

1950, Armen L. Takhtajan, Phylogenetic principles of the system of higher plants, *Botanicheskii Zhurnal* 35(2): 113-139; 1953, English translation by D. I. Lalkow, *the Botanical Review* 19(1): 1-45. Translated into Chinese in 1954.

1954 年，胡先骕译，**高等植物系统的系统发育原理**，54 页；中国科学院，北京。

1950 年，吴长春著，**浙江植物名录**，上册：1-276 页，下册：277-458 页；浙江师范学院，杭州（手写油印本）。

1950, Chang-Chun WU, '***Checklist of Zhejiang Plants***', 1: 1-276 p, 2: 277-458 p; Zhejiang Normal College, Hangzhou (handwritten mimeo).

1950—1953 年，庐山森林植物园更名为中国科学院植物分类研究所庐山工作站，陈封怀任主任。

1950-1953, Lushan Forestry Botanical Garden was renamed Lushan Station of the Institute of Plant Taxonomy, Chinese Academy of Sciences, with Feng-Hwa CHEN as director.

392 本文献的中英文分开发表，且时间、刊物及页码不同。

393 The work was published in Chinese and English in different journals at different times so that the names of the journals and the page numbers are not the same.

394 胡宗刚，2007，1949 年前后的中国植物分类学，科学文化评论 4（3）：107-110。

1950—1954 年，原中央研究院植物研究所高等植物分类室从上海迁往南京九华山，成立中国科学院植物分类研究所华东工作站，裴鉴任主任。

1950-1954, The Higher Plant Department, Institute of Botany, Academia Sinica, was moved from Shanghai to Jiuhua Shan, Nanjing, and named East Station of the Institute of Plant Taxonomy, Chinese Academy of Sciences, with Chien PEI as director.

1950—1958 年，云南农林植物研究所与北平研究院植物学研究所昆明工作站合并成立中国科学院植物分类研究所昆明工作站，蔡希陶任主任；中国西北植物调查所归属中国科学院植物分类研究所，改称西北工作站；夏纬瑛[395]（1896—1987）负责。

1950-1958, Yunnan Botanical Institute and Kunming Station of the Institute of Botany, National Academy of Peiping [Beijing], were merged into the Kunming Station of the Institute of Plant Taxonomy, Chinese Academy of Sciences, with Hse-Tao TSAI as director. The Botanical Survey of North-Western China was renamed Northwest Station of the Institute of Plant Taxonomy, Chinese Academy of Sciences, with Wei-Ying HSIA (1896-1987) in charge.

395 河北赵州人；1916 年考入北京农业专科学校（1920 年改为北平大学农学院）工作，1923 年为实验室助理，1925 年为生物系助理，1928 年为生物系助教，1929 年协助刘慎谔筹建北平研究院植物学研究所，1930 年正式调入植物研究所工作。抗日战争期间西迁陕西武功，任中国西北植物调查所负责人，并先后在河南大学和国立西北农学院任教。1956 年调到中国科学院植物研究所，1960 年调到中国科学院自然科学史研究所。一生从事分类学、生物史和农史研究。参见：志尚惠，1988，沉痛悼念夏纬瑛教授，农业考古 1：387。刘昌芝，1988，怀念夏纬瑛教授，植物杂志 3：37-39。

1951

1951 年 7 月 24 日至 28 日，中国植物学会新一届代表大会（第六届全国代表大会）于北京举行，出席 52 人，理事长为钱崇澍。[396]

1951.07.24-07.28, New National Congress (Sixth National Congress) of the Botanical Society of China was held in Beijing; 52 attendees, with Sung-Shu CHIEN as president.

1951 年，北京师范大学生物学系编，**河北植物名录**，1-114 页，115-125 页，125-126 页；北京师范大学生物学系，北京（内部印制）。

1951, Department of Biology, Beijing Normal University, '***Checklist of Hebei Plants***', 1-114 p, 115-125 p, 125-126 p, Department of Biology, Beijing Normal University, Beijing (internal publication).

1951 年，东北农学院，**东北农学院植物调查研究所丛刊**，Vol. 1。

1951, Northeast Agriculture College, '***Journal of Institute of Botanical Survey, Northeast Agriculture College***', vol. 1.

1951 年，侯定[397]（1921—2008）赴美国圣路易斯华盛顿大学攻读植物学，1952 年获硕士学位，1955 年获博士学位，1955 年进入哈佛大学阿诺德树木园参加胡秀英领衔的中国植物志项目，1956 年应邀赴荷兰莱顿参加马来西亚植物志项目，直至 1986 年退休。

1951, Ding HOU (Ting HOU, 1921-2008) studied botany at Washington University,

Ding HOU (1921-2008) (Photo provided by Naturalis, Nationaal Herbarium Nederland)
侯定（1921—2008）（相片提供者：荷兰国家标本馆）

---

396 中国植物学会五十年编写组，中国植物学会五十年，中国科技史料 6（2）：53。

397 江西新干人，1941—1945 年在江西国立中正大学生物系学习，毕业后在母校担任两年助教，1947—1951 年任台湾大学助教。

New National Congress (Sixth National Congress) of the Botanical Society of China, Beijing, 24-28 July 1951 (Photo provided by Ping MA)

1951 年 7 月 24 日至 28 日，中国植物学会第一届代表大会（第六届全国代表大会）于北京举行（相片提供者：马平）

St. Louis, Missouri, United States of America, received a master's degree in 1952 and a Ph. D. degree in 1955[398], then joined *the Flora of China* project led by Dr. Shiu-Ying HU at the Arnold Arboretum of Harvard University in 1955; joined the *Flora Malesiana* project at Leiden, the Netherlands in 1956, worked there until retirement in 1986[399].

1951 年，胡先骕著，**种子植物分类学讲义**，423 页；中华书局，上海。

1951, Hsen-Hsu HU, '***Textbook of Taxonomy of Seed Plants***', 423 p; Chinese Books, Shanghai.

1951+，中国科学院植物分类研究所 / 中国科学院植物研究所、中国植物学会，**植物分类学报**，vols. 1+, 1951+；季刊，1984 年改为双月刊；1951—1952 年（第 1-2 卷 1 期）刊名为拉丁文 ***Acta Phytotaxonomica***，1953 年（第 2 卷 2 期）刊名为 ***Acta Phytotaxonomica Sinica***；2008 年（第 46 卷）刊名由 ***Acta Phytotaxonomica Sinica*** 改为英文 ***Journal of Systematics and Evolution***（但中文名称不变）；2009 年改为英文版（无中文名称）。本刊 1952 年只出版第 2 卷 1 期，第 2 卷的 2–4 期 1953—1954 年出版；另外，该刊 1959—1963 年和 1966—1973 年两度停刊；1951—1979 年钱崇澍[400]，1979—1982 年秦仁昌，1982—1989 年王文采（1926—），1989—1999 年洪德元（1937—），1999—2004 年杨亲二（1964—），2005—2014 年陈之端（1965—）和仇寅龙（1964—），2014 年至今葛颂（1961—）和文军（1963—）任主编。[401]

1951+, Institute of Plant Taxonomy / Institute of Botany, Academia Sinica / Chinese Academy of Sciences, and Botanical Society of China, ***Acta Phytotaxonomica***, vols. 1-2(1), 1951-1952, ***Acta Phytotaxonomica Sinica***, vols. 2(2)-45, 1953-2007; ***Journal of Systematics and Evolution*** (vols. 46+), 2008+; quarterly 1951-1983, bimonthly 1984+; Chinese edition 1951-2008, English edition, 2009+; discontinued 1959-1963 and 1966-1973; Editors in Chief: Sung-Shu CHIEN, 1951-1979[402], Ren-Chang CHING, 1979-1982, Wen-Tsai WANG (1926-), 1982-1989, De-Yuan HONG (1937-), 1989-1999, Qin-Er YANG (1964-), 1999-2004, Zhi-Duan CHEN (1965-) and Yin-Long QIU (1964-), 2005-2014, Song GE (1961-) and Jun WEN (1963-) since 2014.

1951—1953 年，西藏工作队崔友文（1907—1980）、钟补求进藏采集；崔友文于 1951 年 8 月在昌都附近采集了 706 号标本；钟补求于 1952 年 6 月进藏，1954 年 3 月才返回北京，其路线西迄定结、

398 Ph.D. dissertation: Ding HOU, 1955, A revision of the genus *Celastrus*, *Annals of the Missouri Botanical Garden* 42: 215-302; Advisor: Robert E. Woodson, Jr. (1904-1963).

399 Pieter Baas and Frits Adema, 2001, Dr. Ding HOU 80 years young, *Blumea* 46(2): 201-205; 2008, In Memoriam Ding HOU (1921-2008), *Blumea* 53(2): 233-234.

400 主任、主任编辑或在京常务编委。

401 引自该刊介绍（http://www.plantsystematics.com/cn/ljbwh.asp, 2018 年进入）。

402 As director, editor, or standing member of editorial committee in Beijing.

南至亚东，共采标本 2 437 号（PE）。[403]

1951-1953, Expedition to Xizang by You-Wen TSUI (Yu-Wen TSUI, 1907-1980) and Pu-Chiu TSOONG. You-Wen TSUI collected 706 numbers of plant specimens in Chamdo [Qamdo, Changdu] in August 1951. Pu-Chiu TSOONG explored in the area west to Dinggyê [Dingjie] and south to Yadong between June 1952 and March 1954 and collected 2,437 numbers of plant specimens (PE).

1951—1957 年、1957—1958 年，四川大学派出的采集队共计 15 次，参加人员 20 人次以上，在四川共采集到近 4 万号标本（SZ）；同时还和中国科学院植物研究所合作采集近万号标本（PE、SZ）。[404]

1951-1957 and 1957-1958, Sichuan University sent more than 15 teams and more than 20 personnel to collect in Sichuan. Nearly 40,000 numbers of plant specimens (SZ) were made. Meanwhile, collaborated with staff members from the Institute of Botany, Chinese Academy of Sciences, Beijing, collected more than 10,000 numbers of plant specimens (PE, SZ).

403 吴征镒，1983，西藏植物志，第 1 卷，前言，iii；科学出版社，北京。

404 潘杰，1994，四川植物标本采集史初探，四川文物（S1）：42-48。

1952

1952, John M. Cowan (ed.), with assistance of members of the staff of the Royal Botanic Garden, Edinburgh, and Euan H. M. Cox, ***The Journeys and Plant Introductions of George Forrest V.M.H.***, 252 p; Oxford University Press[405], London.

1952 年，John M. Cowan 编辑，爱丁堡植物园人员和 Euan H. M. Cox 协助，**维多利亚荣誉勋章获得者 George Forrest 的旅行和植物引种**，252 页；牛津大学出版社，伦敦。

1952 年，裴鉴、单人骅，**华东水生维管束植物**，128 页；中国科学院，北京。

1952, Chien PEI and Ren-Hwa SHAN, '***Aquatic Vascular Plants of East China***', 128 p; Chinese Academy of Sciences, Beijing.

1952 年，钱崇澍、胡先骕、林镕、俞德浚、吴征镒、汪发缵、唐进、匡可任、刘慎谔，美军飞机在朝鲜北部和中国东北撒布两种朝鲜南部特产树叶的报告，科学通报（反细菌战特刊）：132-135。

1952, Sung-Shu CHIEN, Hsen-Hsu HU, Yong LING, Te-Tsun YU, Cheng-Yih WU, Fa-Tsuan WANG, Tsin TANG, Ko-Zen KUANG (Ko-Jen KUANG), and Tchen-Ngo LIOU, 'Report of two kinds of endemic species of South Korean plants spread by American air plane into North Korea and Northeast China', *Chinese Science Bulletin* (Special Issue of Anti-Bacteriological Warfare): 132-135.

405 For RHS.

1953

1953 年 1 月，中国科学院植物分类研究所发展为综合研究所，更名为中国科学院植物研究所。

1953.01, Institute of Plant Taxonomy, Chinese Academy of Sciences, developed as a comprehensive institute and renamed Institute of Botany, Chinese Academy of Sciences.

1953 年，吴征镒，中国植物学历史发展的过程和现况，科学通报 2：12-20，生物学通报 3-4：115-123，新华月报，3：205-210，植物学报 2（2）：335-348。[406]

1953, Cheng-Yih WU, 'The process and status of historical development of Chinese botany', *Chinese Science Bulletin* 2: 12-20, *Bulletin of Biology* 3-4: 115-123, *Xinhua Monthly* 3: 205-210, *Acta Botanica Sinica* 2(2): 335-348.

1953—1954 年，中国科学院在黑龙江省哈尔滨设立林业研究所筹备处，刘慎谔任副主任。

1953-1954, Preparatory of Institute of Forestry, Chinese Academy of Sciences, founded in Harbin, Heilongjiang Province, with Tchen-Ngo LIOU as vice director.

1953、1954 年，中国科学院植物研究所（集体编写，39 人），中国植物科属检索表（上、下），植物分类学报 2（3、4）：173-338、339-473；索引，474-536。

1953 and 1954, Institute of Botany, Chinese Academy of Sciences (team work with 39 authors), Claves Familiarum Generumque Plantarum Sinicarum, *Acta Phytotaxonomica Sinica* 2 (3 and 4): 173-338 and 339-473, Index 474-536.

1953-1955, ***Memoirs of the Hong Kong Biological Circle***, vols. 1-3, Newspaper Enterprise Ltd., Hong Kong; 1960-, ***Memoirs of the Hong Kong Natural History Society***, vols. 4-; The Hong Kong Natural History Society, Hong Kong; published irregularly with articles on biology and plant taxonomy[407].

1953—1955 年, ***Memoirs of the Hong Kong Biological Circle***, vols. 1-3, Newspaper Enterprise Ltd., 香港; 1960—, ***Memoirs of the Hong Kong Natural History Society***, vols. 4-;The Hong Kong Natural History Society，香港。这是香港有关生物方面的刊物，不定期发行。该刊物有植物分类学的相关内容。

---

406 此文在不同刊物前后发表四次。

407 John Hodgkiss, 1988, A guide to articles in the Hong Kong Naturalist (1930-1941) and in the Memoirs of the Hong Kong Natural History Society (1953-1988), *Memoirs of the Hong Kong Natural History Society* 18: 67-71.

1953—1957 年，中国科学院植物研究所庐山工作站更名为中国科学院庐山植物园，陈封怀任主任。

1953-1957, Lushan Station of the Institute of Botany, Chinese Academy of Sciences, renamed Lushan Botanical Garden, Chinese Academy of Sciences, with Feng-Hwa CHEN as director.

1953-1957, Dr. Shiu-Ying HU of the Arnold Arboretum of Harvard University, with support from the Continental Development Foundation via China International Foundation, began the *Flora of China* project, with assistance from other Chinese scholars, such as Dr. Ding HOU and Dr. Chi-Wu WANG, who were then in the United States of America. The project was to be completed in three stages: compilation of an index to the Chinese flora; publication of an enumeration of the flora of China; and descriptions and keys to the families, genera and species and citations of important literature and specimens examined. Unfortunately, only family 153, Malvaceae, by Shiu-Ying HU, was published.[408]

1953—1957 年，哈佛大学阿诺德树木园胡秀英，通过中国国际基金在大地基金会的资助下开始启动中国植物志项目，其他参与者包括当时在美国的中国学者侯定和王启无博士等。项目预计分三个阶段，包括编制中国植物索引，出版中国植物名录，做好出版具有描述、检索表、重要文献和检视标本引证的分科植物志的准备等。该项目最后仅胡秀英撰写的第 153 科锦葵科出版。

1953—1977 年，邱炳云[409]（1906—1989）于云南昆明近郊（西山）、嵩明、西双版纳、丽江、禄劝、景东、元江、普文、金平、元谋、武定、富民、江川、下关、洱源、剑川、路南等地采集植物标本。所采标本共计 1 万余号（KUN）。

1953-1977, Ping-Yun Chiu (1906-1989) collected over 10,000 numbers of plant specimens (KUN) in Eryuan, Fumin, Jianchuan, Jiangchuan, Jingdong, Jinping, Kunming (Western Hills), Lijiang, Lunan, Luquan, Puwen, Songming, Xiaguan, Xishuangbanna, Yuanjiang, Yuanmou and Wuding, Yunnan.

---

408 Shiu-Ying HU, 1955, *Flora of China, family 153*, Malvaceae, 80 p; Arnold Arboretum of Harvard University, Cambridge, Massachusetts.

409 四川宜宾人，1932 年跟随蔡希陶赴云南，终生为中国科学院昆明植物研究所采集员。此处仅记载大规模的集中采集，而非邱炳云一生的全部采集；余同。

1954

1954年，中国科学院聘南京师范学院陈邦杰为兼职研究员，同时配备助手，拨出科研专款，在南京建立第一个苔藓植物标本室。该标本室于1984年移至北京的中国科学院植物研究所（PE）。[410]

1954, Chinese Academy of Sciences hired a concurrent research professor Pan-Chieh CHEN of Nanjing Normal College and provided assistants and research funds. Pan-Chieh CHEN established the first bryological herbarium in Nanjing. The specimens were moved to the herbarium (PE) of the Institute of Botany, Chinese Academy of Sciences, Beijing, in 1984.

1954年，秦仁昌，中国蕨类植物科属名词及分类系统，植物分类学报3（1）：93-99。

1954, Ren-Chang CHING, Systematic arrangements of families and genera of Chinese pteridophytes with corresponding names in Chinese, *Acta Phytotaxonomica Sinica* 3(1): 93-99.

1954年，秦仁昌，中国蕨类植物研究的发展概况，植物分类学报3（3）：257-272。

1954, Ren-Chang CHING, A review of the progress of our knowledge of Chinese pteridophytes, *Acta Phytotaxonomica Sinica* 3(3): 257-272.

1954年，傅书遐编著，**中国蕨类植物志属**，1-203页；中国科学院，北京。

1954, Shu-Hsia FU, '***Genera of Chinese Pteridophytes***', 203 p; Chinese Academy of Sciences, Beijing.

1954, Hui-Lin LI, *Davidia* as the Type of a New Family Davidiaceae, *Lloydia* 17(4): 329-331.

1954年，李惠林，珙桐属作为新科珙桐科的模式，*Lloydia* 17（4）：329-331。

1954, Genkei Masamune, ***A List of Vascular Plants of Taiwan***, 172+20 p; Plant Taxonomic Laboratory, Faculty of Science, Taipei 'National' University, Taipei [Taibei].

1954年，正宗严敬，**台湾植物目录**，172+20页；"国立"台湾大学理学院植物分类室，台北。

1954, Boris K. Schischkin (editor in chief), Andrej A. Fedorov and Moisey E. Kirpicznikov, ***Vademecum Methodi Systematis Plantarum Vascularium***—*abbreviations, designationes institutae, nomina geographica,*

410 吴继农，1992，陈邦杰在苔藓科学领域的开拓性研究，南京师大学报（社会科学版），3：39-43；吴鹏程，1998，中国科学院植物研究所苔藓标本室的足迹，植物杂志4：13-14。

109 p; Editio Academiae Scientiarum URSS, Mosqua and Leningrad. Translated into Chinese in 1958.

1958 年, 匡可任译, **高等植物分类学参考手册 ——** 缩写、符号、地名, 131 页; 科学出版社, 北京。

1954, Armen L. Takhtajan, ***Origins of Angiospermous Plants***, Proiskhozhdenie Pokrytosemennykh Rasteniĭ, 96 p; Nauka, Moska. Translated into Chinese in 1955.

1955 年, 朱澂、汪劲武译, 王伏雄校, 76 页; **被子植物的起源**, 科学出版社, 北京。

1954 年, 中国科学院编译局, **种子植物名称**, 160 页; 中国科学院, 北京。

1954, Compilation and Translation Bureau of Chinese Academy of Sciences, ***'Names of Seed Plants'***, 160 p; Chinese Academy of Sciences, Beijing.

1954—1957 年, 由中国科学院组织所属各有关研究所及院外有关大专院校共 34 个单位共同成立了由 231 名人员组成的中国科学院黄河中游水土保持综合考察队。考察队对甘肃、内蒙古、陕西、山西等地进行了自然资源综合考察, 共采集标本 3 万号 (PE、WUK)。

1954-1957, Water and Soil Conservation Comprehensive Expedition with 231 personnel from 34 institutions of the Chinese Academy of Sciences, local universities and colleges explored the middle reaches of the Huang He (Yellow River) in Gansu, Inner Mongolia, Shaanxi and Shanxi and collected more than 30,000 numbers of plant specimens (PE, WUK).

1954—1960 年, 中国科学院植物研究所接管南京总理陵园纪念植物园, 并与中国科学院植物研究所华东工作站合并, 更名为中国科学院南京中山植物园, 裴鉴任主任。

1954-1960, Botanical Garden Memorial Sun Yat-Sen was accepted by the Chinese Academy of Sciences and merged with the East China Station, Institute of Botany, Chinese Academy of Sciences, as the Nanjing Botanical Garden Memorial Sun Yat-Sen, Chinese Academy of Sciences, with Chien PEI as director.

1954—1965 年, 中国科学院将哈尔滨林业研究所筹备处、东北土壤研究所筹备处和长春综合研究所农产化学研究室合并, 在沈阳成立中国科学院林业土壤研究所, 同时设立植物标本室 (IFP), 朱济凡[411] (1912—1987) 任所长, 刘慎谔任副所长兼植物研究室主任。

1954-1965, Preparatory of Institute of Forestry of Harbin, Preparatory of Institute of Pedology and Agricultural and Chemical Products Laboratory of Changchun Research Institute, Chinese Academy of Sciences, were merged into the Institute of Forestry and Pedology, Chinese Academy of Sciences, and the herbarium (IFP) founded at the same time, Shenyang, Liaoning, with Chi-Fan CHU (1912-1987) as director and Tchen-Ngo LIOU as vice director and chairman of the department of botany.

411 浙江诸暨人; 1954 年受命组建中国科学院林业土壤研究所, 并担任所长兼党委书记。1981—1984 年调任南京林产工业学院 (后改为南京林学院) 党委书记。

1954—1968 年，中山大学农林植物研究所改隶于中国科学院，更名为中国科学院华南植物研究所，陈焕镛任所长(1954—1971)。原广西大学经济植物研究所更名为中国科学院华南植物研究所分所，陈焕镛兼任分所所长，钟济新任分所副所长。

1954-1968, The Botanical Institute, Sun Yat-Sen University, was transferred to the Chinese Academy of Sciences and named South China Institute of Botany, Chinese Academy of Sciences, with Woon-Young CHUN as director (1954-1971). The original Economic Institute of Botany, Guangxi University, became a branch, with Woon-Young CHUN concurrently as director of the branch, and Chi-Hsing CHUNG as vice director of the branch.

1955

1955 年 8 月，中国科学院植物研究所招收第一届研究生，植物分类学陈艺林入学，导师为林镕。

1955.08, The first class of graduate students was admitted to the Institute of Botany, Chinese Academy of Sciences, Yi-Ling CHEN, a student in plant taxonomy, advisor: Yong LING.

1955 年 9 月，中国科学院编译局，**孢子植物名称**，163 页；中国科学院，北京。

1955.09, Compilation and Translation Bureau, Chinese Academy of Sciences, '***Names of Cryptogamic Plants***', 163 p; Chinese Academy of Sciences, Beijing.

1955 年 9 月，王忠魁[412]（1922—2016）留学美国纽约州立大学森林学院，1958 年获硕士学位后返台并于东海大学任教，1962 年 8 月再至美国并于 1963 年 10 月获得雪城大学博士学位；1963—1966 年于美国德克萨斯州达拉斯任教于 Bishop College；1966 年 8 月返台继续在东海大学任教，教学与研究生涯前后长达 30 年；期间担任过东海大学生物系系主任以及东海大学理学院院长；1992 年 7 月从东海大学退休；1995 年移居美国亚利桑那州土桑市。

1955.09, Chung-Kuei WANG (1922-2016) went to the United States, studied at the College of Forestry, New York State University, Syracuse, New York, received a master's degree in 1958, then returned to teach at Tunghai Unviersity, returned to the College of Forestry at Syracuse University in August 1962 and obtained a Ph.D. degree in October 1963[413], as teacher in Bishop College, Dallas, Texas, between 1963-1966, then back to Tunghai University as professor in 1966 (including as dean of Department of Biology, and chairman of the School of Science), and retired in July 1992. He moved to Tucson, Arizona in 1995.

1955 年，陈邦杰在南京师范学院开办苔藓植物研究进修班，首届学员 9 人：江西师范学院龚明暄[414]（1923—1975）、复旦大学顾其敏（1932— ）、华东师范大学胡人亮（1932—2009）、东北师范学院郎奎昌（1927—1996）、中国科学院植物研究所黎兴江（1932—，1965 年 12 月调入中国科学院昆明植物研究所）、福建师范学院梁良弼（1928—）、山东大学仝治国[415]（1925—1979）、云南大学徐文宣

412 山东德州人，1945 年 7 月毕业于国立西北农学院森林系；后于青岛市农林事务所担任技士；1948 年 7 月任台湾省政府农林处嘉义山林管理所专员；1949 年 7 月转任台湾林业试验所担任技士；1955 年取得公费留学考植物学第一名，于 9 月奉准从林业试验所离职出国进修。

413 Ph.D. dissertation: Phytogeography of the Mosses of Formosa; Advisor: Edwin H. Ketchledge (1924-2010).

414 江西师范大学阳文静老师根据江西大学保存的个人档案提供；特此致谢。

415 河北定县人，1950 年毕业于山东大学水产系；1957 年调入内蒙古大学。

Pan-Chieh CHEN and his students of the first bryological training class, spring 1955; front row (left-right): Chi-Kuo TUNG, Pan-Chieh CHEN, and Kui-Chang LANG; middle row: Xing-Jiang LI, Ying-Hua ZHAO, Qi-Min GU, and Liang-Bi LIANG; back row: Wen-Xuan XU, Ren-Liang HU, and Ming-Xuan GONG) (Photo provided by Yu-Huan WU)

1955 年春陈邦杰在南京师范学院开办苔藓植物研究进修班，首届学员合影；前排（从左至右）：仝治国、陈邦杰、郎奎昌；中排：黎兴江、赵英华、顾其敏、梁良弼；后排：徐文宣、胡人亮、龚明暄（相片提供者：吴玉环）

（1919—1985）和南京大学赵英华[416]（1929—2011）。后期陆续招收进修生如下：1956 年中国科学院植物研究所吴鹏程（1935—）、1959 年贵阳师范学院钟本固（1929—）、1960 年中国科学院植物研究所罗健馨（1935—）、东北林学院敖志文（1932—2014）、1961 年中国科学院植物研究所郭木森（1940—）、中国科学院华南植物研究所林邦娟（1936—）、中国科学院西北生物土壤研究所张满祥（1934—）、1963 年中国科学院华南植物研究所林尤兴（1934—，“文化大革命”后调到北京）和中山大学李植华

416 辽宁盖县（今盖州市）人（女）；中国植物学文献目录第三和第四册的记载显示从事孢子植物研究，与 2011 年故去时南京大学生物系的讣告相符；但与陈邦杰弟子后期记载的赵蕴华不符；且赵蕴华在中国植物学文献中并没有记载。

Pan-Chie CHEN and his students, 1962; front row (left-right): Xing-Jiang LI, Tsung-Ling WAN, Pan-Chieh CHEN, and Man-Hsiang CHANG; back row: Jian-Xin LUO, Mu-Sen GUO, Pan-Cheng WU, Chien GAO, Ren-Liang HU, and Pang-Juan LIN (Photo provided by Zong-Gang HU)

1962 年陈邦杰与学生合影；前排（从左至右）：黎兴江、万宗玲、陈邦杰、张满祥；后排：罗健馨、郭木森、吴鹏程、高谦、胡人亮、林邦娟（相片提供者：胡宗刚）

Pan-Chie CHEN and his students, Nanjing, summer 1963; front row (left-right): Man-Hsiang CHANG, Tsung-Ling WAN, Pan-Chieh CHEN, and Jian-Xin LUO; middle row: Ren-Liang HU, Pang-Juan LIN, Chi-Hua LI, You-Xing LIN and Mu-Sen GUO; back row: Zhi-Ping WEI, Chi-Kuo TUNG, Huai-Lan QIN, and Pan-Cheng WU (Photo provided by Yu-Huan WU)

1963 年夏，陈邦杰与学生于南京合影；前排（从左至右）：张满祥、万宗玲、陈邦杰、罗健馨；中排：胡人亮、林邦娟、李植华、林尤兴、郭木森；后排：魏志平、仝治国、秦怀兰、吴鹏程（相片提供者：吴玉环）

Pan-Chieh CHEN in the field, 1957 (Photo provided by Pan-Cheng WU)
1957 年，陈邦杰野外采集（相片提供者：吴鹏程）

Pan-Chieh CHEN in his office in Nanjing, 1950-1960s (Photo provided by Pan-Cheng WU)
20 世纪 50–60 年代陈邦杰于南京办公室（相片提供者：吴鹏程）

（1935— ）。1957 年中国科学院林业土壤研究所高谦[417]（1929—2016）作为研究生入学，1962 年毕业。此外还有 1964 年作为研究生入学的丁恒山（江苏南通，1941— ）等。先后参加培训班的还有南京师范学院的助教臧穆[418]（1930—2011）、秦怀兰及中国科学院西北生物土壤研究所魏志平等。

1955, The first national bryological training class was held at Nanjing Normal College and taught by Pan-Chieh CHEN. Nine students from throughout China participated. They included: Ming-Xuan GONG (1923-1975), Jiangxi Normal College, Qi-Min GU (1932-), Fudan University, Ren-Liang HU (1932-2009), East China Normal University, Shin-Chiang LEE (Xing-Jiang LI, 1932-), Institute of Botany, Chinese Academy of Sciences (moved to Kunming Institute of Botany in December 1965), Kui-Chang LANG (1927-

417 祖籍河北献县，1929 年生于黑龙江宁安，1948—1952 年就读于东北大学（毕业于东北师范大学）生物系，1952—1956 年分别在沈阳科学仪器厂和武汉教学标本模型厂负责技术和质量工作，1956 年考取中国科学院林业土壤研究所刘慎谔和陈邦杰二位教授的副博士研究生，从事苔藓植物学研究，1962 年毕业后留所任植物研究室助理研究员，1979—1994 年任植物研究室任副研究员、研究员和研究室主任。著名苔藓植物学家。参见：Wei LI, 2017, Obituary: Chien Gao (1929—2016), *The Bryological Times* 144: 8-9。

418 山东烟台人，1949 年至 1953 年就读于东吴大学（后改为江苏师范学院）生物系，1954—1973 年于南京师范学院任教，1973 年 6 月调至中国科学院昆明植物研究所，历任该所副研究员、研究员。从事真菌系统学、生态地理学、外生菌根及其应用等领域的研究。创建了中国科学院昆明植物研究所隐花植物标本馆，曾任馆长、中国科学院真菌地衣开放实验室副主任、中国真菌学会副理事长。主编和编著数部，发表论文 150 余篇；擅长书画，喜欢集邮，爱好古董。参见黎兴江主编，2013，臧穆纪念册，277 页；昆明（内部印制），中国科学院昆明植物研究所臧穆研究员生平（http://www.kmb.cas.cn/djgz/qtgz/201111/t20111111_3394967.html，2017 年进入）。

1996), Northeast Normal College, Liang-Bi LIANG (1928-), Fujian Normal College, Chi-Kuo TUNG (1925-1979), Shandong University (transferred to Inner Mongolia University in 1957), Wen-Xuan XU (1919-1985), Yunnan University, and Ying-Hua ZHAO (1929-2011), Nanjing University. Students who joined the training class later included Pan-Cheng WU (Peng-Cheng WU, 1935-), Institute of Botany, Chinese Academy of Sciences in 1956; Ben-Gu ZHONG (1929-), Guiyang Normal College in 1959; Jian-Shing LUO (1935-), Institute of Botany, Chinese Academy of Sciences, and Chi-Wen AUR (1932-2014), Northeast Forestry College in 1960; Mu-Sen GUO (1940-), Institute of Botany, Chinese Academy of Sciences, Pang-Juan LIN (1936-), South China Institute of Botany, Chinese Academy of Sciences, and Man-Hsiang CHANG (1934-), Northwest Institute of Biology and Pedology, Chinese Academy of Sciences, in 1961; and You-Xing LIN (1934-), South China Institute of Botany, Chinese Academy of Sciences (transferred to Beijing after cultural revolution) and Chi-Hua LI (1935-), Sun Yat-Sen University, in 1963; Chien GAO (1929-2016), Institute of Forestry and Pedology, Chinese Academy of Sciences, was admitted by CHEN as a graduate student in 1957 and graduated in 1962. Heng-Shan DING (from Nantong, Jiangsu, 1941-), was also admitted by CHEN as a graduate student in 1964. In addition, Mu ZANG (1930-2011) and Huai-Lan QIN, assistant teachers, Nanjing Normal College, and Zhi-Ping WEI, Northwest Institute of Biology and Pedology, Chinese Academy of Sciences, also attended the training class.

1955 年，云南大学秦仁昌调任中国科学院植物研究所，任研究员兼分类室主任。

1955, Ren-Chang CHING, Yunnan University, transferred to Beijing to become research professor and director of department of plant taxonomy, Institute of Botany, Chinese Academy of Sciences.

1955 年，方文培著，**有花植物分类系统的比较**，75 页；四川大学出版社，成都。

1955, Wen-Pei FANG, '***Comparison of the Systems of Flowering Plants***', 75 p; Sichuan University, Chengdu.

1955, Hui-Lin LI, Classification and Phylogency of Nymphaeaceae and Allied Families, *The American Midland Naturalist* 54(1): 33-41, with two new families: Euryalaceae and Barclayaceae.

1954 年，李惠林，睡莲科及其近缘科的分类和系统发育，*The American Midland Naturalist* 54（1）: 33-41，两新科：芡实科和合瓣莲科。

1955 年，刘慎谔等，**东北木本植物图志**，568 页，图版 I—CLXX；科学出版社，北京。

1955, Tchen-Ngo LIOU et al., ***Illustrated Flora of Ligneous Plants of N. E. China***, 568 p, Plates I-CLXX; Science Press, Beijing.

1955 年，中国科学院选出首届学部委员，其中生物地学部委员 83 人，植物分类学家有吴征镒、秦仁昌、林镕、张肇骞、钱崇澍、戴芳澜、陈焕镛、郑万钧等。

1955, First group of academicians elected to the Chinese Academy of Sciences, and 83 members were from biology and geology departments, including plant taxonomists Chao-Chien CHANG, Wan-Chun CHENG, Sung-Shu CHIEN, Ren-Chang CHING, Woon-Young CHUN, Yong LING, Fang-Lan TAI and Cheng-Yih WU.

1955 年，中国科学院在陕西杨陵成立西北农业生物研究所，将中国科学院植物研究所西北工作站纳入其中，后该所改名为中国科学院西北生物土壤研究所。

1955, Northwest Agricultural Institute of Biology, Chinese Academy of Sciences, founded in Yangling, Shaanxi, with Northwest Station of Institute of Botany, Chinese Academy of Sciences, merged with it and renamed Northwest Institute of Biology and Pedology, Chinese Academy of Sciences.

1955 年，中国科学院植物研究所编辑[419]，**中国主要植物图说 —— 豆科**，726 页；科学出版社，北京。

1955, Institute of Botany, Chinese Academy of Sciences, '***Flora Illustralis Plantarum Primarum Sinicarum—Leguminosae***', 726 p; Science Press, Beijing.

1955 年，中国科学院植物研究所钟补求将其父钟观光教授采集的 5 000 多份标本和 5 000 多部中文古籍赠给中国科学院植物研究所。1957 年中国科学院院长郭沫若（1892—1978）委托植物研究所所长钱崇澍颁发奖状。[420]

1955, Pu-Chiu TSOONG, Institute of Botany, Chinese Academy of Sciences, son of Professor Kuan-Kuang TSOONG, donated his father's collection of over 5,000 plant specimens and more than 5,000 ancient and classic Chinese books to the Institute of Botany, Chinese Academy of Sciences. In 1957, he was issued an award by Mo-Ruo GUO (1892-1978), president of the Chinese Academy of Sciences, via Sung-Shu CHIEN, director of the Institute of Botany, Chinese Academy of Sciences.

1955—1956 年，中山大学、中国科学院华南植物研究所和广西分所组成广福林区调查队，由钟济新和何椿年[421]（1906—1968）领队，进入广西临桂考察，采集标本数千号（IBK），并发现裸子植物新属银杉属（松科）。[422]

1955-1956, Sun Yat-Sen University, South China Institute of Botany, Chinese Academy of Sciences and Guangxi Branch, sent Guangfu Forestry Exploring Team, led by Chi-Hsing CHUNG and Chun-Nien

419 具体负责人是汪发缵、唐进，参原书编辑例言第Ⅳ页。

420 详细参见：杨汉碧、金存礼、洪德元，1984，世界马先蒿权威——钟补求教授，植物杂志 5：41-42；胡宗刚，2011，北平研究院植物学研究所史略，61 页；上海交通大学出版社，上海。

421 广东南海人，1932 年毕业于国立中山大学生物系，1932—1938 年先后任国立中山大学生物系助教、广州商检局技士；1941—1947 年为国立中山大学农林植物研究所研究实习员、农场技士；1947 年后任讲师兼副研究员；1954 年起任中国科学院华南植物研究所副研究员；1956 年起负责筹建华南植物园。专长植物分类学，曾参加广州植物志的编写，并合作发表中国红树科志、中国无患子科志、华南五加科补志等论文。

422 傅立国，1998，林海珍珠放异彩——银杉发现与发表始末，植物 4：42-43。

獎　狀

鍾补求先生將其尊严鍾观光教授生前收藏的珍貴图书五千册及历年蒐集的奇異标本五千件捐献我院。这种热爱祖国科学事业的热情，为学术界树立了良好的典范。特予褒奖。

中国科学院院長　郭沫若

1957年1月21日

Pu-Chiu TSOONG donates his father's collection of plant specimens and ancient and classic Chinese books to the Institute of Botany, Chinese Academy of Sciences, 1955. He was issued an award by Mo-Ruo GUO, via Sung-Shu CHIEN, 1957 (Photo provided by Institute of Botany)

1955 年，钟补求将其父钟观光教授采集的标本和中文古籍赠给中国科学院植物研究所。1957 年郭沫若委托钱崇澍颁发奖状（相片提供者：植物研究所）

HO (1906-1968), to Lingui, Guangxi. Thousands of plant specimens were collected (IBK), and *Cathaya* (Pinaceae), a new genus of Gymnosperms, was discovered.

1955-1957, Joint Chinese and Soviet Yunnan Tropical and Subtropical exploration team, from Institute of Botang, Chinese Academy of Sciences, and Botanical Institute, Soviet Academy of Sciences, including botanists Chieh CHEN (1928-2011), Ling-Zhi CHEN (1933-2015), Tai-Ping CHU (1930-), Andrej A. Fedorov (Андрей Александрович Фёдоров; 1908-1987), Kuo-Mei FENG, Moisey E. Kirpicznikov (Моисей Э́льевич Кирпичников; 1913-1995), Heng LI (1929-), Hsi-Wen LI (1931-), Yan-Hui LI (1930-), Igor A. Linczevsky (Игорь Александрович Линчевский; 1908-1997), Ping-I MAO (1925-), Hse-Tao TSAI, Wen-Tsai WANG, Cheng-Yih WU, Shu-Kung WU (Su-Gong WU, 1935-2013), Te-Lin WU (1934-) and Wen-Qing YIN (1927-), surveyed in southern Yunnan to locate suitable areas for rubber cultivation. They collected over 10,000 numbers of plant specimens (KUN, PE) in southern and southeastern Yunnan and on Emei Shan, Sichuan. *Tetrameles* (Tetramelaceae) and *Crypteronia* (Crypteroniaceae) were recorded from China.

1955—1957 年，中国科学院植物研究所和苏联科学院植物研究所组成中苏联合云南热带和亚热带考察队，成员包括植物学家陈介（1928—2011）、陈灵芝（1933—2015）、朱太平（1930—）、Andrej A. Fedorov（1908—1987）、冯国楣、Moisey E. Kirpicznikov（1913—1995）、李恒（1929—）、李

Andrej A. Fedorov (1908-1987) (Photo provided by Komarov Botanical Institute)

Andrej A. Fedorov（1908—1987）相片提供者：（科马洛夫植物研究所）

Moisey E. Kirpicznikov (1913-1995) (Photo provided by Komarov Botanical Institute)

Moisey E. Kirpicznikov（1913—1995）（相片提供者：科马洛夫植物研究所）

锡文(1931—)、李延辉(1930—)、Igor A. Linczevsky (1908—1997)、毛品一(1925—)、蔡希陶、王文采、吴征镒、武素功[423] (1935—2013)、吴德邻(1934—) 和尹文清(1927—) 等。考察队调查了云南南部橡胶宜林地，在云南南部、东南部以及四川峨眉山采集植物标本约1万号(KUN, PE)，并首次记录国产四数木科四数木属和隐翼科隐翼属。

Igor A. Linczevsky (1908–1997) (Photo provided by Komarov Botanical Institute)
Igor A. Linczevsky (1908—1997)(相片提供者：科马洛夫植物研究所)

1955、1958年，胡先骕著，**植物分类学简编**，430页，1955；高等教育出版社，北京；胡先骕著，**植物分类学简编**(修改本)[424]，454页，1958；科学技术出版社，上海。

1955 and 1958, Hsen-Hsu HU, '***Abridged Compilation of Plant Taxonomy***', 430 p, 1955; Higher Education Press, Beijing; Hsen-Hsu HU, '***Abridged Compilation of Plant Taxonomy***', revised ed., 454 p, 1958; Science and Technology Press, Shanghai.

1955—1958年，中国科学院植物研究所组织的中国科学院甘青综合考察队考察甘肃和青海，李沛琼(1934—)、李世英(1921—1985)、钟补求以及来自兰州大学生物系的师生，采到植物标本3 000号(PE)。

1955-1958, Gan-Qing Expedition, Chinese Academy of Sciences, organized by the Institute of Botany, Chinese Academy of Sciences, explored in Gansu and Qinghai. Pei-Chun LI (1934-), Shi-Ying LI (1921-1985), Pu-Chiu TSOONG and teachers and students from the department of biology, Lanzhou University, collected more than 3,000 numbers of plant specimens (PE).

---

423 山西太古人；中国科学院昆明植物研究所研究员，蕨类植物学家，著名青藏高原考察组织者和领导者。自幼失去双亲；中学未毕业便于1951年参军，1955年后入中国科学院植物研究所任见习员，1956年开始先后参加中苏云南生物资源考察，并先后领导了南水北调(藏东南—滇西北)、青藏高原、横断山、喀喇昆仑和昆仑山、可可西里等大规模综合科学考察活动，采集植物标本数万号。详细参见：方瑞征，2014，跋涉于"世界屋脊"的层峦榛莽间——记武素功先生对青藏高原植物资源的考察，科学66(4)：54-56；吕春朝，2014，悠长的思念、深切的缅怀——悼念武素功老师，*Sinopteris* 中国蕨20：7-8。

424 该书第1版，特别是书中关于评论李森科的内容，导致作者遭受批判，该书被禁止发行；1958年，经作者修改后，该书重新出版修改本。

1956

1956 年 9 月 13 日，中国科学院在湖北武汉成立武汉植物园筹备处；1956—1958 年章文才[425]（1904—1998）任主任。

1956.09.13, Preparatory of Wuhan Botanical Garden, Chinese Academy of Sciences, Wuhan, Hubei, was established; with director Wen-Tai CHANG (1904-1998), 1956-1958.

1956 年，陈焕镛，纪念植物学家梅尔博士，科学通报 12：73、33 页。

1956, Woon-Young CHUN, 'In memory of botanist Dr. Elmer D. Merrill', *Chinese Science Bulletin* 12: 73 and 33.

1956, Joseph Lanjouw, Charles Baehni, Walter Robyns, Reed C. Rollins, Robert Ross, Jacques Rousseau, Georg M. Schulze, Albert C. Smith, Roger de Vilmorin and Franz A. Stafleu, ***International Code of Botanical Nomenclature*** **(Paris Code)**, adopted by the Eighth International Botanical Congress, Paris, July 1954, *Regnum Vegetabile* 8:1-338 p. Translated into Chinese in 1961.

1961 年，刘慎谔，**国际植物命名法规**，76 页；中国科学院林业土壤研究所，沈阳（蓝色油印本，内部印制）。[426]

1956, Armen L. Takhtajan, ***Telomophyta I*** Psilophytales-Coniferales, 488 p; Editio Academiae Scientiarum URSS, Mosqua and Leningrad. Translated into Chinese in 1963.

1963 年，匡可任等，**高等植物**（I），486 页；科学出版社，北京。

1956 年 9 月—1959 年 3 月，中国科学院植物研究所派遣首批赴苏联留学人员 7 人，包括植物分

---

425 浙江杭州人，1922 年考入之江大学，1923 年转入金陵大学农学院，1927 年毕业后留校任教；1929 年与许复七、吴耕民、胡昌炽等人发起成立中国园艺学会；1931 年在福建厦门集美农林专科学校担任果树教员兼校长；1933 年任国立浙江大学农学院园艺系讲师兼湘湖实验农场场长；1935 年 4 月赴美留学，同年 7 月进入英国伦敦大学研究院攻读博士学位；1937 年获得博士学位后赴美国康奈尔大学和加州大学研究柑橘；1938 年回国，先后任职多所大学；1950 年任武汉大学园艺系主任；1954 年后任华中农业大学园艺系主任和华中农业大学柑橘研究室主任；著名果树学家、园艺教育家、柑橘专家、中国柑橘学科奠基人之一。参见：中国柑橘学会、湖北省园艺学会、华中农业大学园艺林学学院编，2005，章文才先生诞辰百年（1904—2004）纪念文集，132 页；中国林业出版社，北京。

426 刘慎谔文集（1985）中的著作目录记载为刘慎谔与曹新孙合译。另，西北农学院讲学印刷稿 40 页（双开），西北农学院印刷厂（油印本）。

类学者戴伦凯（1930—）和郑斯绪[427]（1931—1967）等。

1956.09-1959.03, Institute of Botany, Chinese Academy of Sciences, sent first group of seven students, including plant taxonomists Sze-Hsue CHENG (1931-1967) and Lun-Kai DAI (1930-), to USSR for further study.

1956—1959 年，中国科学院组织所属研究所及大专院校、新疆有关部门 90 余人成立新疆综合科学考察队，简焯坡（1916—2003）为考察队副队长。考察队分为 5 个专业组，其中植物组组长为秦仁昌，成员有胡式之、李安仁（1927—）、李世英、刘国钧（1933—2004）、毛祖美（1933—）、王义凤、张佃民等。考察队用 4 年时间先后考察了阿勒泰地区、天山南北坡及昆仑山北坡，行程近 2 万千米，共采集标本 24 000 余号（PE，XJBI）。

1956-1959, Xinjiang Comprehensive Expedition from institutions and colleges and universities organized by Chinese Academy of Sciences, with Cho-Po TSIEN (1916-2003) as vice leader. The botanical team, one of five teams, included Ren-Chang CHING (leader), Shi-Zhi HU, An-Jen LI (1927-), Shi-Ying LI, Guo-Jun LIU (1933-2004), Zu-Mei MAO (1933-), Yi-Feng WANG, Dian-Min ZHANG, et al. They spent about four years in the Altai Shan, Tian Shan and northern Kunlun areas, traveling nearly 20,000 km and collecting more than 24,000 numbers of plant specimens (PE, XJBI).

1956—1960 年，中国科学院与苏联科学院联合组织黑龙江流域综合考察队，考察大兴安岭、小兴安岭和长白山林区，中方参加人员有冯宗炜[428]（1932—2016）、高宪斌、贾成章[429]（1894—1970）、刘慎谔（领队）、邵钧、王战[430]（1910—2000）、朱济凡等。

1956-1960, Sino-Soviet expedition from the Chinese Academy of Sciences and the Soviet Academy of Sciences explored forest areas in the Heilongjiang valley, included the Greater and Lesser Khingan [Xingan] regions and Changbai Shan. Chinese scientists, including Zong-Wei FENG (1932-2016), Xian-Bin GAO, Cheng-Zhang JIA (1894-1970), Tchen-Ngo LIOU (leader), Jun SHAO, Chan WANG[431] (1910-2000), Ji-Fan ZHU et al.

427 江苏徐州人，郑万钧长子，1953 年 9 月于南京大学生物系毕业后分配到中国科学院植物研究所；回国后先后担任植物研究所分类室党支部书记、室副主任等职。

428 浙江嘉兴人，1954 年毕业于南京林学院，1957—1958 年在苏联科学院进修，森林生态学和环境生态学家，1999 年当选中国工程院院士。曾任中国科学院林业土壤研究所、中国科学院生态环境研究中心副所长，中国科学院生态环境研究中心研究员，生态学报主编等职。

429 字佛生。安徽合肥人。著名林学家，中国近代林业开拓者之一。

430 辽宁东沟（安东）人，1932 年入北平大学农学院森林系，1936 年毕业后留校任教；1938 年随校迁入陕西武功，先后任教于国立西北联合大学、国立西北农学院；1943 年转入重庆中央林业试验所任技正；1945 年任教于东北大学、沈阳农学院、东北农学院；1954 年后入中国科学院林业土壤研究所任副研究员、室主任、副所长。著名森林生态学家和植物分类学家。参见：Guo-Fan SHAO, Qi-Jing LIU, Hong QIAN, Ji-Quan CHEN, Jin-Shuang MA, Zheng-Xiang TAN, 2000, Zhan Wang (1911-2000), *Taxon* 49(3): 593-601.

431 Guo-Fan SHAO, Qi-Jing LIU, Hong QIAN, Ji-Quan CHEN, Jin-Shuang MA and Zheng-Xiang TAN, 2000, Zhan Wang (1911-2000), *Taxon* 49(3): 593-601.

1957

1957年5月9日至17日，秦仁昌及侯学煜[432]（1912—1991）赴苏联参加全苏植物学第二届代表大会。

1957.05.09-05.17, Ren-Chang CHING and Hsueh-Yu HOU (1912-1991) attended second All USSR Botanical Congress.

1957年5月31日，中国科学院学部在1955年基础上增选学部委员，胡先骕再次落选。[433]

1957.05.31, Chinese Academy of Sciences re-elected academicians based on elections of 1955, but Hsen-Hsu HU was again unsuccessful.

1957年，傅书遐编著，**中国主要植物图说——蕨类植物门**，280页；科学出版社，北京。

1957, Shu-Hsia FU, '***Flora Illustralis Plantarum Primarum Sinicarum—Pteridophytes***', 280 p; Science Press, Beijing.

1957年，耿煊[434]（1923—2009）赴美国留学，1959年获加州大学伯克利分校博士，并成为宾夕法尼亚大学莫里斯树木园研究人员；1960年赴马来西亚马来亚大学担任植物学讲师；1966年任高级讲师；1984年以新加坡国立大学副教授身份退休。

1957, Hsuan KENG (1923-2009) went to the United States of America to study at the University of California, Berkeley, California; received Ph. D. degree in 1959[435], then became a research scientist at Morris Arboretum, University of Pennsylvania, then to the University of Malaya, Malaysia, as lecturer in 1960, then senior lecturer in 1966, retired as associate professor of the National University of Singapore in 1984.

1957年，乐天宇、徐纬英著，钱崇澍、陈嵘校，**陕甘宁盆地植物志**，274页；中国林业出版社，北京。

1957, Tien-Yu LO and Wei-Ying HSU (eds.), Sung-Shu CHIEN and Yung CHEN (proofreaders), '***Basin Flora of Shaanxi-Gansu-Ningxia***', 274 p; China Forestry Publishing House, Beijing.

---

432 安徽和县人。1937年毕业于国立中央大学农业化学系；1945—1949年留学美国宾夕法尼亚州立大学研究院，获硕士和博士学位。中国科学院植物研究所研究员。长期从事植物学、植被制图、植物生态学等研究工作。1980年当选为中国科学院学部委员。

433 胡宗刚，2005，胡先骕落选学部委员考，自然辩证法通讯 27（5）：67-72；郭金海，2011，1957年中国科学院学部委员的增聘，中国科技史杂志 32（4）：501-521。

434 江苏泰县（今泰州市姜堰区）人，国立中正大学1947年农学学士，后随李惠林赴台湾并于1949年在台湾大学植物研究所取得硕士学位；1950—1957年间担任台湾大学植物系助教及植物学讲师。

435 Ph. D. Dissertation: Morphological Studies of Theaceace; Advisor: Lincoln Costance.

1957、1958、1965 年，吴征镒、王文采，云南热带亚热带地区植物区系研究的初步报告（Ⅰ），植物分类学报 6（2）：183-254，1957；云南热带亚热带地区植物区系研究的初步报告（Ⅰ）（续），植物分类学报 6（3）：267-300，1957；吴征镒、王文采，关于“云南热带亚热带地区植物区系研究的初步报告（Ⅰ）”的一些订正，植物分类学报 7（2）：193-196，1958；吴征镒、李锡文，**云南热带亚热带植物区系研究报告**，第 1 集：146 页，1965；科学出版社，北京。

1957, 1958 and 1965, Cheng-Yih WU and Wen-Tsai WANG, ‘Preliminary reports on studies of the plants of tropical and subtropical regions of Yunnan’, *Acta Phytotaxonomica Sinica* 6(2): 183-254, 1957; ‘Preliminary reports on studies of the plants of tropical and subtropical regions of Yunnan (continued)’, *Acta Phytotaxonomica Sinica* 6(3): 267-300, 1957; ‘Corrections for the preliminary reports on studies of the plants of tropical and subtropical regions of Yunnan’, *Acta Phytotaxonomica Sinica* 7(2): 193-196, 1958; Cheng-Yih WU and Hsi-Wen LI, ‘***Reports on studies of the plants of tropical and subtropical regions of Yunnan***’, 1: 146 p, 1965; Science Press, Beijing.

1958

1958 年 1 月，竹内亮，中国东北裸子植物研究资料，149 页，中国林业出版社，北京。

1958.01, Makoto Takenouchi, '***Research Materials of Gymnosperms of Northeast China***', 149 p; China Forestry Publishing House, Beijing.

1958 年春，中国科学院华南植物研究所陈焕镛和黄观程[436]（1930— ）赴苏联考察 3 个月，其间一个半月在列宁格勒的科马洛夫植物研究所研究，另一个半月参观访问苏联各类植物学机构。

Woon-Young CHUN and Guan-Cheng HUANG in Leningrad Botanical Garden, spring 1958 (left-right): Andrei L. Kursanov (Андрей Львович Курсанов; 1902-1999), academician and director of Plant Physiology Institute, Evgenii M. Lavrenko (Евгений Михайлович Лавренко; 1900–1987), academician, ecologist, Woon-Young CHUN, Pavel A. Balanov (Павел Александрович Баланов; 1892–1962), academician and director of Komarov Botanical Institute, and Guan-Cheng HUANG (Photo provided by Guan-Cheng HUANG)
1958 年春，陈焕镛和黄观程在苏联列宁格勒植物园（从左至右）：植物生理研究所所长库尔萨诺夫院士、生态学家拉夫连科院士、陈焕镛、柯马罗夫植物所所长巴拉诺夫院士、黄观程）（相片提供者：黄观程）

436 印度尼西亚归国华侨，二十世纪五十年代毕业于北京外国语学院俄语专业后一直在中国科学院华南植物研究所（园）从事翻译、情报、编辑以及信息等工作；1955—1957 年作为翻译参加中苏联合云南热带和亚热带考察。此次赴苏联为陈焕镛的专职秘书兼翻译。

Spring 1958, Woon-Young CHUN and Guan-Cheng HUANG (1930-), South China Institute of Botany, Chinese Academy of Sciences, went to USSR for three months, studied for one and a half months at Komarov Botanical Institute, Leningrad [Saint Petersburg], then visited various botanical institutions in USSR for another one and a half months.

1958年8月，中国科学院和苏联科学院合作，苏联科学家塔赫他间（1910—2009）访华，历时3个月，访问城市包括北京、广州等。

1958.08, Through collaboration between the Chinese and Soviet academies of sciences, Armen L. Takhtajan[437] (Армен Леонович Тахтаджян, 1910-2009) visited China (including Beijing and Guangzhou) for three months.

1958年10月，中国科学院西北生物土壤研究所植物组编，**黄河中游黄土区植物名录**，110页；西安第一印刷厂，西安（内部印制）。

1958.10, Plant Group of Northwest Institute of Biology and Pedology, Chinese Academy of Sciences, '***Plant Checklist of the Loess Region in the Middle Yellow Valley***', 110 p; Xi'an First Press, Xi'an (internal publication).

1958年10月，中国科学院植物研究所、北京大学生物系，**河北原产与习见栽培植物名录**，134页；中国科学院植物研究所、北京大学生物系，北京（内部印制）。

1958.10, Institute of Botany, Chinese Academy of Sciences, Department of Biology, Peking University, ***A List of Native and Common Cultivated Plants from Hopei***, 134 p; Institute of Botany, Chinese Academy of Sciences, Department of Biology, Peking University (internal publication).

1958年11月，中国科学院武汉植物园经过两年的筹备，正式于湖北武汉武昌磨山建立；1958—1962年陈封怀任主任。

1958.11, Wuhan Botanical Garden, Chinese Academy of Sciences, founded after two years of preparation, in Moshan, Wuchang, Wuhan, Hubei, with Feng-Hwai CHEN as director, 1958-1962.

1958年12月，西双版纳热带植物园在云南勐腊县勐仑镇由罗梭江环绕而成的葫芦岛上正式建立，隶属中国科学院昆明植物研究所。1959—1970年蔡希陶任主任。

437 Armen Leonovich Takhtajan or Takhtajian (Surname also transliterated as Takhtadjan, Takhtadzhian or Takhtadzhian, June 10, 1910, Shusha, Armenia-November 13, 2009, Saint Petersburg) was a Soviet-Armenian botanist and one of the most important figures in 20th century studies of plant evolution, systematics and biogeography. His other interests included morphology of flowering plants, paleobotany and the flora of the Caucasus. See Robert Thorne, 2010, Armen Leonovich Takhtajan (1910-2009), *Taxon* 59(1): 317.

1958.12, Xishuangbanna Tropical Botanical Garden was established on Hu Lu Dao (Gourd Island) in the Luosuo River, Menglun, Mengla, Yunnan, under the Kunming Institute of Botany, Chinese Academy of Sciences, with Hse-Tao TSAI named director, 1959–1970.

1958年，甘肃兰州西北师范学院在植物标本室（NWTC）基础上成立植物分类研究室，由孔宪武主持，时为中国高校系统中唯一的植物分类学专业研究室；1980年更名为植物研究所，即今日西北师范大学生命科学学院植物研究所。

1958, Research Laboratory of Plant Taxonomy was established in Northwest Normal College, Lanzhou, Gansu, based on the herbarium (NWTC), under leadership of Hsien-Wu KUNG (1897–1984). It was the only Research laboratory of plant taxonomy within the colleges and universities of China. It was renamed Institute of Botany in 1980 and is today's Institute of Botany, College of Life Sciences of Northwest Normal University.

1958年，崔鸿宾、汤彦承，十年内完成中国植物志，科学通报3（10）：296–297。

1958, Hung-Pin TSUI and Yen-Cheng TANG, 'To accomplish *Flora Reipublicae Popularis Sinicae* in ten years', *Chinese Science Bulletin* 3(10): 296–297.

1958, Albert N. Steward, ***Manual of Vascular Plants of the Lower Yangtze Valley, China***, 621 p; Oregon State College, Corvallis, Oregon.

1958年，Albert N. Steward，**中国扬子江下游维管束植物手册**，621页；俄勒冈州立学院，Corvallis, 俄勒冈。

1958年，中国科学院新疆分院成立生物、土壤和地理研究室，并建立植物标本室（XJIB）；1958—1961年刘国钧任植物标本室主任。

1958, Departments of Biology, Pedology and Geography and the herbarium (XJIB) founded under Xinjiang Branch, Chinese Academy of Sciences, with Guo-Jun LIU as director of the Herbarium, 1958–1961.

1958年，中国科学院编译出版委员会名词室编订，**孢子植物名词及名称**，149页；科学出版社，北京。

1958, Names Department of Compilation and Translation Publishing Committee, Chinese Academy of Sciences, '***Terms and Names of Cryptogamic Plants***', 149 p; Science Press, Beijing.

1958年，为满足编写中国植物志的要求，中国科学院植物研究所开办植物科学画训练班，学员

20 多人。他们后来成为中国植物志的绘图骨干。[438]

1958, To meet the demand for illustrations of *Flora Reipublicae Popularis Sinicae*, the Institute of Botany, Chinese Academy of Sciences, organized a plant illustration class of more than 20 students. It became a major force for completing the illustrations for *Flora Reipublicae Popularis Sinicae*.

Plant illustrators at the Institute of Botany, Chinese Academy of Sciences, Beijing (top to bottom and left to right): 1. Li-Sheng WANG, Wei-Jiang JU, Chao-Zhen JI, Rong-Sheng ZHANG, Feng-Xiang WANG and Bao-Heng ZHAO; 2. Shu-Qin CAI, Tai-Li ZHANG, Mu-Sen GUO and Qi-Ming YANG; 3. Gui-Zhen JI, Ting-Rui GUAN, Jing-Mian LIU, Jin-Feng WANG and Ji-Qing LIU; 4. Zhang-Hua WU, Chun-Rong LIU, Rong-Hou ZHANG and Jin-Yong FENG (Photo provided by Ai-Li LI)

中国科学院植物研究所绘图组全体合影（从上至下、从左至右），第 1 排：王利生，鞠卫江，冀朝祯，张荣生，王凤翔，赵宝恒；第 2 排：蔡淑琴，张泰利，郭木森，杨启明；第 3 排：冀桂珍，关庭瑞，刘敬勉，王金凤，刘济清；第 4 排：吴樟华，刘春荣，张荣厚，冯晋庸（相片提供者：李爱莉）

1958、1982、1991 年，侯宽昭编，**中国种子植物科属辞典**，553 页，1958；科学出版社，北京。吴德邻、高蕴璋、陈德昭等修订，**中国种子植物科属词典（修订版）**[439]，632 页，1982；科学出版社，北京。

438 冯晋庸、蒋祖德，1994，中国植物科学画史，中国植物学史，第 366-373 页；科学出版社，北京。

439 本书的中文名称两个版本不完全相同（**辞典与词典**）：第 1 版为中国种子植物辞典，而第 2 版为中国种子植物词典。

吴德邻、高蕴璋、陈德昭等修订，**中国种子植物科属词典（繁体版）**，628 页，1991；南天书局，台北。

1958, 1982, 1991, Foon-Chew HOW, '***A Dictionary of the Families and Genera of Chinese Seed Plants***', 553 p, 1958; Science Press, Beijing; Te-Lin WU, Wan-Cheung KO, Te-Chao CHEN[440] et al., ***A Dictionary of the Families and Genera of Chinese Seed Plants***, ed. 2, 632 p, 1982; Science Press, Beijing; Te-Lin WU, Wan-Cheung KO, Te-Chao CHEN et al., ***A Dictionary of the Families and Genera of Chinese Seed Plants***, ed. 2, Traditional Chinese Edition, 628 p, 1991; Southern Media Publishing Co., Taipei [Taibei].

1958-1959, Sino-East German team explored in Guangdong, Guangxi and Hainan, joined by Chu-Hao WANG (1923-2008, IBSC) et al.; 2,100 numbers of plant specimens were collected (IBSC).

1958—1959 年，中国科学院和德意志民主共和国联合考察广东、广西和海南，中国科学院华南植物研究所王铸豪（1923—2008）等参加，采集植物标本 2 100 号（IBSC）。[441]

1958 年 10 月至 1960 年 10 月，中国科学院植物研究所汤彦承[442]（1926—2016）赴苏联科学院科马洛夫植物研究所研修，归国后继续在中国科学院植物研究所从事植物分类研究。[443]

1958.10-1960.10, Yen-Cheng TANG[444] (1926-2016), Institute of Botany, Chinese Academy of Sciences, studied at the Komarov Botanical Institute, Soviet Academy of Sciences, resumed research on plant taxonomy at the Institute of Botany, Chinese Academy of Sciences, after returning to China.

1958—1961 年，中国科学院华南植物研究所广西分所更名为中国科学院广西植物研究所，陈焕镛兼任所长。

1958-1961, Guangxi Branch, South China Institute of Botany, Chinese Academy of Sciences, renamed Guangxi Institute of Botany, Chinese Academy of Sciences, with Woon-Young CHUN concurrent as director.

1958 and 1962, Woon-Young CHUN and Ko-Zen KUANG (Ko-Jen KUANG), A new genus of Pinaceae—*Cathaya* Chun et Kuang, gen. nov. from the southern and western China, *Botanicheskii Zhurnal SSSR* 43(4): 461-470, 1958; Woon-Young CHUN and Ko-Zen KUANG, De genere *Cathaya* Chun et Kuang, *Acta Botanica Sinica* 10(3): 245-246, 1962.[445]

440 As Te-Choa CHEN in origin.

441 胡启明、曾飞燕，2011，广东植物志，10：327 页；广东科技出版社，广州（附录 2，华南植物研究所（园）植物标本采集简史）。

442 浙江萧山人，1950 年清华大学毕业后到中国科学院植物分类研究所工作；1972—1981 年，曾任植物分类学和植物地理学研究室主任。

443 期间拍摄植物标本照片 678 张，云南大学植物标本室有收藏；署名为汤彦承、郑斯绪。

444 Also as Yan-Cheng TANG or Yan-Chen TANG.

445 The scientific name in the original description by Chun and Kuang in *Botanicheskii Zhurnal* (Moscow and Leningrad) 43: 464. 1958, was invalid since without designation of a type. See Alexander B. Doweld, 2016, (2420) Proposal to conserve the name *Cathaya* Chun & Kuang against *Cathaya* Karav. (*Gymnospermae: Pinales*), *Taxon* 65(1):187-188.

1958、1962 年，陈焕镛、匡可任，中国松科一新属 —— 银杉，*Botanicheskii Zhurnal SSSR* 43（4）：461-470，1958；银杉 —— 我国特产的松柏类植物，植物学报 10（3）：245-246，1962。

1958—1962 年，中国科学院于四川成都建立农业生物研究所，同时成立植物标本室（CDBI）；黄国英（1904—1987）任所长。

1958-1962, Chengdu Institute of Agricultural Biology (and herbarium CDBI), Chinese Academy of Sciences, established in Chengdu, Sichuan, with Kuo-Ying HUANG (1904-1987) as director.

1958—1962 年，中国科学院庐山植物园更名为江西省庐山植物研究所；1958—1960 年温成胜（?—1994）、1960—1962 年刘昌标（1912—？）任所长。

1958-1962, Lushan Botanical Garden, Chinese Academy of Sciences, renamed Jiangxi Lushan Institute of Botany, with Cheng-Sheng WEN (?-1994), 1958-1960 and Chang-Biao LIU (1912-?) 1960-1962, as directors.

1958—1962 年，中国科学院植物研究所昆明工作站更名为中国科学院昆明植物研究所，吴征镒任所长。

1958-1962, Kunming Station of Institute of Botany, Chinese Academy of Sciences, renamed Kunming Institute of Botany, Chinese Academy of Sciences, with Cheng-Yih WU as director.

1958—2005 年，中国科学院林业土壤研究所 / 辽宁省林业土壤研究所 / 中国科学院沈阳应用生态研究所编，刘慎谔主编，**东北草本植物志**，12 卷本；第 1 卷：75 页，1958；第 2 卷：120 页，1959；第 3 卷：242 页，1975；第 4 卷：239 页，1980；第 5 卷：187 页，1976；第 6 卷：308 页，1977；第 7 卷：267 页，1981；第 8 卷：246 页，2005[446]；第 9 卷：447 页，2004[447]；第 10 卷：329 页，2004[448]；第 11 卷：220 页，1976；第 12 卷：292 页，1998[449]；科学出版社，北京。

1958-2005, Institutum Sylviculiturae et Pedologiae Academiae Scientiarum Sinicae / Liaoning Institute of Forestry and Pedology / Institute of Applied Ecology, Chinese Academy of Sciences, Tchen-Ngo LIOU (editor in chief), ***Flora Plantarum Herbacearum Chinae Boreali-Orientalis***, vols. 1-12: 1: 75 p, 1958; 2: 120 p, 1959; 3: 242 p, 1975; 4: 239 p, 1980; 5: 187 p, 1976; 6: 308 p, 1977; 7: 267 p, 1981; 8: 246 p, 2005; 9: 447 p, 2004; 10: 329 p, 2004; 11: 220 p, 1976; 12: 292 p, 1998; Science Press, Beijing.

446 李书心、刘淑珍、曹伟编著。

447 李冀云主编。

448 秦忠时主编。

449 傅沛云主编。

1959

1959年3月，中国科学院植物研究所，**中国植物照片集**[450]，第1册：1–818页，第2册：819–1 750页；学名索引：1–18页；中国科学院植物研究所，北京（内部印制）。[451]

1959.03, The Institute of Botany, Chinese Academy of Sciences, '***The Photographs of Chinese Plants***', Parts I and II: 1–1,750 p, Beijing (internal publication).[452]

1959年11月，中国科学院青海分院生物研究所筹备处在青海西宁成立。

1959.11, Preparatory of Institute of Biology, Qinghai Branch of Chinese Academy of Sciences, Xining, Qinghai, established.

1959年11月13日，中国植物志编辑委员会第一届编辑委员会成立[453]，主编：中国科学院植物研究所钱崇澍、中国科学院华南植物研究所陈焕镛；秘书：中国科学院植物研究所秦仁昌；编辑委员会21人：中国科学院武汉植物研究所陈封怀、张肇骞，中国林业科学研究院林业研究所陈嵘、郑万钧，四川大学方文培，南京大学耿以礼，中国科学院植物研究所胡先骕、简焯坡、姜纪五（1903—1975）、匡可任[454]（1914—1977）、林镕、秦仁昌、唐进、汪发缵、俞德浚、钟补求，华南农学院蒋英，西北师范学院孔宪武，中国科学院林业土壤研究所刘慎谔，中国科学院南京中山植物园裴鉴，中国科学院昆明植物研究所吴征镒。

1959.11.13, First Editorial Committee of Flora Reipublicae Popularis Sinicae founded, with Sung-Shu CHIEN (PE) and Woon-Young CHUN (IBSC) as editors in chief, and Ren-Chang CHING (PE) as secretary; twenty-one editorial committee members: Chao-Chien CHANG (IBSC), Feng-Hwai CHEN (IBSC), Yung CHEN (CAF), Wan-Chun CHENG (CAF), Ren-Chang CHING (PE), Wen-Pei FANG (SZ), Hsen-Hsu HU

---

450 原书说明："中国植物照片集第一批共二巨册，约计7 000种，系从中国科学院植物研究所收藏的大量照片中选出的，包括裸子植物和被子植物，其中大部分为模式标本，也有一些是正确鉴定的标本。"

451 陈家瑞，1994，中国植物学会编，中国植物学史，第161页；中国植物学会五十年编写组，1985，中国植物学会五十年，中国科技史料6（2）：50–55。这些资料记载，除了秦仁昌从欧洲，还有方文培从哈佛、蒋英和傅书遐从台湾拍摄过类似的中国植物相片；但没有提及具体照片的数量以及存放地等。

452 With about 7,000 photos in total.

453 本书所载中国植物志第一至第七届编辑委员会信息均来自中国植物志网址（http://frps.iplant.cn/bwh，2019年进入）。

454 江苏宜兴人，1934年毕业于宜兴高级农林学校，1935年赴日本留学，1937年从日本北海道帝国大学农学部回国参加教师战地贵州服务团，1941年起先后任职于云南农林植物研究所、国立中国医药研究所、北平研究院植物学研究所、中国科学院植物分类研究所和中国科学院植物研究所，1952年任副研究员，1956年晋升为研究员；精通植物拉丁文，编译数部有关植物分类学专著，并与陈焕镛共同发表活化石银杉属。参见：路安民，1987，忆匡可任教授，植物杂志1：38–39。

First editorial committee meeting of *Flora Reipublicae Popularis Sinicae*, Beijing, 1959 (Photo provided by Zong-Gang HU)

1959 年，中国植物志第一次编委会会议于北京举行（相片提供者：胡宗刚）

The Second Editorial Committee meeting of *Flora Reipublicae Popularis Sinicae,* Beijing, 6 September 1961; front row (left-right): Chu SHIH, Dian-Min ZHANG, Li-Kuo FU, Zhi-Yu ZHANG, Lun-Kai DAI, Han-Pi YANG, Pei-Chun LI, Ling-Ti LU, Song-Jun LIANG, Tsue-Chih KU, Jun-Rong TAO, Ching-Chiang LEE, Ge-Ling CHU, Pan-Cheng WU, Yen-Cheng TANG, Wan-Fu JIANG, Chun-Li JIN; middle row: Pu-Chiu TSOONG, Yu-Wen TSUI, Chien PEI, Hsien-Wu KUNG, Ke-Chien KUAN, Yong LING, Ren-Chang CHING, Chao-Chien CHANG, Feng-Hwai CHEN, Hsen-Hsu HU, Woon-Young CHUN, Sung-Shu CHIEN, Yung CHEN, Tchen-Ngo LIOU, Yi-Li KENG, Wen-Pei FANG, Tsin TANG, Wan-Chun CHENG, Pang-Chieh CHEN, Ji-Wu JIANG, Ying TSIANG; and back row: Sing-Chi CHEN, Chieh CHEN, Shiew-Hung WU, An-Jen LI, Yi-Ling CHEN, Te-Tsun YU, Shu-Kang LEE, Ching-Yung CHENG, Ko-Zen KUANG, Cen-Jian QIAO, Hung-Ta CHANG, Cheng-Yih WU, Yu-Chuan MA, Chang-Chun WU, Fa-Tsuan WANG, Zhong-Xun WANG, Jin-Yong FENG, Rong-Hou ZHANG, Chun-Rong LIU, Sze-Hsue CHENG, Cheng-Gung MA (Photo provided by Wu Zhengyi Science Foundation)

1961 年 9 月 6 日，中国植物志第二次编委会（扩大）会议于北京；前排（从左至右）：石铸、张甸民、傅立国、张芝玉、戴伦凯、杨汉碧、李佩琼、陆玲娣、梁松钧、谷翠芝、陶君容、黎兴江、朱格麟、吴鹏程、汤彦承、姜万福、金存礼；中排：钟补求、崔友文、裴鉴、孔宪武、关克俭、林镕、秦仁昌、张肇骞、陈封怀、胡先骕、陈焕镛、钱崇澍、陈嵘、刘慎谔、耿以礼、方文培、唐进、郑万钧、陈邦杰、姜纪五、蒋英；后排：陈心启、陈介、吴兆洪、李安仁、陈艺林、俞德浚、李树刚、诚静容、匡可任、乔曾鉴、张宏达、吴征镒、马毓泉、吴长春、汪发缵、王宗训、冯晋镛、张荣厚、刘春荣、郑斯绪、马成功（相片提供者：吴征镒科学基金会）

(PE), Ji-Wu JIANG (1903–1975, PE), Yi-Li KENG (N), Hsien-Wu KUNG (NWTC), Ke-Zen KUANG (Ko-Jen KUANG, 1914–1977, PE), Yong LING (PE), Tchen-Ngo LIOU (IFP), Chien PEI (NAS), Tsin TANG (PE), Ying TSIANG (CANT), Cho-Po TSIEN (PE), Pu-Chiu TSOONG (PE), Fa-Tsuan WANG (PE), Cheng-Yih WU (KUN) and Te-Tsun YU (PE).

1959 年，徐炳声编著，**上海植物名录**，138 页；科技卫生出版社，上海。

1959, Ping-Sheng HSU, ***Enumeratio Plantarum Civitatis Shanghai***, 138 p; Science and Technology and Hygiene Press, Shanghai.

1959 年，南京大学生物学系、中国科学院植物研究所合编，耿以礼主编，**中国主要植物图说——禾本科**，1 181 页；科学出版社，北京。

1959, Department of Biology, Nanjing University and Institute of Biology, Chinese Academy of Sciences, Yi-Li KENG (editor in chief), ***Flora Illustralis Plantarum Primarum Sinicarum—Gramineae***, 1,181 p; Science Press, Beijing.

1959 年，王伏雄、秦仁昌，十年来的中国植物学，植物学报 8（3）：177–183；生物学通报 10：436–440。

1959, Fu-Hsiung WANG, Ren-Chang CHING, Ten years of botany in China, *Acta Botanica Sinica* 8(3): 177–183; *Bulletin of Biology* 10: 436–440.

1959 年，中国科学院华南植物研究所编，**广东植物名录**，229 页；中国科学院华南植物研究所，广州（内部印制）。

1959, South China Institute of Botany, Chinese Academy of Sciences, '***Checklist of Guangdong Plants***', 229 p; South China Institute of Botany, Chinese Academy of Sciences, Guangzhou (internal publication).

1959 年，中国科学院编译出版委员会名词室编订，**俄拉汉种子植物名称**（试用本），488 页；科学出版社，北京。

1959, Names Department of Compilation and Translation Publishing Committee, Chinese Academy of Sciences, '***Russian-Latin-Chinese Names of Seed Plants***' (trial edition), 488 p; Science Press, Beijing.

1959 年，中国科学院编译出版委员会名词室编订，**拉汉种子植物名称（补编）**，184 页；科学出版社，北京。

1959, Names Department of Compilation and Translation Publishing Committee, Chinese Academy of Sciences, '***Latin-Chinese Names of Seed Plants (Supplement)***' , 184 p; Science Press, Beijing.

1959 年，中国科学院编译出版委员会名词室编订，**拉汉孢子植物名称（补编）**，165 页；科学出版社，北京。

1959, Names Department of Compilation and Translation Publishing Committee, Chinese Academy of Sciences, '***Latin-Chinese Names of Cryptogamic Plants (Supplement)***', 165 p; Science Press, Beijing.

1959—1961 年，基于全国中药资源普查的 4 卷本**中药志**由人民卫生出版社出版。

1959-1961, '***Traditional Chinese Medicinal Materials***', vols. 1-4, published by People's Medical Publishing House based on a national survey of traditional Chinese medicine.

1959—1961 年，由中国科学院植物研究所分类室及生态研究室组成的南水北调综合考察队，张永田（1936—）、姜恕（1925—2019）、应俊生（1934—）、赵从福等人，在四川西部、云南西北部等地采集标本 1 万余号（KUN、PE）。

1959-1961, South-to-North Water Diversion Expedition[455] organized by the departments of plant taxonomy and plant ecology, Institute of Botany, Chinese Academy of Sciences; Yong-Tian CHANG (1936-), Shu JIANG (1925-2019), Tsun-Shen YING (Jun-Sheng YING, 1934-), Cong-Fu ZHAO et al. collected around 10,000 numbers of plant specimens (KUN, PE) in western Sichuan and northwestern Yunnan.

1959—1961 年，1979+，东北林学院 / 东北林业大学，**东北林学院植物标本室汇刊**，年刊，1–3 期，1959—1961；1979 年复刊，易名为**东北林学院植物研究室汇刊**，1979 年发行第 4–5 期，1980 年发行第 6–9 期，1981 年易名为**植物研究**，并按季刊发行且卷数重排；2000 年第 3 期和第 4 期中文名称为**木本植物研究**，2001 年又恢复**植物研究**；2006 年改为双月刊；1985[456]—1991 年周以良（1922—2005）、1992—2005 年聂绍荃（1933—2005）、2006—2013 年祖元刚（1954—）、2014 年至今付玉杰（1967—）任主编。

1959-1961, 1979+, Northeast Forestry College / Northeast Forestry University, ***Bulletin of the Herbarium of North-Eastern Forestry Academy***, Parts I-III, 1959-1961; ***Bulletin of Botanical Laboratory of North-Eastern Forestry Institute***, Parts IV-IX, 1979-1980; ***Bulletin of Botanical Research***, vols. 1+, 1981+; quarterly, 1981-2005, bimonthly since 2006; Editors in Chief: Yi-Liang CHOU (1922-2005), 1985[457]-1991, Shao-Quan NIE (Shou-Chuan NIE, 1933-2005), 1992-2005, Yuan-Gang ZU (1954-), 2006-2013, and Yu-Jie FU (1967-) since 2014.

455 英文翻译自中国南水北调官网（http://www.nsbd.gov.cn/zx/mtgz/201309/t20130910_287488.html，2017 年进入）。

456 1985 年之前没有主编记录。

457 No Editor in Chief before 1985.

1959、1995 年，中国科学院林业土壤研究所植物室编，刘慎谔主编，**东北植物检索表**，655 页，1959；科学出版社，北京；中国科学院沈阳应用生态研究所编，傅沛云主编，**东北植物检索表**，第 2 版，1 007 页，1995，图版 456 幅；科学出版社，北京。

1959 and 1995, Institutum Sylviculiturae et Pedologiae Academiae Scientiarum Sinicae, Tchen-Ngo LIOU (editor in chief), ***Claves Plantarum Chinae Boreali-Orientalis***, 655 p, 1959; Science Press, Beijing; Institute of Applied Ecology, Academia Sinica, Pei-Yun FU (editor in chief), ***Claves Plantarum China Boreali-Orientalis*** ed. 2, 1,007 p, 1995; Science Press, Beijing.

1959—2004 年，中国科学院中国植物志编辑委员会编，**中国植物志**[458]，1–80 卷（126 册）；科学出版社，北京。秦仁昌，1959，中国植物志第 2 卷，406 页。该书出版持续 45 年之久，多达 80 卷，此为率先出版的第一卷（http://frps.iplant.cn/，2019 年 5 月进入）。

1959–2004，Editorial Committee of Flora Reipublicae Popularis Sinicae, ***Flora Reipublicae Popularis Sinicae***[459], vols. 1–80 (126 parts), Science Press, Beijing. Ren-Chang CHING, 1959, vol. 2: 406 p, became the first of 80 volumes in 45 years (http://frps.iplant.cn/, accessed May 2019).

1959, Hui-Lin LI, ***The Garden Flowers of China***, 273p; The Ronald Press Company, New York.

1959, 李惠林，**中国的园林花卉**，273 页；The Ronald Press Company，纽约。

458 “文化大革命”前出版的前 3 卷（2、11、68）主编为钱崇澍、陈焕镛；以后出版的均没有主编，只有编辑。详细参见：胡宗刚、夏振岱，2016，中国植物志编纂史，322 页；上海交通大学出版社，上海；婕妤，2010，《中国植物志》荣耀背后的甘苦，创新科学 2：40-43；黄普华，2017，编研《中国植物志》的回顾与体会，植物科学学报 35（3）：465-468。

459 Jin-Shuang MA and Steve Clemants, 2006, A history and Overiew of the *Flora Reipublicae Popularis Sinicae* (Flora Reipublicae Popularis Sinicae, Flora of China, Chinese edition, 1959–2004), *Taxon* 55(2): 451–460.

1960

1960 年，中国科学院和水利部组成西南地区南水北调综合考察队，涉及怒江、澜沧江和金沙江流域，分设雅砻江分队和滇西北分队。滇西北分队由中国科学院昆明植物研究所、云南大学生物系和云南林学院组成；吕春朝（1939— ）、陶德定（1937—2013）、王守正（1940— ）、武素功、俞绍文（1936— ），历时 3 个月采集标本 1 200 余号 5 000 余份（KUN）。

1960, South-to-North Water Diversion Expedition organized by Chinese Academy of Sciences and Ministry of Water Resources, with two collecting trips to the Yalong River, Sichuan, and northwestern Yunnan. The northwestern Yunnan team was organized by the Kunming Institute of Botany, Chinese Academy of Sciences, Department of Biology, Yunnan University and Yunnan Forestry College. Chun-Chao LÜ [460](1939-), De-Ding TAO (1937-2013), Shou-Zheng WANG (1940-), Shu-Kung WU and Shao-Wen YU (1936-) collected more than 1,200 numbers and more than 5,000 plant specimens (KUN) in three months.

1960+, Academia Sinica, ***Botanical Bulletin of Academia Sinica***, semi-annual, changed to quarterly from volume 29, 1988, and changed to ***Botanical Studies**, An International Journal* from volume 47, 2006.

1960+，中央研究院，**中央研究院植物学汇报**，半年刊，1988 年第 29 卷改为季刊，2006 年第 47 卷起更名为 ***Botanical Studies***，*An International Journal*。

1960—1970 年，中国科学院南京中山植物园易名为中国科学院南京植物研究所；1960—1968 年裴鉴任主任，1968—1970 年王振亚任副主任。

1960-1970, Nanjing Botanical Garden Memorial Sun Yat-Sen, Chinese Academy of Sciences, renamed Nanjing Institute of Botany, Chinese Academy of Sciences, with director Chien PEI, 1960-1968, and Zhen-Ya WANG as vice director, 1968-1970.

1960 年代，庄清漳（1931—1976）[461] 赴加拿大留学，1971 年获不列颠哥伦比亚大学博士学位；1972 年入加拿大印第安人与北部发展事务部从事植物学研究，1973 年入不列颠省博物馆任助理馆员，从事国家公园的植物调查，特别是苔藓，直到 1976 年 11 月 22 日突然故去。详细参见：Sheila Newnham, 1979, Ching-Chang Chuang (1931-1976), *Syesis* 12: 1-2.

1960s, Ching-Chang CHUANG (1931-1976) went to Canada for study bryophyte, and received Ph.D.

460 Chun-Chao LYN in English.

461 台湾台北人，1958 年获台湾大学学士学位，1963 年获台湾大学硕士学位；之后任助教与讲师；主要从事分类学研究，特别是苔藓。

degree in 1971[462]; joined the staff of the Canadian Department of Indian Affairs and Northern Development as a Plant Researcher in 1972, then as Assistant Curator of the Herbarium, Provincial Museum, Victoria, British Columbia, dealing with inventory of provincial parks floras, especially on bryophytes, until his passed away suddenly on November 22, 1976. See Sheila Newnham, 1979, Ching-Chang Chuang (1931–1976), *Syesis* 12: 1–2.

462 Ching-Chang Chuang, 1973, A moss flora of Taiwan exclusive essentially pleurocarpous families. *The Journal of the Hattori Botanical Laboratory* 37: 419–509; Advisor: Wilfred B. Schofield (1927–2008).

1961

1961 年，许建昌[463]（1932—，日文名许田仓园）赴日本留学，1964 年获日本东京大学理学部生物系博士学位，1964—1967 年在东京大学理学部生物系从事博士后研究，1967—1971 年返回台湾任台湾大学理学院植物系副教授，1971—1976 年任教授，1976—1978 年任日本进化生物研究所客座教授，1978—1980 年任日本东京玉川大学农学部非常勤讲师，1980—1983 年玉川大学农学部客员教授，1983—1996 年玉川大学农学部教授。1996 年自玉川大学退休后返台湾定居。

1961, Chien-Chang HSU (1932-, Japanese name: Kyoda Kurasono) went to Japan to study botany; obtained Ph. D. degree from Department of Biology, Faculty of Science, University of Tokyo, 1964[464]; as postdoctoral research fellow in Department of Biology, Faculty of Science, University of Tokyo, 1964-1967; returned to Taiwan as associate professor, Department of Botany, Taiwan University, 1967-1971, then professor, 1971-1976; as visiting professor of the Research Institute of Evolutionary Biology, Japan, 1976-1978; part time lecturer in Faculty of Agriculture, Tamagawa University, 1978-1980; guest professor, Faculty of Agriculture, Tamagawa University, 1980-1983; professor, Faculty of Agriculture, Tamagawa University, 1983-1996; returned to Taiwan after retirement from Tamagawa University in 1996.

1961, Joseph Lanjouw, Charles Baehni, Walter Robyns, Robert Ross, Jacques Rousseau, James M. Schopf, Georg M. Schulze, Albert C. Smith, Roger de Vilmorin and Frans A. Stafleu, ***International Code of Botanical Nomenclature* (Montreal Code)**, adopted by the Ninth International Botanical Congress, Montreal, August 1959, *Regnum Vegetabile* 23:1-372 p. Translated into Chinese in 1964 and 1965.

1964 年，耿伯介译，耿以礼校，**植物命名国际法规**，108 页；南京大学出版社，南京（内部印制）。1965 年，匡可任译，**国际植物命名法规**，262 页；科学出版社，北京（内部印制）。

1961 年，中国科学院在中国科学院新疆分院生物研究室、土壤研究室、地理研究室和新疆综合考察队的基础上，于乌鲁木齐建立中国科学院新疆水土生物资源综合研究所。1961—1966 年杨旭民任所长。

1961, Comprehensive Institute of Water, Soil and Biological Resources, Chinese Academy of Sciences, based on Biological Research Department, Soil Research Department, Geographical Research Department,

463 台湾苗栗人，1951—1955 年台湾师范大学理学院博物系本科毕业获学士学位，1955—1957 年任台湾师范大学理学院博物系助教，1957—1959 年任台湾大学理学院植物系助教，1959—1961 年台湾大学理学院植物研究所毕业获硕士学位。著名禾草专家，台湾植物志禾本科和日文版世界植物的禾本科作者；在日本翻译并出版日文**云南杜鹃花**（1981）、**云南山茶花**（1981）、**云南植物**（3 卷本，1985）等著作。

464 Ph.D. Dissertation: The classification of the genus *Panicum* and its allies, with special reference to their lodicule, style-base and lemma; advisor: Hiroshi Hara (1911-1986).

and Comprehensive Exploration Team, Xinjiang Branch, Chinese Academy of Sciences, Urumqi, Xinjiang, established; with director Xu-Min YANG, 1961-1966.

1961 年，中国植物志第 11 卷（莎草科一部分）出版。这是中国植物志“文化大革命”前出版的三卷的第二卷。

1961, *Flora Reipublicae Popularis Sinicae*, Volume 11 (Cyperaceae in part) published. This was the second of three volumes published before the cultural revolution.

1961—1970 年，中国科学院在青海西宁的青海生物研究所基础上成立西北高原生物研究所，次年正式建立植物标本室（HNWP）[465]；1965—1970 年冯浪（1916—1986）任所长。

1961-1970, Northwest Institute of Plateau Biology, Chinese Academy of Sciences, based on the previous Institute of Biology, was officially established in Xining, Qinghai; herbarium (HNWP) founded the following year, with Lang FENG (1916-1986) as director, 1965-1970.

1961 年 7 月 —1996 年 7 月，中国科学院广西植物研究所改由广西领导，并更名为广西植物研究所，1962—1967 年钟济新、1969—1972 年莫一凡（1912—1983）、1973—1978 年田旭（1917—1987）、1978—1980 年倪兴仁（1917—1998）、1980—1982 年李晓南（1923—2004）、1982—1984 年梁畴芬[466]（1921—1998）、1984—1988 年黄正福（1929—）、1988—1996 年金代钧（1938—）任所长。

1961.07-1996.07, Guangxi Institute of Botany, Chinese Academy of Sciences, renamed Guangxi Institute of Botany, under Guangxi government administration, with directors Chi-Hsing CHUNG, 1962-1967; Yi-Fan MO (1912-1983), 1969-1972; Xu TIAN (1917-1987), 1973-1978; Xing-Ren NI (1917-1998), 1978-1980; Xiao-Nan LI (1923-2004), 1980-1982; Chou-Fen LIANG (1921-1998), 1982-1984; Zheng-Fu HUANG (1929-), 1984-1988; Dai-Jun JIN (1938-), 1988-1996.

1961、2012 年，中华人民共和国商业部土产废品局、中国科学院植物研究所主编，**中国经济植物志**，上册：1-1 246 页、下册：1 247-2 273 页，1961；科学出版社，北京（内部印制）。中华人民共和国商业部土产废品局、中国科学院植物研究所主编，**中国经济植物志**，上册：1-1 246 页，下册：1 247-2 236 页，2012；科学出版社，北京。[467]

1961 and 2012, Bureau of Local and Special Products, Ministry of Commerce of the People's Republic of China and Institute of Botany, Chinese Academy of Sciences (Editors in Chief), '***Economic Flora of***

465 中国科学院西北高原生物研究所志编纂委员会，2012，中国科学院西北高原生物研究所志，第 674 页；青海人民出版社，西宁。

466 广东开平人；1949 年毕业于广西大学农学院园艺专业，后应陈焕镛之邀加入植物分类学研究，1963 年任广西植物研究所资源室副主任，1975 年任植物分类室副主任，1981 年任副所长，1982—1984 年任所长。

467 重印说明参见 2012 年版上册和下册第 i-ii 页。

China', vol. 1: 1-1,246 p and 2: 1,247-2,273 p, 1961; Science Press, Beijing (printed internally). Bureau of Local and Special Products, Ministry of Commerce of the People's Republic of China and Institute of Botany, Chinese Academy of Sciences (editors in chief), '***Economic Flora of China***', vol. 1: 1-1,246 p and 2: 1,247-2,236 p, 2012; Science Press, Beijing.

1962

1962年2月17日，人民日报发表胡先骕水杉歌；英文稿由作者自译，1966年发表于香港 *Eastern Horizon (Hong Kong)* 5（4）：26-28。

1962.02.17, On *Metasequoia* by Hsen-Hsu HU published in *the People's Daily*. The English version was published by the author in *Eastern Horizon (Hong Kong)* 5(4): 26-28, 1966, in Hong Kong.

Tsan-Iang CHUANG (1933-1994) (Photo provided by Huei-Mei HSU, CHUANG's wife, and his son Carl CHUANG)
庄灿旸（1933—1994）（相片提供者：庄夫人许惠美及其子 Carl）

1962年8月，庄灿旸[468]（1933—1994）赴美国留学，1966年获加州大学伯克利分校博士学位，同时应聘至罗德岛大学金斯顿校区任教一学年，1967年受聘伊利诺伊州立大学任植物学助理教授兼植物标本馆馆长，1972年任副教授，1977年任教授，直至1994年。

1962.08, Tsan-Iang CHUANG (1933-1994) studied botany at the University of California, Berkeley, received Ph.D. degree in 1966[469], spent one year teaching in the University of Rhode Island, Kingston, Rhode Island, 1966; in 1967 he went to Illinois State University, Normal, Illinois, and was successfully assistant professor and Curator of the Illinois State Univrersity Herbarium, then associate professor in 1972 and full professor of botany in 1977 until 1994.

468 台湾新竹人（1933年4月24日至1994年5月24日），省立新竹中学高中毕业考入台湾省立师范学院植物系，1956年毕业获学士学位，1959年于台湾大学植物研究所硕士班毕业获硕士学位，1959—1962年任“中央”研究院植物研究所任助理研究员，并于1960年创建植物标本馆（HAST）同时兼任馆长；其中1960年下半年至1961年年底服兵役一年半（曾驻金门）；在台期间曾发表数篇有关植物分类学的文章，包括金门植物小志。感谢庄先生旅美的新竹同乡、加州大学伯克利分校的校友陈仲钦先生，旅居加州湾区的庄夫人许惠美及其子 Carl 提供详细信息与照片，感谢中国科学院西双版纳植物园杨玺提供哈佛大学阿诺德树木园的相关信件，感谢加州大学伯克利分校植物标本馆档案室提供庄先生的博士论文导师 Lincoln Constance 撰写的回忆文章（1994；未刊行）、庄先生的完整简历以及出版物详细目录。

469 Ph.D. dissertation: Tsan-Iang Chuang and Lincoln Constance, 1969, A systematic study of *Perideridia* (Umbelliferae-apioideae), *University of California Publications in Botany* 55: 1-74; Advisor: Lincoln Constance.

Ching-Chieu HUANG, Feng-Hwai CHEN, and Chi-Ming HU in Guangzhou, 1982 (left-right) (Photo provided by Royal Botanic Garden, Edinburgh)
黄成就、陈封怀、胡启明（从左至右）1982 年于广州（相片提供者：爱丁堡植物园）

1962 年 11 月，陈封怀离开中国科学院武汉植物园到广州，协助陈焕镛主持中国科学院华南植物研究所植物园，并兼任中国科学院华南植物研究所副所长。

1962.11, Feng-Hwai CHEN moved from Wuhan Botanical Garden, Chinese Academy of Sciences, to Guangzhou to assist Woon-Young CHUN in charge of South China Botanical Garden, South China Institute of Botany, Chinese Academy of Sciences, and served as vice director of South China Institute of Botany, Chinese Academy of Sciences.

1962 年，"中央"研究院植物研究所在台湾台北南港复所[470]，而植物标本馆（HAST）则在一年前就已经成立；1962—1971 年李先闻（1902—1976）、1972—1977 年郭宗德（1933— ）、1977—1983 年邬宏潘（1933— ）、1983—1989 年陈庆三（1935— ）、1989—1996 年周昌弘（1942— ）、1996—1996 年杨祥发（1932—2007）、1997—2003 年萧介夫（1948— ）、2003—2008 年贺瑞华（1948— ）、2008—2009 年林纳生（1953— ）、2010—2012 年黄焕中（1948— ）、2013 年至今陈荣芳（1953— ）任

470 "中央"研究院植物研究所，1992，回顾与展望——"中央"研究院植物研究所在台复所三十周年纪念，102 页；"中央"研究院植物研究所（内部印制），台北；"中央"研究院植物研究所，1992，"中央"研究院植物研究所在台复所三十周年纪念出版品及著作目录，217 页；"中央"研究院植物研究所，台北（内部印制）。实际上复所的准备工作早在 1954 年便已经开始，只是当时没有地方，不得不借助台湾大学等地办公，直至新的大楼落成才正式复所。详细参见：http://ipmb.sinica.edu.tw/index.html/?q=node/774（2018 年进入）。

所长。2005 年“中央”研究院植物研究所更名为植物暨微生物学研究所；与此同时，相关的动植物研究人员则组成“中央”研究院生物多样性研究中心，包括植物标本馆（HAST）。

Institute of Plant and Microbial Biology, Academia Sinica, Taipei, 2016 (Photo provided by Jin-Shuang MA)
2016 年台北的“中央”研究院植物暨微生物学研究所（相片提供者：马金双）

Biodiversity Research Center, Academia Sinica, Taipei, 2016 (Photo provided by Jin-Shuang MA)
2016 年台北的“中央”研究院生物多样性研究中心（相片提供者：马金双）

1962, Institute of Botany, Academia Sinica, restored in Nankang [Nangang], Taipei [Taibei], Taiwan, with the herbarium (HAST) established previous year, and directors Hsien-Wen LI (1902–1976), 1962–1971; Tsong-Teh KUO (1933–), 1972–1977; Hong-Pang WU (1933–), 1977–1983; Ching-San CHEN (1935–), 1983–1989; Chang-Hung CHOU (1942–), 1989–1996; Shang-Fa YANG (1932–2007), 1996–1996; Jei-Fu SHAW (1948–), 1997–2003; Tuan-Hua David HO (1948–), 2003–2008; Na-Sheng LIN (1953–), 2008–2009; Anthony HUANG (1948–), 2010–2012; and Long-Fang Oliver CHEN (1953–), since 2013. Institute of Botany renamed Institute of Plant and Microbial Biology in 2005; at the sametime, the related scientists from institutes of Zoology and Botany were formed as Biodiversity Research Center, Academia Sinica, including Herbarium (HAST).

1962—1970 年，中国科学院昆明植物研究所更名为中国科学院植物研究所昆明分所；吴征镒任分所所长。

1962-1970, Kunming Institute of Botany, Chinese Academy of Sciences, became a branch of the Institute of Botany, Chinese Academy of Sciences, with director Cheng-Yih WU.

1962 年 9 月 —1970 年，中国科学院农业生物研究所更名为中国科学院西南生物研究所；1963 年、1966 年黄国英、1968—1970 年邓国彪（1913—1983）任所长。

1962.09-1970, Chengdu Institute of Agricultural Biology, Chinese Academy of Sciences, renamed Southwest Institute of Biology, Chinese Academy of Sciences, with directors Guo-Ying HUANG, 1963 and 1966, and Guo-Biao DENG (1913-1983), 1968-1970.

1962—1970 年，江西省庐山植物研究所更名为中国科学院庐山植物园；1962—1965 年沈洪欣（1923—2006）、1965—1967 年温成胜、1968—1970 年慕宗山（1922—1994）任园主任。

1962-1970, Jiangxi Lushan Institute of Botany renamed Lushan Botanical Garden, Chinese Academy of Sciences, with directors Hong-Xin SHEN (1923-2006), 1962-1965; Cheng-Sheng WEN, 1965-1967; and Zong-Shan MU (1922-1994), 1968-1970.

The main gate of Lushan Botanical Garden, Chinese Academy of Sciences, 1962-1970 (Photo provided by Zong-Gang HU)
1962—1970 年中国科学院庐山植物园大门（相片提供者：胡宗刚）

1962—1975 年、1984—1987 年、1993 年，北京师范大学生物系编，**北京植物志**，上册，1-532 页，1962；中册，533-1965 页，1964，下册[471]，1-349 页，1975；人民出版社，北京；贺士元、尹祖棠编，**北京植物志**，第 2 版，上册，1-710 页，1984；下册，711-1 476 页，1987；北京出版社，北京；贺士元、尹祖棠编，**北京植物志**，1992 年修订版，上册，1-710 页；下册，711-1 510 页，1993；北京出版社，北京。

1962-1975, 1984-1987, 1993, Department of Biology, Beijing Normal University, '***Flora of Beijing***', vols. 1-3; 1: 1-532 p, 1962, 2: 533-1,965 p, 1964, 3: 1-349 p, 1975; Beijing People's Press, Beijing; Shih-

471 名为：北京地区植物志——单子叶植物。

Yuen HO and Zu-Tang YIN ***Flora of Beijing***, ed. 2, vols. 1: 1–710 p, 1984; 2: 711–1,476 p, 1987; Beijing Press, Beijing; Shih-Yuen HO and Zu-Tang YIN, ***Flora of Beijing***, 1992 revised ed., vols. 1: 1–710 p; 2: 711–1,510 p, 1993; Beijing Press, Beijing.

1962, 1965, 1966, 1974, 1978, 1993, 2002, 2004, 2012, Colonial Herbarium / Hong Kong Herbarium / Hong Kong Herbarium / Hong Kong Government Agriculture and Fishery Department / Hong Kong Government Agriculture, Fisheries and Conservation Department, ***Check List of Hong Kong Plants***, 22 p, 1962; ed. 2, 61 p, 1965; ed. 3, 67 p, 1966; ed. 4, 115 p, 1974; ed. 5, 142 p, 1978; ed. 6, 159 p, 1993; revised ed. 6[472], 407 p, 2002; ed. 7, 198 p, 2004; ed. 8, 219 p, 2012; Colonial Herbarium, Hong Kong / Hong Kong Herbarium, Hong Kong / Hong Kong Government Agriculture and Fishery Department, Hong Kong / Hong Kong Government Agriculture, Fisheries and Conservation Department, Hong Kong.

1962、1965、1966、1974、1978、1993、2002、2004、2012 年，殖民地植物标本室 / 香港植物标本室，**香港植物名录**，22 页，1962；第 2 版，61 页，1965；第 3 版，67 页，1966；第 4 版，115 页，1974；第 5 版，142 页，1978；第 6 版，159 页，1993；第 6 版增订本[473]，407 页，2002；第 7 版，198 页，2004；第 8 版，219 页，2012；殖民地植物标本馆，香港 / 香港特别行政区渔农自然护理署，香港植物标本室，香港。

472 South China Institute of Botany, Chinese Academy of Sciences, and Hong Kong Herbarium, Agriculture, Fisheries and Conservation Department, Te-Lin WU (editor in chief).

473 中国科学院华南植物研究所、渔农自然护理署香港植物标本室，吴德邻主编。

1963

1963 年 10 月 16 日至 26 日，中国植物学会三十周年年会（第七届全国代表大会）于北京举行，全国 200 多位代表出席；理事长为钱崇澍。[474]

1963.10.16-10.26, Thirtieth anniversary meeting (Seventh National Congress) of Botanical Society of China held in Beijing; more than 200 attended from throughout China. Sung-Shu CHIEN was elected president.

1963 年，中国植物志第 68 卷（玄参科一部分）出版。这是“文化大革命”前出版 3 卷的最后 1 卷。

1963, *Flora Reipublicae Popularis Sinicae*, Volume 68 (Scrophulariaceae in part) published. This was the last of the three volumes published before the cultural revolution.

1963、1974 年，中国科学院编译出版委员会名词室，关克俭、陆定安，**英拉汉植物名称**（试用本），986 页，1963；科学出版社，北京；关克俭、诚静容编；**拉汉种子植物名称**，第 2 版，543 页，1974；科学出版社，北京。

1963, 1974, Names Department of Compilation and Translation Publishing Committee, Chinese Academy of Sciences, Ke-Chien KUAN and Ding-An LU, '***English-Latin-Chinese Names of Plants***' (trial edition), 986 p, 1963; Science Press, Beijing; Ke-Chien KUAN and Ching-Yung CHENG, '***Latin-Chinese Names of Chinese Seed Plants***', ed. 2, 543 p, 1974; Science Press, Beijing.

1963、2009 年，冯国楣，云南植物调查和采集的历史回顾与前瞻，学术研究（自然科学版）4：34-42，1963；冯国楣原著、李德仁改写，云南植物调查采集的历史回顾，昆明文史资料选集 32[475]：4 863-4 869，2009；云南科技出版社，昆明。

1963 and 2009, Kuo-Mei FENG, 'Review and expectation on history of plant exploration in Yunnan', *Academic Research (Natural Science)* 4: 34-42, 1963; Kuo-Mei FENG (original author), De-Ren LI (revised), 'Review on history of plant exploration in Yunnan', *Selected Works of Literature and History of Kunming* 32: 4,863-4,869, 2009; Yunnan Science and Technology Press, Kunming.

1963+, Valery I. Grubov, ***Plantae Asiae Centralis (Plants of Central Asia, Plants collections from China and Mongolia, Plantae Asiae Centralis)***, Volume 1+, Izd-vo Akademii nauk SSSR, Moskva and

474 中国植物学会编，1963，中国植物学会三十周年年会论文摘要汇编，382 页；中国科学技术情报研究所，北京。中国植物学会五十年编写组，1985，中国植物学会五十年，中国科技史料 6（2）：54。

475 即昆明文史资料集萃第 6 集。

Leningrad; Vol. 1: 167 p, 1963[476]/188 p, 1999[477]; 2: 134 p, 1966/165 p, 2000; 3: 118 p, 1967/149 p, 2000; 4: 246 p, 1968/315 p, 2001; 5: 208 p, 1970/241 p, 2002; 6: 82 p. 1971/87 p, 2002; 7: 137 p, 1977/169 p, 2003; 8a: 125 p, 1988/170 p, 2003; 8b: 88 p, 1998/108 p, 2003; 8c: 180 p, 2000/260 p, 2004; 9: 149 p, 1989/197 p, 2005; 10: 118 p, 1994/139 p, 2005; 11: 122 p, 1994/136 p, 2007; 12: 169 p, 2001/190 p, 2007; 13: 129 p, 2002/149 p, 2007; 14a: 153 p, 2003/176 p, 2007; 14b: 222 p, 2008/XXX; 15: 142 p, 2006/XXX[478]; 16: 134 p, 2007/XXX.

1963+，Valery I. Grubov，**中亚植物（中国和蒙古的植物）**[479]，vols. 1+; Izd-vo Akademii nauk SSSR，Moskva and Leningrad；第 1 卷：167 页[480]，1963/188 页，1999；2：134 页，1966/165 页，2000；3：118 页，1967/149 页，2000；4：246 页，1968/315 页，2001；5：208 页，1970/241 页，2002；6：82 页，1971/87 页，2002；7：137 页，1977/169 页，2003；8a：125 页，1988/170 页，2003；8b：88 页，1998/108 页，2003；8c：180 页，2000/260 页，2004；9：149 页，1989/197 页，2005；10：118 页，1994/139 页，2005；11：122 页，1994/136 页，2007；12：169 页，2001/190 页，2007；13：129 页，2002/149 页，2007；14a：153 页，2003/176 页，2007；14b：222 页，2008/XXX；15：142 页，2006/XXX；16：134 页，2007/XXX。

1963—1972 年，1963—1969 年李春祥、1969—1972 年王绍志、1969—1972 年蔡德炳任中国科学院武汉植物园副主任。

1963-1972, Vice directors of Wuhan Botanical Garden, Chinese Academy of Sciences: Chun-Xiang LI, 1963-1969; Shao-Zhi WANG, 1969-1972; and De-Bing CAI, 1969-1972.

1963、1978 年，陈邦杰主编，**中国藓类植物属志**，上册：304 页，1963，下册：331 页，1978；科学出版社，北京。

1963 and 1978, Pan-Chieh CHEN (editor in chief), ***Genera Muscorum Sinicorum***, 1: 304 p, 1963; 2: 331 p, 1978; Science Press, Beijing[481].

476 Russian edition.

477 English edition.

478 Not published; same as below.

479 全书计划出版 20 卷，截至 2008 年，俄文版已出版 16 卷，英文版已出版 14 卷。

480 李世英译，1976，第 1 卷，亚洲中部植物概论，生物学译丛，3：39-94；青海生物研究所，西宁（内部印制）。

481 Howard A. Crum, 1964, *The Bryologist* 67(3): 383; Benito T. TAN, 1979, *The Bryologist* 82(4): 638-641.

1964

1964 年 2 月，北京大学生物系植物教研室，**河北省植物检索表**，396 页；北京大学生物系植物教研室，北京（油印本）。

1964.02, Botany Group, Department of Biology, Peking University, '***Key to Plants of Hebei Province***', 396 p; Botany Group, Department of Biology, Peking University, Beijing (mimeo printed).

1964 年，徐炳声、金德孙，新分类学与植物命名，生物科学动态 10: 1-6。

1964, Ping-Sheng HSU and De-Sun JIN, 'New systematics and botanical nomenclature', *Trends of Biological Science* 10: 1-6.

1964 年，崔友文、李培元，P. Giraldi 在陕西采集植物地点的考证，植物分类学报 9(3): 308-312。

1964, You-Wen TSUI and Pei-Yuan LI, Notes on the localities of Giraldi's plant-collecting in Shen-Si, *Acta Phytotaxonomica Sinica* 9(3): 308-312.

1964 年，林镕，中国植物学三十年来的回顾，生物学通报 2: 1-8。

1964, Yong LING, 'Review on the Chinese botany in the past thirty years', *Bulletin of Biology* 2: 1-8.

1964 年，谢万权[482]（1929—）赴日本东京教育大学留学，1966 年获得硕士学位，返台后任教于中兴大学；1972 年获日本国立东京教育大学理学博士；退休前为中兴大学教授。

1964, Wang-Chueng SHIEH (1929-) went to Japan for study in Tokyo University of Education, received Master's degree in 1966, then back to Chung Hsing University for teaching, and received Ph.D. degree[483] from Tokyo University of Education in 1972; he was professor of Chung Hsing University before retired.

1964—1977 年，中国科学院华南植物研究所、广东省植物研究所编辑，陈焕镛主编，**海南植物志**，4 卷本；第 1 卷：517 页，1964；第 2 卷：470 页，1965；第 3 卷：629 页，1974；第 4 卷：644 页，1977；科学出版社，北京。

1964-1977, South China Institute of Botany, Chinese Academy of Sciences, and Guangdong Institute of

482 台湾桃园人，台湾省立农学院毕业。

483 Ph.D. dissertation: A synopsis of the fern family Pteridaceae sensu Copeland in Taiwan, supervised by Hiroshi Ito (1909-2006), Tokyo University of Education, D.Sc., 1972. Wang-Chueng SHIEH, 1973, A synopsis of the fern family Pteridaceae (sensu Copeland) in Taiwan, *Journal of Science & Engineering* 5: 191-232.

Botany, Woon-Young CHUN (editor in chief), ***Flora Hainanica***, vols. 1-4; 1: 517 p, 1964, 2: 470 p, 1965, 3: 629 p, 1974, 4: 644 p, 1977; Science Press, Beijing.

1965

1965, Hsen-Hsu HU, The major groups of living beings: A new classification, *Taxon* 14(8): 254-261.

1965 年，胡先骕，生物的主要类群：一个新系统，*Taxon* 14（8）：254-261。

1965 年，吴征镒，中国植物区系的热带亲缘，科学通报 1：25-33。

1965, Cheng-Yih WU, 'The tropical affinity of the Flora of China', *Chinese Science Bulletin* 1: 25-33.

1965 年，位于新疆乌鲁木齐的中国科学院新疆水土生物资源综合研究所分建为新疆生物土壤研究所和新疆地质地理研究所；1974—1984 年徐文哲任新疆生物土壤研究所所长。

1965, Xinjiang Comprehensive Research Institute of Water, Soil and Biological Resources, Chinese Academy of Sciences, Urumqi, was split into Institute of Biology and Pedology, and Institute of Geology and Geography, Chinese Academy of Sciences. Director of Institute of Biology and Pedology: Wen-Zhe XU, 1974-1984.

1965 年 4 月 7 日 —1967 年，中国科学院西北农业生物研究所的植物研究室独立，成立中国科学院西北植物研究所；苏建中为负责人。

1965.04.07-1967, Department of Botany, Northwest Agricultural Institute of Biology, Chinese Academy of Sciences, stood alone as independent Northwest Institute of Botany, Chinese Academy of Sciences, person in charge: Jian-Zhong SU.

1966

1966 年，林镕任中国科学院植物研究所代所长。

1966, Yong LING named acting director, Institute of Botany, Chinese Academy of Sciences.

1967, Vernon H. Heywood, ***Plant Taxonomy***, 60 p, 1967; St. Martin's Press, New York; Vernon H. Heywood, ***Plant Taxonomy***, ed. 2, 63 p, 1976; Edward Arnold, London. Translated into Chinese in 1976.

1976 年，柯植芬[484]译，洪德元校，**植物分类学**，86 页；科学出版社，北京。

1966, 1984, Sheila Pim, ***The Wood and the Trees****—a biography of Augustine Henry*, 256 p, 1966; Macdonald: London; ***The Wood and the Trees***, ed. 2, 252 p, 1984; Boethius Press, Kilkenny, Ireland.

1966、1984 年，Sheila Pim，**林木与树木**，Augustine Henry 传记，256 页，1966；Macdonald，伦敦；**林木与树木**，第 2 版，252 页，1984；Boethius Press，Kilkenny，爱尔兰。

1966, 1973, 1983, 1992, William T. Stearn, ***Botanical Latin***, 566 p, 1966, Nelson, London; ***Botanical Latin***, ed. 2, 566 p, 1973; David and Charles, Newton Abbot Devon; ***Botanical Latin***, ed. 3, 566 p, 1983; David and Charles, Newton Abbot Devon; ***Botanical Latin***, ed. 4, 546 p, 1992; Timber Press, Portland, Oregon. Translated into Chinese in 1978 and 1980.

1978、1980 年，秦仁昌译，俞德浚、胡昌序校，**植物学拉丁文**，上册：712 页，1978；下册：344 页，1980；科学出版社，北京。[485]

---

484 取自科学院植物所分类室三个首字（科植分）的谐音。

485 译自第 2 版。

1967

1967 and 1968, John Hutchinson, ***Key to the Families of Flowering Plants of the World***, revised and enlarged for use as a supplement to the *Genera of Flowering Plants*, 117 p, 1967; Clarenden Press, Oxford; reprinted with corrections, 1968. Translated into Chinese in 1977 and 1983.

1977、1983 年，宁夏农学院植物教研组译，**世界有花植物分科检索表**，212 页，1977；宁夏农学院，银川（铅字油印本，内部印刷）。洪涛译[486]，李扬汉[487]校，**世界有花植物分科检索表**，173 页，1983；农业出版社，北京。

486 据 1968 年重印版翻译。

487 江西南昌人（1913—2004），1929—1935 年就读于江西九江同文中学，1935—1936 年于南京金陵大学农学院（主系农艺，辅系园艺）学习，1936—1939 年就读于四川成都金陵大学农学院（转主系植物系，辅系植物病理系），1938—1939 年任金陵大学植物系见习助教，1940—1945 年任金陵大学植物系助教、讲师，1945—1946 年赴美国耶鲁大学林学研究院进修，1946—1952 年任南京金陵大学农学院植物系副教授、教授，其中 1946—1948 年兼中央林业试验所技正（荐任），1952—1956 年任南京农学院教授兼生物教研组主任，1956—1972 年任南京农学院农学系教授兼系副主任、系主任，1972—1978 年任江苏农学院农学系教授兼系主任，1979—2004 年任南京农学院（1984 年更名为南京农业大学）农学系教授兼系主任、杂草研究室主任。著名杂草专家。

1968

1968, Andrei Baranov, ***Basic Latin for Plant Taxonomists***, 146 p; Impex India, New Delhi. Translated into Chinese in 1988.

1988 年，赵能[488]译，**植物分类学基础拉丁语**，202 页；四川科学技术出版社，成都。

1968, 1982, Charles Jeffery, ***An Introduction to Plant Taxonomy***, 128 p, 1968; Churchill, London; Charles Jeffery, ***An Introduction to Plant Taxonomy***, ed. 2, 154 p, 1982; Cambridge University Press, Cambridge and New York. Translated into Chinese in 1990.

1990 年，胡征宇、李建强、傅运生译[489]，毕列爵校，**植物分类学入门**，150 页；上海科技教育出版社，上海。

1968—1978 年，中国科学院华南植物研究所改隶广东省，名称先后更改为广东省农林水科技服务站经济作物队（1968—1972）和广东省植物研究所（1972—1978）。

1968-1978, South China Institute of Botany, Chinese Academy of Sciences, transferred to administration of Guangdong Province and renamed Economic Crop Team, Scientific Service Station of Guangdong Agriculture, Forestry and Hydrology (1968-1972) and Guangdong Institute of Botany (1972-1978).

1968 年，李贤祉（1945—）赴美国加州大学 Santa Cruz 分校攻读博士，1973 年获得博士学位；1973—1975 年于哈佛大学阿诺德树木园从事 Mercer Fellow 研究；1975 年返港加入香港高尔夫球会从事草地草的工作；1976 年加入香港赛马会工作直至 1996 年，参与了沙田马场的建造，以及香港三个高尔夫球场、一些体育场馆及广东中山第一个高尔夫球场的建造；1996 年赴新加坡从事赛马跑道的建造及营运；2002 年回香港后加入香港中文大学生命科学学院成为其高级名誉研究员至今。

1968, Yin Tse LEE[490] (1945-, Eric Y. T. LEE) went to University of California, Santa Cruz, California, received Ph. D. degree, 1973[491]; Mercer Fellow in the Arnold Arboretum of Harvard University, 1973-1975; returned to Hong Kong in 1975, started career as turf professional in the Hong Kong Golf Club, 1975-1976, and Hong Kong Jockey Club, 1976-1996. Became esteemed turf expert and consulted on construction of three new golf courses plus many sports fields locally and also first golf course in China in Zhongshan,

488 又作赵良能。

489 据李建强确认，译自 1982 年版。

490 Eric Y. T. LEE received B. Sc. degree in Biology from Chinese University of Hong Kong in 1966.

491 Ph.D. Dissertation: Yin-Tse LEE and Jean H. Langenheim, 1975, Systematics of genus *Hymenaea* L. (Leguminosae, Caesalpinioideae, Detarieae); *University of California Publications in Botany* 69: 1-109; Advisor: Jean H. Langenheim (1925-).

Guangdong; joined consulting company and Turf Club in Singapore for construction and management of horse racing tracks for new racecourse, 1996; returned to Hong Kong in 2002 and became Senior Honorary Research Fellow in School of Life Sciences in Chinese University of Hong Kong to today.

1969

1969 年 12 月，谢阿才、杨再义，**新撰台湾植物名汇**，1 082 页；台湾大学农学院、中国农村复兴联合委员会，台北。

1969.12, A-Tsai HSIEH and Tsai-I YANG, ***Nomenclature of Plants in Taiwan***, 1,082 p; Agriculture College, Taiwan University and Agriculture Recovery Committee of China, Taipei [Taibei].

1969, Daniel J. Foley, ***The Flowering World of "Chinese" Wilson***, 334 p; The MacMillan Co., New York.

1969 年，Daniel J. Foley，**"中国"威尔逊的有花植物世界**，334 页；The MacMillan Co., 纽约。

1969—1973 年，第二次全国中草药资源普查；全国中草药汇编编写组，**全国中草药汇编**，上册：1 009 页，1975；下册：1 020 页，1978；人民卫生出版社，北京。

1969-1973, Second national survey of traditional Chinese medicine carried out. National Herbal Compendium Editorial Group, '***National Herbal Compendium***', volumes 1: 1,009 p, 1975, and 2: 1,020 p, 1978, published by People's Medical Publishing House, Beijing.

1970

1970—1973 年，广西植物研究所编，**广西植物名录**，3 卷本；第 1 册：70 页，1970；第 2 册：841 页，1971；第 3 册：199 页，1973；广西植物研究所，雁山，广西（内部印制）。

1970-1973, Guangxi Institute of Botany, '***Checklist of Guangxi Plants***', vols. 1-3; 1: 70 p, 1970; 2: 841 p, 1971; 3: 199 p, 1973; Guangxi Institute of Botany, Yanshan, Guangxi (internal publication).

1970 年 7 月至 1978 年，中国科学院西双版纳热带植物园更名为云南省热带植物研究所；蔡希陶

Xishuangbanna Tropical Botanical Garden, 1970s (Photo provided by Zong-Gang HU)
20 世纪 70 年代的西双版纳热带植物园（相片提供者：胡宗刚）

任主任。[492]

1970.07-1978, Xishuangbanna Tropical Botanical Garden, Chinese Academy of Sciences, renamed Yunnan Institute of Tropical Botany, with Hse-Tao TSAI as director.

1970 年 12 月至 1978 年，中国科学院西南生物研究所更名为四川省生物研究所；邓国彪任所长。

1970.12-1978, Southwest Institute of Biology, Chinese Academy of Sciences, renamed Sichuan Institute of Biology, with Guo-Biao DENG as director.

1970 年 7 月至 1978 年 3 月，中国科学院林业土壤研究所更名为辽宁省林业土壤研究所；陶炎（1923—1999）任所长。

1970.07-1978.03, Institute of Forestry and Pedology, Chinese Academy of Sciences, renamed Liangning Institute of Forestry and Pedology, with Yan TAO (1923-1999) as director.

1970—1978 年，中国科学院昆明植物研究所改名为云南植物研究所；吴征镒任所长。

1970-1978, Kunming Institute of Botany, Chinese Academy of Sciences, renamed Yunnan Institute of Botany, with Cheng-Yih WU as director.

1970—1979 年，中国科学院西北高原生物研究所更名为青海省生物研究所；常韬（1915—2012）任所长。

1970-1979, Northwest Institute of Plateau Biology, Chinese Academy of Sciences, renamed Qinghai Institute of Biology, with Tao CHANG (1915-2012) as director.

1970—1983 年，中国科学院西北植物研究所更名为陕西省中国科学院西北植物研究所，1972—1976 年许世越、1978—1983 年王育英任所长。

1970-1983, Northwest Institute of Botany, Chinese Academy of Sciences, renamed Shaanxi Institute of Northwest Botany, Chinese Academy of Sciences, with directors Shi-Yue XU, 1972-1976 and Yu-Ying WANG, 1978-1983.

1970—1983 年，中国科学院南京植物研究所更名为江苏省植物研究所；1970—1978 年王振亚任副主任，1978—1983 年单人骅任主任。

1970-1983, Nanjing Institute of Botany, Chinese Academy of Sciences, renamed Jiangsu Institute of Botany, Zhen-Ya WANG as vice director, 1970-1978; and Ren-Hwa SHAN as director, 1978-1983.

492 1974 年 11 月 11 日蔡希陶在西双版纳突发脑血栓，并在昆明医治；1977 年在北京再次复发，返昆明后于 1981 年 3 月 9 日病逝。

1971

1971 年，野田光藏，**中国东北区（满洲）植物志**[**中国东北区（满洲）の植物志**]，1 613 页；风间书房，东京。

1971, Mitsuzo Noda[493], ***Flora of the N.-E Provinces (Manchuria) of China***, 1,613 p; Kazama Shobo, Tokyo.

1971—1995 年，中国科学院庐山植物园更名为江西省庐山植物园，1971—1978 年慕宗山、1979—1983 年秦治平（1918—2014）、1983—1989 年徐祥美（1942—）、1989—1994 年杨涤清（1940—）、1994—1995 年吴炳文（1938—）任主任（或副主任）。

1971-1995, Lushan Botanical Garden, Chinese Academy of Sciences, renamed Jiangxi Lushan Botanical Garden, with directors (or vice director) Zong-Shan MU, 1971-1978; Zhi-Ping QIN (1918-2014), 1979-1983; Xiang-Mei XU (1942-), 1983-1989; Di-Qing YANG (1940-), 1989-1994; and Bing-Wen WU (1938-), 1994-1995.

1971 and 2001, Wilfrid Blunt, ***The Compleat Naturalist - A Life of Linnaeus,*** 256 p, 1971; Viking Press, New York, and Collins, London; Wilfrid Blunt, ***Linnaeus - The Compleat Naturalist,*** 264 p, 2001, with an introduction by William T. Stearn; Princeton University Press, Princeton, New Jersey, and Frances Lincoln, London. Translated into Chinese in 2017.

2017 年，徐保军译，**林奈传**——才华横溢的博物学家，287 页[494]；商务印书馆，北京。

493 Laboratory of Phytotaxo-Morphology, Faculty of Science, Niigata University, 1975, *Prof. Mitsuzo Noda (1909-1995) commemorative publication on his retirement from the Faculty of Science, Niigata University*, 107 p.

494 译自 2001 年重印版。

1972

1972 年 4 月至 11 月，中国科学院北京植物研究所和青海生物研究所在西藏军区后勤部协助下，与当地医务工作者组成西藏队，在拉萨地区及日喀则地区对西藏中草药进行了调查；鲍显诚（1934—）、郎楷永（1937—）、马成功（1936—1992）、王金亭（1932—）共采得标本 6 000 余号（PE）。

1972.04–11, Xizang Expedition organized by Beijing Institute of Botany, and Qinghai Institute of Plateau Biology, Chinese Academy of Sciences, and local medical staff, assisted by local military in Xizang, surveyed for medicinal plants in Lhasa and Shigatse [Rigaze] districts, Xian-Cheng BAO (1934–), Kai-Yung LANG (1937–), Cheng-Gung MA (1936–1992) and Chin-Ting WANG (1932–) collected 6,000 numbers of plant specimens (PE).

1972 年，张满祥，**秦岭苔藓植物名录**，58 页；中国科学院西北植物研究所，陕西，武功（内部印制）。

1972, Man-Hsiang CHANG, '***Checklist of Bryophytes of Qinling***', 58 p; Northwest Institute of Botany, Chinese Academy of Sciences, Wugong, Shaanxi (internal publication).

1972 年，林镕、钟补求，植物分类学的国际动态（内部文件），林镕文集，895–902 页，2013。

1972, Yong LING and Pu-Chiu TSOONG, International Status of Plant Taxonomy (internal document), *Collected Works of Ling Yong (Yong LING)*, 895–902 p, 2013.

1972, Frans A. Stafleu, Charles E. B. Bonner, Rogers McVaugh, Robert D. Meikle, Reed C. Rollins, Robert Ross, James M. Schopf, Georg M. Schulze, Roger de Vilmorin and Edward G. Voss, ***International Code of Botanical Nomenclature*** **(Seattle Code)**, adopted by the Eleventh International Botanical Congress, Seattle, August 1969. Translated into Chinese in 1974, 1975 and 1976.

1974、1975、1976 年，谢万权译，国际植物命名规约，中华林学季刊 7（2）：87–95，1974；森林学报（中兴大学森林系）4：1–17，1975；5：15–35，1976；廖日京译，**国际植物命名法规**，74 页，1975；台湾大学农学院森林系，台北（油印本）；1976 年重印。[495]

1972 年，中国科学院植物研究所编，**河北植物名录**，264 页；中国科学院植物研究所，北京（内部印制）。

1972, Institute of Botany, Chinese Academy of Sciences, '***Checklist of Hebei Plants***', 264 p; Institute of

495 参见：刘棠瑞、廖日京译，1981，树木学，下册：1 013–1 101；台湾商务印书馆，台北；黄增泉，1983，高等植物分类学原理，132–151；"国立"编译馆，台北。

Botany, Chinese Academy of Sciences, Beijing (internal publication).

1972+，福建省亚热带植物研究所，**三胶通讯**（内部发行），1974 年起更名为**亚热带植物通讯**，1983 年起公开发行，2000 年起由半年刊改为季刊，2000 年第 4 期起更名为**亚热带植物科学**；1983—1992 年李来荣（1908—1992）、1993—2011 年黄维南（1933—）、2012 年至今苏明华（1951— ）任主编。

1972+, Fujian Institute of Subtropical Botany, '***Three Gums Communication***' (internal publication), changed to '***Subtropical Plants Newsletter***' in 1974, then officially published in 1983, changed from semiannual to quarterly in 2000 and renamed ***Subtropical Plant Science*** from issue number four in 2000; Editors in Chief: Lai-Rong LI (1908-1992), 1983-1992, Wei-Nan HUANG (1933-), 1993-2011, and Ming-Hua SU (1951-), since 2012.

1972—1976 年，1982 年、1983 年，中国科学院植物研究所主编，**中国高等植物图鉴**，5 册本，第 1 册：1 157 页，1972；第 2 册：1 312 页，1972；第 3 册：1 083 页，1974；第 4 册：932 页，1975；第 5 册：1 146 页，1976；科学出版社，北京；1982、1983，中国科学院植物研究所主编，**中国高等植物图鉴 补编**，第 1 册：806 页，1982；第 2 册：879 页，1983；科学出版社，北京。

1972-1976, 1982, 1983, Institute of Botany, Chinese Academy of Sciences (editor in chief), ***Iconographia Cormophytorum Sinicorum***, vols. 1-5; 1: 1,157 p, 1972; 2: 1,312 p, 1972; 3: 1,083 p, 1974; 4: 932 p, 1975; 5: 1,146 p, 1976; Science Press, Beijing; 1982, 1983, Institute of Botany, Chinese Academy of Sciences (editor in chief), ***Iconographia Cormophytorum Sinicorum Supplementum***, vols. 1-2, 1: 806 p, 1982, 2: 879 p, 1983; Science Press, Beijing.

1972 年 7 月 1 日至 1978 年 11 月 1 日，中国科学院植物研究所更名为中国科学院北京植物研究所。

1972.07.01-1978.11.01, Institute of Botany, Chinese Academy of Sciences, renamed Beijing Institute of Botany, Chinese Academy of Sciences.

1972—1978 年，中国科学院武汉植物研究所更名为湖北省植物研究所；李春祥任副所长。

1972-1978, Wuhan Institute of Botany, Chinese Academy of Sciences, renamed Hubei Institute of Botany, Chun-Xiang LI as vice director.

1972—1986 年、1992—2002 年，云南省热带植物研究所 / 中国科学院云南热带植物研究所 / 中国科学院西双版纳热带植物园，**热带植物研究**，1–47 期（不定期，内部发行）。

1972-1986 and 1992-2002, Yunnan Institute of Tropical Botany / Yunnan Institute of Tropical Botany, Chinese Academy of Sciences / Xishuangbanna Tropical Botanical Garden, Chinese Academy of Sciences, '***Tropical Plants Research***', Parts 1-47 (issued irregularly and internally).

1972，毕培曦[496]（1948—）自香港赴美国留学，1974年获加州大学伯克利分校硕士学位，1977年获加州大学伯克利分校博士学位；返港后任教于香港中文大学，曾兼任香港中文大学中药研究中心主任。

1972, Paul Pui-Hay BUT (1948-), from Hong Kong went to United States of America, for study, received masters' degree in 1974 and Ph.D. degree in 1977[497], from University of California, Berkeley; back to Hong Kong for teaching in Chinese University of Hong Kong, and once was director of Center of Chinese Medicine, Chinese University of Hong Kong.

496 1968—1972年香港中文大学本科，学士毕业。

497 Ph.D. dissertation: Paul P. H. BUT, 1977, Systematics of *Pleuropogon* R. Br. (Poaceae); Advisor: Lincoln Constance.

1973

1973 年 2 月 19 日至 3 月 7 日，中国科学院在广东广州召开中国动植物志会议（简称“三志”，即中国植物志、中国动物志和中国孢子植物志）；中国植物志编写重新开始。来自 26 个省（市、区）的有关科研机构、高等院校、科技管理部门和其他单位的代表共计 181 人，列席代表 48 人，其中中国植物志编写人员 56 人，工作人员 3 人。[498]

1973.02.19-03.07, Meeting of Floras of China and Fauna Sinica (i.e. San Zhi, Two floras and one fauna, i.e. *Flora Reipublicae Popularis Sinicae*, *Cryptogamic Flora of China* and *Fauna Sinica*) organized by Chinese Academy of Sciences held in Guangzhou, Guangdong; *Flora Reipublicae Popularis Sinicae* restarted. One hundred eighty one voting representatives and 48 non-voting representatives, including 56 botanists working on *Flora Reipublicae Popularis Sinicae*, plus 3 additional office staff members, from research institutes, colleges, universities and scientific management organizations from 26 provinces attended the meeting.

1973 年 6 月，新乡师范学院生物系植物教研室编，**河南省维管植物目录**，166 页，新乡师范学院，河南新乡（油印本）。

1973.06, Botany Group, Department of Biology, Xinxiang Normal College, '***Checklist of Vascular Plants of Henan Province***', 166 p; Xinxiang Normal College, Xinxiang, Henan (mimeo printed).

1973 年，林善雄，台湾产苔类科属检索表[499]，东海学报 14：1-33。

1973, Shan-Hsiung LIN[500], Keys to the families and genera of the mosses of Taiwan, *Journal of Science Tunghai University*[501] 14: 1-33.

1973, Tang-Shui LIU and Ming-Jou LAI, A census of the genera and families of bryophytes and lichens of Taiwan, *Sylva* 7: 136-144.

1973 年，刘棠瑞、赖明洲，台湾苔藓类及地衣类植物之科、属，台大森林学刊 7：136-144。

1973 年，中国植物志第二届编辑委员会成立，主编为中国科学院植物研究所林镕，副主编为中

498 胡宗刚、夏振岱，2016，中国植物志编纂史，142-144 页；上海交通大学出版社，上海；http://www.cas.cn/jzzky/ysss/bns/200909/t20090928_2529193.shtml (accessed December 2017).

499 请读者注意，台湾的林善雄先生等使用的“苔藓类”表述与大陆目前所使用的相反，即苔藓称为“藓苔”；还有苔类称为“藓类”；藓类称为“苔类”；而角苔称为“角藓”。

500 As Sang-Hsiung LIN.

501 *Tunghai Journal.*

国科学院昆明植物研究所吴征镒、中国科学院植物研究所崔鸿宾[502]（1928—1994）、简焯坡、洪德元；编辑委员会19人：中国科学院华南植物研究所陈封怀，中国科学院植物研究所陈心启（1931—）、戴伦凯、汤彦承、徐全德（1921—）、俞德浚、钟补求，中国科学院西北植物研究所崔友文、何业祺（1932—），江苏省植物研究所丁志遵（1927—）、单人骅，四川大学方文培，中国科学院林业土壤研究所傅沛云（1926—2017）、刘慎谔，华南农学院蒋英，西北师范学院孔宪武，中国科学院昆明植物研究所李锡文，厦门大学曾沧江（1928—1982），中国林业科学院林业研究所郑万钧。

1973, Second Editorial Committee of Flora Reipublicae Popularis Sinicae established, with Yong LING (PE) as Editor in Chief and De-Yuan HONG (PE), Cho-Po TSIEN (PE), Hung-Pin TSUI (1928-1994, PE) and Cheng-Yih WU (KUN) as Vice Editors in Chief, with 19 members of the editorial committee: Feng-Hwai CHEN (IBSC), Sing-Chi CHEN (1931-, PE), Wan-Chun CHENG (CAF), Lun-Kai DAI (PE), Wen-Pei FANG (SZ), Pei-Yun FU (1926-2017, IFP), Yeh-Chi HO (1932-, WUK), Hsien-Wu KUNG (NWTC), Hsi-Wen LI (KUN), Tchen-Ngo LIOU (IFP), Ren-Hwa SHAN (NAS), Yen-Cheng TANG (PE), Chih-Chi TING (1927-, NAS), Chang-Chiang TSENG (1928-1982, AU), Ying TSIANG (CANT), Pu-Chiu TSOONG (PE), You-Wen TSUI (Yu-Wen TSUI, WUK), Quan-De XU (1921-, PE) and Te-Tsun YU (PE).

1973—1976年，青藏综合考察队对西藏进行了大规模的综合考察。考察队成员1973年有倪志诚（1942—）、武素功；1974年增加了陈书坤（1936—2018）、程树志、顾立民、何关福（1935—）、郎楷永、南勇，以及西藏医院的洛桑西挠和西藏军区卫生处肖永会等，并且由黄荣福（1940—）、陶德定、杨永昌（1927—）、臧穆等组成补点组对山南地区进行补点；1975年有郎楷永、倪志诚、武素功；1976年有黄永福、郎楷永、倪志诚、陶德定、武素功，又再次进藏并增加了苏志云（1936—）、尹文清。4年的考察共采得标本15 000余号。1975年和1976年陈书坤、杜庆（1938—）、管开云（1953—）、吴征镒、杨崇仁（1943—）、臧穆等两次到西藏，共采得标本4 000余号。与此同时，考察队的植被组陈伟烈（1939—）、李渤生（1946—）、王金亭等也采得标本14 000余号，林业组采得标本4 500余号，草场组采得标本2 000余号。中国科学院西北高原生物研究所的郭本兆[503]（1920—1993）、潘锦堂[504]（1933—2004）、刘尚武、王为义、徐朗然（1936—）、周立华（1934—2013）等也数次到西藏考察采集，先后获得万余号标本。这些从西藏获得的约7万号植物标本（HNWP，KUN，PE），成为编研5卷本西藏植物志的主要依据。[505]

502 北京人，1951年南开大学毕业到中国科学院植物研究所工作，长期担任中国植物志秘书、副主编等职。详细参见：崔鸿宾（遗稿），2008，我所经历的《中国植物志》三十年，中国科技史杂志 29（1）：73-89。

503 河南滑县人；1948年毕业于国立西北农学院，同年进入北平研究院植物学研究所西北工作站，1950年入中国科学院西北农业土壤研究所，1953年赴南京大学进修禾本科，1956年回中国科学院西北农业土壤研究所，1965年调入中国科学院西北高原生物研究所，先后任植物室主任、研究员、副所长；著名禾本科专家。

504 甘肃天水人，1956年入兰州大学，1961年毕业后先后在青海农牧学院、西宁市实验学校任教；1963年调至中国科学院西北高原生物研究所植物研究室工作，先后任助理研究员、副研究员、研究员，植物研究室副主任，1993年11月退休。虎耳草科专家。

505 吴征镒，1983，西藏植物志，第一卷，前言，iii - iv；科学出版社，北京。

Su-Kung WU (left) and Cheng-Yih WU (right) in Zhangmu, Nyalam, Xizang, 1975.(Photo provided by Wu Zhengyi Science Foundation)

1975年，武素功（左）和吴征镒（右）于西藏聂拉木县樟木镇（相片提供者：吴征镒科学基金会）

1973-1976, Qinghai-Xizang Expedition conducted comprehensive survey in Xizang, Chi-Cheng NI (1942-) and Shu-Kung WU in 1973, joined by Shu-Kun CHEN (1936-2018), Shu-Zhi CHENG, Li-Min GU, Guan-Fu HE (1935-), Kai-Yung LANG, Yong NAN, Yong-Hui XIAO (Sanitation Department of Xizang Military Bureau), Luoshang XINAO (Xizang Hospital), as well as Supplement Team comprising Rong-Fu HUANG (1940-), De-Ding TAO, Yong-Chang YANG (1927-) and Mu ZANG to Shannan District in 1974, Kai-Yung LANG, Chi-Cheng NI, and Shu-Kung WU in 1975 and Yong-Fu HUANG, Kai-Yung LANG, Chi-Cheng NI, De-Ding TAO and Shu-Kung WU went to Xizang with Zhi-Yun SU (1936-) and Wen-Qing YIN in 1976. More than 15,000 numbers of plant specimens were collected in four years. Shu-Kun CHEN, Qing DU (1938-), Kai-Yun GUAN (1953-), Cheng-Yih WU, Chong-Ren YANG (1943-) and Mu ZANG surveyed Xizang again and collected around 4,000 numbers of plant specimens in 1975-1976. At the same time, the vegetation team, including Wei-Lie CHEN (1939-), Bo-Sheng LI (1946-), Chin-Ting WANG et al., collected more than 14,000 numbers of plant specimens. The forestry team collected around 4,500 numbers of plant specimens and the grassland team collected more than 2,000 numbers of plant specimens. An additional 10,000 numbers of plant specimens were collected by Pung-Chao KUO (1920-1993), Shang-Wu LIU, Jin-Tang PAN (1933-2004), Lang-Rang XU (1936-), Wei-Yi WANG and Li-Hua ZHOU (1934-2013) of Northwest Institute of Plateau Biology, Chinese Academy of Sciences. In total about 70,000 numbers of plant specimens (HNWP, KUN, PE) from Xizang (Tibet) became the major vouchers for the five volumes of *Flora Xizangica*.

1973 年 10 月、1982 年 4 月，杨再义，**台湾植物名汇**，929 页，1973；中正科学技术研究讲座基金董事会、中国农村复兴联合委员会，台北；杨再义，**台湾植物名汇**，1 632 页，1982；天然书社，台北。

1973.10 and 1982.04, Tsai-I YANG, ***A List of Plants in Taiwan***, 929 p, 1973; Chung Cheng Science and Technology Foundation and Agriculture Recovery Committee of China, Taipei [Taibei]; Tsai-I YANG, ***A List of Plants in Taiwan***, 1,632 p, 1982; Natural Publishing Co. Ltd., Taipei [Taibei].

1973—2000 年，中国植物志编辑委员会办公室，**中国植物志编辑委员会编写工作简讯**，第 1–90 期（内部印制）。[506]

1973-2000, Office of the Editorial Committee of Flora Reipublicae Popularis Sinicae, '***Newsletter of Flora Reipublicae Popularis Sinicae***', parts 1-90 (internal publication).

506 胡宗刚、夏振岱，2016，中国植物志编纂史，266 页；上海交通大学出版社，上海。

1974

1974.8.27-9.23, The United States Plant Studies Delegation, organized under the auspices of the Committee on Scholarly Communication with the People's Republic of China of the National Academy of Sciences, the Social Sciences Research Council and the American Council of Learned Societies, spent four weeks in China (Hong Kong - Beijing - Jilin - Liaoning - Xi'an - Chengdu - Nanjing - Shanghai - Suzhou - Guangzhou - Hong Kong). The delegation consisted of ten plant scientists: Richard L. Bernard, Norman E. Borlaug, Nyle C. Brady, Glenn W. Burton, John L. Creech, Jack R. Harlan, Arthur Kelman, Henry M. Munger, George F. Sprague and Sterling Wortman. Alexander P. DeAngelis, from the National Academy of Sciences and a specialist in the study of Chinese literature, who served as secretary and translator for the delegation, and Philip A. Kuhn, Asian specialist and professor of Chinese history at the University of Chicago, also served as a translator.

1974年8月27日—9月23日，由国家科学院、社会科学研究理事会和美国学术团体理事会资助的中华人民共和国学术交流委员会美国植物研究代表团访问中国（行程经由香港、北京、吉林、辽宁、西安、成都、南京、上海、苏州、广州、香港），代表团成员包括10位植物学家：Richard L. Bernard、Norman E. Borlaug、Nyle C. Brady、Glenn W. Burton、John L. Creech、Jack R. Harlan、Arthur Kelman、Henry M. Munger、George F. Sprague和Sterling Wortman。美国科学院的中文文学专家Alexander P. DeAngelis任代表团秘书兼中文翻译，芝加哥大学的亚洲专家与中国历史学教授Philip A. Kuhn也是代表团翻译。

1974，Stephanne B. Sutton, ***In China's Border Provinces*** *– The Turbulent Career of Joseph Rock, Botanist-Explorer*, 334 p; Hastings House, New York. Translated into Chinese in 2013.

2013年，李若虹译，**苦行孤旅**——约瑟夫F. 洛克传，436页；上海辞书出版社，上海。

1974年，中国科学院新疆生物土壤研究所更名为新疆生物土壤沙漠研究所。1984—1992年夏训诚（1935— ）、1992—1994年李述刚（1934— ）、1994—1998年李崇舜（1939— ）任所长。

1974, Xinjiang Institute of Biology and Pedology, Chinese Academy of Sciences, renamed Xinjiang Institute of Biology, Pedology and Desert, with directors Xun-Cheng XIA (1935-), 1984-1992; Shu-Gang LI (1934-), 1992-1994; and Chong-Shun LI (1939-), 1994-1998.

1974—1985年、2013年、2013年，中国科学院西北植物研究所编著，**秦岭植物志**，3卷本1（1-5）、

Participants at first editorial meeting, *Flora Tsinlingensis,* 1961 (Photo provided by Northwest A&F University)
1961 年，秦岭植物志编审委员会第一次会议留影（相片提供者：西北农林科技大学）

2、3（1）册[507]；第 1 卷第 1 册：476 页，1976；第 1 卷第 2 册：647 页，1974；第 1 卷第 3 册：500 页，1981；第 1 卷第 4 册：421 页，1983；第 1 卷第 5 册：442 页，1985；第 2 卷：246 页，1974；第 3 卷第 1 册：329 页，1978；科学出版社，北京。李思锋、黎斌主编，**秦岭植物志增补**[508]，419 页，图版 19，2013；科学出版社，北京。郭晓思、徐养鹏编著，**秦岭植物志**，第 2 版，第 2 卷：石松类和蕨类植物，298 页，彩图 48，2013；科学出版社，北京。

1974–1985, 2013, 2013, Instituto Botanico Boreali-Occidentali Academiae Sinicae Edita, ***Flora Tsinlingensis***, vols. 1(1–5), 2 and 3(1); 1(1): 476 p, 1976, 1(2): 647 p, 1974, 1(3): 500 p, 1981, 1(4): 421 p, 1983, 1(5): 442 p, 1985, 2: 246 p, 1974, 3(1): 329 p, 1978; Science Press, Beijing; Si-Feng LI and Bin LI (editors in chief), ***Flora Tsinlingensis Supplementum***[509], 419 p; plates 1–19, 2013; Science Press, Beijing; Xiao-Si GUO and Yang-Pong HSU, ***Flora Tsinlingensis***, ed. 2, vol. 2: Lycophyta et Monilophyta, 298 p, color photos 48, 2013; Science Press, Beijing.

507 原计划第 3 卷第 2 册的苔类和第 4 卷的地衣至今未出版。

508 种子植物。

509 Spermatophyte.

1974—1995 年，中国科学院中国植物志编辑委员会，**中国植物志参考文献目录**（中国植物志参考资料），第 1-44 期[510]；中国科学院中国植物志编辑委员会，北京（内部印制）。该文献涵盖起止时间为 1958—1994 年。

1974-1995, Editorial Committee of Flora Reipublicae Popularis Sinicae, '***Bibliography of the Reference Literatures for Flora Reipublicae Popularis Sinicae*** (*Reference Materials for Flora Reipublicae Popularis Sinicae*)', parts 1-44; Editorial Committee of Flora Reipublicae Popularis Sinicae, Beijing (internal publication). The bibliography covered the period from 1958 through 1994.

1974+，中国科学院植物研究所、中国植物学会、高等教育出版社，**植物学杂志**，季刊，1976 年改为双月刊；1977 年改名为**植物杂志**，2004 年第 2 期改名为**生命世界**，2005 年改为月刊；1979[511]—1983、1983 年曹宗巽（1920—2011）[512]、1983—1993 年高信曾（1926—）、1994—1999 年胡玉熹（1937—）、1999—2003 年傅德志（1952—）、2004 年至今崔金钟（1959—）任主编。

1974+, Institute of Botany, Chinese Academy of Sciences, Botanical Society of China, Higher Education Press, '***Botanical Magazine***', quarterly, bimonthly since 1976, renamed '***Plant Magazine***' in 1977, as ***Plants*** from 1983 to 2004, then ***Life World*** from the second issue of 2004, and monthly since 2005; Editor in Chief: Tsung-Hsun TSAO (1920-2011), 1979-1983[513], Xin-Zeng GAO (1926-), 1983-1993, Yu-Xi HU (1937-), 1994-1999, De-Zhi FU (1952-), 1999-2003, and Jin-Zhong CUI (1959-), since 2004.

---

510 该出版物不仅不规则而且内容广泛，特列出详细目录如下：**1**，30 页，1974，关于种的划分问题；**2**，680 页，1974，1958—1972 目录（上、下册）；**3**，5+9 页，1974，关于种的划分问题；**4**，59 页，1974，四川、云南、贵州地名考；**5**，49 页，1975，高等植物分类学中常用拉丁缩写及苏联植物志术语缩写解说；**6**，85 页，1975，中国植物志植物中文名称（8 开本）；**7**，22 页，1975，我国本草学的发展与儒法斗争的探讨；**8**，155 页，1975，1973 年目录；**9**，14 页，1975，关于种的划分问题；**10**，14 页，1975，广西新旧地名的初步校订；**11**，48 页，1975，谈谈分类学中的一些哲学问题；**12**，156 页，1975，1974 年目录；**13**，64 页，1976，开门编志经验交流；**14**，28 页，1976，植物分类学译丛；**15**，24 页，1976，学习自然辩证法参考资料；**16**，163 页，1977，1975 年目录；**17**，76 页，1977，植物分类学译丛；**18**，46 页，1977，西藏自治区地名考；**19**，233 页，1977，1976 年目录；**20**，76 页，1977，植物分类学译丛；**21**，85 页，1978，青海地名初编；**22**，80 页，1978，植物分类学译丛；**23**，233 页，1978，1977 年目录；**24**，135 页，1979，植物学家人名、植物学书名、植物学期刊名缩写；**25**，16+8 页，1979，西藏自治区地名考（二），P. Giraldi 在陕西采集植物地名的考证；**26**，145 页，1979，1978 年目录；**27**，219 页，1980，1979 年目录；**28**，185 页，1981，1980 年目录；**29**，39 页，1981，中国植物采集简史；**30**，188 页，1982，1981 年目录；**31**，71 页，1983，国际植物命名法规简介；**32**，165 页，1983，1982 年目录；**33**，160 页，1984，1983 年目录；**34**，162 页，1985，1984 年目录；**35**，212 页，1986，1985 年目录；**36**，172 页，1987，1986 年目录；**37**，142 页，1988，1987 年目录；**38**，104 页，1989，1988 年目录；**39**，143 页，1990，1989 年目录；**40**，128 页，1991，1990 年目录；**41**，132 页，1992，1991 年目录；**42**，157 页，1993，1992 年目录；**43**，150 页，1994，1993 年目录；**44**，132 页，1995，1994 年目录。

511 1974 年创刊至 1978 年没有主编记录。

512 山东济南人（女），1936 年入国立清华大学，1940 年毕业于国立西南联合大学生物学并留校任教，1945 年赴美国，1948 年获威斯康星大学麦迪逊分校植物学与生物化学博士学位，后在 University of Texas 和 Atlanta University 任教，1951 年回国后任教于清华大学，1952 年院系调整后一直任教于北京大学。著名植物生理学家。

513 No Editor in Chief recorded before 1979.

1975

1975.07.12-08.07, Shiu-Ying HU, Arnold Arboretum of Harvard University, joined the staff of the Faculty of Science, Chinese University of Hong Kong, made her first visit to China since leaving for the United States in 1946 (visited Hong Kong, Shenzhen, Hangzhou, Shanghai, Suzhou, Nanjing, Shenyang, Dalian, Beijing, and Guangzhou), met many Chinese botanists (especially in Beijing, Guangzhou, Nanjing and Shenyang) and wrote a general account of Chinese botany.[514]

1975 年 7 月 12 日至 8 月 7 日，哈佛大学阿诺德树木园胡秀英参加香港中文大学理学院的大陆访问团，此为其 1946 年离开大陆后首次回国访问（行程经由香港、深圳、杭州、上海、苏州、南京、沈阳、大连、北京、广州），其间会见很多植物学家（特别是在北京、广州、南京和沈阳），返美国后撰写了相关见闻。

1975，浅間一男，**被子植物の起源**，400 页，三省堂，东京（Kazuo Asama, ***The Origin of the Angiosperms***, 400 p, Sanseido, Tokyo）。Translated into Chinese in 1988.

1988 年，谷祖纲、珊林译，**被子植物的起源**，220 页；海洋出版社，北京。

1975 年，中华植物学会于台湾台北成立。

1975, Chinese Society of Botany established in Taipei [Taibei], Taiwan.

1975, Harold R. Fletcher, ***A Quest of Flowers****—The plant explorations of Frank Ludlow and George Sherriff told from their diaries and other occasional writings*, 387 p; Edinburgh University Press, Edinburgh.

1975 年，Harold R. Fletcher，**寻求花** ——Frank Ludlow 与 George Sherriff 日记与其他偶见作品所讲述的植物考察故事，387 页；爱丁堡大学出版社，爱丁堡。

1975 年，刘棠瑞、赖明洲，台湾藓苔植物之分类，台湾省立博物馆科学年刊 18：57-91。

1975, Tang-Shui LIU and Ming-Jou LAI, The classification of Bryophyta of Taiwan, *Annual Bulletin of National Taiwan Museum* 18: 57-91.

1975 年，吴鹏程、罗健馨，苔藓植物的研究概况，植物分类学报 13（1）：131-136。

---

514 Shiu-Ying HU, 1975, The tour of a botanist in China, *Arnoldia* 35(6): 265-295; Shiu-Ying HU, 1977, More about tours of botanists and gardeners in China, *Arnoldia* 37(3): 157-163.

1975, Pan-Cheng WU and Jian-Shing LUO, Present status of bryology research, *Acta Phytotaxonomica Sinica* 13(1): 131–136.

1975, Ulrich Schweinfurth and Heidrun Schweinfurth-Marby, ***Exploration in the Eastern Himalayas and the River Gorge Country of Southeastern Tibet*** *– Francis (Frank) Kingdon Ward (1885–1958) – an annotated bibliography with a map of the area of his expeditions*, 114 p; Steiner, Wiesbaden.

1975 年，Ulrich Schweinfurth and Heidrun Schweinfurth-Marby，Francis (Frank) Kingdon Ward（1885—1958）**东喜马拉雅和藏东南河流峡谷的考察 ——** 附说明的文献目录及考察地区图，114 页；Steiner，Wiesbaden。

1975 年，厦门水产学院养殖系水生生物教研组编，**水生维管束植物图册**，224 页；厦门水产学院，厦门。

1975, Aquatic Biology Teaching Group, Department of Cultural Breeding, Xiamen Aquatic College, '***Icones of Aquatic Vascular Plants***', 224 p; Xiamen Aquatic College, Xiamen.

1975 年，中国科学院新疆生物土壤沙漠研究所编，**新疆植物名录**，72 页；中国科学院新疆沙漠土壤研究所生物研究室（铅印本），乌鲁木齐。

1975, Xinjiang Institute of Biology, Pedology and Desert, Chinese Academy of Sciences, '***Checklist of Xinjiang Plants***', 72 p; Department of Biology, Xinjiang Institute of Biology, Pedology and Desert, Chinese Academy of Sciences (internal publication), Urumqi.

1975—1977 年，中国植物志第三届编辑委员会成立，主编：中国科学院植物研究所林镕，副主编：中国科学院昆明植物研究所吴征镒，中国科学院植物研究所崔鸿宾、简焯坡、洪德元；编辑委员会 18 人：中国科学院植物研究所曹子余（1935— ）、陈心启、戴伦凯、冀朝祯（1936— ）、汤彦承、徐全德、俞德浚、钟补求，中国科学院林业土壤研究所李书心（1926—2007），中国科学院成都生物研究所傅发鼎（1936— ），中国科学院华南植物研究所阮云珍（1936— ），杭州植物园姚昌豫（1936— ），中国科学院西北植物研究所何业祺、唐昌林（1934— ），南京大学王正平（1929— ），中国科学院新疆生物土壤沙漠研究所杨戈（1938— ），广西植物研究所张本能（1938— ），中国林业科学院林业研究所郑万钧。

1975–1977, Third Editorial Committee of Flora Reipublicae Popularis Sinicae founded, with Yong LING (PE) as Editor in Chief and De-Yuan HONG (PE), Cho-Po TSIEN (PE), Hung-Pin TSUI (PE) and Cheng-Yih WU (KUN) as Vice Editors in Chief, with 18 editorial members: Zi-Yu CAO (1935–, PE), Ben-Neng CHANG (1938–, IBK), Sing-Chi CHEN (PE), Wan-Chun CHENG (CAF), Lun-Kai DAI (PE), Yeh-Chi HO (WUK), Chao-Zhen JI (1936–, PE), Shu-Xin LI (1926–2007, IFP), Fa-Ding PU (1936–, CDBI), Yun-Zhen RUAN (1936–, IBSC), Chang-Lin TANG (1934–, WUK), Yen-Cheng TANG (PE), Pu-Chiu TSOONG

(PE), Cheng-Ping WANG (1929-, N), Quan-De XU (PE), Ge YANG (1938-, XJBI), Chang-Yu YAO (1936-, HHBG) and Te-Tsun YU (PE).

1975-1979, 1993-2003, 2012, Editorial Committee of the Flora of Taiwan, Hui-Lin LI (editor in chief)[515], ***Flora of Taiwan***, vols. 1-6; 1: 562 p, 1975, 2: 722 p, 1976, 3: 1,000 p, 1977, 4: 994 p, 1978, 5: 1,166 p, 1978, 6: 665 p, 1979; Epoch Publishing Co., Taipei [Taibei]. Tseng-Chieng HUANG (editor in chief), ***Flora of Taiwan***, ed. 2, vols. 1-6[516]; 1: 648 p, 1994, 2: 855 p, 1996, 3: 1,084 p, 1993, 4: 1,217 p, 1998, 5: 1,143 p, 2000, 6: 343 p, 2003; Department of Botany, Taiwan University, Taipei [Taibei]. Taiwan Normal University, 2012, ***Flora of Taiwan*, second edition, *Supplement*** *Gymnospermae and Angiospermae*, 414 p; Taiwan Normal University Press, Taipei [Taibei]. The Engler Medal in Sliver was awarded by the International Association of Plant Taxonomy on completion of the second edition in 2004.

1975—1979 年、1993—2003 年、2012 年，台湾植物志编辑委员会，李惠林主编[517]，**台湾植物志**，6 卷本；第 1 卷：562 页，1975；第 2 卷：722 页，1976；第 3 卷：1 000 页，1977；第 4 卷：994 页，1978；第 5 卷：1 166 页，1978；第 6 卷：665 页，1979；现代关系出版社，台北。台湾植物志第 2 版编辑委员会，黄增泉主编，**台湾植物志**，第 2 版，6 卷本；第 1 卷：648 页，1994；第 2 卷：855 页，1996；第 3 卷：1 084 页，1993；第 4 卷：1 217 页，1998；第 5 卷：1 143 页，2000；第 6 卷：343 页，2003；台湾大学植物学系，台北。台湾师范大学，2012，**台湾植物志**，**第 2 版**，**补遗**，裸子植物和被子植物，414 页；台湾师范大学出版社，台北。台湾植物志第 2 版于 2004 年获得国际植物分类协会恩格勒银质奖章。

---

515 Hui-Lin LI had been elected an academician of the Academia Sinica in 1964 and actively raised funds to edit the *Flora of Taiwan*. Dr. Tetsuo Koyama (1933-) at the New York Botanical Garden was informed that the U.S. government would provide support through funds for academic research in Taiwan. Dr. Tetsuo Koyama and Prof. Tseng-Chieng HUANG initiated the editing of the Flora and Professor Hui-Lin LI was appointed the main convener. He also invited professors Tang-Shui LIU and Charles E. DeVol (1902-1989) to serve as editors. In 1969, two sides formally signed an academic cooperation agreement that resulted in funding for the project from the United States National Science Foundation.

516 A Checklist of the Vascular Plants of Taiwan, vol. 6: 15-139, 2003.

517 李惠林教授自 1964 年当选“中央”研究院院士始，即积极为编辑台湾植物志筹募经费；时在纽约植物园的小山铁夫（Tetsuo Koyama，1933—）博士得知美国政府欲资助台湾学术研究，便与黄增泉教授发起编辑台湾植物志，推举李惠林教授为主要召集人，并邀刘棠瑞教授及棣慕华 [Charles E. DeVol，1902—1989，美国传教士与植物学家，生于江苏六合一个美国医生与传教士家庭，大半生在中国传教（包括台湾）；印第安纳大学植物学博士，曾在台湾大学任教] 教授等共 5 位担任编辑。1969 年，双方签订学术合作协议，由美国国家科学基金会资助，1975—1979 年出版第 1 版 6 卷本。

1976

1976 年 5 月 5 日至 13 日，中国科学院中国孢子植物志编辑委员会第二届会议于北京中国科学院微生物研究所举行，来自北京、广东、江苏、浙江、云南、山东、湖北、辽宁等省市的编辑委员会和应邀代表共 46 人出席会议。中国孢子植物志包括中国海藻志、中国淡水藻志、中国真菌志、中国地衣志和中国苔藓志。

1976.05.05-05.13, Second Consilio Florarum Cryptogamarum Sinicarum, Academiae Sinicae Edita held at the Institute of Microbiology, Chinese Academy of Sciences, Beijing. Forty six attendees from Beijing, Guangdong, Hubei, Jiangsu, Liaoning, Shandong, Yunnan and Zhejiang. The *Cryptogamic Flora of China* covers *Flora Algarum Marinarum Sinicarum, Flora Algarum Sinicarum Aquae Dulcis, Flora Fungorum Sinicorum, Flora Lichenum Sinicorum, and Flora Bryophytorum Sinicorum.*

1976 年，洪德元，试论分类学的基本原理，植物分类学报 14（2）：90-100。

1976, De-Yuan HONG, Discussion on the basic principle of taxonomy, *Acta Phytotaxonomica Sinica* 14(2): 90-100.

1976, Ming-Jou LAI, A Key to the families of Bryophyta in Taiwan, *Taiwania* 21(1): 73-82.

1976 年，赖明洲，台湾苔藓植物科之检索表，*Taiwania* 21（1）：73-82。

1976, Ming-Jou LAI and Jen-Rong WANG-YANG, Index Bryoflorae Formosensis, *Taiwania* 21(2): 159-203.

1976 年，赖明洲、王杨贞容，台湾苔藓类目录，*Taiwania* 21（2）：159-203。

1976—1978 年，中国科学院武汉植物研究所和中国科学院植物研究所联合其他单位组成神农架植物资源考察队，参加约 140 人次，获得 10 000 号 5 万多份植物标本（HIB、PE）。[518]

1976-1978, Shennongjia Expedition of Plant Resources consists by about 140 persons, from Wuhan Institute of Botany and Institute of Botany, Chinese Academy of Sciences, and from other institutions, more than 10,000 numbers and more than 50,000 plant specimens were collected (HIB, PE).

518 中国科学院武汉植物研究所编著，神农架植物，467 页，照片 63，1980；湖北人民出版社，武汉。

1976, Benito C. TAN[519] (1946-2016, Chinese Filipinos) went to Canada for study of bryology, received Ph.D. degree from University of British Columbia in 1981[520], assistant professor of University of the Philippines at Los Banos, 1980-1988, curator of bryophyte and pteridophytes, University of Philippine at Los Banos Museum of Natural History, 1981-1988, curatorial assistant at New York Botanical Garden, 1988-1989, research associate in bryology at Farlow Herbarium (FH), Harvard University, 1990-1997, senior lecturer in botany, National University of Singapore, 1997-1998, associate professor in biology, National University of Singapore, 1998-2007, keeper of herbarium, Singapore Botanic Gardens, 2007-2010; research associate of University and Jepson Herbarium, University of California, Berkeley, 2013-2016. Ben is world famous bryologist, and joined many field expedition and projects in China; member of the editorial committee of *Bryoflora of China* (English version) from 1991-2016, and one of chief editors of *Chenia*, 1993-2016.

1976年, Benito C. TAN[521]（1946—2016，华裔菲律宾人）赴加拿大学习苔藓学, 1981年获得加拿大不列颠哥伦比亚大学博士学位; 1980—1988年任菲律宾大学Los Banos分校助理教授, 1981—1988年任菲律宾大学Los Banos分校自然历史博物馆苔藓与蕨类植物馆馆长, 1988—1989年

Benito C. TAN (1946-2016) (Photo provided by Boon-Chuan HO)
Benito C. TAN（1946—2016）(相片提供者：何文钏）

519 Benito (Ben) Ching TAN (August 30, 1946-December 23, 2016), an internationally recognized bryologist and Research Associate at the University and Jepson Herbaria died in December 2016. Among his many honors, Ben was a fellow of the California Academy of Sciences and recipient of the prestigious Richard Spruce Award of the International Association of Bryologists. Ben specialized in the systematics and biogeography of mosses, especially the family Sematophyllaceae, and published many studies on East Asian bryophytes. He was considered the world's foremost expert on tropical bryophytes and was often sought after to identify rainforest taxa (http://ucjeps.berkeley.edu/people/Tan_obit.html, accessed by 2018). For details, see Boon-Chuan HO and James R. Shevock, 2018, A tribute to Benito C. TAN (1946-2016), distinguished musicologist, *Philippine Journal of Systematic Biology*, 12(1): iv-XXVIII; Peng-Cheng WU, 2018, Dr. Benito C. TAN, our friend and editor—In memory of his great contribution to Chinese bryology, *Chenia* 13: 172-173.

520 Ph.D. Dissertation: A moss flora of the Selkirk and Purcell Mountain Ranges, southeastern British Columbia, 928 p; Advisor: Wilfred B. Schofield (1927-2008).

521 华裔菲律宾人，出生于马尼拉；1967年本科毕业于菲律宾远东大学，1974年硕士毕业于菲律宾Diliman分校；中文名字“贝尼托陈”曾在隐花植物生物学第1期（1993）中记载；另一个中文名字“陈班头”曾在云南植物志第19卷（苔藓植物藓类纲，2005）中使用。

任纽约植物园研究助理，1990—1997 年任哈佛大学隐花植物标本馆研究馆员，1997—1998 年任新加坡大学植物学高级讲师，1998—2007 年任新加坡大学植物学副教授，2007—2010 年任新加坡植物园植物标本馆馆长；2013—2016 年任加州大学伯克利分校植物标本馆附属研究员；世界著名苔藓学家，多次参加中国的野外考察和研究项目；1991—2016 年任英文版中国藓类植物志编辑委员会委员，1993—2016 年任隐花植物生物学主编之一。

1976 年、1979 年，2001—2002 年，湖北省植物研究所、中国科学院武汉植物研究所编，**湖北植物志**，2 卷本，第 1 卷：508 页，1976；第 2 卷：522 页，1979；湖北人民出版社，武汉。中国科学院武汉植物研究所编著，傅书遐主编，**湖北植物志**，4 卷本[522]；第 1 卷：508 页，2001；第 2 卷：510 页，2002；第 3 卷：746 页，2002；第 4 卷：692 页，2002；湖北科学技术出版社，武汉。

1976 and 1979, 2001-2002, Hubei Institute of Botany, Wuhan Institute of Botany, Chinese Academy of Sciences, ***Flora Hupehensis***, vols. 1-2; 1: 508 p, 1976, 2: 522 p, 1979; Hubei People's Press, Wuhan. Wuhan Institute of Botany, Chinese Academy of Sciences, Shu-Hsia FU (editor in chief), ***Flora Hubeiensis***, vols. 1-4; 1: 508 p, 2001, 2: 510 p, 2002, 3: 746 p, 2002, 4: 692 p, 2002; Hubei Science and Technology Press, Wuhan.

1976—1980 年，四川省卫生厅、商业厅和省供销社组织，四川省中药研究所对全省各地县进行药用植物普查，共采集 20 余万份标本，成为现今重庆市中药研究院（原四川省中药研究所，SM）标本馆馆藏标本的主要来源。

1976-1980, Under the leadership of the Health Bureau, Commerce Bureau and Supply and Marketing Cooperative of Sichuan Province, the Sichuan Institute of Chinese Materia Medica organized a survey of all counties in the province for medicinal plants and collected more than 200,000 plant specimens. The specimens became the major collection of today's Chongqing Academy of Chinese Materia Medica (SM).

522 该志前后正式出版印刷两次，内容一致，但署名与西文标题不同（前者为韦氏拉丁化，而后者为拼音拉丁化）；且出版社也不同。

1977

1977.06.29-07.08, Chinese-American botanical expedition in Taiwan: Ilan, Hualian, Nantou, Taibei, Taoyuan. David E. Boufford (1941-, MO) and Emily W. Wood (1949-, MO), Chang-Fu HSIEH, Ming-Jou LAI (1949-2007) and Ching-I PENG (1950-2018, TAI), collected 247 numbers of vascular plants (David E. Boufford et al. nos. 19202-19448, CM, MO).

1977年6月29日至7月8日，中美植物联合考察台湾的宜兰、花莲、南投、台北、桃园，密苏里植物园 David E. Boufford（1941—）和 Emily W. Wood（1949—），台湾大学谢长富、赖明洲（1949—2007）和彭镜毅（1950—2018），采集维管束植物247号（CM、MO）。

Field Trip in Taiwan (left-right): David E. Boufford, Chang-Fu HSIEH and Emily W. Wood at Hehuan Shan, 29 June-8 July 1977 (Photo provided by David E. Boufford)
1977年6月29日至7月8日，中美植物学家考察台湾合欢山（从左至右）：David E. Boufford、谢长富、Emily W. Wood（相片提供者：鲍棣伟）

1977年，辽宁省林业土壤研究所编，**东北藓类植物志**，404页；科学出版社，北京。

1977, Liaoning Institute of Forestry and Pedology, ***Flora Muscorum Chinae Boreali-Orientalis***, 404 p; Science Press, Beijing.

1977年、1982年，2013—2017年，

江苏省植物研究所编，**江苏植物志**，2卷本：上册：1-502页，1977，江苏人民出版社，南京；下册：1-1 010页，1982，江苏科学技术出版社，南京。刘启新主编，**江苏植物志**，第2版，5卷本：第1卷：

573 页，彩 32，2017；第 2 卷：507 页，彩 32，2013；第 3 卷：528 页，彩 32，2015；第 4 卷：540 页，彩 32，2015；第 5 卷：473 页，彩 32，2015；江苏科学技术出版社，南京[523]。

1977 and 1982, 2013-2017, Jiangsu Institute of Botany, '***Flora of Jiangsu***', 1: 1-502 p, 1977; Jiangsu People's Press, Nanjing; 2: 1-1,010 p, 1982; Jiangsu Science and Technology Press, Nanjing. Qi-Xin LIU (editor in chief), ***Flora of Jiangsu***, ed. 2, 5 vols. 1: 573 p, color plates 32, 2017; 2: 507 p, color plates 32, 2013; 3: 528 p, color plates 32, 2015; 4: 540 p, color plates 32, 2015; 5: 473 p, color plates 32, 2015; Jiangsu Science and Technology Press, Nanjing.

1977—2010 年，云南省植物研究所、中国科学院昆明植物研究所，吴征镒主编，**云南植物志**，21 卷本、总索引；第 1 卷：870 页，1977；第 2 卷：889 页，1979；第 3 卷：795 页，1983；第 4 卷：823 页，1986；第 5 卷：809 页，1991；第 6 卷：910 页，1995；第 7 卷：888 页，1997；第 8 卷：778 页，1997；第 9 卷：807 页，2003；第 10 卷：929 页，2006；第 11 卷：754 页，2000；第 12 卷：884 页，2006；第 13 卷：918 页，2004；第 14 卷：885 页，2003；第 15 卷：874 页，2003；第 16 卷：876 页，2006；第 17 卷：650 页，2000；第 18 卷：525 页，2002；第 19 卷：681 页，2005；第 20 卷：785 页，2006；第 21 卷：477 页，2005；总索引，682 页，2010；科学出版社，北京。

1977-2010, Yunnan Institute of Botany, Kunming Institute of Botany, Chinese Academy of Sciences, Cheng-Yih WU (editor in chief), ***Flora Yunnanica***, vols. 1-21 and Index; 1: 870 p, 1977; 2: 889 p, 1979; 3: 795 p, 1983; 4: 823 p, 1986; 5: 809 p, 1991; 6: 910, 1995; 7: 888 p, 1997; 8: 778 p, 1997; 9: 807 p, 2003; 10: 929 p, 2006; 11: 754 p, 2000; 12: 884 p, 2006; 13: 918 p, 2004; 14: 885 p, 2003; 15: 874 p, 2003; 16: 876 p, 2006; 17: 650 p, 2000; 18: 525 p, 2002; 19: 681 p, 2005; 20: 785 p, 2006; 21: 477 p, 2005; Index: 682 p, 2010; Science Press, Beijing.

523 马克平，2017，《江苏植物志》修订再版，让人耳目一新，生物多样性 25（8）：914；夏仟仟，2017 年 8 月 8 日，一本“内外兼修”的地方植物志——“中国最美图书”这样打造（http://www.bookdao.com/article/399429/，2018 年进入）。

1978

1978.05.04-06.20, Peter S. Green (1920-2009), Deputy Director and Keeper of the Herbarium, and John B. Simmons (1937-), Curator, Royal Botanic Gardens, Kew, visited Guangzhou, Hainan, Lushan, Hangzhou, Beijing, Guilin. They visited five botanical gardens and six herbaria. They collected about 150 specimens in Hong Kong and Hainan.[524]

1978 年 5 月 4 日至 6 月 20 日，英国皇家植物园邱园副主任兼植物标本馆馆长 Peter S. Green（1920—2009）和 John B. Simmons（1937—）研究员访问中国，行程经广州、海南、庐山、杭州、北京、桂林。他们访问了 5 个植物园和 6 个植物标本馆。两人在香港和海南采集 150 号植物标本。

1978.05.20-06.18, Botanical Society of America Delegation to the People's Republic of China; ten members lead by Lawrence Bogorad (1921-2003), Harvard University, Bruce M. Bartholomew (1946-), Botanical Garden of University of California, Berkeley, Thomas S. Elias (1942-), Cary Arboretum, New York Botanical Garden, Richard H. Hageman (1917-2002), University of Illinois, Urbana, Richard A. Howard (1917-2003), Arnold Arboretum of Harvard University, J. William Schopf (1941-), University of California, Los Angeles, Jane Shen-Miller (1933-), U.S. National Science Foundation, Richard C. Starr (1924-1998), University of Texas, Austin, William TAI (1932-2006), Michigan State University, East Lansing and Anitra Thorhaug, Florida International University. They visited major botanical institutes, botanical gardens and universities in Beijing, Guangzhou, Hangzhou, Hong Kong, Kunming, Nanjing, Shanghai, Suzhou and Wuhan to meet Chinese botanists and discuss current developments in botanical research[525].

1978 年 5 月 20 日至 6 月 18 日，美国植物学会代表团 10 人访华，代表团成员是哈佛大学 Lawrence Bogorad（1921—2003，领队）、加州大学伯克利植物园 Bruce M. Bartholomew（1946—）、纽约植物园 Cary Arboretum 的 Thomas S. Elias（1942—）、伊利诺伊大学 Urbana 的 Richard H. Hageman（1917—2002）、哈佛大学阿诺德树木园 Richard A. Howard（1917—2003）、加州大学洛杉矶分校 J. William Schopf（1941—）、美国国家自然科学基金委 Jane Shen-Miller（沈育培，1933—）[526]、德州大学

524 Peter S. Green and John B. Simmons, 1978, *Report of a visit to the People's Republic of China*, 33 pages, Royal Botanic Gardens, Kew.

525 Richard A. Howard, 1978, Botanical impressions of the People's Republic of China, *Arnoldia* 38(6): 218-237; Anitra Thorhaug (editor), 1978, Botany in China: Report of the Botanical Society of America Delegation to the People's Republic of China, May 20-June 18, 1978, United States-China Relations Report No. 6; Center for Research in International Studies, Stanford University, 1978, 154 p, Stanford, California; Jane Shen-Miller, 1979, The Botanical Society of American delegation trip to the People's Republic of China, *BioScience* 29(5): 300-305; Bruce Bartholomew, Richard A. Howard, Thomas S. Elias, 1979, Phytotaxonomy in the People's Republic of China, *Brittonia* 31(1): 1-25.

526 Jane Yu-Pei SHEN, Ph.D., Michigan State University (1959).

Botanical Society of America Delegation to China at Institute of Botany, Chinese Academy of Sciences, Beijing, 20 May–18 June 1978 (left-right): J. William Schopf, Jia-Ge LI, Richard C. Starr, Jen HSU, Yan-Feng FU, Thomas S. Elias, Su-Xuan WU, Lawrence Bogorad, Richard A. Howard, Anitra Thorhaug, Richard H. Hageman, Pei-Sung TANG, Bruce M. Bartholomew, Te-Tsun YU, William TAI, Jane Shen-Miller, Xian-Bo WANG, Xu-Chuan DUAN, Xue-Yu HOU (Photo provided by Bruce M. Bartholomew)

1978 年 5 月 20 日至 6 月 18 日，美国植物学家代表团访问北京的中国科学院植物研究所（从左至右）：J. William Schopf、李佳格、Richard C. Starr、徐仁、付燕凤、Thomas S. Elias、吴素萱、Lawrence Bogorad、Richard A. Howard、Anitra Thorhaug、Richard H. Hageman、汤佩松、Bruce M. Bartholomew、俞德浚、戴威廉、沈育培、王献博、段续川、侯学煜（相片提供者：Bruce M. Bartholomew）

Austin 的 Richard C. Starr(1924—1998)、密执安州立大学 East Lansing 分校的 William TAN[527](戴威廉，1932—2006)，以及佛罗里达国际大学的 Anitra Thorhaug。代表团参观了北京、广州、杭州、昆明、南京、上海、苏州、武汉、香港等地的主要植物学机构、植物园和大学，会见了中国植物学家，并讨论当前植物学研究的进展。

1978 年 8 月，彭镜毅[528]（1950—2018）赴美国留学，1982 年 5 月获美国圣路易斯华盛顿大学生物学博士学位；1982 年 12 月至 1987 年 7 月任“中央”研究院植物研究所副研究员，1983 年任“中央”研究院植物研究所植物标本馆（HAST）馆长，1987 年 8 月至 2004 年 12 月任“中央”研究院植物研究所研究员，2005—2015 年任“中央”研究院生物多样性研究中心研究员，2007—2015 年兼任生物多样性研究博物馆（含植物标本馆等）馆长，2015 年退休。

Ching-I PENG (1950-2018) in 2016 (Photo provided by Jin-Shuang MA)
2016 年的彭镜毅（1950—2018）（相片提供者：马金双）

1978.08, Ching-I PENG (1950-2018) entered at Washington University, St. Louis, Missouri, United States of America, received Ph. D. in 1982[529]; employed by the Institute of Botany, Academia Sinica, 1983, as associate curator, 1982-1987; director of the herbarium (HAST), Institute of Botany, Academia Sinica, then curator, 1987-2004; curator of Biodiversity Center, Academia Sinica, 2005-2015; director of

527 江苏扬州人，毕业于上海震旦中学和扬州中学，1949 年随父母赴台，1961 年中兴大学农艺学系毕业后赴美国，1963 和 1967 年在犹他州立大学分别获硕士和博士学位（植物遗传学），1967—1969 年在斯坦福大学作为博士后从事研究工作并与任教于该校的 Peter H. Raven 相识，后任教于密执安州立大学，1988 年应 Peter H. Raven 之邀入密苏里植物园主持英文版中国植物志项目，并频繁穿梭于中美两国，为促成英文版中国植物志项目和早期的工作做出不可替代的贡献（英文版中国植物志第 13 卷特别纪念他）；1998 年主持马里兰大学全球华人事务研究中心并继续穿梭于中美之间；2002 年发起建立华盛顿特区国家树木园中国园倡议并于 2004 年两国签署备忘录之后创建中国园之友会。

528 1968 年 9 月至 1972 年 6 月中兴大学植物学系学士，1972—1974 年服兵役，1974 年 9 月至 1976 年 6 月台湾大学植物学研究所硕士；1976 年 7 月至 1977 年 12 月台湾省畜产试验所杨梅分所研究助理，1978 年 1 月至 1978 年 7 月“中央”研究院植物研究所助理研究员。

529 Ph.D. dissertation, A biosystematics study of *Ludwigia* sect. *Microcarpium* (Onagraceae), Advisor: Peter H. Raven. Ching-I PENG, 1989, The systematics and evolution of *Ludwigia* sect. *Microcarpium* (Onagraceae), *Annuals of the Missouri Botanical Garden* 76(1): 221-302.

the Biodiversity Research Museum, Academia Sinica, 2007-2015; retired 2015.

1978 年 9 月，赖明洲[530]（1949—2007）赴芬兰留学，1981 年 9 月获博士学位；1981—1982 年任职于台湾博物馆，1983 年任职台湾自然科学博物馆并兼任东海大学副教授，1983—1985 年任东海大学景观系助理教授，1985 年任副教授，1989—1992 年任系主任，同时在辅仁大学和中国文化大学兼职。1988 年开始访问大陆，并先后受聘于香港中文大学、中国科学院植物研究所、中国科学院沈阳应用生态研究所、上海自然博物馆、华东师范大学、上海师范大学等，在芬兰成立芬中植物学基金，自费资助中国植物学出版有关学术专著（如 S. C. CHEN et al.，*Bibliography of Chinese Systematic Botany*, 1993），资助中国苔藓研究人员赴海外研究（如 1990 年中国 9 位苔藓学者赴芬兰开会并于会后在芬兰进行研究），资助年轻的中国植物学人员赴芬兰攻读博士学位，如吴声华（1961— ）、何小兰（1965— ）、饶鹏程（1963— ），以及帮助中国苔藓学者支付国际学会的会费等，1990 年通过芬中植物学基金创办亚洲苔藓与地衣学报。[531]

1978.09, Ming-Jou LAI (1949-2007), studied bryology in Finland, received Ph. D.

Ming-Jou LAI in 1981 and 1997 (Photo provided by Acta Bryolichenologica Asiatica)
1981 年和 1997 年的赖明洲（相片提供者：亚洲苔藓与地衣学报）

530 台湾基隆人，1967—1971 年台湾大学林学学士，1971—1973 年台湾大学树木学硕士兼林学系标本馆助理，1973—1975 年服役于海军，1975—1976 年和 1976—1978 年分别任台湾大学植物系助理和讲师。

531 Timo J. Koponen, 2010, Ming-Jou LAI (1949-2007), bryologist, lichenologist, ornamental horticulturist and generous benefactor, *Acta Bryolichenologica Asiatica* 3 (Dr. Ming-Jou LAI Memorial Volume): 3-18; Chien GAO and Tong CAO, 2010, In memory of Dr. Ming-Jou LAI and fruitful cooperation in bryological research, *Acta Bryolichenologica Asiatica* 3 (Dr. Ming-Jou LAI Memorial Volume): 19-22.

degree in 1981[532]; returned to Taiwan as curator of Taiwan Museum, 1981-1982; curator of Taiwan Museum of Natural Science, Taichung, 1983; assistant and associate professors at Tunghai University, 1983 and 1985; dean of Landscape Architecture Department and professor at Fu Jen Catholic University and Chinese Culture University, 1989-1992; visited mainland China, 1988; honorary professor, Chinese University of Hong Kong, Institute of Botany, Chinese Academy of Sciences, Institute of Applied Ecology, Chinese Academy of Sciences, Shanghai Natural History Museum, East China Normal University, and Shanghai Normal University; established and financed Finnish-Chinese Botanical Foundation to support publication of botanical monographs (such as S. C. CHEN et al., *Bibliography of Chinese Systematic Botany*, 1993) and travel expenses of Chinese bryologists to visit Finland (such as nine Chinese bryologists to attend meeting in 1990 and to study there after the meeing) and young Chinese students, such as Sheng-Hua WU (1961-), Xiao-Lan HE (1965-), Peng-Cheng RAO (1963-), to study abroad for their Ph.D.; paid registration fees for Chinese bryologists to attend international conference and issued *Acta Bryolichenologica Asiatica* in 1990.

1978 年 9 月，四川省生物研究所更名为中国科学院成都生物研究所；1978—1979 年邓国彪、1980—1982 年伍义泽（1917—2012）、1982—1987 年仇镛（1930—2008）、1987—1991 年张永地（1932—2005）、1991—1995 年刘照光（1934—2001）、1995—2008 年李伯刚（1947—）、2008—2012 年吴宁（1964—）、2013 年至今赵新全（1959—）任所长。

1978.09, Sichuan Institute of Biology renamed Chengdu Institute of Biology, Chinese Academy of Sciences, with directors Guo-Biao DENG, 1978-1979; Yi-Ze WU (1917-2012), 1980-1982; Yong QIU (1930-2008), 1982-1987; Yong-Di ZHANG (1932-2005), 1987-1991; Zhao-Guang LIU (1934-2001), 1991-1995; Bo-Gang LI (1947-), 1995-2008; Ning WU (1964-), 2008-2012; and Xin-Quan ZHAO (1959-), since 2013.

1978 年 10 月 5 日 —10 月 14 日，中国植物学会成立四十五周年年会（第八届全国代表大会）于云南昆明举行，269 人出席[533]；理事长为汤佩松[534]（1903—2001）。

1978.10.05-10.14, Forty fifth anniversary meeting (Eighth National Congress) of Botanical Society of China held in Kunming, 269 attendees, with Pei-Sung TANG (1903-2001) as president.

1978 年 11 月 27 日，中国科学院北京植物研究所改名为中国科学院植物研究所。1978—1983 年汤佩松、1983—1987 年钱迎倩（1932—2010）、1987—1990 年路安民（1939—）、1990—1998 年张新时（1934—）、1998—2006 年韩兴国（1959—）、2006—2010 年马克平（1958—）、2010—2016 年方

532 Ph.D. dissertation: Ming-Chou LAI, 1981, Taxonomic and floristic studies on the mosses and lichens of Taiwan, China. *Publications from the Department of Botany, University of Helsinki* 9: 1-40; Advisor: Teuvo T. Ahti (1934-).

533 邹安寿，1979，征程万里奋勇攀登 —— 记中国植物学会四十五周年年会，植物杂志 1：1-3。

534 湖北蕲水人；植物生理学家、中国植物生理学的奠基人之一；参见娄成后主编，1993，汤佩松论文选集，747 页；中国世界语出版社，北京。

Forty fifth anniversary of Botanical Society of China in Kunming, 5 October 1978 (left-right) : Cheng-Yih WU, Te-Tsun YU, Hsiang-Chung SUN, Yen-Chin YANG, Hung-Ta CHANG, Rui-Qi ZHANG, Mien CHENG, Ren-Hwa SHAN, Su-Xuan WU, Jen HSU, Hao-Jan ZHU, Yuan-Zhen ZHAO, Chen-Ying CHOU, Dong-Lai TAO. Xing-Wu ZHAO (Photo provided by Wu Zhengyi Science Foundation)

1978 年 10 月 5 日，中国植物学会四十五周年年会于昆明（从左至右）：吴征镒、俞德浚、孙祥钟、杨衔晋、张宏达、张瑞琪、郑勉、单人骅、吴素萱、徐仁、朱浩然、赵元桢、周贞英、陶东来、赵星武（相片提供者：吴征镒科学基金会）

精云（1959—）、2016 年至今汪小全（1968—）[535] 任所长。

1978.11.27, Beijing Institute of Botany, Chinese Academy of Sciences, renamed Institute of Botany, Chinese Academy of Sciences, with directors Pei-Sung TANG, 1978-1983; Ying-Chien CHIEN (1932-2010), 1983-1987; An-Ming LU (1939-), 1987-1990; Hsin-Shi CHANG (1934-), 1990-1998; Xing-Guo HAN (1959-), 1998-2006; Ke-Ping MA (1958-), 2006-2010; Jing-Yun FANG (1959-), 2010-2016; and Xiao-Quan WANG (1968-), since 2016.

1978 年，秦仁昌，中国蕨类植物科属的系统排列和历史来源（一、续），植物分类学报 16（3）：1-19、16（4）：16-37。

1978, Ren-Chang CHING, The Chinese fern families and genera: systematic arrangement and historical origin (Ⅰ and cont.), *Acta Phytotaxonomica Sinica* 16(3): 1-19 and 16(4):16-37.

535 2016—2017 年任副所长主持工作。

1978 年，路安民，张芝玉，对于被子植物进化问题的述评，植物分类学报 16（4）：1-15。

1978, An-Ming LU and Chin-Yu CHANG, A brief review of the problems of the evolution of angiosperms, *Acta Phytotaxonomica Sinica* 16(4): 1-15.

1978, Frans A. Stafleu, Vincent Demoulin, Werner R. Greuter, Paul H. Hiepko, Igorj A. Linczevski, Rogers McVaugh, Robert D. Meikle, Reed C. Rollins, Robert Ross, James M. Schopf and Edward G. Voss, ***International Code of Botanical Nomenclature*** **(Leningrad Code)**, adopted by the Twelfth International Botanical Congress, Leningrad [Saint Petersburg], July 1975, *Regnum Vegetabile* 97: 1-457 p. Translated into Chinese in 1982 and 1984.

1982、1984 年，汤彦承编，**国际植物命名法规简介**[536]，69 页，1982；西北植物研究所分类室，陕西武功（油印本）；赵士洞译，俞德浚、耿伯介校，**国际植物命名法规**，295 页，1984；科学出版社，北京。

1978, Herbert E. Street, ***Essays in Plant Taxonomy***, 304 p; Academic Press, London, New York, San Francisco. Translated into Chinese in 1986.

1986 年，石铸、李娇兰、曾建飞译，秦仁昌、何关福校，**植物分类学简论**，334 页；科学出版社，北京。

1978, Armen L. Takhtajan, ***Floristicheskie Oblasti Zemli***, 246 p; Nauka, Leningrad [Saint Petersburg]; English edition, ***Floristic Regions of the World***, translated by Theodore J. Crovello with the assistance and collaboration of the author and under the editorship of Arthur J. Cronquist, 522 p, 1986; University of California Press, Berkeley, California. Translated into Chinese in 1988.

1988 年，黄观程译，张宏达校，**世界植物区系区划**，311 页；科学出版社，北京。

1978 年，温都苏，**种子植物名称**，303 页；内蒙古人民出版社，呼和浩特。

1978, Wendusu[537], '***Names of Seed Plants***', 303 p; Inner Mongolia People's Press, Hohhot[538].

1978 年，云南植物研究所更名为中国科学院昆明植物研究所；1978—1983 年吴征镒、1983—1990 年周俊（1932—）、1990—1995 年孙汉董（1939—）、1995—1996 年许再富（1939—）、1997—2005 年郝小江（1951—）、2006—2015 年李德铢（1963—）、2015 年至今孙航（1963—）任所长。

536 副标题：在西北植物研究所座谈会上的发言稿（1982 年 3 月）。此时悉尼法规已经通过，但内容依然是列宁格勒法规（参见原书第 8 页）。

537 People of Mongolia often have only a single name; to render 'Wendusu' in Chinese requires three *hanzi* (characters), 温都苏 .

538 Huhhot or Huhehaote in pinyin transliteration.

1978, Yunnan Institute of Botany renamed Kunming Institute of Botany, Chinese Academy of Sciences, with directors Cheng-Yih WU, 1978-1983; Jun CHOW (1932-), 1983-1990; Han-Dong SUN (1939-), 1990-1995; Zai-Fu XU (1939-), 1995-1996; Xiao-Jiang HAO (1951-), 1997-2005; De-Zhu LI (1963-), 2006-2015; and Hang SUN (1963-), since 2015.

1978—1981 年，云南省热带植物研究所更名为中国科学院云南热带植物研究所；蔡希陶任所长。

1978-1981, Yunnan Institute of Tropical Botany renamed Yunnan Institute of Tropical Botany, Chinese Academy of Sciences, with director Hse-Tao TSAI.

1978—1986 年，中国植物志编辑委员会成立第四届编辑委员会，主编：中国科学院植物研究所俞德浚；副主编：中国科学院昆明植物研究所吴征镒、中国科学院植物研究所崔鸿宾；编辑委员会 37 人：新疆八一农学院安争夕[539]（1928—），中国科学院华南植物研究所陈德昭（1926—）、陈封怀、黄成就，中国科学院昆明植物研究所陈介、李锡文、裴盛基[540]（1938—），北京医学院诚静容[541]（1913—2012），中国科学院植物研究所陈心启、傅立国（1932—）、关克俭[542]（1913—1982）、李安仁、谢瑛（1926—）、张瑞琪（1908—1988），中国科学院南京植物研究所 / 江苏省植物研究所陈守良（1921—2013）、单人骅，四川大学方文培，中国科学院西北植物研究所傅坤俊（1912—2010），中国科学院林业土壤研究所傅沛云，中国科学院武汉植物研究所傅书遐[543]（1916—1986），南京大学耿伯介[544]（1917—1997）、王正平，中国科学院西北高原生物研究所郭本兆，北京师范大学贺士元[545]（1927—2015），庐山植物园赖书绅（1931—），湖南师范学院李丙贵（1921—2009），华南农学院

539 此为安峥皙的简写。

540 四川梓潼人，中国著名民族植物学家；参见：*裴盛基文集*，上、下卷，1 219 页，2018；云南科技出版社，昆明。

541 辽宁辽阳人（女），锡伯族；1934 年入国立清华大学，1938 年转入国立四川大学，毕业后留校任教；1947 年赴美国留学，1948 年于田纳西大学获硕士学位，1949 年入哈佛大学，1952 年获硕士学位，回国组建北京大学药学系植物学教研室，先后任教授和室主任直至 1990 年退休；著名药用植物学家。参见：马金双、陈虎彪，2012，缅怀诚静容教授，植物分类与资源学报 34（6）：633-634。

542 北京人，1934 年国立清华大学生物系毕业，1934—1953 年先后任教于国立北平师范大学、北京大学、东北大学、湖南师范学院、湖南大学、天津第一军医大学，1954 年调入中国科学院植物研究所工作。

543 江西南昌人，20 世纪 30 年代末靠自学考入国立四川大学园艺系，1943 年入位于江西泰和国立中正大学的静生生物调查分所工作，为胡先骕当年学生之一；1946 年随静生生物调查所复员北平，1947 年受胡先骕的委派赴台湾核对标本，返回后继续在静生生物调查所、中国科学院植物研究所从事分类学研究；1958 年奉陈封怀之邀调入武汉植物园，主持标本馆建设并编写*湖北植物志*、*中国植物志*景天科等。

544 江苏南京人，耿以礼之子，1942 年毕业于国立中央大学生物系，跟随其父耿以礼在国立中央大学 / 国立南京大学 / 南京大学从事教学与禾本科研究；著名禾本科专家。

545 北京人，满族，1947 年入国立北平师范大学，1952 年毕业留校工作，先后任助教、讲师、副教授和教授，主编北京植物志、北京植物检索表和河北植物志。参见刘全儒、马金双，2015，深切怀念贺士元教授，植物分类与资源学报 37（4）：491-492。

Kunming Institute of Botany, Chinese Academy of Sciences, 2018: 1. Administration, 2. Main Gate, 3. Laboratory, 4. Germplasm Bank, 5. Greenhouse, 6. Herbarium (Kunming Institute of Botany) (Photo provided by Kunming Institute of Botany)
2018 年的中国科学院昆明植物研究所：1. 行政楼，2. 大门，3. 实验室，4. 种质库，5. 温室，6. 植物标本馆 (KUN)（相片提供者：昆明植物研究所）

李秉滔（1936—），广西植物研究所李树刚[546]（1915—1998），兰州大学彭泽祥（1924—），浙江自然博物馆韦直（1929—），复旦大学徐炳声[547]（1924—2016），东北林学院杨衔晋，中山大学张宏达[548]（1914—2016），福建省亚热带植物研究所张永田，云南大学朱维明（1930—），厦门大学曾沧江，华东师范大学郑勉[549]（1899—1987）。

1978-1986, Fourth Editorial Committee of Flora Reipublicae Popularis Sinicae established, Te-Tsun YU (PE) as Editor in Chief, Hung-Pin TSUI (PE) and Cheng-Yih WU (KUN) as Vice Editors in Chief; 37 editorial members: Cheng-Hsi AN (1928-, XJA), Hung-Ta CHANG (1914-2016, SYS), Yong-Tian CHANG (FJSI), Chieh CHEN (KUN), Feng-Hwai CHEN (IBSC), Shou-Liang CHEN (1921-2013, NAS), Sing-Chi CHEN (PE), Te-Chao CHEN (1926-, IBSC), Ching-Yung CHENG (1913-2012, PEM), Mien CHENG (1899-1987, HSNU), Wei-Ming CHU (1930-, PYU), Wen-Pei FANG (SZ), Kun-Tsun FU (1912-2010, WUK), Li-Kuo FU (1932-, PE), Pei-Yun FU (IFP), Shu-Hsia FU (1916-1986, HIB), Shih-Yuen HO (1927-2015, BNU), Ping-Sheng HSU (1924-2016, FUS), Ching-Chieu HUANG (IBSC), Pai-Chieh KENG (1917-1997, N), Ke-Chien KUAN (1913-1982, PE), Pung-Chao KUO (HNWP), Shu-Shen LAI (1931-, LBG), Shu-Kang LEE (Shu-Kang LI, 1915-1998, IBK), An-Jen LI (PE), Bing-Gui LI (1921-2009, HNNU), Hsi-Wen LI (KUN), Ping-Tao LI (1936-, CANT), Sheng-Ji PEI (1938-, KUN), Tse-Hsiang PEN (1924-, LZU), Ren-Hwa SHAN (NAS), Chang-Chiang TSENG (AU), Cheng-Ping WANG (N), Zhi WEI (1929-, ZM), Ying XIE (1926-, PE), Yen-Chin YANG (NEFI), Rui-Qi ZHANG (1908-1988, PE).

1978—1987年，辽宁省林业土壤研究所更名为中国科学院林业土壤研究所；1978—1980年陶炎、1980—1984年曾昭顺、1984—1987年高拯民（1931—1992）任所长。

1978-1987, Liaoning Institute of Forestry and Pedology renamed Institute of Forestry and Pedology, Chinese Academy of Sciences, with directors Yan TAO, 1978-1980; Chao-Shun TSENG, 1980-1984; Tseng-Min KAO (1931-1992), 1984-1987.

1978、1981、1993年，北京师范大学生物系植物组，**北京地区植物检索表**，399页，1978；人民

546 广东台山人，1933年考入广东省中医专门学校（后因父亲病故而辍学），1934年入国立中山大学农林植物研究所任见习采集员，1935年转入广西大学植物研究所工作，1941年入国立广西大学农学院森林系，1946年毕业，再回广西植物研究所，先后任助理研究员、副研究员、研究员、室主任、副所长等；1981年创办广西植物，并承担主编工作16年之久。

547 浙江绍兴人，1948年毕业于南通学院农科，后入复旦大学任教直至退休。1951年创建复旦大学植物标本室（FUS），1993年标本储藏量已达到10万号；主编上海植物志，并培养众多弟子，其中包括洪德元等人。参见：徐栋，2017，缅怀外公——复旦大学著名植物分类学家徐炳声教授，植物科学学报35（4）：631-636。

548 广东揭西人，1935年入国立中山大学，1939年毕业，历任助教、讲师、副教授、教授、植物教研室主任、生物系主任，专于植物分类学及植物生态学，特别是山茶属系统与分类；提出著名的华夏植物区系理论。详细参见：叶创兴，2011，张宏达与他的植物学理论，中山大学学报（社会科学版）51（2）：51-63页。

549 江苏武进人，1919—1923年就读于日本东京文理科大学，1923年回国后历任江苏第四师范兼特设师范专科教员、南京工业学校专科教师、江苏太仓师范校长、南京市立师范校长、江苏中学校长，1939年之后历任成都光华大学、西北政治学院、国立西北大学教授，1946—1951年任同济大学教授兼生物系主任，1951年后任华东师范大学教授。

出版社，北京；北京师范大学生物系植物组，贺士元等，修订本，**北京植物检索表**，414 页，1981；1992 年增订版，468 页，1993；北京出版社，北京。

1978, 1981, 1993, Plant Group of Department of Biology, Beijing Normal University, '***Keys to Plants of Beijing Area***', 399 p, 1978; Beijing People's Press, Beijing; Plant Group of Department of Biology, Beijing Normal University, Shih-Yuen HO et al., revised ed., ***Claves Plantarum Pekinensis***, 414 p, 1981; and enlarged ed. 1992, 468 p, 1993; Beijing Press, Beijing.

1978—1985 年、1989—1998 年，内蒙古植物志编辑委员会，马毓泉主编，**内蒙古植物志**，8 卷本；1：294 页，1985；2：390 页，1978；3：309 页，1978；4：223 页，1979；5：442 页，1980；6：355 页，1982；7：282 页，1984；8：372 页，1985；内蒙古人民出版社，呼和浩特；[550] **内蒙古植物志**，第 2 版，5 卷本；1：408 页，1998；2：759 页，1991；3：716 页，1990；4：907 页，1992；5：634 页，1994；内蒙古人民出版社，呼和浩特。

1978-1985, 1989-1998, Commissione Redactorum Florae Intramongolicae, Yu-Chuan MA (editor in chief), ***Flora Intramongolica***, vols. 1-8; 1: 294 p, 1985, 2: 390 p, 1978, 3: 309 p, 1978[551], 4: 223 p, 1979, 5: 442 p, 1980, 6: 355 p, 1982, 7: 282 p, 1984, 8: 372 p, 1985; Inner Mongolia People's Press, Hohhot; ***Florae Intramongolica***, ed. 2, vols. 1-5; 1: 408 p, 1998, 2: 759 p, 1991, 3: 716 p, 1990, 4: 907 p, 1992, 5: 634 p, 1994; Inner Mongolia People's Press, Hohhot.

1978—2003 年，湖北省植物研究所更名为中国科学院武汉植物研究所；1978—1984 年孙祥钟、1984—1988 年胡鸿钧（1934—）、1988—1990 年郑重（1936—）、1990—1992 年侯嵩生（副所长，1931—）、1992—1997 年胡鸿钧、1997—2003 年[552]黄宏文（1957—）任所长。

1978-2003, Hubei Institute of Botany renamed Wuhan Institute of Botany, Chinese Academy of Sciences, with directors Hsiang-Chung SUN, 1978-1984; Hong-Jun HU (1934-), 1984-1988; Zhong ZHENG (Chong CHENG, 1936-), 1988-1990; Song-Sheng HOU (vice director, 1931-), 1990-1992; Hong-Jun HU, 1992-1997; and Hong-Wen HUANG (1957-), 1997-2003.

550 格鲁波夫，1988，评《内蒙古植物志》，植物学杂志 73（12）：1 769-1 771；张玉钧译，1991，干旱区资源与环境 5（3）：115-117。

551 As *Flora Chinae Intramongolicae*.

552 其中 1997—2000 年为常务副所长并主持工作。

Participants at meetings of Flora *Intramongolica* (upper, first edition, 10 January 1981, and lower, second edition, 25 September 1994) (Photo provided by Ping MA)

内蒙古植物志会议合影（上，1981 年 1 月 10 日第一版；下，1994 年 9 月 25 日第二版）（相片提供者：马平）

1979

1979 年 5 月 1 日至 6 月 1 日，中国植物学会组成 10 人代表团访美，作为 1978 年美国植物学家代表团访华的回访，代表团成员有中国科学院植物研究所汤佩松（团长）、中国科学院上海植物生理研究所殷宏章（1908—1992，副团长）、中国科学院昆明植物研究所吴征镒（副团长）、中国科学院植物研究所徐仁[553]（1910—1992）、中国科学院植物研究所俞德浚、江苏植物研究所 南京中山植物园盛诚桂、中国科学院南京地质古生物研究所李星学（1917—2010）、中国科学院上海药物研究所方圣鼎、

Chinese botanical delegation to Washington D.C., the United States of America, 1 May-1 June 1979 (left-right): Cheng-Kuei SHENG, Sheng-Ding FANG, Hsing-Hsueh LEE, Feng-Lin SU, Te-Tsun YU, Hung-Chang YIN, Cheng-Yih WU, Jen HSU, Pei-Sung TANG, Bing-Jun QIU (Photo provided by Wu Zhengyi Science Foundation)

1979 年 5 月 1 日至 6 月 1 日，中国植物学家代表团访问美国于华盛顿（从左至右）：盛诚桂、方圣鼎、李星学、苏凤林、俞德浚、殷宏章、吴征镒、徐仁、汤佩松、邱秉钧（相片提供者：吴征镒科学基金会）

553 安徽当涂人，古植物学家，详细参见：朱为庆主编，2000，徐仁著作选集，323 页；地震出版社，北京。

中国科学院外事局苏凤林（处长）和邱秉钧（翻译）。代表团参观访问了主要的植物学机构与学校。[554]

1979.05.01-06.01, In return for the 1978 delegation from the United States of America to China, the Botanical Society of China sent a reciprocal delegation to the United States of America. Participants included Sheng-Ding FANG, Shanghai Institute of Materia Medica, Chinese Academy of Sciences, Jen HSU (1910-1992), Institute of Botany, Chinese Academy of Sciences, Hsing-Hsueh LEE (1917-2010), Nanjing Institute of Geology and Palaeontology, Chinese Academy of Sciences, Cheng-Kuei SHENG, Jiangsu Institute of Botany, Nanjing Botanical Garden Memorial Sun Yat-Sen, Pei-Sung TANG (leader), Institute of Botany, Chinese Academy of Sciences, Cheng-Yih WU (vice leader), Kunming Institute of Botany, Chinese Academy of Sciences, Hung-Chang YIN (1908-1992, vice leader), Shanghai Institute of Plant Physiology, Chinese Academy of Sciences and Te-Tsun YU, Institute of Botany, Chinese Academy of Sciences. They were accompanied by Feng-Lin SU (leader of foreign department) and Bing-Jun QIU (translator) from the Foreign Affairs Bureau, Chinese Academy of Sciences. They visited major botanical institutions and universities.

1979 年 10 月，中国植物学会首次蕨类植物学学术讨论会在浙江天目山举行[555]，全国 14 个省（市、区）的代表出席；中国科学院植物研究所、学部委员秦仁昌教授致大会开幕词。[556]

1979.10, First national symposium of the Botanical Society of China on ferns held at Tianmu Shan, Zhejiang. Representatives from 14 provinces attended; professor Ren-Chang CHING, academician, Institute of Botany, Chinese Academy of Sciences, gave the opening remarks.

1979 年，秦仁昌，二十年来我国的植物分类学，植物分类学报 17（4）：1-6。

1979, Ren-Chang CHING, Twenty years of Chinese systematic botany, *Acta Phytotaxonomica Sinica* 17(4): 1-6.

1979 年，秦仁昌，发展中的中国植物分类学，植物杂志 4：1-3。

1979, Ren-Chang CHING, ‘On the development of Chinese plant taxonomy’, *Plant Magazine* 4: 1-3.

1979 年，林英、程景福，**维管束植物鉴定手册**[557]，373 页；江西人民出版社，南昌。

1979, Ying LIN, Jing-Fu CHENG, ‘***Identification Handbook of Vascular Plants***’, 373 p; Jiangxi People’s Publishing House, Nanchang.

---

554 俞德浚，1979，美国植物园巡礼，植物杂志，6：1-2、38；吴征镒，1980，访问 借鉴 赶超 —— 赴美观感，植物杂志，1：1-2。

555 薛惠，1980，中国植物学会举行学术讨论会，植物杂志 1：38-39。

556 鲁侨，1980，全国首次蕨类植物学学术讨论会在我省召开，今日科技 2：7。

557 作者在前言中提到在 1965 年版基础上修订，但编者未见。

1979，林善雄[558]（1942—）留学美国研究苔藓，1982年获宾夕法尼亚州立大学生物系博士，1982年返台后于东海大学任副教授，1987年任教授；主要从事苔藓学研究。

1979, Shan-Hsiung LIN (1942-) went to United States of America to study bryology, received Ph.D. degree from the Pennsylvania State University in 1982[559], returned to Taiwan as associate professor and professor in Tunghai University since 1982 and 1987.

1979年，刘和义[560]（1952—）赴美国德州大学植物系留学，1980年转入俄亥俄州立大学植物系攻读博士，1986年毕业获博士学位；1987—1989年任台湾自然科学博物馆副研究员，1989年至今为中山大学生物系副教授；2014年至今任台湾植物分类学会理事长。

1979, Ho-Yih LIU (1952-) studied botany at the University of Texas, then transferred to the Department of Botany, Ohio State University in 1980; obtained his Ph. D. degree in 1986[561]; returned to become associate curator, Taiwan Museum of Natural Science, Taichung [Taizhong] 1987, then associate professor, Sun Yat-Sen University in Taiwan from 1989 to the present. He has served as president of the Taiwan Society of Plant Systematics since 2014.

1979年，青海省生物研究所更名为中国科学院西北高原生物研究所；1979—1984年夏武平（1918—2009）、1984—1987年李家藻（1925—）、1987—1991年王祖望（1933—）、1991—1992年樊乃昌（1940—）、1992—1995年杜继曾（1938—）、1996—2000年张宝琛（1941—）、2000—2008年赵新全、2008年至今张怀刚（1962—）任所长。

1979, Qinghai Institute of Biology renamed Northwest Institute of Plateau Biology, Chinese Academy of Sciences, with directors Wu-Ping XIA (1918-2009), 1979-1984; Jia-Zao LI (1925-), 1984-1987; Zu-Wang WANG (1933-), 1987-1991; Nai-Chang FAN (1940-), 1991-1992; Ji-Zeng DU (1938-), 1992-1995; Bao-Chen ZHANG (1941-), 1996-2000; Xin-Quan ZHAO, 2000-2008; Huai-Gang ZHANG (1962-), since 2008.

1979年，吴征镒，论中国植物区系的分区问题，云南植物研究 1：1-22。

---

558 生于日本名古屋，1946年随家入台；1965年获中兴大学植物学系学士，1965年服兵役，1966—1969年于中兴大学和三义中学从事教学以及“中央”研究院植物研究所从事研究工作，1969年后任东海大学助教，1974—1976年台湾大学植物研究所硕士；1976—1979年任东海大学讲师。

559 Ph.D. dissertation: A taxonomic revision of Phyllogoniaceae (Bryopsida), 283 p; Advisor: Ronald A. Pursell (1930-2014); Shan-Hsiung LIN, 1983 & 1984, A Taxonomic Revision of Phgllogoniacae (Bryopsida), Part I & II, *Journal of Taiwan Museum,* 36(2): 37-86 & 37(2): 1-54.

560 祖籍福建安溪，生于台湾新北市；1971年入台湾大学植物系，1975年本科毕业获学士学位；1975年入台湾大学植物系硕士班，1977年毕业获硕士学位；1977—1978年任东海大学生物系助教，1978—1979年任国防医学院药学系研究助理。

561 Ho-Yih LIU, 1986, A monograph of the genus *Aeonium* (Crassulaceae-Sempervivoideae); Advisor: Tod F. Stuessy (1943-).

1979, Cheng-Yih WU, The regionalization of Chinese flora, *Acta Botanica Yunnanica* 1: 1-22.

1979 年，中国科学院植物研究所主编，**中国高等植物科属检索表**，733 页；科学出版社，北京。

1979, Institute of Botany, Chinese Academy of Sciences (editor in chief), ***Claves Familiarum Generumque Cormophytorum Sinicorum***, 733 p; Science Press, Beijing.

1979 年，中国植物学会植物分类学术交流会在四川成都召开[562]，代表 120 人，提交文章 146 篇。[563]

1979, Symposium on Plant Taxonomy of Botanical Society of China held in Chengdu, Sichuan, 120 attendees, 146 papers submitted.

1979+，中国科学院昆明植物研究所，**云南植物研究**，1-32 卷，1979—2010；**植物分类与资源学报**，33-37 卷，2011—2015；***Plant Diversity***，英文版，38 卷 +，2016—；季刊，2002 年改为双月刊。1975—1978 年间为内部发行，共 10 期。其中 1975—1977 年每年 2 期，而 1978 年 3 期外加增刊 1 期。内部发行中的第 5 期（1977 年第 1 期）为云南植物志第 1 卷分属与分种检索表；另外，1988—2009 年期间出版增刊 I-XVI；1979—2005 年吴征镒、2006—2015 年李德铢、2016 年至今 Sergei Volis（1961— ）和周浙昆（1956— ）任主编。

1979+, Kunming Institute of Botany, Chinese Academy of Sciences, ***Acta Botanica Yunnanica***, vols. 1-32, 1979-2010; ***Plant Diversity and Resources***, vols. 33-37, 2011-2015; ***Plant Diversity***, English edition, vols. 38+, 2016-; quarterly 1979-2001; bimonthly since 2002; Editors in Chief: Cheng-Yih WU, 1979-2005, De-Zhu LI, 2006-2015, Sergei Volis (1961-) and Zhe-Kun ZHOU (1956-), since 2016.

1979 年 9 月 14 日至 1992 年 2 月 23 日，王弼昭[564]（1953—1992）于台湾采集蕨类植物标本 19 000 余号（HAST）。

1979.09.14-1992.02.23, Bi-Jao WANG (1953-1992) collected more than 19,000 ferns in Taiwan (HAST).

1979—2003 年，广东省植物研究所更名为中国科学院华南植物研究所；1979—1983 年陈封怀、1983—1987 年郭俊彦（1926—2018）、1987—1995 年屠梦照（1934— ）、1995—2000 年梁承邺（1938— ）、2000—2001 年赵南先（1959—，常务副所长）、2001—2003 年彭少麟（1956— ）任所长。

562 中国植物学会五十年编写组，1985，中国植物学会五十年，中国科技史料 6（2）：54。

563 薛惠，1980，中国植物学会举行学术讨论会，植物杂志 1：38-39。

564 台湾高雄人，1973—1977 年台湾师范大学生物系学习，毕业后先后服兵役并任中学教师，1981 年开始接手家族油漆生意；1992 年 3 月 15 日于屏东普安山登山采集时遇难。参见：台湾省特有生物研究保育中心、"中央"研究院植物研究所，1999，台湾杰出的蕨类采集者王弼昭纪念集（1953—1992），207 页；台北。

1979-2003, Guangdong Institute of Botany renamed South China Institute of Botany, Chinese Academy of Sciences, with directors Feng-Hwai CHEN, 1979-1983; Tsun-Yan KUO (1926-2018), 1983-1987; Meng-Zhao TU (1934-), 1987-1995; Cheng-Ye LIANG (1938-), 1995-2000; Nan-Xian ZHAO (1959-, Standing Vice Director), 2000-2001; Shao-Lin PENG (1956-), 2001-2003.

1979—2008 年，中国科学院华南植物研究所叶华谷（1956—）在广东、香港、澳门以及西藏，采集植物标本 1.4 万余号（IBSC）。[565]

1979-2008, Hua-Gu YE (1956-), South China Institute of Botany, Chinese Academy of Sciences, collected about 14,000 numbers of plant specimens in Guangdong, Hong Kong, Macao and Xizang (Tibet) (IBSC).

565 胡启明、曾飞燕，2011，广东植物志，10：319；广东科技出版社，广州［附录 2，华南植物研究所（园）植物标本采集简史］。

1980

1980.08.15–11.15, Sino-American botanical expedition to western Hubei: *Metasequoia* region of Lichuan and Shennongjia Forest District. American botanists: Bruce M. Bartholomew (UC, leader), David E. Boufford (CM), Theodore R. Dudley (1936–1994, NA) and James L. Luteyn (1948–, NY), Stephen A. Spongberg (1942–, A) and Chinese botanists: Ao-Lo CHANG (Ao-Luo ZHANG, 1935–, KUN), Zhong ZHENG (HIB), Shan-An HE (1932–, NAS), Yen-Cheng TANG (leader, PE) and Tsun-Shen YING (PE). They collected 2,085 numbers of vascular plants (A, HIB, KUN, NA, NAS, NY, PE, US) and made 621 collections of living plants and seeds[566]. American botanists also visited botanical institutions in Beijing, Hangzhou, Kunming, Nanjing, Shanghai and Wuhan.

1980年8月15日至11月15日，中美植物联合考察湖北水杉原产地和神农架林区，美方成员为加州大学伯克利分校 Bruce M. Bartholomew（领队），匹斯堡卡内基自然历史博物馆 David E. Boufford，国家树木园 Theodore R. Dudley（1936—1994），纽约植物园 James L. Luteyn（1948—）和哈佛大学阿诺德树木园 Stephen A. Spongberg(1942—)；中方成员为中国科学院植物研究所汤彦承(领队)、应俊生，江苏植物研究所贺善安（1932—），武汉植物研究所郑重和昆明植物研究所张敖罗（1935—）；考察队采集植物标本2 085号（A、HIB、KUN、NA、NAS、NY、PE、US），以及活植物和种子等621号。除此之外，美方成员还访问了北京、杭州、昆明、南京、上海和武汉的有关机构。

1980年9月，中国科学院中国孢子植物志编辑委员会第三届编辑委员会成立，代主编王云章，副主编饶钦止、魏江春[567]（1931—），编辑委员会29人。

1980.09, Third Consilio Florarum Cryptogamarum Sinicarum, Academiae Sinicae Edita established, Yun-Chang WANG, acting Editor in Chief, Chin-Chih JAO and Jiang-Chun WEI (1931–), Vice Editors in Chief, with 29 editorial members.

---

566 Bruce Bartholomew, David E. Boufford, Chang Ao-Luo, Cheng Zhong, Theodore R. Dudley, He Shan-An, Jin Yi-Xing, Li Qing-Yi, James L. Luteyn, Stephen A. Spongberg, Sun, S. C., Tang Yan-Cheng, Wan Jia-Xiang and Ying Tsun-Sheng, 1983, The 1980 Sino-American botanical expedition to western Hubei province, People's Republic of China. *Journal of Arnold Arboreum* 64 (1): 1–103; Bruce Bartholomew, David E. Boufford and Stephen A. Spongberg, 1983, *Metasequoia glyptostroboides*–Its present status in central China. *Journal of Arnold Arboretum* 64 (1): 105–128; Michael Dosmann and Peter Del Tredici, 2003, Plant introduction, distribution and survival: A case study of the 1980 Sino-American botanical expedition, *BioScience* 53(6): 588–597.

567 陕西咸阳人，1955年于西北农学院（今西北农林科技大学）毕业后进入中国科学院西北农业生物研究所工作，1956年调至中国科学院应用真菌研究所，1958年赴苏联留学，1962年毕业于苏联科学院研究生院，获副博士学位，同年回到中国科学院微生物研究所工作，1997年入选中国科学院院士；中国地衣学主要奠基者和学术带头人。参见：中国菌物学会，2011，贺魏江春院士八十华诞，菌物学报30（6）文前。

Sino-American botanical expedition to Hubei, 15 August–15 November 1980: 1. American botanists under Metasequoia tree in Modaoxi (left-right): David E. Boufford, James L. Luteyn, Bruce M. Bartholomew, Stephen A. Spongberg, Theodore R. Dudley; 2. Chinese botanists under Metasequoia tree in Modaoxi (left-right): Kai-Yun GUAN, Zhong ZHENG, Wen-Jun HE, Shan-An HE, Yi-Xing JIN, Tsun-Shen YING, Yen-Cheng TANG, Ao-Lo CHANG, Qing-Yi LI, Xu WANG, Jia-Xiang WAN; 3. Expedition en route to Shennongjia (Photo provided by David E. Boufford)

1980 年 8 月 15 日至 11 月 15 日，中美植物学家联合考察湖北：1. 美国植物学家于磨刀溪水杉树下（从左至右）：David E. Boufford、James L. Luteyn、Bruce M. Bartholomew、Stephen A. Spongberg、Theodore R. Dudley；2. 中国植物学家于磨刀溪水杉树下（从左至右）：管开云、郑重、何文俊、贺善安、金义兴、应俊生、汤彦承、张敖罗、李清义、王旭、万家祥；3. 赴神农架考察路上（相片提供者：鲍棣伟）

1980 年 9 月，中国科学院成都生物研究所植物分类室，**植物分类研究**，报道石蒜科芒苞草属（仅 1 期）。

1980.09, Department of Plant Taxonomy, Chengdu Institute of Biology, Chinese Academy of Sciences, ***Phytotaxonomic Research*** with description of *Acanthochlamys* P. C. Kao (one issue only).

1980 年，张宏达，华夏植物区系的起源与发展，中山大学学报（自然科学版）1：89-98。

1980, Hung-Ta CHANG, The origin and development of the Cathaysian flora, *Acta Scientiarum Naturalium Universitatis Sunyatseni* 1: 89–98.

1980, Kristin S. Clausen and Shiu-Ying HU, Mapping the collecting localities of E. H. Wilson in China, *Arnoldia* 40(3): 139–145.

1980 年，Kristin S. Clausen、胡秀英，图注威尔逊在华采集地点，*Arnoldia* 40(3): 139-145。

1980 年，方文培编著，**拉丁文植物学名词及术语**，211 页；四川人民出版社，成都。本书 1957 年曾以**植物学名词及术语**印刷，126 页；四川大学，成都（内部印制）。

1980, Wen-Pei FANG, '***Latin Names and Terminology of Botany***', 211 p; Sichuan People's Publishing House, Chengdu. The title of the book was printed as '***Names and Terminology of Botany***', 126 p in 1957 by Sichuan University (internal publication).

1980, Richard A. Howard, E. H. Wilson as a botanist (Part I and Ⅱ), *Arnoldia* 40(3): 102-138, 40(4): 154-193.

1980 年，Richard A. Howard，植物学家威尔逊（Ⅰ、Ⅱ），*Arnoldia* 40（3）：102-138，40（4）：154-193。

1980 年，日本京都大学蕨类植物学家岩槻邦男（1936—）访问云南中国科学院昆明植物研究所，由臧穆陪同赴大理苍山考察。

1980, Kunio Iwatsuki (1936-), pteridologist from Kyoto University, Japan, visited Kunming Institute of Botany, Chinese Academy of Sciences, Kunming, Yunnan, and conducted field studies on Cang Shan, Dali, Yunnan, accompanied by Mu ZANG.

1980 年，甘伟松等校订，**中国高等植物图志**，1 211 页；宏业书局，台北。

1980, Wei-Sung KAN et al., '***Illustration Flora of Chinese Higher Plants***', 1,211 p; Hongye Books, Taipei [Taibei].

1980，潘富俊（1951—）[568] 赴美留学，1985 年获得夏威夷大学博士学位，1985—1994 年任台湾林业试验所副研究员，1994—2004 年任台湾林业试验所森林生物系主任，2004 年至今任台湾林业试验所研究员。

1980, Fuh-Jiunn PAN (1951-) went to United States of America to study, received Ph.D. degree from University of Hawaii in 1985[569], associate professor of Taiwan Forestry Research Institute, 1985-1994, director of department of forestry, Taiwan Forestry Research Institute, 1994-2004, and professor of Taiwan Forestry Research Institute since 2004.

568 私立中国文化大学森林学系学士，台湾大学森林研究所硕士。资料来自网址介绍（http://www2.pccu.edu.tw/crtdla/ 潘富俊 .html, 2018 年进入）。

569 Ph.D. dissertation: Systematics and Genetics of the *Leucaena diversifolia* (Schlecht.) Benth. Complex; Advisor: James L. Brewbaker (1926-).

1980, Clive A. Stace, ***Plant Taxonomy and Biosystematics***, 279 p; University Park Press, Baltimore; ***Plant Taxonomy and Biosystematics***, 2nd Edition, 1989, 264 p; Edward Arnold, London. Translated into Chinese in 1986.

1986 年，韦仲新、缪汝槐、谢翰铁译，**植物分类与生物系统学**，328 页；科学出版社，北京。

1980, Armen L. Takhtajan, Outline of the classification of flowering plants (Magnoliophyta), *The Botanical Review* 46(3): 225-359. Translated into Chinese in 1986.

1986 年，黄云晖译，王伯荪校，**有花植物（木兰植物）分类大纲**，224 页；中山大学出版社，广州。

1980, Pan-Cheng WU, Present state of bryology in China, *Taxon* 29(2/3): 369-370.

1980 年，吴鹏程，中国苔藓现状，*Taxon* 29（2/3）：369-370。

1980 年，西藏植物名录编辑组编，**西藏植物名录**，463 页；西藏自治区科学技术委员会，拉萨（内部印制）。

1980, Editorial Group of Plant Checklist of Xizang (Tibet), ***An Enumeration of the Vascular Plants of Xizang (Tibet)***, 463 p; Xizang Science and Technology Committee, Lhasa (internal publication).

1980.10-1981.01, Shiu-Ying HU invited by State Department of Education Administration and Botanical Society of Heilongjiang, visited and lectured in Beijing, Chengdu, Guangzhou, Hankou, Harbin, Kunming, Lanzhou, Lushan, Nanjing, Shanghai, Simao, Xishuangbanna, Xi'an and Xuzhou. Gave lectures at Anhui University, Institute of Forestry and Pedology of Chinese Academy of Sciences, Nanjing University, Northeast Forestry College, Northwest Normal College, Sichuan University, Sun Yat-Sen University and Yunnan Institute of Tropical Botany, Chinese Academy of Sciences; audience came from nearly all of China. Among her lectures were 14 on plant taxonomy at South China Agricultural College, Guangzhou.

1980 年 10 月至 1981 年 1 月，胡秀英应教育部和黑龙江省植物学会的邀请，到北京、成都、广州、哈尔滨、汉口、兰州、庐山、昆明、南京、上海、思茅、西安、西双版纳、徐州访问，分别在安徽大学、东北林学院、南京大学、西北师范学院、四川大学、中山大学、中国科学院沈阳林业土壤研究所和云南热带植物研究所讲学，听众来源几乎覆盖全国所有的机构。其中在华南农学院做了 14 场有关植物学，尤其是植物分类学的报告。[570]

570 胡秀英，1981，植物学学术讲座（一、二、三、四），华南农学院学报，2（1）：22-29；2（2）：93-103；2（3）：68-76；2（4）：93-98。

Participants at Dr. Shiu-Ying HU's lecture, Northeast Forestry College, Harbin, 7–12 November 1980 (Photo provided by Arnold Arboretum of Harvard University)

胡秀英博士 1980 年 11 月 7 日至 12 日于哈尔滨东北林学院讲学留念（相片来源：哈佛大学阿诺德树木园）

1981

1981.04.20-06.03, Sino-British botanical exploration in Yunnan: Cang Shan. David F. Chamberlain (1941-, E), Peter A. Cox, Peter Hutchison, Charles Roy Lancaster[571] (1937-), Robert J. Mitchell (1936-, STA), from the U.K.; Rhui-Cheng FANG (1932-, KUN), Kuo-Mei FENG (KUN), Kai-Yun GUAN (KUN), Chun-Chao LÜ (KUN), Zheng-Wei LÜ (1931-2001, KUN), Tian-Lu MING (1937-, KUN), De-Ding TAO (KUN) and Gaby Lock (German student), from Kunming Institute of Botany, Chinese Academy of Sciences.[572]

1981年4月20日至6月3日，中英植物联合考察云南苍山，英方成员包括苏格兰爱丁堡植物园 David F. Chamberlain（1941— ）、圣安德鲁大学植物园 Robert J. Mitchell（1936— ），以及 Peter A. Cox、Peter Hutchison、Charles Roy Lancaster（1937— ）等6人；中方成员包括中国科学院昆明植物研究所方瑞征（1932— ）、冯国楣、管开云、闵天禄（1937— ）、吕春朝、吕正伟（1931—2001）、陶德定以及德国留学生 Gaby Lock[573] 等。

1981.08.21-08.28, Thirteenth International Botanical Congress held in Sydney, Australia, attended by 33 Chinese delegates. Reports on the congress from 26 delegates were given to 300 attendees from throughout China in Hefei, Anhui, October 11-17, 1981.

1981年8月21日至28日，第十三届国际植物学大会在澳大利亚悉尼举行，中国植物学家33人[574]代表出席[575]。同年10月11至17日于安徽合肥市举行“第十三届国际植物学会议传达报告会”，26位参加第十三届国际植物学会议的代表做了传达报告[576]；全国300多人参加。

1981年10月1日至1982年10月10日，4位1980年参加中美植物学考察队的中方人员分别

571 British plantsman, gardener, author and broadcaster, is most widely known for his work on the long running BBC TV programme, Gardeners' World. He has also regularly appeared on the BBC Radio show Gardeners' Question Time and is also a freelance writer and lecturer. Formerly first curator of the Hillier Arboretum (now Sir Harold Hillier Gardens), he has travelled the world on plant finding expeditions, including many times to China, particularly to southwestern China.

572 Robert J. Mitchell with others, 1984, Cang Shan – The Report of the Sino-British Exploration to China – 1991, 80 p; The University of St. Andrews Printing Department, College Gate, St. Andrews, KY 16 9AL.

573 联合培养，中方指导教授：吴征镒等。

574 另外台湾有10人参加；参见：本报记者，1982，让植物学更好地为祖国四化服务 —— 第十三届国际植物学会议传达报告会侧记，植物杂志1：10-12。

575 汪向明，1983，澳大利亚植物科学概况见闻，武汉植物学研究1（1）：129-133。

576 参见：本报记者，1982，让植物学更好地为祖国四化服务 —— 第十三届国际植物学会议传达报告会侧记，植物杂志1：10-12；中国植物学会、河北省植物学会、河北省植物生理学会，第十三届国际植物学会议传达报告汇编，194页，1981年12月（铅印本）；其中第58页为出席会议的中国（大陆）33名代表名单。

Sino-British expedition to Cang Shan, 20 April-3 June 1981 (Photo provided by Kunming Institute of Botany)
1981 年 4 月 20 日至 6 月 3 日，中英植物学家考察苍山（相片提供者：昆明植物研究所）

赴美国研修与考察，其中江苏植物研究所贺善安和中国科学院昆明植物研究所张敖罗赴华盛顿特区的美国国家树木园、中国科学院植物研究所应俊生和武汉植物研究所郑重赴哈佛大学阿诺德树木园；而没有参加考察队的中国科学院植物研究所陈心启代替汤彦承赴纽约植物园。

1981.10.01-1982.10.10, Four Chinese members of the 1980 Sino-American botanical expedition to China spent one year in the United States of America for research and field work; Ao-Lo CHANG (Ao-Luo ZHANG), Kunming Institute of Botany, Chinese Academy of Sciences, and Shan-An HE, Jiangsu Institute of Botany, were at the National Arboretum in Washington, D.C., Zhong ZHENG, Wuhan Institute of Botany, Chinese Academy of Sciences, and Tsun-Shen YING, Institute of Botany, Chinese Academy of Sciences were at the Arnold Arboretum of Harvard University. Sing-Chi CHEN, Institute of Botany, Chinese Academy of Sciences, who was not on the 1980 expedition, but sent as a substitute for Yen-Cheng TANG, was at the New York Botanical Garden.

1981 年 11 月 26 日，国务院学位委员会批准博士学位授予点和首批博士生导师，其中植物分类学导师包括北京中国科学院植物研究所秦仁昌、俞德浚，北京中国林业科学研究院郑万钧，北京医学院诚静容，广东广州中山大学张宏达，黑龙江哈尔滨东北林学院杨衔晋，辽宁沈阳中国科学院林业土壤研究所王战，湖北武汉武汉大学孙祥钟，四川成都四川大学方文培。

1981.11.26, Academic Degrees Committee of the State Council approved the first doctoral program in China. First doctoral advisors of plant taxonomy included: Hung-Ta CHANG, Sun Yat-Sen University, Guangzhou, Guangdong; Wan-Chun CHENG, Chinese Academy of Forestry, Ching-Yung CHENG, Beijing Medical College, Ren-Chang CHING and Te-Tsun YU, Institute of Botany, Chinese Academy of Sciences, Beijing; Wen-Pei FANG, Sichuan University, Chengdu, Sichuan; Hsiang-Chung SUN, Wuhan University, Wuhan, Hubei; Chan WANG, Institute of Forestry and Pedology, Chinese Academy of Sciences, Shenyang, Liaoning; Yen-Chin YANG, Northeast Forestry College, Harbin, Heilongjiang.

1981 年，赵淑妙[577]（1954—）赴美国路易斯安那州新奥尔良杜兰大学留学，1985 年获博士学位；返台湾后 1985—1987 年任“中央”研究院植物研究所助理研究员、1987—1993 年任副研究员、1993—2004 年任研究员、2005 年任“中央”研究院生物多样性研究中心研究员；其中 1999—2003 年兼任“中央”研究院植物研究所植物标本馆（HAST）馆长。现任“中央”研究院生物多样性研究中心主任、特聘研究员。

1981, Shu-Miaw CHAW (1954-) went to Tulane University, New Orleans, Louisiana, United States of America to study botany, and received Ph. D. degree in 1985[578]; employed by the Institute of Botany, Academia Sinica, as project research associate, 1985-1987, associate research fellow, 1987-1993; research fellow (with tenure), 1993-2004, and she also as curator of the herbarium (HAST), Institute of Botany,

577 台中人（女），私立东海大学生物系本科毕业，台湾大学植物系硕士毕业。

578 Ph.D. dissertation: A taxonomic study of the Old World species of *Antirhea* Commerson ex Jussieu (Rubiaceae), 231 p; Advisor: Steven P. Darwin (1949-).

Five Chinese botanists in USA (all in spring and summer of 1982), 1 October 1981–10 October 1982: 1. Botanical Garden of University of North Carolina (left-right): Ao-Luo ZHANG, C. Ritchie Bell, Sing-Chi CHEN, Zhong ZHENG, Tsun-Shen YING, Shan-An HE; 2. Clay County, North Carolina: Elizabeth A. Kellogg, Stephen A. Spongberg, Sing-Chi CHEN, Tsun-Shen YING, Shan-An HE; 3. Heggie's Rock, Georgia: Tsun-Shen YING, Ao-Luo ZHANG, Sing-Chi CHEN, American botanist (female), Shan-An HE, American botanist (male), Zhong ZHENG; 4. Lyndon State College, Vermont: Zhong ZHENG, Sing-Chi CHEN, Ao-Luo ZHANG, Tsun-Shen YING, Shan-An HE; 5. White Mountains, New Hampshire: Tsun-Shen YING, Sing-Chi CHEN, David S. Conant, Ao-Luo ZHANG, Shan-An HE, Zhong ZHENG; 6. White Mountains, New Hampshire: Zhong ZHENG, Sing-Chi CHEN, David S. Conant, Shan-An HE, Ao-Luo ZHANG, Tsun-Shen YING (Photo provided by David E. Boufford)

1981 年 10 月 1 日至 1982 年 10 月 10 日，五位中国植物学家在美国（1982 年春和夏）：1. North Carolina 大学植物园（从左至右）：张敖罗、C. Ritchie Bell、陈心启、郑重、应俊生、贺善安；2. Clay County，North Carolina：Elizabeth A. Kellogg、Stephen A. Spongberg、陈心启、应俊生、贺善安；3. Heggie Rock，Georgia：应俊生、张敖罗、陈心启、美国植物学家（女），贺善安、美国植物学家（男）、郑重；4. Lyndon 州立学院，Vermont：郑重、陈心启、张敖罗、应俊生、贺善安 5. White Mountains，New Hampshire：应俊生、陈心启、David S. Conant、张敖罗、贺善安、郑重；6. White Mountains，New Hampshire：郑重、陈心启、David S. Conant、贺善安、张敖罗、应俊生（相片提供者：鲍棣伟）

Academia Sinica, 1999-2003; research fellow of Biodiversity Center of Academia Sinica, 2005, currently director and distinguished research fellow.

1981 年，高谦、张光初著，**东北苔类植物志**，220 页；科学出版社，北京。

1981, Chien GAO and Kuang-Chu CHANG, ***Flora Hepaticarum Chinae Boreali-Orientalis***, 220 p; Science Press, Beijing.

1981 年，路安民，现代有花植物分类系统初评，植物分类学报 19（3）：279-290。

1981, An-Ming LU (An-Min LU), A preliminary review of the modern classification systems of the flowering plants, *Acta Phytotaxonomica Sinica* 19(3): 279–290.

1981+，广西壮族自治区中国科学院广西植物研究所、广西植物学会，**广西植物**，vols. 1+, 1981+；季刊，2002 年改为双月刊，2016 年又改为月刊。本刊 1975—1978 年名称为**植物研究通讯**（内部发行）；其中 1975—1976 年每年两期，1977—1978 年每年 4 期，而 1979 年未印刷，1980 年试刊改为现名并发行 4 期（其中第 3 和 4 期为合刊）；1981 年正式发行并于第 4 期加入外文刊名 *Guihaia*；1987—1996 年李树刚、1996—2000 年金代钧、2001—2010 年李锋（1953—）、2010 年至今李先琨（1967—）任主编。[579]

1981+, Guangxi Institute of Botany, Guangxi Zhuang Autonomous Region and Chinese Academy of Sciences, and Guangxi Society of Botany, ***Guihaia***, vols. 1+, quarterly, bimonthly 2002 and monthly since 2016, as ***Botanical Research Bulletin*** (irregularly issued internally), between 1975-1978, Guilin, Guangxi; Editors in Chief: Shu-Kang LEE, 1987-1996, Dai-Jun JIN, 1996-2000, Feng LI (1953-), 2001-2010, and Xian-Kun LI (1967-) since 2010.[580]

1981+，四川植物志编辑委员会，方文培主编，**四川植物志**；1：509 页，1981；2：250 页，1983；3：309 页，1985；4：493 页，1988；5（1）：427 页，2017[581]；5（2）：457 页，1988；6：410 页，1988；7：416 页，1991；8：571 页，1990；9：544 页，1989；10：687 页，1992；11：185 页，1994；12[582]：338 页，1998；13：308 页，1999；14：181 页，1999；15：199 页，1999；16[583]：415 页，2005；17[584]：347

579 本刊 1981—1986 年间无主编，参见编辑委员会介绍：http://www.guihaia-journal.com/ch/common_item.aspx?parent_id=20150604020647001&menu_id=20160128030429581&is_three_menu=0（2018 年进入）。

580 No Editor in Chief from 1981 to 1986 according to the introduction at: http://www.guihaia-journal.com/ch/common_item.aspx-?parent_id=20150604020647001&menu_id=20160128030429581&is_three_menu=0 (accessed at 2018).

581 主编：高宝莼，周永红，唐亚，何兴金。

582 本卷内容简介中提到四川植物志记载四川高等植物，全书共分 26 卷出版。

583 李伯刚主编。第 16 卷后记记载，第 1-15 卷共计 129 科 4 000 多种，1 700 多幅图版，700 余万字。

584 高宝莼主编。第 17 卷后记记载，第 1-17 卷共计 134 科 5 000 多种，2 000 余幅图版，800 余万字。

页，2007；21[585]：549 页，2012；四川人民出版社（1981—1983）、四川科学技术出版社（1985—1988、1994、2012、2017+）、四川民族出版社（1991—2007），成都。

1981+, Editorial Committee of Flora Sichuanica, Wen-Pei FANG (editor in chief), ***Flora Sichuanica***, 1: 509 p, 1981; 2: 250 p, 1983; 3: 309 p, 1985; 4: 493 p, 1988; 5(1): 427 p, 2017; 5(2): 457 p, 1988; 6: 410 p, 1988; 7: 416 p, 1991; 8: 571 p, 1990; 9: 544 p, 1989; 10: 687 p, 1992; 11: 185 p, 1994; 12[586]: 338 p, 1998; 13: 308 p, 1999; 14: 181 p, 1999; 15: 199 p, 1999; 16: 415 p, 2005; 17: 347 p, 2007[587]; 21: 549 p, 2012; Sichuan People's Publishing House (1981-1983), Sichuan Science and Technology Press (1985-1988, 1994, 2012, 2017+) and Sichuan Nationality Press (1991-2007), Chengdu.

1981+，中国科学院西北植物研究所 / 西北农林科技大学，**西北植物学报**，vols. 1+，1981—1984：半年刊；1985—1993：季刊；1994 年改为双月刊，2003 年又改为月刊。本刊 1980—1984 年中文名为**西北植物研究**，1985 年改为现名；1981—1983 年李中宪（1912—1982）、1983—1994 年李振声（1931—）、1994—2012 年胡正海（1930—2018）、2012 年至今赵忠（1958—）任主编。

1981+, Northwest Institute of Botany, Chinese Academy of Sciences / Northwest A&F University, ***Acta Botanica Boreali-Occidentalia Sinica***, vols. 1+, semi-annually 1981-1984, quarterly 1985-1993, bimonthly 1994-2002, monthly since 2003; Editors in Chief: Zhong-Xian LI (1912-1982), 1981-1983, Chen-Sheng LI (1931-), 1983-1994, Zheng-Hai HU (1930-2018), 1994-2012, and Zhong ZHAO (1958-), since 2012.

1981 年 8 月至 1982 年 4 月，广西科委组织金秀县大瑶山综合考察，广西植物研究所等 10 多家机构参加，莫新礼（1934—1988）、韦发南（1941—）、覃民府、沈开福、文和群（1957—2013）、钟树华（1958—）、叶华谷等采集植物标本 3 600 多号（IBK, IBSC）。

1981.08-1982.04, Dayao Shan Expedition organized by Guangxi Scientific Committee to Jinxiu included Xin-Li MO (1934-1988, GXF), Kai-Fu SHEN, Min-Fu QIN, Fa-Nan WEI (1941-, IBK), He-Qun WEN (1957-2013, IBK), Shu-Hua ZHONG (1958-, IBK), Hua-Gu YE (IBSC) et al. from Guangxi Institute of Botany and more than ten institutions, collected more than 3,600 numbers of plant specimens (IBK, IBSC).

1981—1983 年，中国科学院组织横断山综合科学考察队，植物组成员包括植物研究所陈伟烈、李渤生、李良千（1953—）、李沛琼、郎楷永、王金亭、汪楣芝（1947—），昆明植物研究所成晓（1957—）、费勇（1960—2001）、李恒、武素功、俞宏渊（1963—）、张长芹（1953—）对滇西北、川西和藏东昌都地区进行调查，采集标本约 40 000 号（KUN、PE），成为编研两卷本横断山区维管植物（1993、

585 高宝莼主编。

586 There were about to be 26 volumes in total for *Flora Sichuanica* according to the information provided in volume 17.

587 The published volumes 1-17 included treatments for 134 families description of over 5,000 species, and more than 2,000 illustrations and included over 8 million words (in Chinese) based on the summary in volume 17 (2007).

1994）的主要依据。

1981-1983, The Hengduan Mountain Expedition of Chinese Academy of Sciences in northwestern Yunnan, western Sichuan and Chamdo [Qamdo, Changdu] in eastern Xizang. Participants, Institute of Botany, Beijing: Wei-Lie CHEN, Bo-Sheng LI, Liang-Qian LI (1953-), Pei-Chiong LI, Kai-Young LANG, Chin-Ting WANG, Mei-Zhi WANG (1947-); Kunming Institute of Botany: Xiao CHENG (1957-), Yong FEI (1960-2001), Heng LI, Shu-Kung WU, Hong-Yuan YU (1963-) and Chang-Qin ZHANG (1953-). The approximately 40,000 numbers of plant specimens (KUN, PE) collected became the basis for the two volume *Vascular Plants of the Hengduan Mountains* (1993 and 1994).

1981 年 12 月至 1986 年 12 月，裴盛基任中国科学院云南热带植物研究所所长。

1981.12-1986.12, Sheng-Ji PEI named director of Yunnan Institute of Tropical Botany, Chinese Academy of Sciences.

1981—1992 年，江苏省植物研究所 南京中山植物园，**南京中山植物园研究论文集**[588]；1992 年由**植物资源与环境取代**。

1981-1992, Jiangsu Institute of Botany, Nanjing Botanical Garden Memorial Sun Yat-Sen, ***Bulletin of Nanjing Botanical Garden Memorial Sun Yat-Sen***, superseded by ***Journal of Plant Resources and Environment*** from 1992.

1981—1988 年、1997—1998 年，丁宝章、王遂义、高增义（第 1 卷），丁宝章、王遂义（第 2–4 卷）主编，**河南植物志**，vols. 1-4；1：632 页，1981；2：670 页，1988；3：781 页，1997；4：581 页，1998；河南人民出版社（第 1 卷）、河南科学技术出版社（第 2–4 卷），郑州。

1981-1988 and 1997-1998, Bao-Zhang DING, Sui-Yi WANG and Zeng-Yi GAO (vol. 1, editors in chief), Bao-Zhang DING and Sui-Yi WANG (vols. 2-4, editors in chief), '***Flora of Henan***', vols. 1 and 2-4; 1: 632 p, 1981, 2: 670 p, 1988, 3: 781 p, 1997, 4: 581 p, 1998; Henan People's Publishing House (vol. 1) and Henan Science and Technology Press (vols. 2-4), Zhengzhou.

1981 年，林赞标[589]（1948—）赴美国俄勒冈州立大学留学，1985 年获博士学位；1985—1987 年于密执安州立大学从事博士后研究；返台后任台湾林业试验所副研究员和研究员，2000—2017 年任台湾大学植物系教授；专长兰科。

1981, Tsan-Piao LIN (1948-) studied at the Oregon State University, Oregon, United States of America;

588 此刊物没有卷或集的数目，具体期刊年代（与出版年代）如下：1980（1981）、1981（1983）、1982（1984）、1983（1985）、1984—1985（1986）、1986（1987）、1987（1988）、1988—1989（1990）、1990（1991）、1991（1992）。

589 台湾宜兰人；1966 年入中兴大学植物系，1970 年本科毕业获学士学位；1971 年入台湾大学植物研究所硕士班，1973 年毕业获硕士学位；1973—1978 年任台湾林业试验所任研究助理，1978 年任副研究员。

obtained his Ph.D. degree[590] in 1985; as postdoc research fellow of Michigan State University, 1985-1987; returned to become research scientist of Taiwan Forestry Research Institute, then professor of department of botany, Taiwan University, 2000-2017; specializing in Orchidaceae.

590 Tsan-Piao LIN, 1985, Acid Hydrolases and Seed Shriveling in Triticale Seed; Oregon State University, 78 pages.

1982

1982.09.23-09.25, Biogeographical Relationships between Temperate Eastern Asia and Temperate Eastern North America: The twenty-Ninth Annual Systematics Symposium held at Missouri Botanical Garden, St. Louis, Missouri, the United States of America, more than 20 papers published in *Annals of the Missouri Botanical Garden* 70 (3): 421-576 and 70(4): 577-749, 1983, about half authored by Chinese botanists, David H. S. CHANG (Hsin-Shi CHANG), Sing-Chi CHEN, Shan-An HE, De-Yuan HONG, Hsueh-Yu HOU, Jen HSU, Cheng-Yih WU, Tsun-Shen YING and Zhong ZHENG.

1982 年 9 月 23 日至 9 月 25 日，美国密苏里植物园第二十九届年会：温带东亚和北美东部生物地理关系研讨会举行，1983 年 *Annals of the Missouri Botanical Garden* 分两期 70（3）：421-576、70（4）：577-749 发表 20 多篇相关文章，约一半来自中国植物学者：张新时、陈心启、贺善安、洪德元、侯学煜、徐仁、吴征镒、应俊生、郑重。

1982 年 9 月，陈晓亚[591]（1955—）赴英国留学，1985 年 10 月获雷丁大学博士学位，1985—1991 年于南京大学生物系任讲师和副教授，1991—1994 年先后在德国图宾根大学和美国普渡大学从事博士后研究，1994 年 10 月至今任中国科学院上海植物生理生态研究所研究员，2010 年至今兼职上海辰山植物园园长和中国科学院上海辰山植物科学研究中心主任，先后担任植物生理生态研究所所长、中国科学院上海生命科学研究院院长。现从事植物分子生理学，早年从事分类学研究；2005 年当选中国科学院院士，2008 年当选发展中国家科学院院士。

1982.09, Xiao-Ya CHEN (1955-) entered graduated shool in England, received his Ph.D. degree from University of Reading in October 1985[592]; lecturer and associate professor in Department of Biology, Nanjing University, 1985 to 1991; postdoctoral research fellows in Universität Tübingen, Germany, and Purdue University, United States of America, 1991 to 1994; professor of Institute of Plant Physiology and Ecology, Chinese Academy of Sciences, Shanghai, since October 1994, and director of Shanghai Chenshan Botanical Garden and president of Shanghai Chenshan Plant Science Research Center, Chinese Academy of Sciences, since 2010. He was director of Institute of Plant Physiology and Ecology, Chinese Academy of Sciences, and president of Shanghai Institute for Biological Sciences, Chinese Academy of Sciences. His research focuses is on plant molecular physiology but previously on plant taxonomy. He was elected an academician of the Chinese Academy of Sciences in 2005 and of the World Academy Sciences for the advancement of science in developing countries in 2008.

---

591 江苏扬州人，1978 年 2 月至 1982 年 2 月就读于南京大学生物学系植物学专业，获学士学位。

592 Ph.D. thesis: Systematic studies in *Angelica* L. and related genera (Northeast China), 249 p; Advisor: Vernon H. Heywood (1927-).

1982, Ren-Chang CHING and Chu-Hao WANG, A brief report on the progress of pteridological research in China, *American Fern Journal* 72(1): 1-4.

1982 年，秦仁昌、王铸豪，中国蕨类植物研究进展简报，*American Fern Journal* 72（1）：1-4。

1982 年，黄生[593]（1941—）赴美留学，1984 年美国圣路易斯密苏里大学获硕士学位、1989 年再次赴美，1992 年获圣路易斯密苏里大学获博士学位；1976—1986 年任台湾师范大学讲师，1986-1995 年任副教授，1996—2006 年任教授，2006 年退休；2005 年 4 月至 2006 年 2 月曾代理台湾师范大学校长。

1982, Shong HUANG (1941-) went to United States of America to study botany, received master's degree from University of Missouri St. Louis in 1984, and went to United States of America again in 1989, and received Ph.D. degree from University of Missouri St. Louis in 1992; as lecturer of Taiwan Normal University, 1976-1986, associate professor 1986-1995, and professor 1996-2006, and retired in 2006. He was acting president of Taiwan Normal University from April 2005 to Feburary 2006.

1982 年，韩振乾主编，**汉朝植物名称词典**，1 739 页；辽宁人民出版社，沈阳。

1982, Zhen-Qian HAN (editor in chief), '***Chinese-Korean Dictionary of Plant Names***', 1,739 p; Liaoning People's Publishing House, Shenyang.

1982, Hui-Lin LI, ***Contributions to Botany*** *Studies in Plant Geography, Phylogeny and Evolution, Ethnobotany and Dendrological and Horticultural Botany*, 527 p; Epoch Publishing Co., Ltd., Taipei [Taibei].

1982 年，李惠林，**植物学论丛** 植物地理、演化及系统、民族植物学及树木与园艺植物学论文选[594]，527 页；时代出版社，台北。

1982 年，卜莱斯著，台湾植物采集记，中华林学丛刊，第 2 号，247 页；中华林学会，台北。

1982, William R. Price, Plant Collecting in Formosa, *General Technical Report*, No. 2, 247 p; The Chinese Forestry Association, Taipei [Taibei].

1982 年，任波涛，**植物学拉丁文基础课本**，164 页；北京林学院，北京（内部印制）。

1982, Po-Tao REN, ***Elementa Linguae Latinae Botanicae***, 164 p; Beijing Forestry College, Beijing (internal publication).

---

593 北京人，1960 年入台湾师范大学博物系（后改为生物系），1965 年毕业（五年制，含最后一年实习）。

594 该书是李惠林诞辰七十周年文集，收载 1932—1981 年论著 42 篇；书前有李惠林女儿所撰写的传记和李惠林所发表的论著目录（共 270 篇、部）。

1982 年，邢公侠编，秦仁昌校，**蕨类名词及名称**，112 页；科学出版社，北京。

1982，Kung-Hsia SHING, Ren-Chang CHING (proofreader), ***A Glossary of Terms and Names of Ferns***, 112 p; Science Press, Beijing.

1982, Stephen A. Spongberg and David E. Boufford, *Acta Phytotaxonomica Sinica*—A bibliographic summary of published volumes, *Taxon* 31(4): 705-707.

1982 年，Stephen A. Spongberg、David E. Boufford，植物分类学报 —— 出版卷册目录总结，*Taxon* 31（4）：705-707。

1982，邹稚华[595]（1957—）赴美留学，1990 年获得纽约大学博士学位，返台后于“中央”研究院植物研究所从事研究。

1982, Chih-Hua TSOU (1957-) went to United States of America to study botany, received Ph.D. degree from the City University of New York in 1990[596], research associate fellow in Institute of Botany, Academia Sinica after back to Taiwan.

1982 年，杨再义，**台湾植物名汇**，1 281+351 页；天然书社，台北。

1982, Tsai-I YANG, ***A List of Plants in Taiwan***, 1,281+351 p; Natural Publishing Co. Ltd., Taipei [Taibei].

1982+，新疆八一农学院编著，**新疆植物检索表**，5 册本，第 1 册，516 页，1982；第 2 册，642 页，1983；第 3 册，469 页，1983；第 4 册和第 5 册，XXX；新疆人民出版社，乌鲁木齐。

1982+, Xinjiang August 1 Agricultural College, ***Claves Plantarum Xinjiangensium***, vols. 1-5, 1, 516 p, 1982, 2, 642 p, 1983, 3, 469 p, 1983; 4 & 5, XXX; Xinjiang People's Press, Urumqi.

1982+，浙江省林业研究所 / 浙江省林业研究院 / 国家竹类研究中心，**竹子研究汇刊**，vols. 1+，1982+；半年刊，1987 年改为季刊；2016 年更改为**竹子学报**；1982—1996 年周重光（1913—2000）、1997—2014 年王树东（1950—）、2015 年于辉（1971—）、2016 年至今王玉魁（1962—）任主编。

1982+, Zhejiang Forestry Institute / Zhejiang Academy of Forestry / China National Bamboo Research Center, ***Journal of Bamboo Research***, vols. 1+, semiannual 1982-1986, quarterly since 1987; Editors in Chief: Chung-Kuang CHOU (1913-2000), 1982-1996, Shu-Dong WANG (1950-), 1997-2014, Hui YU (1971-), 2015, Yu-Kui WANG (1962-), since 2016.

---

595 1975—1979 年台湾大学植物系学士（女），1979—1982 年台湾大学植物研究所硕士。

596 Ph.D. dissertation: The embryology, reproductive morphology, and systematics of Lecythidaceae; Advisor: Scott A. Mori (1941-).

1982—1995 年，1991 年，福建植物志编写组编著，林来官主编，**福建植物志**，6 卷本；第 1 卷：631 页，1982；第 2 卷：417 页，1985；第 3 卷：556 页，1988；第 4 卷：669 页，1990；第 5 卷：470 页，1993；第 6 卷：724 页，1995；福建科学技术出版社，福州；福建植物志编写组编著，林来官主编，**福建植物志**，修订版：第 1 卷：645 页，1991；福建科学技术出版社，福州。

1982-1995, 1991, Editorial Group of Flora of Fujian, Lai-Kuan LING (editor in chief), ***Flora Fujianica***, vols. 1-6; 1: 631 p, 1982, 2: 417 p, 1985, 3: 556 p, 1988, 4: 669 p, 1990, 5: 470 p, 1993, 6: 724 p, 1995; Fujian Science and Technology Publishing House, Fuzhou; Editorial Group of Flora of Fujian, Lai-Kuan LING (editor in chief), ***Flora Fujianica***, revised ed., vol. 1: 645 p, 1991; Fujian Science and Technology Publishing House, Fuzhou.

1982—1999 年，南京林产工业学院 / 南京林学院 / 南京林业大学，**竹类研究**，vols. 1-9，1982—1999；1982—1985 年为半年刊，1986 年改为季刊，1999 第 2 期停刊。本刊是我国竹类综合研究类刊物；其中 1975—1979 年曾作为内部资料发行，共 16 期。

1982-1999, Nanjing Technological College of Forest Products / Nanjing Forestry College / Nanjing Forestry University, ***Bamboo Research***, vols. 1-9, semiannual between 1982-1985, quarterly in 1986, suspended after part 2, 1999.

1982—2002 年，中国科学院西北高原生物研究所，**高原生物学集刊**，vols. 1-15（不定期出版）。

1982-2002, Northwest Institute of Plateau Biology, Chinese Academy of Sciences, ***Acta Biologica Plateau Sinica***, vols. 1-15 (published irregularly).

1982—2004 年，贵州植物志编辑委员会编，李永康（1–9 卷）、陈谦海（10 卷）主编，**贵州植物志**，10 卷本；第 1 卷：393 页，1982；第 2 卷：700 页，1986；第 3 卷：484 页，1990；第 4 卷：758 页，1989；第 5 卷：688 页，1988；第 6 卷：643 页，1989；第 7 卷：771 页，1989；第 8 卷：701 页，1988；第 9 卷：410 页，1989；第 10 卷：607 页，2004；贵州人民出版社（第 1–3 卷），贵阳；四川民族出版社（第 4–9 卷），成都；贵州科技出版社，贵阳（第 10 卷）。

1982-2004, Editorial Committee of ***Flora Guizhouensis***, Yong-Kang LI (editor in chief, vols. 1-9) and Qian-Hai CHEN (editor in chief, vol. 10), '***Flora of Guizhou***' / ***Flora Guizhouensis***, vols. 1-10; 1: 393 p, 1982, 2: 700 p, 1986, 3: 484 p, 1990, 4: 758 p, 1989, 5: 688 p, 1988, 6: 643 p, 1989, 7: 771 p, 1989, 8: 701 p, 1988, 9: 410 p, 1989, 10: 607 p, 2004; Guizhou People's Press (vols. 1-3), Guiyang, Sichuan Minorities Press (vols. 4-9), Chengdu and Guizhou Science and Technology Publishing House (vol. 10), Guiyang.

1982、1983 年，李丙贵、刘林翰、万绍宾，湖南植物志资料（第一集），湖南师范学院学报，自然科学版，1982 年增刊（总第 9 期）：1–115 页，1982；李丙贵、彭寅斌、万绍宾、杨保民、刘林翰，湖南植物志资料（第二集），湖南师范学院学报，自然科学版，1983 年增刊（总第 13 期）：1–150 页，

1983。

1982 and 1983, Bing-Gui LI, Lin-Han LIU, Shao-Bin WAN, Materials for Flora of Hunan (I), *Journal of Hunan Teacher's College, Natural Science Edition*, 1982 Supplement (No. 9): 1–115 p, 1982; Bing-Gui LI, Yin-Bin PENG, Shao-Bin WAN, Bao-Min YANG, Lin-Han LIU, Materials for Flora of Hunan (II), *Journal of Hunan Teacher's College, Natural Science Edition*, 1983 Supplement (No. 13): 1-150 p, 1983.

1983

1983 年 2 月，杨远波[597]（1943—）赴美国留学，1988 年获美国圣路易斯大学生物学博士学位；1989 年 7 月至 1992 年 7 月任台湾林业试验所副研究员，1992 年 8 月至 1999 年 7 月任中山大学生物系副教授，1998 年 8 月至 2008 年 7 月任教授，其中 2001 年 8 月至 2004 年 7 月兼任中山大学生物系主任。2008 年 7 月至 2013 年 7 月任大叶大学生物技术与资源学院教授。

1983.02, Yuen-Po YANG (1943-) studied at St. Louis University, St. Louis, Missouri, United States of America, received Ph. D. degree in 1988[598]; employed by the Taiwan Forestry Research Institute, as associate research professor, July 1989 to July 1992; associate professor in Department of Biological Sciences, Sun Yat-Sen University, August 1992 to July 1999; professor, August 1998 to July 2008; also dean of Department of Biological Sciences, August 2001 to July 2004; Professor of School of Biotechnology and Bioresources, Dayeh University, July 2008 to July 2013.

1983 年 5 月，顾红雅（1960—）[599] 赴美国留学，1987 年 12 月获美国圣路易斯华盛顿大学理学博士，1988 年 4 月至 1989 年 9 月和 1989 年 9 月至 1990 年 6 月先后于中国科学院植物研究所和北京大学生物系从事博士后研究，1990 年 7 月至 1992 年 11 月任北京大学生命科学学院副教授，1992 年 12 月至今任北京大学生命科学学院教授；主要从事系统与进化植物学、植物分子生物学、植物遗传多样性和自然类群的起源，基因家族的功能和演化等研究。

1983.05, Hong-Ya GU (1960-) went to United States of America to study, received her Ph.D. degree from Washington University, St. Louis, Missouri, in December 1987[600], postdoctoral research fellow in the Institute of Botany, Chinese Academy of Sciences, from April 1988 to September 1989, Department of Biology, Peking University, from September 1989 to June 1990, associate professor of School of Life Sciences, Peking University from July 1990 to November 1992, and professor of School of Life Sciences, Peking University, since December 1992, research focus on systematic and evolutionary botany and plant molecular biology, genetic diversity and origin of natural taxa, and function and evolution of certain plant gene families.

---

597 1964 年 8 月至 1969 年 7 月台湾师范大学生物学系学士；1969 年 8 月至 1972 年 7 月台湾大学植物学研究所硕士；1973 年 7 月至 1975 年 3 月任台湾林业试验所生物系研究助理；1975 年 4 月至 1981 年 6 月任台湾林业试验所生物系技士；1981 年 7 月至 1989 年 6 月任台湾林业试验所助理研究员。

598 Ph.D. dissertation: Systematics of Subgenus *Bladhia* of *Ardisia* (Myrsinaceae), 316 p; Adviser: John D. Dwyer (1915-2005).

599 江苏扬州人（女），1978 年 2 月至 1982 年 2 月南京大学生物系植物学学士，1982 年 3 月至 1983 年 5 月南京大学生物系教育部出国代培研究生。

600 Ph.D. dissertation: A Biosystematic Study of the Genus *Kalimeris*; Advisor: Peter H. Raven. Hong-Ya GU and Peter C. Hoch, 1997, Systematics of *Kalimeris* (Asteraceae: Astereae), *Annuals of the Missouri Botanical Garden* 84(4): 762-814.

1983.05.23-05.28, World Conference of Bryology held at National Museum of Nature and Science, Tokyo, attended by Man-Hsiang CHANG, Chien GAO, Ming-Jou LAI, Pang-Juan LIN, Shan-Hsiung LIN, Chia-Li WU and Pan-Cheng WU from China. This was the first international bryological meeting attended by Chinese bryologists.[601]

1983年5月23日至5月28日，世界苔藓植物学大会在东京国立科学博物馆召开，中国代表高谦、赖明洲、林邦娟、林善雄、吴嘉丽、吴鹏程、张满祥等参加。这是中国苔藓学者首次参加国际苔藓学大会。[602]

1983.09.07-09.26, Zennoske Iwatsuki (1929-2015)[603], Botanical Institute, Hiroshima University, Japan, visited China with his wife and gave presentations. Visited Institute of Forestry and Pedology, Shenyang, Liaoning, Institute of Botany, Beijing, and South China Institute of Botany, Guangzhou, Guangdong, Chinese Academy of Sciences. Collected bryophytes at Qian Shan, Liaoning, Changbai Shan, Jilin, and Dinghu Shan, Guangdong.[604]

1983年9月7日至26日，日本广岛大学植物研究所岩月善之助（1929—2015）夫妇访问中国。他们走访了沈阳的中国科学院林业土壤研究所、北京的中国科学院植物研究所和广州的中国科学院华南植物研究所并做报告，同时赴辽宁千山、吉林长白山和广东鼎湖山等地考察并采集苔藓。

1983.09-10, Paul L. Redfearn, Jr. (1926-, SMS), Department of Biology, Southwest Missouri State University, visited China with his wife. Visited Institute of Botany, Beijing, Kunming Institute of Botany, Kunming, Yunnan, Chengdu Institute of Biology, Chengdu, Chinese Academy of Sciences, Yunnan University, Kunming and Sichuan University, Chengdu; had discussions with Chinese bryologists and mycologists, including Ren-Liang HU, Tsung-Ling WAN[605] (1907-2006, Pan-Chieh CHEN's widow), Pan-Cheng WU, Ruo-Li XIONG (1928-1985), Wen-Xuan XU and Mu ZANG; also briefly visited Emei Shan, Sichuan.[606]

1983年9月至10月，西南密苏里州立大学生物系 Paul L. Redfearn，Jr.（1926—）和夫人，先后访问位于北京的中国科学院植物研究所、云南昆明的中国科学院昆明植物研究所、成都的中国科学院成都生物研究所、云南大学和四川大学，并与中国苔藓学家和菌物学家交流，包括胡人亮、万宗

601 Stephan R. Gradstein, 1983, On the state of affairs in I. A. B., a report on the IAB council meeting in Tokyo, 1983, *The Bryological Times* 22: 1-3.

602 Pan-Cheng WU, 1983 (1984), Out and about in the People's Republic of China, *The Bryological Times* 24: 3.

603 Hironori Deguchi, 2016, Obituary: Zennoske Iwatsuki (1929-2015), *Hattoria* 7: 1-8.

604 Zennosuke Iwatsuki, 1984, A bryological trip to China, *The Bryological Times* 26: 3-5.

605 As Thung-Ling WAN in IPNI.

606 Paul L. Redfearn, Jr. 1984, A visit to the People's Republic of China, *The Bryological Times* 27: 3-5.

玲[607]（1907—2006，陈邦杰遗孀）、吴鹏程、熊若莉（1928—1985）、徐文宣和臧穆等；还对四川的峨眉山进行简短考察。

1983年10月1日至6日，中国植物学会第九届会员代表大会暨五十周年年会于山西太原举行，与会代表400多人；理事长汤佩松。本次会议确认以往历届大会序号。[608]

1983.10.01-10.06, Ninth National Congress and 50th anniversary meeting of the Botanical Society of China held in Taiyuan, Shanxi; more than 400 attendees. Pei-Song TANG was elected president. The sequence number of previous congresses of Botanical Society of China was officially confirmed.

Fiftieth anniversary meeting of Botanical Society of China, Taiyuan (Photo provided by Zong-Gang HU)
中国植物学会五十周年年会于太原（相片提供者：胡宗刚）

607 四川永川人，毕业于国立中央大学生物系，抗战时曾任教于重庆大学等多个大中专院校，1954年调入中国科学院植物研究所任助理研究员，在南京师范学院的苔藓植物研究室协助陈邦杰开展研究工作，并参加野外考察和研究生培养等。苔藓植物研究室搬回北京后退休（陈家幼女陈佐芳于美国圣地亚哥通过李植华先生于2017年8月25日提供，特此致谢）。

608 植边，1983，中国植物学会成立五十周年年会暨第九届会员代表大会在太原举行，植物杂志6：1-2。

年　年　會　一九八三年十月於太原

Council Member of Botanical Society of China, Taiyuan, 1983 (Photo provided by Zong-Gang HU)
1983 年中国植物学会理事会于太原（相片提供者：胡宗刚）

The editorial committee members, *Flora Reipublicae Popularis Sinicae*, Taiyuan, 1983 (Photo provided by Zong-Gang HU)
1983 年，中国植物志编辑委员会于太原（相片提供者：胡宗刚）

1983 年，董仁威、邱沛篁，**绿海探宝** —— 林奈与方文培，121 页；四川少年儿童出版社，成都。

1983, Ren-Wei DONG and Pei-Huang QIU, '***Lü Hai Tan Bao***—*Carl Linnaeus and Wen-Pei FANG*', 121 p; Sichuan Children's Publishing House, Chengdu.

1983, Rolf M. T. Dahlgren, General aspects of angiosperm evolution and macrosystematics, *Nordic Journal of Botany* 3(1): 119–149. Translated into Chinese in 1987.

1987 年，张芝玉译，何关福校，被子植物的进化和大系统的概况，生物科学参考资料 22：21–53；科学出版社，北京。

1983 年，洪德元，中国植物分类学进展，植物杂志 2：1–3。

1983, De-Yuan HONG, 'Advances in Chinese plant taxonomy', *Plants* 2: 1–3.

1983 年，李亮恭著，**中国生物学发展史**，216 页；中央文物供应社，台北。

1983, Liang-Kung LI, '***Development History of Chinese Biology***', 216 p; Central Heritage Provider, Taipei [Taibei].

1983, Robert F. Thorne, Proposed new realignments in the angiosperms, *Nordic Journal of Botany* 3(1): 85-117. Translated into Chinese in 1987.

1987 年，张金泉译，王铸豪校，关于被子植物重新组合的新观点，生物科学参考资料 22：185-210；科学出版社，北京。

1983 年，吴鹏程、罗健馨、汪楣芝，一个原始的苔类的目 —— 藻苔目在中国的发现，植物分类学报 21（1）：105-107。

1983, Pan-Cheng WU, Jian-Shing LUO and Mei-Zhi WANG, The primitive liverwort Takakiales is discovered in Xizang (Tibet), *Acta Phytotaxonomica Sinica* 21(1): 105-107.

1983 年，颜素珠编著，**中国水生高等植物图说**，335 页；科学出版社，北京。

1983, Su-Zhu YAN, '***Icones of Aquatic Higher Plants of China***', 335 p; Science Press, Beijing.

1983 年，中国科学院武汉植物研究所编著，王宁珠等编，**中国水生维管束植物图谱**，683 页；湖北人民出版社，武汉。

1983, Wuhan Institute of Botany, Chinese Academy of Sciences, Ning-Zhu WANG et al. (eds.), '***Icones of Aquatic Vascular Plants of China***', 683 p; Hubei People's Press, Wuhan.

1983 年，中国植物学会植物分类专业委员会：中国种子植物分类学的回顾和展望；秦仁昌：五十年来的中国蕨类植物学；苔藓植物组（高谦等）：中国苔藓植物学的研究概况及其展望。中国植物学会五十周年年会学术报告及论文摘要汇编，30-39、40-43、44-53 页；中国植物学会，北京（内部印制）。

1983, Committee of Phytotaxonomy of Botanical Society of China: *Vistas in Taxonomy of Spermatophytes in China*; Ren-Chang CHING: *Fifty years of Pteridology in China*; Committee of Bryology Section (Chien GAO et al.): *The Historical Survey and Prospects of Chinese Bryology*; Reports and abstracts presented at a meeting commemorating the 50th anniversary of the Botanical Society of China, 30-39, 40-43 and 44-53 p; Botanical Society of China, Beijing (internal publication).

1983, Edward G. Voss, Herve M. Burdet, William G. Chaloner, Vincent Demoulin, Paul H. Hiepko, John McNeill, Robert D. Meikle, Dan H. Nicolson, Reed C. Rollins, Paul C. Silva and Werner R. Greuter, ***International Code of Botanical Nomenclature* (Sydney Code)**, adopted by the Thirteenth International Botanical Congress, Sydney, August 1981, *Regnum Vegetabile* 111: 1-472 p. Translated into Chinese in

1983-1985.

1983年，1983—1985年，汤彦承译，**国际植物命名法规简介**，71页，1983，中国植物志参考文献目录第31册；中国植物志编辑委员会,北京（油印本）[609]；汤彦承[610],国际植物命名法规简介（I-IX），植物学通报，1（1）：55-59，1（2）：57-59，1983；2（1）：53-55，2（2-3）：87-92，2（4）：51-57，2（5）：58-61，2（6）：49-54，1984；3（1）：59-62，3（2）：53-56，1985[611]。

1983+，中国科学院武汉植物研究所/中国科学院武汉植物园、湖北省植物学会，**武汉植物学研究**，vols. 1+，不规则发行，1985年改为季刊，2000年改为双月刊，2011年更名为**植物科学学报**；1983—1988年孙祥钟、1989—1991年郑重、1992—1993年侯嵩生、1993—2004年郑重、2005—2009年黄宏文、2010—2014年李绍华（1957—）、2015年至今王青锋（1968—）任主编。

1983+, Wuhan Institute of Botany / Wuhan Botanical Garden, Chinese Academy of Sciences and Botanical Society of Hubei, ***Wuhan Botanical Research***, changed to ***Journal of Wuhan Botanical Research*** in 1985, changed to ***Plant Science Journal*** since 2011; vols. 1+, irregularlly, quarterly since 1985, bimonthly in 2000; Editors in Chief: Hsiang-Chung SUN, 1983-1988, Zhong ZHENG, 1989-1991, Song-Sheng HOU, 1992-1993, Zhong ZHENG, 1993-2004, Hong-Wen HUANG, 2005-2009, Shao-Hua LI (1957-), 2010-2014, Qing-Feng WANG (1968-) since 2015.

1983, Shan-Hsiung LIN, A Taxonomic Revision of Phyllogoniaceae (Bryopsida) Part I, *Journal of Taiwan museum* 36 (2): 37-86, with a new family Orthorrynchiaceae.

1983年，林善雄，带叶苔科之分类学修订Ⅰ，台湾省立博物馆半年刊36（2）：37-86，含新科直喙藓科。

1983、1984年，毕列爵，从19世纪到建国之前西方国家对我国进行的植物资源调查，武汉植物学研究1（1）：119-128，1983；毕列爵、李建强，从19世纪到建国之前西方国家对我国进行的植物资源调查（续），武汉师范学院学报1：77-84，1984。

1983, 1984, Lie-Jue BI, The botanical exploration in China by western countries during the 149 years (1800-1949) before liberation, *Wuhan Botanical Research* 1(1): 119-128, 1983; Lie-Jue BI and Jian-Qiang LI, The botanical exploration in China by western countries during the 149 years (1800-1949) before liberation (cont.), *Bulletin of Wuhan Teaching College* 1: 77-84, 1984.

609 1984年12月江西植物志编辑委员会曾重新印刷（标题相同）。

610 中国植物志参考文献目录原稿是1982年在西北植物研究所座谈会上的简介，参考的是列宁格勒法规的内容；本次印刷修改稿参考了悉尼法规；详细参见该文献的几点说明。遗憾的是，之后的植物学通报连续9次报道中并没有交代具体的依据版本，而且在简史部分还提到“本法规由悉尼法规取代”。据汤先生的学术严谨程度，推断其依据的应为悉尼法规（尽管没有具体说明）。

611 汤彦承、郑儒永合译。

1983、1995 年，中国植物学会编，王宗训主编，**中国植物学文献目录**，1：1–620 页，2：621–1 226 页，3：1 227–1 793 页，1983；4：1–1 463 页，1995；科学出版社，北京。

1983 and 1995, Botanical Society of China, Zong-Xun WANG (Tsung-Hsun WANG, editor in chief), ***Bibliography of Chinese Botany***, 1: 1-620 p, 2: 621-1,226 p, 3: 1,227-1,793 p, 1983; 4: 1-1,463 p, 1995; Science Press, Beijing.

1983、1986 年，中国科学院青藏高原综合科学考察队编，**青藏高原研究 —— 横断山考察专集**，1：1–291 页，1983；云南人民出版社，昆明；2：1–623 页，1986；北京科学技术出版社，北京。

1983 and 1986, Comprehensive Scientific Expedition to Qinghai-Xizang Plateau, Chinese Academy of Sciences[612], ***Studies in Qinghai-Xizang (Tibet) Plateau—Special Issue of Hengduan Mountains Scientific Expedition***, 1: 1-291 p, 1983; Yunnan People's Publishing House, Kunming, 2: 1-623 p, 1986; Beijing Science and Technology Press, Beijing.

1983—1987 年，第 3 次全国中药资源普查，先后出版中国中药资源、中国中药资源志要、中国常用中药材、中国中药区划、中国药材资源地图集等。

1983-1987, Third national survey of traditional Chinese medicinal materials carried out; *Chinese Resources of Traditional Chinese Medical Materials*, *Records of China Traditional Chinese Medical Materials*, *Chinese Common Traditional Chinese Medical Materials*, *Division of China Traditional Chinese Medical Materials* and *Atlas of China Traditional Chinese Medical Resources* published sequentially.

1983—1987 年，中国科学院青藏高原综合科学考察队，吴征镒主编，**西藏植物志**，5 卷本；第 1 卷：791 页，1983；第 2 卷：956 页，1985；第 3 卷：1 047 页，1986；第 4 卷：1 021 页，1985；第 5 卷：955 页，1987；科学出版社，北京。[613]

1983-1987, Comprehensive Expedition to Qinghai-Xizang Plateau, Chinese Academy of Sciences, Cheng-Yih WU (editor in chief), ***Flora Xizangica***, vols. 1-5; 1: 791 p, 1983, 2: 956 p, 1985, 3: 1,047 p, 1986, 4: 1,021 p, 1985, 5: 955 p, 1987; Science Press, Beijing.

1983—1993 年，陕西省中国科学院西北植物研究所更名为西北植物研究所；1983—1984 年李振声、1985—1989 年于兆英（1934—1993）、1989—1993 年白守信（1933—）任所长。

1983-1993, Shaanxi Institute of Northwest Botany, Chinese Academy of Sciences, renamed Northwest Institute of Botany, with directors Chen-Sheng LI, 1983-1984, Zhao-Ying YU (1934-1993), 1985-1989, and Shou-Xin BAI (1933-), 1989-1993.

612 As Academia Sinica.

613 李晖、于顺利、土艳丽、央金卓嘎，2009,《西藏植物志》亟待修订，西藏科技 5（194）：67-68。

1983—1993 年，贺善安任江苏省植物研究所 南京中山植物园主任。

1983-1993, Shan-An HE named director, Jiangsu Institute of Botany, Nanjing Botanical Garden Memorial Sun Yat-Sen.

1983—1994 年，中国科学院植物研究所，**植物学集刊**，vols. 1-7（不定期出版）。

1983-1994, Institute of Botany, Academia Sinica, ***Botanical Research***, *Contributions from the Institute of Botany, Academia Sinica*, vols. 1-7 (published irregularly).

1983—1995 年，中国科学院华南植物研究所，**中国科学院华南植物研究所集刊**，vols. 1-10（不定期出版）。

1983-1995, South China Institute of Botany, Academia Sinica, ***Acta Botanica Austro-Sinica***, vols. 1-10 (published irregularly).

1983、2001 年，关克俭等，**拉汉英种子植物名称**，1 036 页，1983；科学出版社，北京；朱家柟主编，**拉汉英种子植物名称**，第 2 版，1 393 页，2001；科学出版社，北京。

1983, 2001, Ke-Chien KUAN et al., ***Dictionary of Seed-Plants Names Latin-Chinese-English***, 1,036 p, 1983; Science Press, Beijing; Jia-Nan ZHU (editor in chief), ***Dictionary of Seed-Plants Names Latin-Chinese-English***, ed. 2, 1,393 p, 2001; Science Press, Beijing.

1983—2004 年，中国树木志编辑委员会编，郑万钧主编，**中国树木志**，4 卷本；第 1 卷：1-929 页，1983；第 2 卷：931-2 398 页，1985；第 3 卷：2 399-3 969 页，1997；第 4 卷：3 972-5 429 页，2004；中国林业出版社，北京。

1983-2004, Editorial Committee of Sylva Sinica, Wan-Chun CHENG (editor in chief), ***Sylva Sinica***, vols. 1-4; 1: 1-929 p, 1983, 2: 931-2,398 p, 1985, 3: 2,399-3,969 p, 1997, 4: 3,972-5,429 p, 2004; China Forestry Publishing House, Beijing.[614]

1983—2010 年、2015+，广东省植物学会，**广东省植物学会会刊**，1-19、20-[615]；广东省植物学会，广州（内部发行）。

1983-2010, 2015+, Botanical Society of Guangdong, ***Bulletin of Botanical Society of Guangdong***, parts 1-19, 20-; Botanical Society of Guangdong, Guangzhou (internal publication).

---

614 Jin-Shuang MA, 2005: Book Reviews: An indispensable work in Chinese plant taxonomy "*Sylva Sinica*", by Wan-Chun Cheng, p. 1-5,429, China Forestry Publishing House, *Taxon* 54(1): 262-263.

615 目前只见到网络版（http://bxh.scbg.cas.cn/xhkw/201507/P020150703550134421932.pdf，2018 年进入）。

1984

1984.06-07, 1984 Sino-American botanical expedition to Yunnan: Anning, Chuxiong, Dali (Cang Shan), Kunming, Lufeng, Lunan, Nanhua, Songming, Weishan, Xiangyun, Yangbi. Bruce M. Bartholomew (CAS), David E. Boufford (A/GH), Hsi-Wen LI (KUN), Cheng-Gung MA (PE), Dan H. Nicolson (1933-2016, US)[616], Tsun-Shen YING (PE), Shao-Wen YU (KUN), Si HE (1959-, PE), Paul L. Redfearn, Jr. (MO, SMS) and Yong-Ge SU (KUN) collected 2,006 numbers of vascular plant specimens (Sino-American Botanical Expedition nos. 1-2,006, A, KUN, PE) and 2,328 numbers of bryophytes specimens (Paul L. Redfearn. Jr., nos. 1-2,328 (KUN, MO, PE).[617]

1984 年 6 月至 7 月，1984 年中美植物联合考察云南安宁、楚雄、大理（苍山）、昆明、禄丰、路南、南华、嵩明、巍山、祥云、漾濞。加州科学院 Bruce M. Bartholomew，哈佛大学 David E. Boufford，史密森学会的 Dan H. Nicolson（1933—2016），中国科学院昆明植物研究所李锡文、苏永革、俞绍文，中国科学院植物研究所何思（1959—）、马成功、应俊生，西南密苏里州立大学、密苏里植物园 Paul L. Redfearn，Jr. 等，采集维管束植物标本 2 006 号（A、KUN、PE）、苔藓植物标本 2 328 号（KUN、MO、PE）。

1984 年 8 月，张祺编、朱家柟校，**植物命名手册**，286 页；北京农业大学，北京（内部印制）。

1984.08, Qi ZHANG (ed.), Jia-Nan ZHU (proofreader), ‘***Handbook of Botanical Nomenclature***’, 286 p; Beijing Agricultural University, Beijing (internal publication).

1984 年 9 月，沈中桴[618]（1957—）赴美留学，1992 年获得纽约大学博士学位；返台后从事植物地理以及台湾生物相研究。

1984.09, Chung-Fu SHEN (1957-) went to United States of America for study, received Ph.D. degree from the City University of New York in 1992[619]. After back to Taiwan, he works on plant geography and Taiwan Flora.

1984 年 10 月 21 日，中国植物学会于湖北武汉举办植物分类学原理讨论会，洪德元、徐炳声、

616 Dan H. Nicolson (1933-2016), *The Plant Press* 19(3): 6-7, 2016.

617 David E. Boufford and Bruce Bartholomew, 1986, The 1984 Sino-American Botanical Expedition to Yunnan, China, *Arnoldia* 46(4): 15-36.

618 1974—1979 年台湾大学森林系本科学士，1979—1981 年台湾大学森林系硕士研究生，1981—1983 年服兵役，1983—1984 年完成台湾大学森林系硕士学位。

619 Ph.D. Dissertation: A monograph of the genus *Fagus* Tourn. ex L. (Fagaceae); Advisor: Arthur J. Cronquist (1919-1992).

En route to Huadianba, Sino-American expedition to Yunnan (Cang Shan), June–July, 1984 (Photo provided by David E. Boufford)
1984 年 6 月至 7 月，中美植物学家联合考察云南（苍山）花甸坝路上（相片提供者：鲍棣伟）

Participants at National Symposium on Principles of Plant Taxonomy, Wuhan, 21 October 1984 (Photo provided by Ping MA)
1984 年 10 月 21 日，全国植物分类学原理讨论会于武汉（相片提供者：马平）

汤彦承和徐克学等主讲，近百人出席。

1984. 10. 21, National Symposium on Principles of Plant Taxonomy, organized by Botanical Society of China, held in Wuhan, Hubei; lectures by De-Yuan HONG (PE), Ping-Sheng HSU (FUS), Yen-Cheng TANG and Ke-Xue XU (PE); nearly 100 attendees.

1984 年，张宏达，种子植物系统分类提纲，广东省植物学会会刊 2：17-18。

1984, Hung-Ta CHANG, Outline of spermatophyta classification, *Bulletin of Botanical Society of Guangdong* 2: 17-18.

1984，Isabel S. Cunningham, ***Frank N. Meyer—Plant Hunter in Asia***, 317 p; Iowa State University Press, Ames, IA. Translated into Chinese in 2015 and 2018.

2015[620] 和 2018 年，谭继清、白史且、谭志艰、李平等译，**佛兰克 梅尔——美国现代植物探险家**，260 页、232 页；自印，重庆，佛兰克 梅尔传书稿编著者印制。[621]

1984 年，路安民，诺・达格瑞被子植物分类系统介绍和评注，植物分类学报 22（6）：497-508。

1984, An-Ming LU, Introduction and notes to Rolf M. T. Dahlgren's system of classification of the angiosperms, *Acta Phytotaxonomica Sinica* 22(6): 497-508.

1984 年，徐炳声，被子植物系统发育研究的现状与展望，云南植物研究 6（1）：1-10。

1984, Ping-Sheng HSU, Present day aspects and perspectives of the study of angiosperm phylogeny, *Acta Botanica Yunnanica* 6(1): 1-10.

1984, Shiu-Ying HU invited to visit China by South China Institute of Botany, Chinese Academy of Sciences, Guangzhou; visited Beijing, Guangdong, Guangxi, Hubei, Inner Mongolia, Jiangsu, Ningxia, Qinghai and Shanghai; presented many lectures, especially at South China Agricultural College, South China Institute of Botany and Northwest Institute of Plateau Biology, Chinese Academy of Sciences.

1984 年，胡秀英应中国科学院华南植物研究所的邀请，在北京、广东、广西、湖北、内蒙古、江苏、宁夏、青海、上海，特别是在华南农学院、中国科学院华南植物研究所、中国科学院西北高原生物研究所等，做了 59 个 3 小时的报告，开了 40 个座谈会（其中，6 月 21 日至 7 月 3 日在中国科学院华南植物研究所做了 5 次报告）。[622]

---

620 首页时间为 2015 年，而译后记时间则为 2016 年 2 月春节。

621 没有书号与出版社。

622 胡秀英，1987，国外植物学家二三事，武汉植物学研究 5（2）：197-204; Shiu-Ying HU, 1987, Remarks of a foreign student in professor Fernald's last class at Harvard University, *Journal of Wuhan Botanical Research* 5(2): 197-204；陈梦玲，1996，美籍华人之光——胡秀英博士，植物杂志 5：36-38。

1984, Timo J. Koponen, A historical review of Chinese bryology, p. 283-313, Jirí Vána (ed.), *Proceedings of the Third Meeting of the Bryologists from Central and East Europe, Praha*.

1984 年，Timo J. Koponen，中国苔藓学历史回顾，283–313 页；Jirí Vána 编辑，第三届中东欧苔藓学家会议论文集，布拉格。

1984 年，阳顾平、徐尤清合编，**蕨类藻类名词辞典**，112+159 页；名山出版社，台北。

1984，Chin-Ping YANG and You-Ching HSU (eds.), ***English Chinese Dictionary of Pteridophyte and Algae***, 112 p + 159 p; Mingshan Press, Taipei [Taibei].

1984 年，王文采，被子植物分类系统选介（I、II），植物学通报 2（5）：11-17、33，2（6）：15-20。

1984, Wen-Tsai WANG, On six important systems of classification of the Angiosperms (I and II), *Chinese Bulletin of Botany* 2(5): 11-17, 33, 2(6): 15-20.

1984, Pan-Cheng WU, Fifty years of Chinese bryology, *The Journal of the Hattori Botanical Laboratory* 56: 29-38.

1984 年，吴鹏程，五十年来的中国苔藓植物学，*The Journal of the Hattori Botanical Laboratory* 56：29-38。

1984 年，吴鹏程等著，**苔藓名词及名称**，124 页；科学出版社，北京。

1984, Pan-Cheng WU et al., ***A Glossary of Terms and Names of Bryophytes***, 124 p; Science Press, Beijing.

1984 年，中国科学院昆明植物研究所编，吴征镒主编，**云南种子植物名录**[623]，上册：1-1 070 页，下册：1 071-2 259 页；云南人民出版社，昆明。

1984, Kunming Institute of Botany, Cheng-Yih WU (editor in chief), ***Index Florae Yunnanensis***, 1: 1-1,070 p; 2: 1,071-2,259 p; Yunnan People's Publishing House, Kunming.

1984 年，郑重，哈佛大学植物标本馆湖北木本植物标本志要，武汉植物学研究 2（1）：1-219。

1984, Zhong ZHENG, A conspectus of Hubei woody plants in the Harvard University Herbaria, U. S. A., *Wuhan Botanical Research* 2(1): 1-219.

623 初稿曾于 1959 年分三册内部印制。

1984—1988 年，四川省林业科学研究院森林植物分类研究室，**森林植物研究**，vols. 1-4（不定期，内部发行）。

1984-1988, Forest Plant Taxonomy Department, Sichuan Academy of Forestry, ***Bulletin of Forest Plant Research***, vols. 1-4 (irregular; issued internally).

1985

1985年7月，何思[624]（1959—）赴美国密苏里植物园做访问学者，1986年9月转赴辛辛那提大学攻读博士研究生，1992年6月获辛辛那提大学理科博士学位，1992年7月至1993年10月于密苏里植物园从事博士后研究，1993年11月至1999年8月任密苏里植物园助理研究员、1999年9月至2006年12月任副研究员、2007年至今任研究员；特别是亚洲苔藓学研究，主持英文版中国藓类植物志项目。

1985.07, Si HE (1959-) went to the Missouri Botanical Garden as visiting scholar, then transferred to University of Cincinnati for graduate studies; received Ph.D. degree from University of Cincinnati, June 1992[625]; postdoctoral research fellow at Missouri Botanical Garden, July 1992 to October 1993, then assistant curator, November 1993 to August 1999, associate curator, September 1999 to December 2006, and curator since 2007; specializing in Asian bryophytes, in charge of *Moss Flora of China* project.

1985年9月，文军[626]（1963—）赴美国留学，1991年获美国俄亥俄州立大学博士学位；1992—1995年分别于哈佛大学阿诺德树木园和美国史密森学会国家自然历史博物馆从事博士后研究，1995—2000年任科罗拉多州立大学助理教授兼标本馆馆长，2000—2005年任芝加哥自然博物馆（终身）副研究员，2005—2008年任史密森学会研究科学家、副研究员，2008年至今任史密森学会研究科学家、研究员。

1985.09, Jun WEN (1963-) studied at Ohio State University, Columbus, Ohio, United States of America; received Ph. D. degree in 1991[627]; postdoctoral research fellow, Arnold Arboretum of Harvard University and National Museum of Natural History, Smithsonian Institution, 1992-1995, assistant professor and curator, Colorado State University, 1995-2000; associate curator (with tenure), Field Museum, Chicago, Illinois, 2000-2005; research scientist and associate curator, Smithsonian Institution, 2005-2008, research scientist and curator, since 2008.

1985年11月，国务院学位委员会批准博士生导师制度后的首批植物分类学博士毕业，中山大学

624 广东梅州人，1982年中山大学生物系毕业获学士学位后入中国科学院植物研究所任实习研究员。

625 Ph.D. dissertation: A worldwide taxonomic revision of the genera *Homalia*, *Pendulothecium* (Musci: Neckeraceae) and *Symphyodon* (Musci: Symphyodontaceae); Advisor: Jerry A. Snider (1937-).

626 湖北公安人（女），1980—1984年华中农学院林学系学习获学士学位；1984—1985年于母校任教。

627 Ph.D. dissertation: Systematics of *Aralia*, Araliaceae; Advisor: Tod F. Stuessy.

张宏达指导的叶创兴（1946— ）[628]、武汉大学孙祥钟指导的赵佐成（1941— ）[629]分别获博士学位。

1985.11, First Ph. D. degree in plant taxonomy officially awarded after approval of doctoral program by Academic Degrees Committee of the State Council; Chuang-Xing YE (1946-), advisor: Hung-Ta CHANG, Sun Yat-Sen University, and Zuo-Cheng ZHAO (1941-), advisor: Hsiang-Chung SUN, Wuhan University, were awarded Ph. D. degree.

1985.12.10-1986.01.14, Sino-American bryological expedition to Yunnan: Kunming, Simao, Xishuangbanna. Bruce H. Allen (1952-, MO), Marshall R. Crosby (1943-, MO) and Robert E. Magill (1947-, MO), Paul L. Redfearn, Jr. (MO, SMS), Mei-Zhi WANG (PE) and Pan-Cheng WU (PE), collected 3,000 numbers of specimens of bryophytes (MO, PE); Guangdong: Dinghu Shan, Seven Star Crags and Luofu Shan where joined by Pang-Juan LIN (IBSC), collected 300 numbers of specimens of bryophytes (IBSC, MO, PE).

1985 年 12 月 10 日至 1986 年 1 月 14 日，中美苔藓联合考察云南昆明、思茅、西双版纳，美国密苏里植物园的 Bruce H. Allen（1952— ）、Marshall R. Crosby（1943— ）和 Robert E. Magill（1947— ），西南密苏里州立大学、密苏里植物园的 Paul L. Redfearn, Jr.，中国科学院植物研究所的汪楣芝和吴鹏程，采集了 3 000 号苔藓植物标本（MO、PE）；之后加入的中国科学院华南植物研究所的林邦娟，在广东鼎湖山、七星岩、罗浮山采集了 300 号苔藓植物标本（IBSC、MO、PE）。

1985 年 12 月，国家博士后科研流动站管理协调委员会正式批准了第一批博士后流动站，植物学专业包括北京的北京大学、广州的中山大学、武汉的武汉大学和北京的中国科学院植物研究所。

1985.12, First group of national postdoctoral stations approved by the State Postdoctoral Scientific Research and Moving Management Committee, botany was in Peking University, Beijing, Sun Yat-Sen University, Guangzhou, Wuhan University, Wuhan, and Institute of Botany, Chinese Academy of Sciences, Beijing.

1985, Chi-Ju HSUEH, Reminiscences of collecting the type specimens of *Metasequoia glyptostroboides* H. H. Hu and Cheng, *Arnoldia* 45(4): 10-18.

1985 年，薛纪如[630]，水杉模式标本采集的回忆，*Arnoldia* 45（4）：10-18。

1985 年，刘慎谔文集编辑委员会[631]，**刘慎谔文集**，342 页；科学出版社，北京。

628 博士论文：山茶亚科的系统研究。

629 博士论文：中国水筛 *Blyxa* 植物研究。

630 河北临城人（1921—1999），1941 年入国立中央大学生物系，二年级转入林学系，1945 年毕业后考入国立中央大学研究生部，拜师郑万钧；1946 年受导师委托，两次赴四川万县采集水杉标本；1948 年硕士毕业后到福建研究院植物研究所工作，后赴云南大学、昆明农林学院、西南林学院等地从教；著名树木学与竹类专家。

631 刘慎谔文集编辑组 1977 年由朱济凡、王战、曹新孙、刘媖心和赵大昌等 5 人组成。

1985, Editorial Committee for collected works of Tchen-Ngo LIOU, '***Collected Works of Tchen-Ngo LIOU***', 342 p; Science Press, Beijing.

1985年，王宗训，回忆北平研究院植物研究所，中国科技史料6（2）：16–20。

1985, Zong-Xun WANG, Recollections of botany research workshop in Beiping Research Institute (Memoirs of the Institute of Botany, National Academy of Peiping [Beijing]), *China Historical Materials of Science and Technology* 6(2): 16–20.

1985年，中国科学院青藏高原综合科学考察队编著，黎兴江主编，**西藏苔藓植物志**，581页；科学出版社，北京。

1985, The Comprehensive Scientific Expedition to Qinghai-Xizang Plateau, Chinese Academy of Sciences, Shin-Chiang LEE (editor in chief), ***Bryoflora of Xizang***, 581p; Science Press, Beijing.

1985+, 周以良主编, **黑龙江省植物志**, 11卷本；第1卷: 231页, 1985；第2、3卷: XXX; 第4卷: 483页, 1992；第5卷: 392页, 1992；第6卷: 371页, 1998；第7卷: 424页, 2003；第8卷: 430页, 2001；第9卷: 363页, 1998；第10卷: 404页, 2002；第11卷: 262页, 1993；东北林业大学出版社，哈尔滨。

1985+, Yi-Liang CHOU (editor in chief), ***Flora Heilongjiangensis***, vols. 1-11; 1: 231 p, 1985; 2 & 3: XXX; 4: 483 p, 1992; 5: 392 p, 1992; 6: 371 p, 1998; 7: 424 p, 2003; 8: 430 p, 2001; 9: 363 p, 1998; 10: 404 p, 2002; 11: 262 p, 1993; Northeastern Forestry University Press, Harbin.

1985+, 应绍舜, **台湾高等植物彩色图志**[632], 第1卷: 530页, 1985；第2卷: 742页, 1987；第3卷: 663页, 1988；第4卷: 851页, 1992；第5卷: 647页, 1995；第6卷: 714页；1998；作者自行出版，台北。

1985+, Shao-Shun YING, ***Coloured Illustrated Flora of Taiwan***, Vol. 1: 530 p, 1985; 2: 742 p, 1987; 3: 663 p, 1988; 4: 851 p, 1992; 5: 647 p, 1995, 6: 714 p, 1998; published by the author himself, Taipei.

632 本书名为高等植物，但是目前出版的内容既没有苔藓也没有蕨类，而且作者也没有具体说明系统安排，且自第6卷出版以来，过去的十余年间没有任何有关本书的消息，尽管作者在第6卷自序中提及第7卷在考虑之中。

1985—1992 年，中国科学院兰州沙漠研究所编，刘媖心[633]主编，**中国沙漠植物志**，3 卷本；第 1 卷：546 页，1985；第 2 卷：464 页，1987；第 3 卷：508 页，1992；科学出版社，北京。日文版：中国科学院兰州沙漠研究所编，刘媖心主编，德冈正三 訳·解说，2002，**中国砂漠·沙地植物図鉴（木本编）**，543 页；东方书店，东京。

1985-1992, Lanzhou Institute of Desert Research, Chinese Academy of Sciences, Ying-Hsin LIOU (Ying-Xin LIU, editor in chief), ***Flora in Desertis Reipublicae Populorum Sinarum***, vols. 1-3; 1: 546 p, 1985; 2: 464 p, 1987; 3: 508 p, 1992; Science Press, Beijing. Translated into Japanese in 2002 (woody only).

1985，于兆英，陕西省植物分类学研究的回顾与展望，西北植物学报 5（2）：163–168。

1985, Zhao-Ying YU, Retrospect and prospect for the taxonomic studies in Shaanxi province, *Acta Botanica Boreali-Occidentalia Sinica* 5(2): 163-168.

633 刘慎谔之女，山东牟平人（1919—2012），治沙专家、沙漠植物学家；1937—1938 年在北京中法大学理学院生物系学习，1940—1944 年在国立西北农学院园艺系学习并获农学学士，1944—1947 年在陕西三原女中、郿县中学和宝鸡励行中学任教，1950—1953 年任东北农学院东北植物调查研究所实习研究员，1953—1959 年历任中国科学院林业土壤研究所护林组副组长，章古台治沙试验站副站长，沙坡头治沙试验站业务秘书、副站长和助理研究员，1959—1960 年任中国科学院治沙队业务秘书和植物组组长、沙坡头试验站副站长，1966—1978 年任中国科学院兰州冰川冻土沙漠研究所中国沙漠地区药用植物编写负责人，1978—1987 年任中国科学院兰州冰川冻土沙漠研究所植物治沙（沙漠）室副主任、主任；1982 年晋升为副研究员，1986 年晋升为研究员。刘媖心，1993，中国沙漠植物志编写梗概，中国沙漠 13（3）：14-17。参见：王康富，刘媖心传略，载于刘媖心、黄兆华编，2000，植物治沙与草原治理，238-241 页；甘肃文化出版社，兰州。

1986

1986.07.03-07.15, Sino-Finland bryological expedition to Hainan: Jianfengling. Timo J. Koponen (1939-, H), president, International Association of Bryologists, Pang-Juan LIN (IBSC) and Pan-Cheng WU (PE) collected 500 numbers of bryophyte specimens (H, IBSC, PE).

1986 年 7 月 3 日至 15 日，中芬苔藓联合考察海南尖峰岭，国际苔藓学会主席、赫尔辛基大学 Timo J. Koponen（1939— ），中国科学院华南植物研究所林邦娟和中国科学院植物研究所吴鹏程，采集苔藓植物标本 500 号（H、IBSC、PE）。

1986 年 7 月 10 日至 15 日，第二次全国蕨类植物学学术讨论会在江西庐山召开，全国 15 个省（市、区）34 位代表出席，收到 40 篇论文。[634]

1986.07.10-07.15, Second national symposium on ferns held in Lushan, Jiangxi, with 34 representatives from 15 provinces; 40 papers submitted.

1986.08.19-10.17, Sino-American botanical expedition to Guizhou (Fanjing Shan) Cengong, Jiangkou, Shiqian, Songtao, Yinjiang. Bruce M. Bartholomew (CAS), David E. Boufford (A/GH), Qian-Hai CHEN (HGAS), S. Z. FANG (HGAS), Jin-Gen QI (PE), Stephen A. Spongberg (A), Zhan-Huo TSI (1937-2001, PE), Yu-Lin TU (1941-, GNUB), Pei-Shan WANG (1936-, HGAS), Ying-Hai XIANG (1936-, HGAS) and Tsun-Shen YING (PE) collected 2,474 numbers of vascular plants (Sino-American Guizhou Expedition nos. 1-2,474 (A, CAS, GNUB, HGAS, PE).

1986 年 8 月 19 日至 10 月 17 日，中美植物联合考察贵州（梵净山）岑巩、江口、石阡、松涛、印江，加州科学院 Bruce M. Bartholomew，哈佛大学 David E. Boufford 和 Stephen A. Spongberg，贵州科学院生物研究所陈谦海、S. Z. FANG、王培善（1936— ）和向应海（1936— ），贵州师范大学屠玉麟（1941— ），中国科学院植物研究所齐金根、吉占和（1937—2001）、应俊生，采集植物标本 2 474 号（A、CAS、GNUB、HGAS、PE）。

1986 年 11 月，中国科学院中国孢子植物志编辑委员会第四届编辑委员会成立，主编：曾呈奎[635]

634 张朝芳，1986，学会活动，植物杂志 6：37。

635 福建省厦门人，著名海洋生物学家；详细参见：曾呈奎，1994，曾呈奎文集（上、下卷），1 279 页；海洋出版社，北京。

Sino-American expedition to Guizhou (Fanjing Shan) ,19 August -17 Octorber 1986 (Photo provided by David E. Boufford)
1986 年 8 月 19 日至 10 月 17 日，中美植物学家联合考察贵州（梵净山）（相片提供者：鲍棣伟）

（1909—2005），副主编：黎尚豪[636]（1917—1993）、余永年[637]（1923—2014）、魏江春、吴鹏程，编辑委员会 26 人。

1986.11, Fourth Consilio Florarum Cryptogamarum Sinicarum, Academiae Sinicae Edita established, Cheng-Kui TSENG (1909-2005), Editor in Chief, Shang-Hao LEY (1917-1993), Yung-Nien YU (1923-2014), Jiang-Chun WEI, and Pan-Cheng WU, Vice Editors in Chief, with 26 editorial members.

1986 年，张宏达，种子植物系统分类提纲，中山大学学报（自然科学版）25（1）：1-13。

1986, Hung-Ta CHANG, Outline of Spermatophyta classification, *Acta Scientiarum Naturalium Universitatis Sunyatseni* 25(1): 1-13.

1986 年，陈守良、刘守炉主编，**江苏维管植物检索表**，559 页；江苏科学技术出版社，南京。

1986, Shou-Liang CHEN and Sheo-Lu LIOU (editors in chief), '***Key to Vascular Plants of Jiangsu***', 559 p; Jiangsu Science and Technology Press, Nanjing.

1986 年，丁淦主编，**苔藓名词辞典**，338 页；五洲出版社，台北。

1986, Gan DING (editor in chief), ***English Chinese Dictionary of Bryophytes***, 338 p; Wuzhou Press, Taipei [Taibei].

1986 年，徐炳声、金德孙，论物种的客观真实性，云南植物研究 8（2）：229-238。

1986, Ping-Sheng HSU and De-Sun JIN, The objective reality of species, *Acta Botanica Yunnanica* 8(2): 229-238.

1986，Joseph Needham, ***Science and Civilisation in China***, vol. 6, Biology and Biological Technology, Part 1, Botany. 741 p; Cambridge University Press, Cambridge, UK. Translated into Chinese in 2006.

2006 年，袁以苇等译，**中国科学技术史**，第 6 卷，生物学及相关技术，第 1 分册，植物学，672 页；科学出版社，北京；上海古籍出版社，上海。

1986 年，仇寅龙[638]（1964—）赴美国留学，1993 年获北卡罗来纳大学 Chapel Hill 分校博士学位，

636 广东梅县人，著名藻类学家；详细参见：中国科学院水生生物研究所，2017，黎尚豪先生百年诞辰纪念文集，143 页；中国科学院水生生物研究所，武汉（内部印制）；中国科学院水生生物研究所，2017，黎尚豪论文集，212 页；中国科学院水生生物研究所，武汉（内部印制）。

637 重庆万州人，著名菌物学家；详细参见：余永年著，2003，余永年文选，454 页；学苑出版社，北京。青宁生，2014，真菌学家——余永年，微生物学报 54（12）：1515-1516.

638 江苏常州人，1980—1984 年就读于南京农业大学园艺系，1984—1986 年江苏省植物研究所研究生。

1994—1995年和1996—1997年分别在印第安纳大学Bloomington分校和美国国家卫生研究院从事博士后研究，1998—1999年任瑞士苏黎世大学助理教授，2000—2002年任麻州大学Amherst分校助理教授，2003—2008年任密歇根大学Ann Arbor分校助理教授、2008年至今任副教授；主要研究陆生植物的起源与演化。

1986, Yin-Long QIU (1964-) studied at the University of North Carolina, Chapel Hill, United States of America; received Ph.D. degree in 1993[639], postdoctoral fellow, Indiana University, Bloomington, Indiana, 1994-1995, National Institutes of Health (NIH), 1996-1997; assistant professor, University of Zurich, 1998-1999, assistant professor, University of Massachusetts, Amherst, 2000-2002; assistant professor, University of Michigan, Ann Arbor, Michigan, 2003-2008; associate professor, University of Michigan, Ann Arbor, Michigan, since 2008. Research focus origin and evolution of land plants.

1986, Paul L. Redfearn, Jr. and Pan-Cheng WU, Catalog of the mosses of China, *Annals of the Missouri Botanical Garden* 73(1): 177-208.

1986年, Paul L. Redfearn, Jr.、吴鹏程，中国藓类名录, *Annals of the Missouri Botanical Garden* 73（1）: 177-208。

1986年，丁广奇、王学文编，**植物学名解释**，463页；科学出版社，北京。

1986, Kuan-Chi TING and Xue-Wen WANG, '***Explanation of Plant Names***', 463 p; Science Press, Beijing.

1986年，王芳礼编著，**英汉种子植物名词词典**，411页；湖北辞书出版社，武汉。

1986, Fang-Li WANG, ***English-Chinese Dictionary of Spermatophyte Nouns***, 411 p; Hubei Cishu Press, Wuhan.

1986年，张德山，山东种子植物名录，烟台师范学院学报2（1）：61-96，2（2）：45-79。

1986, De-Shan ZHANG, The directory contents of seed plants in Shan Dong, *Yantai Teacher's College Journal* 2(1): 61-96 and 2(2): 45-79.

1986、1990年，中日植物联合考察西藏色季拉山等地，中国科学院植物研究所郎楷永、李渤生、孙世洲等与日本东北大学及大阪大学学者共采集标本4 000余号（PE）。

1986 and 1990, Sino-Japanese botanical expeditions to Xizang: Shegyla [Sejila] Shan. Kai-Young LANG, Po-Sheng LI, Shi-Zhou SUN (PE) and scientists from Tohoku and Osaka universities, Japan, collected 4,000 numbers of specimens of plants (PE).

639 Ph.D. Dissertation: Molecular divergence between Asian and North American species of *Magnolia* sect. *Rytidospermum* (Magnoliaceae); Advisors: Clifford R. Parks and Mark W. Chase (1951-).

1986—1991 年，河北植物志编辑委员会编，贺士元主编，**河北植物志**，3 卷本；第 1 卷：831 页，1986；第 2 卷：676 页，1989；第 3 卷：698 页，1991；河北科学技术出版社，石家庄。

1986-1991, Editorial Committee of Flora Hebeiensis, Shih-Yuen HO (editor in chief), ***Flora Hebeiensis***, vols. 1-3; 1: 831 p, 1986; 2: 676 p, 1989; 3: 698 p, 1991; Hebei Science and Technology Publishing House, Shijiazhuang.

1986—1992 年，安徽编写组编著，钱啸虎主编，**安徽植物志**，5 卷本；第 1 卷：281 页，1986；第 2 卷：583 页，1987；第 3 卷：695 页，1990；第 4 卷：697 页，1991；第 5 卷：615 页，1992；安徽科学技术出版社，合肥（1986，1991—1992）及中国展望出版社，北京（1987，1990）。

1986-1992, Editorial Group of Flora of Anhui, Hsiao-Hu CHIEN (editor in chief), '***Flora of Anhui***', vols. 1-5; 1: 281 p, 1986; 2: 583 p, 1987; 3: 695 p, 1990; 4: 697 p, 1991; 5: 615 p, 1992; Anhui Science and Technology Publishing House, Hefei (1986, 1991-1992), China Prospect Publishing House, Beijing (1987, 1990).

1986、1988、2007 年，马德滋、刘惠兰编著；**宁夏植物志**，2 卷本；上册：505 页，1986；下册：555 页，1988；宁夏人民出版社，银川；马德滋、刘惠兰、胡福秀主编，**宁夏植物志**，第 2 版，2 卷本；上册：635 页，2007；下册：642 页，2007；宁夏人民出版社，银川。

1986 and 1988, 2007, De-Zi MA and Hui-Lan LIU (eds.), ***Flora Ningxiaensis***, vols. 1-2; 1: 505 p, 1986; 2: 555 p, 1988; Ningxia People's Press, Yinchuan; De-Zi MA, Hui-Lan LIU and Fu-Xiu HU (editors in chief), ***Flora Ningxiaensis***, ed. 2, vols. 1-2; 1: 635 p, 2: 642 p, 2007; Ningxia People's Press, Yinchuan.

1987

1987. 7. 24-8. 1, Fourteenth International Botanical Congress held in West Berlin, Federal Republic of Germany, 69 attendees from China.

1987 年 7 月 24 日至 8 月 1 日，第十四届国际植物学大会于联邦德国西柏林举行，中国代表 69 人参加。[640]

1987 年 8 月 20 日，中国科学院植物研究所在原植物分类研究室基础之上，成立中国科学院系统与进化植物学开放研究实验室；2001 年 11 月 14 日改为中国科学院植物研究所系统与进化植物学重点实验室；2005 年又通过整合植物标本馆（含古植物馆）、植物分类研究室和古植物研究室，成为系统与进化植物学国家重点实验室。[641]

1987.08.20, Laboratory of Systematic and Evolutionary Botany, Chinese Academy of Sciences, Beijing, founded, based on Department of Plant Taxonomy, Institute of Botany, Chinese Academy of Sciences, Beijing; November 14, 2001, Laboratory of Systematic and Evolutionary Botany, Chinese Academy of Sciences, renamed Key Laboratory of Systematic and Evolutionary Botany, Chinese Academy of Sciences; 2005, Key Laboratory of Systematic and Evolutionary Botany, Chinese Academy of Sciences, combined with the herbarium (PE, including Chinese Paleobotany Museum), the plant taxonomy department and the paleobotany department, formed a State Key Laboratory of Systematic and Evolutionary Botany.

Herbarium (IFP), Institute of Applied Ecology, Chinese Academy of Sciences, 2014 (Photo provided by Wei LI)
2014 年中国科学院沈阳应用生态研究所植物标本馆（IFP）（相片提供者：李薇）

1987 年 10 月 27 日，中国科学院林业土壤研究所更名为中国科学院沈阳应用生态研究所；1987—1994 年沈善敏（1933— ）、1994—2000 年孙铁珩（1938—2013）、2000—2008 年何兴元

640 张新时、王金亭，1988，第十四届国际植物学大会在西柏林举行，植物生态学与地植物学学报 12（2）：159-161；王伏雄、胡昌序，1988，第十四届国际植物学大会概况汇报，中国植物学会五十五周年年会学术论文摘要汇编，1-6 页（附出席会议的中国大陆学者名单）。

641 http://www.lseb.cn/Text.aspx?ItemID=24 (accessed 19 September 2017).

Chinese delegates to XIV International Botanical Congress, West Berlin, 24 July – 1 August 1987 (Photo provided by Wu Zhengyi Science Foundation)
1987 年 7 月 24 日至 8 月 1 日，中国代表出席西柏林的第十四届国际植物学大会（相片提供者：吴征镒科学基金会）

Group photo, Department of Plant Taxonomy and Herbarium (PE), 1996 (Photo provided by Institute of Botany)

1996 年植物分类室与标本馆（PE）集体合影（相片提供者：植物研究所）

Institute of Botany, Chinese Academy of Sciences, 2019: 1. Herbarium (PE), 2. Information center, 3. Main gate, 4. Greenhouse (Photos provided by Institute of Botany and Fang-Pu LIU)
2019 年的中国科学院植物研究所：1. 植物标本馆（PE），2. 信息中心，3. 大门，4. 温室（相片提供者：植物研究所，刘方谱）

（1962—）、2008—2015 年韩兴国、2015—2016 年姬兰柱（1960—）、2016 年至今朱教君（1965—）任所长。

1987.10.27, Institute of Forestry and Pedology, Chinese Academy of Sciences, renamed Institute of Applied Ecology, Chinese Academy of Sciences, with directors Shan-Min SHEN (1933-), 1987-1994; Tie-Heng SUN (1938-2013), 1994-2000; Xing-Yuan HE (1962-), 2000-2008; Xing-Guo HAN, 2008-2015; Lan-Zhu JI (1960-), 2015-2016; and Jiao-Jun ZHU (1965-), since 2016.

1987 年，白学良，内蒙古藓类植物初报，内蒙古大学学报（自然科学版）18（2）：311-350。

1987, Xue-Liang BAI, Preliminary report of the mosses in Inner Mongolia, China, *Acta Scientiarum Naturalium Universitatis Intramongolicae* 18(2): 311-350.

1987 年，徐炳声，近二十年来植物分类学的进展，武汉植物学研究 5（1）：77-92。

1987, Ping-Sheng HSU, Progress in plant taxonomy in the past two decades, *Journal of Wuhan Botanical Research* 5(1): 77-92.

1987 年，纪念孔宪武教授逝世三周年筹备小组编，**孔宪武教授纪念文集**，40 页；甘肃省植物学会、西北师范学院植物研究所，兰州（内部印制）。

1987, Preparatory Group on Third Anniversary of Death of Professor Hsien-Wu KUNG, '***Memorial Works of Hsien-Wu Kung***', 40 p; Botanical Society of Gansu and Institute of Botany, Northwest Normal College, Lanzhou (internal publication).

1987 年，林苏娟[642]（1957—）赴日本留学，1989 年获东京大学理学院硕士学位，1992 年获博士学位；1993—1994 年密苏里植物园访问学者，1995—1996 年任东京大学植物园研究员，1996—2000 任年日本立教大学兼职讲师，2001—2002 年任东京大学植物园访问研究员，2002—2016 年任日本岛根大学副教授同时兼职日本国立鸟取大学副教授，2016 年至今任日本岛根大学教授同时兼职日本国立鸟取大学教授，并任岛跟大学生物生命科学专攻副主任；另外，1994—2003 年任中国科学院江苏省植物研究所客座研究员，2010—2011 年美国哈佛大学植物标本馆访问研究员。林苏娟教授为现任日本植物分类学会中唯一的外国人教授，且为日本植物分类学会中 3 位女性教授之一（会员中女性只有 3%），更是国立岛根大学生物资源学部中唯一的外国人教授，第一位女性教授。

1987, Su-Juan LIN (1957-) went to Japan to study botany, received master's degree from Graduated School of Science, University of Tokyo in 1989, and received Ph.D. degree[643] in 1992, visiting researcher in Missouri Botanical Garden, 1993-1994, researcher of Botanical Garden, Faculty of Science, University of

642 福建人（女），1978—1982 年南京大学植物学系本科毕业，1982—1987 年任福建师范大学助理研究员。

643 Ph.D. dissertation: Systematic study of *Dryopteris varia* complex (Dryopteridaceae); adviser: Kunio Iwatsuki.

Tokyo, 1995–1996, part-time lecturer, Faculty of Science, Rikkyo University, 1996–2000, visiting researcher of Botanical Garden, Faculty of Science, University of Tokyo, 2001–2002, associate professor, Department of Biological Science, Faculty of Life and Environmental Science, Graduate School of Shimane University, and associate professor, Tottori University, 2002–2016, and professor, Department of Biological Science, Faculty of Life and Environmental Science, Graduate School of Shimane University, and professor, Tottori University since 2016, at the sametime as vice director of Biological Science and Biotechnology, Graduate School, Shimane University. In additional, she was as research scientist of Jiangsu Institute of Botany, Chinese Academy of Sciences, China, 1994–2000, and visiting researcher of Harvard University Herbaria, the United States of America, 2010–2011. She is the only foreign professor in the Japanese Society of Plant Taxonomy today, and one of the three women professors among the Japanese Society of Plant Taxonomy (about 3% only). She is the only foreign professor, and also the first female professor in Faculty of Life and Environmental Science, Shimane University.

1987年，祁承经主编，**湖南植物名录**，466页；湖南科学技术出版社，长沙。

1987, Cheng-Jing QI (editor in chief), ***The List of Hunan Flora***, 466 p; Hunan Science and Technology Publishing House, Changsha.

1987年，青藏高原综合科学考察丛书（青藏高原隆起及其对自然环境和人类活动影响的综合研究）获国家自然科学一等奖；该丛书包括西藏植物志（5卷）和西藏苔藓植物志，主要获奖人员包括植物分类学家吴征镒和武素功。

1987, Results of the scientific expedition to the Qinghai-Xizang Plateau, including *Flora Xizangica* (five volumes) and *Bryoflora of Xizang*, awarded National Natural Science First Prize, Cheng-Yih WU and Shu-Kung WU represented plant taxonomists.

1987, Armen Takhtajan, ***Systema Magnoliophytorum***, 439 p[644]; Officina Editoria «Nauka» sectio Leninopolitana, Leninopoli [Saint Petersburg]. Partly translated into Chinese in 1989.

1989年，吴征镒、李恒摘译，**木兰植物系统**，中国植物区系研究参考资料第1辑，40页；云南植物研究编辑部，昆明（内部印制）。

1987年，国家环境保护局，中国科学院植物研究所编，**中国珍稀濒危保护植物名录**，1：96页；科学出版社，北京。

1987, National Administrative Bureau of Environment Protection, Institute of Botany, Chinese Academy of Sciences, '***Catalogue of Chinese Rare and Endangered Plants***', 1: 96 p; Science Press, Beijing.

644 Система магнолиофитов (*Sistema Magnoliofitov*, aka *Systema Magnoliophytorum*).

1987 年，中国孢子植物志编辑委员会编著的中国孢子植物志[645]分为中国海藻志、中国淡水藻志、中国真菌志、中国地衣志及中国苔藓志。本纪事只收录了中国苔藓志。中国苔藓志中文版和英文版各计划 12 卷册。

1987, *Cryptogamic Flora of China*[646] divided into *Flora Algarum Marinarum Sinicarum, Flora Algarum Sinicarum Aquae Dulcis, Flora Fungorum Sinicorum, Flora Lichenum Sinicorum,* and *Flora Bryophytorum Sinicorum*, only the last title accounted for in this chronicle. The Chinese and English edition of *Flora Bryophytorum Sinicorum* will be in 12 volumes.

1987 年，中国高等植物图鉴及中国高等植物科属检索表获国家自然科学一等奖，完成人：中国科学院植物研究所王文采、汤彦承及其研究团队。

1987, Wen-Tsai WANG and Yen-Cheng TANG, Institute of Botany, Chinese Academy of Sciences, Beijing, and their research collective awarded National Natural Science First Prize for *Iconographia Cormophytorum Sinicorum* and *Claves Familiarum Generumque Cormophytorum Sinicorum*.

为表彰在自然科学研究中做出的重要贡献，特颁发此证书，以资鼓励。

证书

1987 年国家自然科学奖

项目名称：中国高等植物图鉴及中国高等植物科属检索表

主要研究者：王文采　汤彦承及其研究集体（中国科学院植物研究所等）

获奖等级：一等

中华人民共和国
国家科学技术委员会主任　宋健

证书号：Z87107

1988年8月28日

Wen-Tsai WANG, Yen-Cheng TANG and their research collective awarded National Natural Science First Prize , 1987 (Photo provided by Qi LIN)

1987 年王文采和汤彦承及其研究集体获国家自然科学一等奖（相片提供者：林祁）

645 田金秀，1994，我国孢子植物物种信息库——《中国孢子植物志》，中国科学院院刊 3：263-264；褚鑫等，2015，国家自然科学基金重大项目“中国孢子植物志”编研概述，中国科学基金 1：60-61。

646 Chinese pteridophytes included in *Flora Reipublicae Popularis Sinicae* and *Flora of China*.

1987、1988 年，陈德懋、曾令波，中国植物学发展史略 —— 植物分类学发展简史（Ⅰ, Ⅱ, Ⅲ），华中师范大学学报 21（1）：117-127，21（4）：637-644，1987；22（4）：477-486，1988。

1987 and 1988, De-Mao CHEN[647] and Ling-Bo ZENG, History evolution of China's botany—a brief history of China's botany systematics development (part I, II, III), *Journal of Central China Normal University* 21(1): 117-127, 21(4): 637-644, 1987, 22(4): 477-486, 1988.

1987—1988 年，中国科学院对武陵山地区（川东南、鄂西南、湘西和黔东北）进行生物考察，植物队伍主要参加队员来自北京、广州、昆明、成都和武汉，共采集标本 16 000 号（IBSC、KUN、PE）；其成果为 1995 年出版的武陵山地区维管植物检索表（详细参见 1995）。

1987-1988, Chinese Academy of Sciences biological expedition to Wuling Shan (SE Sichuan, SW Hubei, W Hunan and NE Guizhou); botanical teams from Beijing, Chengdu, Guangzhou, Kunming and Wuhan, collected 16,000 numbers of plant specimens (IBSC, KUN, PE); major results incorporated in *Keys to the Vascular Plants of the Wuling Mountains*, published in 1995 (see 1995 for details).

1987—1992 年，中国植物志第五届编辑委员会成立，主编：中国科学院昆明植物研究所吴征镒，副主编：中国科学院植物研究所崔鸿宾；委员 21 人：中国科学院华南植物研究所陈德昭、黄成就、林有润，中国科学院江苏省植物研究所 南京中山植物园陈守良，中国科学院植物研究所陈心启、陈艺林、戴伦凯、傅立国、李安仁、夏振岱（1935—），中国科学院西北高原生物研究所郭本兆、刘尚武，东北林业大学黄普华（1932—），复旦大学胡嘉琪（1932—），中国科学院成都生物研究所孔宪需（1930—）、李朝銮（1938—1998），广西植物研究所李树刚，中国科学院昆明植物研究所李锡文，福建师范大学林来官[648]（1925—2014），浙江自然博物馆韦直，中国科学院西北植物研究所徐朗然。

1987-1992, Fifth Editorial Committee of Flora Reipublicae Popularis Sinicae established, Cheng-Yih WU (KUN), Editor in Chief, and Hung-Pin TSUI (PE) as Vice Editor in Chief, 21 additional members: Shou-Liang CHEN (NAS), Sing-Chi CHEN (PE), Te-Chao CHEN (IBSC), Yi-Ling CHEN (PE), Lun-Kai DAI (PE), Li-Kuo FU (PE), Pung-Chao KUO (HNWP), Chia-Chi HU (1932-, FUS), Ching-Chieu HUANG (IBSC), Pu-Hwa HUANG (1932-, NEFI), Hsien-Shiu KUNG (1930-, CDBI), Shu-Kang LEE (IBK), An-Jen LI (PE), Chao-Luang LI (1938-1998, CDBI), Hsi-Wen LI (KUN), Yuou-Ruen LING (IBSC), Lai-Kuan LING (1925-2014, FNU), Shang-Wu LIU (HNWP), Zhi WEI (ZM), Zhen-Dai XIA (1935-, PE) and Lang-

647 陈德懋（1941—）于 1982—1990 年发表多篇文章，其拼写应该是 Chen De-Mao，但有时却用 Chen Dao-Men，而且不止一篇文章。

648 1941 年 8 月至 1949 年 8 月在福建省研究院动植物研究所植物分类室任练习生、技佐、技士；1949 年 8 月至 1952 年 8 月在福州大学自然科学研究所植物分类室任技术员；1952 年 8 月至 1969 年 12 月在福建师范学院生物系植物分类教研室任助教、讲师（其间：1963 年 11 月至 1965 年 9 月在中国科学院植物研究所进修植物分类）；1970 年 1 月至 1974 年 8 月在厦门大学教育系农基专业植物教研组任讲师；1974 年 8 月至 1990 年 8 月在福建师范大学生物系任讲师、副教授、教授；1990 年 8 月退休。

Participants at enlarged editorial committee meeting for *Flora Reipublicae Popularis Sinicae*, Nanjing, 5 May 1987 (Photo provided by Wu Zhengyi Science Foundation)
1987 年 5 月 5 日中国植物志编辑委员会扩大会议于南京（相片提供者：吴征镒科学基金会）

Participants at 5th editorial committee meeting (1987-1992) of *Flora Reipublicae Popularis Sinicae*, Beijing, May 1988 (Photo provided by Wu Zhengyi Science Foundation)
1988 年 5 月中国植物志第五届编辑委员会会议（1987—1992）于北京（相片提供者：吴征镒科学基金会）

Rang XU (WUK).

1987 年 1 月至 1996 年 12 月，中国科学院云南热带植物研究所（植物园部分）更名为中国科学院西双版纳热带植物园，主任许再富，并隶属于中国科学院昆明植物研究所。

1987.01-1996.12, Yunnan Institute of Tropical Botany (part of botanical garden) renamed Xishuangbanna Tropical Botanical Garden, Chinese Academy of Sciences, Zai-Fu XU director, under administration of Kunming Institute of Botany, Chinese Academy of Sciences.

1987—2011 年，中国科学院华南植物研究所 / 中国科学院华南植物园编，陈封怀主编（1–2 卷）、吴德邻主编(3–10 卷),**广东植物志**[649],10 卷本；第 1 卷：600 页,1987；第 2 卷：511 页,1991；第 3 卷：511 页, 1995；第 4 卷：446 页, 2000；第 5 卷：498 页, 2003；第 6 卷：445 页, 2005；第 7 卷：543 页, 2006；第 8 卷：431 页，2007；第 9 卷：558 页，2009；第 10 卷：330 页，2011；广东科技出版社，广州。

1987-2011, South China Institute of Botany, Academia Sinica / South China Botanical Garden, Chinese Academy of Sciences, Feng-Hwai CHEN (editor in chief, vols. 1-2), Te-Lin WU (editor in chief, vols. 3-10), ***Flora of Guangdong***[650], vols. 1-10; 1: 600 p, 1987; 2: 511 p, 1991; 3: 511 p, 1995; 4: 446 p, 2000; 5: 498 p, 2003; 6: 445 p, 2005; 7: 543 p, 2006; 8: 431 p, 2007; 9: 558 p, 2009; 10: 330 p, 2011; Guangdong Science and Technology Press, Guangzhou.

1987，狄维忠，**贺兰山维管植物**，378 页；西北大学出版社，西安。

1987, Wei-Zhong DI, ***Plantae Vasculares Helanshanicae***, 378 p; Northwest University Press, Xi'an.

649 包括海南。

650 Including Hainan

1988

1988.06.15-10.10, Sino-Japanese botanical expedition to Karakoram and Kunlun mountains, Qinghai, Xinjiang, Xizang. Yong FEI (KUN), Hideaki Ohba (1943-, TI), Shu-Kung WU (leader, KUN) and Yu-Hu WU (1951-, HNWP), collected 2,212 numbers of plant specimens (HNWP, KUN, TI).

1988 年 6 月 15 日至 10 月 10 日，中日植物联合考察青海、新疆、西藏的喀喇昆仑和昆仑山，中国科学院昆明植物研究所费勇、武素功（领队），东京大学大场秀章（1943— ），中国科学院西北高原生物研究所吴玉虎（1951— ），采集植物标本 2 212 号（HNWP、KUN、TI）。

1988.08.20-09.22, Chinese-American bryological exploration in Sichuan: Dujiangyan (formerly Guanxian), Emei Shan, Hongyuan, Jiuzhaigou and Songpan. Bruce H. Allen (MO), Paul L. Redfearn. Jr. (MO, SMS), Pan-Cheng WU (PE) and one anonymous scientist (CDBI), collected 2,000 numbers of bryophyte specimens (CDBI, MO, PE).

1988 年 8 月 20 日至 9 月 22 日，中美苔藓联合考察四川都江堰（原灌县）[651]、峨眉山、红原、九寨沟和松潘，密苏里植物园 Bruce H. Allen，西南密苏里州立大学、密苏里植物园 Paul L. Redfearn, Jr.，中国科学院植物研究所吴鹏程和中国科学院成都生物研究所 1 人，采集苔藓植物标本 2 000 号（CDBI、MO、PE）。

1988.08.21-09.16, Chinese-American botanical expedition to Sichuan: Dujiangyan. Bruce M. Bartholomew (CAS), David E. Boufford (A/GH), Gang LI (1965-, PE) and Guang-Hua ZHU (1964-2005, PE), collected 992 numbers of vascular specimens (D. E. Boufford and B. Bartholomew, with G. LI and G. H. ZHU nos. 23934-24926, A, CAS, PE).

1988 年 8 月 21 日至 9 月 16 日，中美植物联合考察四川都江堰，加州科学院 Bruce M. Bartholomew，哈佛大学 David E. Boufford，中国科学院植物研究所李岗（1965— ）和朱光华（1964—2005），采集维管束植物 992 号（A，CAS，PE）。

1988.09.05-09.10, International Symposium on Systematic Pteridology held in Beijing on occasion of 90th anniversary of birth of Professor Ren-Chang CHING, 72 attendees from 12 countries.[652]

1988 年 9 月 5 日至 10 日，纪念秦仁昌教授诞辰九十周年国际蕨类植物学学术讨论会在北京举行，

651 1988 年 5 月改为现名。

652 Kung-Hsia SHING and Karl U. Kramer (eds.), 1989, *Proceedings of the International Symposium on Systematic Pteridology*, 330 p; China Science and Technology Press, Beijing.

Chinese-American botanical expedition to Sichuan, 21 August–16 September 1988 (left-right): 1. Rear center: Bruce M. Bartholomew; 2. Standing: Gang LI; 3. Standing: Gang LI and Guang-Hua ZHU; 4. Jun CHEN, Peng ZHAO, Guang-Hua ZHU and Bin LIN (Photo provided by David E. Boufford)

1988 年 8 月 21 日至 9 月 16 日，中美植物学家考察四川（从左至右）：1. Bruce M. Bartholomew（后中）；2. 李岗（站立者）；3. 李岗和朱光华（站立者，左一和左二）；4. 陈军、赵鹏、朱光华、林彬等（相片提供者：鲍棣伟）

来自 12 个国家的 72 名代表出席。[653]

1988.10.01-10.07, Cheng-Yih WU and Peter H. Raven[654] (1936-), representing China and the United States of America respectively as joint editors in chief, signed agreement at the Missouri Botanical Garden, St. Louis, Missouri, to compile *Flora of China*. It was planned that 15 years would be needed to complete the 25 volumes in English. Editorial committee for the *Flora of China* was established; members were: Bruce M. Bartholomew (CAS), David E. Boufford (A/GH), Shou-Liang CHEN (NAS), Sing-Chi CHEN (PE), Hong-Bin CUI (PE), Lun-Kai DAI (PE), Ching-Chiu HUANG (IBSC), Hsi-Wen LI (KUN), Nancy R. Morin (1948-, MO), Peter H. Raven (MO), William TAI (MO) and Cheng-Yih WU (KUN); Secretaries: Chia-Jui CHEN (1935-, PE) and Hong-Ya GU (PE); William TAI and Lun-Kai DAI were to coordinate between the Chinese and American members of the editorial committee[655].

1988 年 10 月 1 日至 7 日，吴征镒和 Peter H. Raven[656]（1936—）作为联合主编分别代表中美双方在密苏里州圣路易斯市密苏里植物园签署英文版中国植物志的编撰工作合作协议，用 15 年时间完成英文版中国植物志 25 卷，并成立编辑委员会：加州科学院 Bruce M. Bartholomew，哈佛大学 David E. Boufford，江苏省植物研究所 南京中山植物园陈守良，中国科学院植物研究所陈心启、崔鸿宾、戴伦凯，中国科学院华南植物研究所黄成就，中国科学院昆明植物研究所李锡文、吴征镒，密苏里植物园 Nancy R. Morin（1948—）、Peter H. Raven、戴威廉；秘书：中国科学院植物研究所陈家瑞（1935—）、顾红雅；同时指定戴威廉和戴伦凯负责中美双方的直接联络。

1988 年 10 月 20 日至 23 日，中国植物学会第十届会员代表大会暨五十五周年学术年会在四川成都举行，参加会议的代表有 450 余人。理事长：王伏雄（1913—1995）。[657]

1988.10.20-10.23, Tenth National Congress and 55th anniversary meeting of Botanical Society of China held in Chengdu, Sichuan; more than 450 attendees; Fu-Hsiung WANG (1913-1995) elected president.

---

653 邢公侠、克拉姆编辑，1989，国际蕨类植物学科学讨论会论文集，330 页；中国科学技术出版社，北京。

654 B.S., University of California, Berkeley, California, 1957, Ph.D., University of California, Los Angeles, California, 1960, Post-doctoral Fellow at British Museum (Natural History), 1960-1961, Taxonomist and Curator, Rancho Santa Ana Botanic Garden, Claremont, California, 1961-1962, Assistant, then Associate Professor, Stanford University, Stanford, California, 1962-1971; Director / President of Missouri Botanical Garden, 1971-2010, President Emeritus of Missouri Botanical Garden since 2010.

655 Related information cited from Cheng-Yih WU, Peter H. Raven, De-Yuan HONG (eds.), 2013, *Flora of China*, volume 1, History of the Flora of China, 1-20 p.

656 美国植物学家，1936 年生于中国上海（次年随父母返回美国），1957 年加州大学伯克利分校本科最高荣誉毕业，1960 年获加州大学洛杉矶分校博士学位，1960—1961 年英国自然历史博物馆博士后（美国国家自然科学基金），1961—1962 年任加州克莱尔蒙特圣塔安娜牧场植物园植物分类学家和研究员，1962—1971 年任斯坦福大学助理教授和副教授，1971—2010 年任密苏里植物园主任，2010 至今为荣誉主任；世界著名植物学家，特别是推动与组织编写英文版中国植物志；1994 年当选为中国科学院首批外籍院士。

657 中国植物学会编，1988，中国植物学会五十五周年年会学术论文摘要汇编，776 页；中国植物学会，北京（内部印制）。

Cheng-Yih WU and Peter H. Raven signing agreement to compile *Flora of China*, Missouri Botanical Garden, St. Louis, MO, 1-7 October 1988 (Photo provided by Wu Zhengyi Science Foundation).

1988 年 10 月 1 日至 7 日，吴征镒和 Peter H. Raven 于密苏里圣路易斯密苏里植物园签署协议编写英文版中国植物志（相片提供者：吴征镒科学基金会）

1988 年，中国林学会主编，**陈嵘纪念集**，98 页；中国林业出版社，北京。

1988, Chinese Society of Forestry (Editor in Chief), '***Memorial Works of Yung CHEN***', 98 p; China Forestry Publishing House, Beijing.

1988, William J. Haas, Transplanting botany to China: The cross-cultural experience of Chen Huanyong (Woon-Young CHUN), *Arnoldia* 48(2): 9–25. Translated into Chinese in 1993.

1993 年，许兆然译，陈焕镛和阿诺德树木园 [658]，植物学通报 10（4）：32–42。

1988 年，耿以礼、耿伯介、王正平、宋桂卿、谢权中合编，**中国种子植物分科检索表及图解**，541 页；南京大学出版社，南京。

1988, Yi-Li KENG, Pai-Chieh KENG, Cheng-Ping WANG, Gui-Qing SONG and Quan-Zhong XIE, '***Key and Illustration for families of Chinese Seed Plants***', 541 p; Nanjing University Press, Nanjing.

1988, Chen-Meng KUO and Tzen-Yuh CHIANG, Index of Taiwan Hepaticae, *Taiwania* 33: 1–46.

1988 年，郭城孟、蒋镇宇，台湾藓类植物名录 [659]，*Taiwania* 33：1–46。

1988 年，林善雄编著，台湾苔类植物名录，*Yushania* 5（4）：1–39。

1988, Shan-Hsiung LIN, List of mosses of Taiwan, *Yushania* 5(4): 1–39.

1988 年，汪子春，中国近现代生物学发展概况，中国科技史料 9（2）：17–35。

1988, Zi-Chun WANG, A concise history of modern biology of China, *China Historical Materials of Science and Technology* 9(2): 17–35.

1988 年，王宗训，中国近代植物学回顾，植物杂志 4：2–5。

1988, Zong-Xun WANG, 'A review of modern Chinese botany', *Plants* 4: 2–5.

1988 年，赵毓棠、吉金祥编，王文采、任波涛审校，**拉汉植物学名辞典**，726 页；吉林科学技术出版社，长春。

1988, Yu-Tang ZHAO and Jin-Xiang JI (eds.), Wen-Tsai WANG and Po-Tao REN (proofreaders), ***Dictionarium Latino-Sinicum Nominum Scientificorum Plantarum***, 726 p; Jilin Science and Technology Press, Changchun.

---

658 附译后记。

659 即苔类植物。

1988 年，中国植物学会，**秦仁昌论文选**，366 页；科学出版社，北京。

1988, Botanical Society of China, ***Selected Papers of Ching Ren Chang*** *(Ren-Chang CHING)*, 366 p; Science Press, Beijing.

1988.07-1989.02, Chi-Ming HU (1935-), South China Institute of Botany, Chinese Academy of Sciences, Guangzhou, traveled to Institute of Biology, Aarhus University, Denmark, to study Myrsinaceae and Primulaceae for ***Flora of Thailand*** [660].

1988 年 7 月至 1989 年 2 月，中国科学院华南植物研究所胡启明（1935— ）赴丹麦 Aarhus 大学生物研究所，参加**泰国植物志**紫金牛科和报春花科的编研。

1988、1992 年，辽宁植物志编辑委员会，李书心主编，**辽宁植物志**，2 卷本；上册：1 439 页，1988；下册：1 245 页，1992；辽宁科学技术出版社，沈阳。

1988 and 1992, Editorial Committee of Flora of Liaoning, Shu-Xin LI (editor in chief), ***Flora Liaoningica***, vols. 1-2; 1: 1,439 p, 1988, 2: 1,245 p, 1992; Liaoning Science and Technology Press, Shenyang.

1988—1996 年，宁夏回族自治区农业现代化基地办公室、宁夏回族自治区畜牧局、陕西省西北植物研究所编著，于兆英、徐养鹏主编（第 1 卷），徐养鹏、王克制、于兆英主编（第 2 卷），徐养鹏、王克制主编（第 3、4 卷），**中国滩羊区植物志**，4 卷本；第 1 卷：177 页，1988；第 2 卷：479 页，1993；第 3 卷：379 页，1996；第 4 卷：385 页，1996；宁夏人民出版社，银川。

1988-1996, Agricultural Modernization Base and Bureau of Animal and Husbandry of Ningxia Hui Autonomous Region and Northwest Institute of Botany, Shaanxi Province, Zhao-Ying YU and Yang-Pong HSU (editors in chief, vol. 1), Yang-Pong HSU, Ke-Zhi WANG and Zhao-Ying YU (editors in chief, vol. 2), Yang-Pong Hsu and Ke-Zhi WANG (editor in chief, vols. 3-4), ***Flora Sinensis in Area Tan-Yang***, vols. 1-4; 1: 177 p, 1988; 2: 479 p, 1993; 3: 379 p,1996; 4: 385 p, 1996; Ningxia People's Press, Yinchuan.

1988、1990、1991、1993 年，梁畴芬，谈谈中国人的姓名如何翻译成西文名，广西植物 8（4）：375-376，1988；评许兆然的中国人姓名西译方案，广西植物 10（1）：93-96，1990；再议人名西译问题，广西植物 11（3）：286-288，1991；人名西译问题研究续报，广西植物 13（2）：143，1993。

1988, 1990, 1991, 1993, Chou-Fen LIANG, 'How to translate the person's name from Chinese to Roman', *Guihaia* 8(4): 375-376, 1988; 'Evaluation of Xu Zhao-Ran's Chinese translation of Chinese

660 Kai Larsen and Chi-Ming HU, 1996, Myrsinaceae. In: Kai Larsen (ed.), *Flora of Thailand*, 6(2): 81-178; The Forest Herbarium, Royal Forest Department, Bangkok; Chi-Ming HU, 1999, Primulaceae. In: Thawatchai Santisuk and Kai Larsen (eds.), *Flora of Thailand*, 7(1): 155-168; The Forest Herbarium, Royal Forest Department, Bangkok.

Chi-Ming HU in Aarhus University, Denmark (Photo provided by Chi-Ming HU)
胡启明在丹麦 Aarhus University（相片提供者：胡启明）

names', *Guihaia* 10(1): 93-96, 1990; 'Reconsidering the issue of translation of names of people', *Guihaia* 11(3): 286-288, 1991; 'Continued report on the translation of name in the West', *Guihaia* 13(2): 143, 1993.

1989

1989年3月，中国科学院成都生物研究所植物分类室，**四川植物研究**；只出版1期，报道芒苞草科，包括芒苞草，但称为“第2期”，实际上是1980年植物分类研究（一）的续刊[661]。

1989.03, Department of Plant Taxonomy, Chengdu Institute of Biology, Chinese Academy of Sciences, ***Acta Botanica Sichuanica***. Only one issue; family Acanthochlamydaceae P. C. Kao was described; following publication in *Phytotaxonomic Research*, 1980, in which P. C. Kao described *Acanthochlamys bracteata*.

1989年3月，向秋云[662]（1962—）赴美国，先后在史密森学会和哈佛大学从事研究，1990年入华盛顿州立大学攻读博士，1995年获得博士学位；1995—1997年在俄亥俄州立大学从事博士后研究，1997—2000年任爱达荷州立大学生物系助理教授兼爱达荷自然历史博物馆研究员，2001—2006年任北卡罗来纳州立大学植物生物学系助理教授、2006—2012年任副教授、2012年至今任教授。

1989.03, Qiu-Yun XIANG (Jenny XIANG, 1962-) traveled to the United States of America as visiting scientist at the Smithsonian Institution and Arnold Arboretum of Harvard University; then studied at Washington State University from 1990, received Ph.D. degree in 1995[663]; postdoctoral research fellow at Ohio State University, 1995-1997; assistant professor, Department of Biology, Idaho State University, and curator of the Idaho State Museum of Natural History, 1997-2000, assistant professor, 2001-2006, associate professor, 2006-2012, and professor since 2012, in the Department of Plant Sciences, North Carolina State University, Raleigh, North Carolina.

1989年3月至1991年3月，中国科学院昆明植物研究所管开云（1953—）[664]赴爱丁堡皇家植物园进修植物系统分类学，1993年4月至1995年3月任中国科学院昆明植物研究所副研究员兼所长助理、外办主任、科研处长，1995年8月至2006年9月任研究员兼昆明植物园主任，2006年10月至2009年12月任中国科学院昆明植物研究所研究员；其间，2001年10月至2002年2月任日本淡路景观园艺学校客座教授，2002年3月日本大阪府立大学应用生命科学专业在职博士生，2007年3月

661 高宝莼，1980，植物分类研究1：1-3；陈心启，1981，植物分类学报19（3）：323-329；高宝莼，1989，四川植物志9：483-516；Zhan-he JI and Alan W. Meerow, *Flora of China* 24:273, 2000。

662 1982年中山大学生物系本科毕业后入中国科学院植物研究所植物分类室工作（女）。

663 Ph.D. dissertation: Molecular systematics and biogeography of *Cornus* L. and putative relatives, 204 p; Advisor: Douglas E. Soltis.

664 云南景谷人，1972年3月至1975年8月云南师范学院外语系英语专业学习，1975年9月至1979年2月任中国科学院昆明植物研究所科技翻译，1979年3月至1980年2月云南大学外语系高级英语进修班学员，1980年3月任中国科学院昆明植物研究所所长秘书并从事植物分类研究；其间担任1980年中美湖北神农架植物联合考察翻译。

获得博士学位；2010 年 1 月至今任中国科学院新疆生态与地理研究所研究员、副所长，兼任吐鲁番沙漠植物园主任，2016 年起同时兼任伊犁植物园园长。

1989.03–1991.03, Kai-Yun GUAN (1953–), Kunming Institute of Botany, Chinese Academy of Sciences, went to Royal Botanic Garden, Edinburgh to study plant systematics and taxonomy, returned to Kunming as associate professor, assistant to director, director of foreign affair and of science department, Kunming Institute of Botany, Chinese Academy of Sciences, April 1993 to March 1995, then professor and director of Kunming Botanical Garden, and professor of Kunming Institute of Botany, Chinese Academy of Sciences, October 2006 to December 2009; visiting professor at Awaji Landscape Planning and Horticulture Academy, Japan, August 1995-September 2006, and Professor of Kunming Institute of Botany, Chinese Academy of Scieuces, October 2001 to February 2002, on the-job Ph.D. candidate in Applied Life Sciences, Osaka Prefecture University, from March 2002, received his Ph.D. degree in March 2007[665]; research professor and vice director of Xinjiang Institute of Ecology and Geography, Chinese Academy of Sciences, since January 2010, concurrently director of Turpan Botanical Garden, and concurrently director of Yili Botanical Garden since 2016.

1989.07.30–08.25, Sino-German botanical expedition to Xizang: Amdo [Anduo], Cona [Cuona], Damxung [Dangxiong], Nyingchi [Linzhi], Mainling [Milin], and Qinghai: Golmud [Geermu], Qumarlêb [Qumalai], Fenghuo Shan, Zhidoi [Zhiduo]. Scientists from Georg-August-Universität Göttingen, Germany, and Rong-Fu HUANG (HNWP), collected 864 numbers of plant specimens (HNWP).

1989 年 7 月 30 日至 8 月 25 日，中德植物联合考察西藏的安多、当雄、林芝、米林、错那，青海的格尔木、曲麻莱、风火山、治多，德国的哥廷根大学科学家，中国科学院西北高原生物研究所黄荣福，采集植物标本 864 号（HNWP）。

1989.08.28–08.30, Joint editorial committee meeting of Flora of China held at South China Institute of Botany, Chinese Academy of Sciences, Guangzhou. It was decided that the English language *Flora of China* would not be a direct translation of *Flora Reipublicae Popularis Sinicae*, but instead a revision co-authored by Chinese authors and non-Chinese collaborators, preferably persons with expertise in the family. Attendees were Bruce M. Bartholomew (CAS), David E. Boufford (A/GH), Paul Pui-Hay BUT (CUHK), Chia-Jui CHEN (PE), Shou-Liang CHEN (NAS), Shu-Kun CHEN (KUN), Sing-Chi CHEN (PE), Hong-Bin CUI (PE), Lun-Kai DAI (PE), Hong-Ya GU (PE), Jian-Guo HAN (PE), Chi-Ming HU (IBSC), Ching-Chieu HUANG (IBSC), Hsi-Wen LI (KUN), Ping-Tao LI (CANT), Yuou-Ruen LING (IBSC), Xin-Zhong LU (IBSC), Nancy R. Morin (MO), Peter H. Raven (MO), Alastair Scott (MO), De-Yan TAN (IBSC), William TAI (MO), Cheng-Yih WU (KUN), Xiu-Wen XIA (IBSC), Zhen-Dai XIA (PE) and Jian-Fei ZENG (1939–, Science Press). Between 1988 and 1993, Paul Pui-Hay BUT, Chang-Chung CHOU (HAST), David S. Ingram (1941–, E) and Laurence E. Skog (1943–, US) joined the editorial committee.

665 Ph.D. dissertation: Studies on diversity and conservation of the genus *Begonia* L. (Begoniaceae) in China; Advisor: Hirofumi Yamaguchi (1946–).

Joint editorial committee meeting of *Flora of China*, Guangzhou, 28-30 August 1989 (Photo provided by Wu Zhengyi Science Foundation)
1989 年 8 月 28 日至 30 日，英文版中国植物志编辑委员会联席会议于广州（相片提供者：吴征镒科学基金会）

1989 年 8 月 28 日至 8 月 30 日，英文版中国植物志编辑委员会联席会议在中国科学院华南植物研究所举行，会议决定英文版并非中文版的直接翻译，而是中外作者共同修订。会议出席者有加州科学院 Bruce M. Bartholomew，哈佛大学 David E. Boufford，香港中文大学毕培曦，中国科学院植物研究所陈家瑞、陈心启、崔鸿宾、戴伦凯、顾红雅、韩建国、夏振岱，江苏省植物研究所 南京中山植物园陈守良，中国科学院昆明植物研究所陈书坤、李锡文、吴征镒，中国科学院华南植物研究所胡启明、黄成就、林有润、路新中、谈德颜、夏秀文，华南农学院李秉滔，密苏里植物园 Nancy R. Morin、Peter H. Raven、Alastair Scott、戴威廉和科学出版社曾建飞（1939—）。编辑委员会 1988—1993 年增加毕培曦、"中央" 研究院植物研究所周昌弘、爱丁堡植物园 David S. Ingram（1941—）和史密森学会的 Laurence E. Skog（1943—）。[666]

1989 年 10 月，吴声华（1961—）[667] 赴芬兰留学，1990 年 11 月获赫尔辛基大学植物系博士学位，

666 本书有关英文版中国植物志编辑委员会的相关内容均引自英文版中国植物志第 1 卷第 18-19 页，2013。
667 台湾新竹人，1979 年 10 月 —1983 年 6 月东海大学生物系学士，1983 年 9 月 —1985 年 6 月东海大学生物研究所硕士，从事苔藓研究，1987 年 9 月 —1989 年 6 月台湾大学植物学研究所博士肄业。

1991 年 1 月任台湾自然科学博物馆副研究员、1995 年 12 月任研究员；主要从事菌类学研究，但也从事其他类群；详细参见：2016 年条目。

1989.10, Sheng-Hua WU (1961-) studied in Finland; received his Ph.D. degree from Department of Botany, University of Helsinki, November 1990[668]; associate curator, January 1991, and curator since December 1995, at Taiwan Museum of Natural Science, Taichung. Major research focus on fungi and also other taxa; for details, see: 2016.

1989 年 10 月 17 日 至 20 日，第一届系统与进化植物学青年研讨会于北京中国科学院植物研究所举行，来自全国 20 个省（市、区）的 84 人出席。

1989.10.17-10.20, First Youth Symposium of Systematic and Evolutionary Botany held in Institute of Botany, Chinese Academy of Sciences, Beijing, 84 attendees from 20 provinces.

1989 年，傅立国主编，**中国珍稀濒危植物**，365 页；上海教育出版社，上海。

1989, Li-Kuo FU (editor in chief), '***Rare and Endangered Plants of China***', 365 p; Shanghai Educational Publishing House, Shanghai.

1989, An-Ming LU, Explanatory notes on R. Dahlgren's system of classification of the angiosperms, *Cathaya* 1: 149-160.

1989 年，路安民，R. Dahlgren 的被子植物系统注释，*Cathaya* 1：149-160。

1989 年，鲁迎青（1963—）[669] 赴美国留学，1995 年获威斯康星大学麦迪逊分校植物系博士学位，1995—1996 年任威斯康星大学麦迪逊分校植物系 Research Fellow，1996—2003 年先后任杜克大学生物系博士后、Solan Fellow 和 Research Associate；2003 年至今任中国科学院植物研究所研究员、研究组组长。

1989, Ying-Qing LU (1963-) went to the United States of America to study in the Department of Botany, University of Wisconsin, Madison, Wisconsin; received Ph.D. degree in 1995[670], Research Fellow, Department of Botany, University of Wisconsin, Madison, Wisconsin, 1995-1996, Postdoctoral, Sloan Fellow and Research Associate, Department of Biology, Duke University, 1996-2003; Professor and Group Leader, Institute of Botany, Chinese Academy of Sciences, Beijing, since 2003.

---

668 Ph.D. dissertation: The Corticiaceae (Basidiomycetes) subfamilies Phlebioideae, Phanerochaetoideae and Hyphodermoideae in Taiwan; Advisor: Tuomo Niemelä (1940-).

669 籍贯山西浑源，生于宁夏青铜峡（女）；1979—1983 年获南京大学生物学本科学士学位，1983—1985 年任中国药科大学生药系助教；1985—1988 年获中国科学院植物研究所研究生硕士学位；1988—1989 年任中国科学院植物研究所系统与进化开放实验室实习研究员。

670 Ph.D. dissertation: Ecological genetics of *Impatiens capensis*: the responses of mating systems to density variation in natural populations and the correlation between mating system and inbreeding depression; advisor: Donald M. Waller.

1989, Charles Lyte, ***Frank Kingdon-Ward —The Last of the Great Plant Hunters***, 218 p; J. Murray, London.

1989 年，Charles Lyte，**Frank Kingdon-Ward—— 最后的伟大植物猎人**，218 页；J. Murray，伦敦。

1989 年，上海自然博物馆编著，徐炳声主编，**长江三角洲及邻近地区孢子植物志**，573 页；上海科学技术出版社，上海。

1989, Shanghai Natural History Museum, Bing-Sheng XU (Ping-Sheng HSU, editor in chief), ***Cryptogamic Flora of the Yangtze Delta and Adjacent Regions***, 573 p; Shanghai Scientific and Technical Publishers, Shanghai.

1989 年，史念海主编，**辛树帜先生诞生九十周年纪念论文集**，503 页；农业出版社，北京。

1989, Nian-Hai SHI (editor in chief), '***Memorial Works on Mr. Shu-Chih HSIN's 90th birth anniversary*** *(S. S. Sin)*', 503 p; Agriculture Press, Beijing.

1989 年，宋朝枢等著，**中国珍稀濒危保护植物**，453 页；中国林业出版社，北京。

1989, Chao-Shu SONG et al., '***Chinese Rare and Endangered Protected Plants***', China Forestry Publishing House, Beijing.

1989 年，苏少泉、郭景春编，傅沛云校，**英拉汉杂草名称**，390 页；农业出版社，北京。

1989, Shao-Quan SU and Jing-Chun GUO (eds.), Pei-Yun FU (proofreader), '***English-Latin-Chinese Names of Weeds***', 390 p; Agriculture Press, Beijing.

1989 年，杨宗愈[671]（1961— ）赴英国雷丁大学植物系攻读博士，1994 年获博士学位后返回台湾。1995 年任台湾自然科学博物馆副研究员，2015 年至今为研究员。1995—2001 年同时兼任中国文化大学副教授。2001—2016 年兼任东海大学副教授，2016 年至今任教授。2005—2012 年兼任台北艺术大学副教授。2009—2016 年兼任中兴大学合聘副教授，2016 至今任教授。

1989, Tsung-Yu YANG (T. Y. Aleck YANG, 1961-) studied at Department of Botany, the University of Reading, England, received his Ph.D. degree in 1994[672]; returned to Taiwan as associate curator and curator of Taiwan Museum of Natural Science, Taichung, 1995-2015, and 2015 to present; concurrently associate

671 别号中宇；祖籍云南，生于台湾高雄；1980—1981 年就读于中国文化大学植物系，1981—1984 年东海大学生物系毕业获学士学位，1984—1987 年台湾大学植物研究所毕业获硕士学位；1986—1989 年任东海大学生物系专任助教。

672 Ph.D. thesis: A revision of *Viorna* group of species (Section *Viorna* sensu Pranti) in the genus *Clematis* L. (Ranunculaceae); Advisors: David M. Moore (1933-2013) and Jeffrey B. Harborne (1928-2002).

professor at Chinese Culture University, Taipei [Taibei], 1995-2001; associate professor and professor of Tunghai University, 2001-2016; associate professor at Taipei University of the Arts, 2005-2012; associate professor at the Chung Hsing University, Taichung, 2009-2016, and professor since 2016.

1989+，傅坤俊主编，**黄土高原植物志**，6卷本；第1卷：648页，2000；第2卷：547页，1992；第5卷：557页，1989；科学出版社，中国林业出版社，科学技术文献出版社，北京。

1989+, Kun-Tsun FU (editor in chief), ***Flora Loess-Plateaus Sinicae***, vols. 1-6; 1: 648 p, 2000; 2: 547 p, 1992; 5: 557 p, 1989; Science Press, China Forestry Publishing House, Science and Technical Documentation Press, Beijing.

1989、1990年，钟本固、熊源新，贵州藓类植物名录（Ⅰ，Ⅱ），贵州师范大学 自然科学专集1（总第10期）：41-51，1989；贵州师范大学学报 自然科学版3（总第14期）：22-32，1990。

1989 and 1990, Ben-Gu ZHONG and Yuan-Xin XIONG, A catalogue of Musci of Guizhou (I and II), *Journal of Guizhou Normal University Natural Science Edition* 1 (No. 10): 41-51, 1989; 3 (No. 14): 22-32, 1990.

1989—1990年，中国科学院考察贵州红水河上游，植物研究所李良千等、华南植物研究所吴德邻等、昆明植物研究所苏志云等，采集植物标本1.1万（IBSC、KUN、PE）。[673]

1989-1990, Expedition of Chinese Academy of Sciences to upper Hongshui River, Guizhou; Liang-Qian LI et al. (PE), Te-Lin WU et al. (IBSC) and Zhi-Yun SU et al. (KUN), collected 11,000 numbers of plant specimens (IBSC, KUN, PE).

1989, 1992, 1998, Leonard Forman and Diane Bridson, ***The Herbarium Handbook***, 214 p, 1999; ed. 2, 303 p; 1992; ed. 3, 334 p; 1998; Royal Botanic Gardens, Kew. Translated into Chinese in 1998.

1998年，姚一建等译，**标本馆手册**，第3版，299页；皇家植物园，克佑。[674]

1989年8月—1991年4月，中山大学陈宝樑[675]（1942—1991）赴荷兰国家植物标本馆研究木兰科。[676]

1989.08-1991.04, Bao-Liang CHEN (1942-1991), Sun Yat-Sen University, studies Magnoliaceae in Leiden, the Netherlands (L).

673 吴德邻主编，1996，红水河上游地区植物调查研究报告集，294页；科学出版社，北京。

674 Kew的音译，即邱园。

675 1961年9月至1968年7月，中山大学生物学本科，1978年10月至1981年10月中山大学生物系硕士研究生，1981年12月留校工作。

676 Bao-Liang CHEN and Hans P. Nooteboom, 1993, Notes on Magnoliaceae III: The Magnoliaceae of China, *Annals of the Missouri Botanical Garden* 80(4): 999-1104.

1989—1993 年，浙江植物志编辑委员会编，**浙江植物志**，8 卷本[677]；总论（章绍尧、丁炳扬主编）：343 页，1993；第 1 卷（张朝芳[678]、章绍尧主编）：411 页，1993；第 2 卷（王景祥主编）：408 页，1992；第 3 卷（韦直、何业祺[679]主编）：541 页，1993；第 4 卷（裘宝林主编）：423 页，1993；第 5 卷（方云亿主编）：355 页，1989，第 2 版[680]，356 页，1992；第 6 卷（郑朝宗主编）：390 页，1993；第 7 卷（林泉主编）：584 页，1993；浙江科学技术出版社，杭州。[681]

1989-1993, Editorial Board of Flora of Zhejiang, ***Flora of Zhejiang***, General and vols. 1-7[682]; General (editors in chief, Shao-Yao ZHANG and Bing-Yang DING): 343 p, 1993, 1 (editors in chief, Chao-Fang ZHANG and Shao-Yao ZHANG): 411 p, 1993, 2 (editor in chief, Jing-Xiang WANG): 408 p, 1992, 3 (editors in chief, Zhi WEI and Ye-Qi HE): 541 p, 1993, 4 (editor in chief, Bao-Lin QIU): 423 p, 1993, 5 (editor in chief, Yun-Yi FANG): 355 p, 1989 and ed. 2[683], 356 p, 1992, 6 (editor in chief, Chao-Zong ZHENG): 390 p, 1993, 7 (editor in chief, Quan LIN): 584 p, 1993; Zhejiang Science and Technology Publishing House, Hangzhou.

1989-1995, Chinese Botanical Society[684], ***Chinese Journal of Botany***[685], vols. 1-7(2).

1989—1995 年，中国植物学会，**植物学杂志**[686]，vols. 1-7（2）。

1989, 2008, Charles Roy Lancaster, ***Roy Lancaster Travels in China****—A Plantsman's Paradise*, 516 p, 1989; Antique Collectors' Club, Woodbridge; Charles Roy Lancaster, ***Plantsman's Paradise****—Travels in China*, ed. 2, 511 p, 2008; Garden Art Press, Woodbridge.

1989、2008 年，Charles Roy Lancaster，**Roy Lancaster 在中国旅行**——园艺爱好者的天堂，516 页，1989；Antique Collectors' Club，Woodbridge；Charles Roy Lancaster，**园艺爱好者的天堂**——在中国旅行，

677 即总论卷和第 1-7 卷。

678 浙江东阳人（1923—2002），1955 年毕业于华东师范大学生物系，分配到浙江师范学院工作，1957 年被划分为右派，1979 年才获得平反，1980 年任讲师，1986 年任副教授，1994 年任教授；1980 年开始跟随秦仁昌从事蕨类植物研究，1987 年倡议成立中国蕨类植物协会并任理事长。详细参见：丁炳扬、张庆勉，2002，为发展蕨类植物科学奋斗到最后一刻——忆中国蕨类植物协会原理事长张朝芳教授，中国花卉协会蕨类植物分会简讯 6：1-4；曾汉元，深切缅怀张朝芳教授，中国花卉协会蕨类植物分会简讯 6：4-6。

679 20 世纪 80 年代初期由中国科学院西北植物研究所调到杭州师范学院。

680 目前第 2 版只有第 5 卷；实为第 1 版其他卷于 1993 年完成后，对 1989 年出版的第 5 卷进行的修订。

681 李卓凡，1996，一部有特色的地方志——评《浙江植物志》，中国图书评论 4：42-43。

682 Total of 8 volumes: general, plus volumes 1-7.

683 Only volume five for the second edition.

684 i.e. Botanical Society of China.

685 *The Chinese Journal of Botany* differs from *Chinese Journal of Botany* published in 1936 by the Chinese Society of Biological Sciences only by the article 'the' preceding the title.

686 本刊刊名与 1936 年中国生物科学学会出版的 *The Chinese Journal of Botany* 除冠词“The”外同名。

第 2 版，511 页，2008；Garden Art Press，Woodbridge。

1989-2008, Laboratory of Systematic and Evolutionary Botany and Herbarium, Institute of Botany, Chinese Academy of Sciences, Beijing, ***Cathaya***, *Annals of the Laboratory of Systematic and Evolutionary Botany and Herbarium, Institute of Botany, Chinese Academy of Sciences*, vols. 1-16 and 17-18, 1989-2004 and 2005-2008.

1989—2008 年，中国科学院植物研究所系统与进化植物学开放研究实验室, ***Cathaya***，中国科学院植物研究所系统与进化植物学开放研究实验室年刊, vols. 1-16, 1989—2004, 17-18, 2005—2008。

1990

1990.03.04-03.07, Chinese-American symposium on bryophytes held in Beijing; decision made to compile a collaborative China and USA *Bryoflora of China* in English.

1990 年 3 月 4 日至 7 日，中美苔藓植物学讨论会于北京召开，决定中美联合编写英文版中国苔藓植物志。[687]

1990 年 5 月 21 日—8 月 23 日，由国家科委、中国科学院、国家环境保护局及青海省人民政府共同组建的可可西里综合科学考察队，进行了历史上第一次全面、综合的自然科学考察，全队 68 名成员分为地理、地质和生物 3 个大组 27 个专业，植物学家中国科学院昆明植物研究所武素功为队长，获得大量宝贵的第一手资料与植物标本。[688]

1990.05.21-08.23, Integrated scientific expedition to Hoh Xil, Qinghai, organized by State Science Committee, Chinese Academy of Sciences, National Administrative Bureau of Environment Protection, and the Government of Qinghai Province was first comprehensive expedition to the region; team was composed of 68 scientits from geography, geology and biology, representing 27 scientific fields, led by Shu-Kung WU, botanist, Kunming Institute of Botany, Chinese Academy of Sciences. Considerable original scientific data as well as valuable specimens were obtained.

1990.08.12-08.19, Congress of East Asiatic Bryology held in Helsinki, Finland; more than sixty international attendees, including Tong CAO (1945-), Si HE, Ming-Jou LAI, Shin-Chiang LEE, Zhi-Hua LI, Pang-Juan LIN, Jian-Xin LUO, Pan-Cheng WU, Sheng-Hua WU and Mu ZANG from China.[689, 690]

1990 年 8 月 12 日至 19 日，东亚苔藓学会议在芬兰赫尔辛基举行，60 多人出席，中国学者包括曹同（1945— ）、何思、赖明洲、李植华、黎兴江、林邦娟、罗健馨、吴鹏程、吴声华、臧穆。

687 吴鹏程老师提供。

688 李炳元，1990，青海可可西里地区综合科学考察初报，山地研究 8（3）：161-166；武素功、张以茀、李炳元，1991，青海可可西里地区综合科学考察再报，山地研究 9（2）: 93-98；可可西里综合科学考察队编，1994，青藏高原腹地——可可西里综合科学考察，108 页；上海科学技术出版社，上海；武素功，1994，可可西里综合科学考察，科学 46：46-49；武素功，2003，青藏高原腹地——可可西里综合科学考察追记，上海科学生活 8：64-71。

689 Jetter Lewinsky, 1990, Congress of east Asiatic bryology, Helsinki, August 12-19, 1990, *The Bryological Times* 57/58: 5-8.

690 Timo J. Koponen and Jaakko Hyvönen (eds.), 1992, Proceedings of the congress of east Asiatic bryology, Helsinki, August 12-19, 1990, *Bryobrothera* 1: i–xii, 332 p.

1

The integrated scientific expedition to Hoh Xil, Qinghai, 21 May–23 August 1990: 1. Leaving Xining; 2. Su-Kung WU, Leader, and three vice leaders; 3. Su-Kung WU studying map; 4. No-Man's land, 5000 m elevation ; 5. Su-Kung WU in the field; 6. On the way; 7. Camp site at 4400 m elevation (Photo provided by Rhui-Cheng FANG)

1990 年 5 月 21 日至 8 月 23 日，可可西里综合考察队：1. 西宁出征；2. 队长武素功和三位副队长；3. 武素功野外研究地图；4. 海拔 5000 m 无人区；5. 武素功于野外；6. 考察途中；7. 海拔 4400 m 宿营（相片提供者：方瑞征）

The Congress of East Asiatic Bryology in Helsinki, Finland, 12-19 August, 1990; 1. Participants in Helsinki University (left-right): Mu ZANG, Jian-Xing LUO, Pang-Juan LIN, Sheng-Hua WU, Chin-Chinh LAI LIANG, Shin-Chiang LEE, Si HE, Tong CAO, Zhi-Hua LI, Ming-Jou LAI, and Pan-Cheng WU; 2. Participants at the field station: Zhi-Hua LI, Pan-Cheng WU, Mu ZANG, Shin-Chiang LEE, Si HE, Ming-Jou LAI, Pang-Juan LIN, Tong CAO, and Marshall Crosby (Photo provided by Si HE)

1990年8月12日至19日，东亚苔藓学会议于芬兰赫尔辛基；1. 赫尔辛基大学留念（从左至右）：臧穆、罗健馨、林邦娟、吴声华、赖梁金枝（赖明洲夫人）、黎兴江、何思、曹同、李植华、赖明洲、吴鹏程；2. 野外台站合影（从左至右）：李植华、吴鹏程、臧穆、黎兴江、何思、赖明洲、林邦娟、曹同、Marshall Crosby（相片提供者：何思）

1990 年 9 月，钱宏[691]（1957—）赴北美留学，1990 年 9 月至 1993 年 8 月在加拿大温哥华不列颠哥伦比亚大学、1993 年 9 月至 1995 年 8 月在美国北卡罗来纳大学 Chapel Hill 分校从事博士后研究，1995 年 9 月至 1999 年 8 月任加拿大不列颠哥伦比亚大学副研究员，1999 年 8 月至 2001 年 4 月于加拿大不列颠哥伦比亚大学攻读生物地理学博士研究生并获博士学位，2001 年 5 月至 2008 年 8 月任美国伊利诺伊州博物馆副研究员，2008 年 8 月至今任研究员；主要从事大尺度生物多样性比较和生物地理学研究。

1990.09, Hong QIAN (1957-) went to North America, as postdoctoral fellow at University of British Columbia, Vancouver, Canada, September 1990 to August 1993, and University of North Carolina, Chapel Hill, North Carolina, United States of America, September 1993 to August 1995, then research associate scientist at University of British Columbia, September 1995 to August 1999, studied at University of British Columbia, August 1999 to April 2001, received Ph.D. degree in 2001[692]; associate scientist, Illinois State Museum, United States of America, May 2001 to August 2008, scientist since August 2008. His major areas of study focus on the large scale biodiversity comparisons and biogeography.

1990 年 9 月，朱光华[693]（1964—2005）赴美国留学，1995 年 9 月获美国密苏里大学博士学位，同时就职于密苏里植物园，从事英文版中国植物志的编辑工作，2001 年任美方英文版中国植物志项目主任；2005 年 11 月 2 日因病过世[694]。

1990.09, Guang-Hua ZHU (1964-2005) went to United States of America to study botany, received Ph.D. degree from University of Missouri St. Louis in September 1995[695], editorial staff of *Flora of China* project in Missouri Botanical Garden immediately afterword, became the project director since 2001; but died by illness in November 2, 2005.

1990.10.01-10.02. Joint editorial committee meeting of Flora of China held at Harvard University, Cambridge, Massachusetts, United States of America. Attendees: Ihsan A. Al-Shehbaz (1939-, MO), Bruce M. Bartholomew (CAS), David E. Boufford (A/GH), Paul Pui-Hay BUT (CUHK), Chia-Jui CHEN (PE), Shou-Liang CHEN (NAS), Sing-Chi CHEN (PE), Robert Cook (A/GH), Hong-Bin CUI (PE), Bryan E. Dutton

691 籍贯浙江宁波，1978 年 2 月 —1982 年 1 月安徽农学院林学系本科，获学士学位，1982 年 2 月 —1985 年 1 月安徽农学院森林植物学研究生，获硕士学位，1985 年 2 月 —1986 年 1 月任安徽农学院助教，1986 年 2 月 —1989 年 1 月中国科学院林业土壤所即中国科学院沈阳应用生态研究所研究生，获博士学位，博士论文：Hong QIAN, 1989, Alpine tundra of Mt. Changbai: phytotaxonomy, florology and phytoecology, Institute of Applied Ecology, Chinese Academy of Sciences, Shenyang, China, 452 p; 指导教授：王战。1989 年 2 月 —1990 年 9 月任中国科学院植物研究所助理研究员。

692 Ph.D. dissertation: Multiscale comparisons of vascular plant diversity and geography between East Asia and North America, University of British Columbia, Vancouver, Canada, 152 p; Advisor: Karel Klinka (1937-2015).

693 内蒙古满洲里人，1981—1985 年和 1985—1988 年于内蒙古师范大学获理学学士和硕士学位，1988 年在中国科学院植物研究所攻读博士学位。

694 参见：http://flora.huh.harvard.edu/china/editors/Guanghua_Zhu_memorial_program.pdf（2017 年 12 月进入）。

695 Ph.D. dissertation: Systematics of *Dracontium* L. (Araceae), 311 p; Advisor: Thomas B. Croat (1938-).

Joint editorial committee meeting of *Flora of China* at Arnold Arboretum of Harvard University, 1-2 October 1990 (Photo provided by Wu Zhengyi Science Foundation)

1990 年 10 月 1 日至 2 日，英文版中国植物志编辑委员会联席会议于哈佛大学阿诺德树木园（相片提供者：吴征镒科学基金会）

(1958-, A/GH), Hong-Ya GU (PEY), Jian-Guo HAN (PE), Ching-Chiu HUANG (IBSC), Hsi-Wen LI (KUN), Peter H. Raven (MO), Orbelia R. Robinson (CAS), William TAI (MO) and Cheng-Yih WU (KUN).

1990年10月1日至2日，英文版中国植物志编辑委员会联席会议在哈佛大学举行，密苏里植物园 Ihsan A. Al-Shehbaz（1939—）、Peter H. Raven、戴威廉，加州科学院 Bruce M. Bartholomew、Orbelia R. Robinson，哈佛大学 David E. Boufford、Robert Cook、Bryan E. Dutton（1958—），香港中文大学毕培曦，中国科学院植物研究所陈家瑞、陈心启、崔鸿宾、韩建国，江苏省植物研究所 南京中山植物园陈守良，北京大学顾红雅，中国科学院华南植物研究所黄成就，中国科学院昆明植物研究所李锡文、吴征镒出席。

1990, Xue-Liang BAI, Conspectus of Flora Hepaticae in Nei Mongol, *Acta Scientiarum Naturalium Universitatis Intramongolicae* 21(2): 264-276.

1990年，白学良，内蒙古苔类植物志大纲，内蒙古大学学报（自然科学版），21（2）：264-276。

1990, George Bishop, ***Travels in Imperial China***—*The Exploration and Discoveries of Père David*, 192 p; Cassell, London.

1990年，George Bishop，**中华帝国之旅行**——David 神父的考察与发现，192页；Cassell，伦敦。

1990, Tong CAO, Chien GAO, Kuang-Chu CHANG and Pang-Juan LIN, Current Chinese bryological literature, *Acta Bryolichenologica Asiatica* 2(1-2): 41-63.

1990年，曹同、高谦、张光初、林邦娟，当代中国苔藓植物文献，*Acta Bryolichenologica Asiatica* 2（1-2）：41-63。

1990年，蒋镇宇（1960—）[696] 赴美国留学，1994年获美国圣路易斯华盛顿大学生物学博士学位；1994—1996年于“中央”研究院植物研究所从事博士后研究，1996—1997年任成功大学生命科学系讲师、1997—2001年任副教授、2001—2006年为教授、2006年至今为特聘教授；其间2009—2012年为系主任兼所长、2012年至今为生态农业与生物资源中心主任。

1990, Tzen-Yuh CHIANG (Tseng-Yu CHIANG, 1960-) studied at Washington University, St. Louis, Missouri, United States of America; received Ph. D. degree in 1994[697]; returned as postdoctoral research fellow at Institute of Botany, Academia Sinica, Taipei [Taibei], 1994 to 1996, then lecturer, 1996-1997,

696 1982年台湾师范大学生物系本科毕业获学士学位，1989年台湾大学植物研究所毕业获硕士学位（以上信息来自成功大学网站：http://o.bio.ncku.edu.tw/files/13-1356-109600.php，2017年12月进入）。

697 Ph. D. Dissertation: Phylogenetics and Morphological Evolution of the Dicnemonaceae (Mosses, Order: Dicranales), Phylogenetics and Evolution of the Hylocomiaceae (Mosses, Order: Hypnales); Advisor: Barbara A. Schaal (1947-).

associate professor, 1997-2001, professor, 2001-2006, distinguished professor since 2006, Department of Biology / Department of Life Sciences, Cheng Kung University; also chair of department, 2009-2012, and director of Research Center of Ecological Agriculture and Bio-Resources since 2012.

1990, Chien GAO and Tong CAO, Research activities on Chinese bryology throughout the 1980s, *Acta Bryolichenologica Asiatica* 1(1, 2): 21-30.

1990 年，高谦、曹同，八十年代中国苔藓植物学术研究动态，*Acta Bryolichenologica Asiatica* 1（1、2）：21-30。

1990 年，苟新京编著，**青海种子植物名录**，284 页；青海种子植物名录编写组，青海（内部印制）。

1990, Xin-Jing GOU, ***Index Florae Qinghaiensis***, 284 p; Group of Qinghai Index Florae Qinghaiensis, Xining (internal publication).

1990, Sinikka S. Piippo, Annotated catalogue of Chinese Hepaticae and Anthocerotae. *The Journal of the Hattori Botanical Laboratory* 68: 1-192.

1990 年，Sinikka S. Piippo，中国苔纲和角苔纲名录，*The Journal of the Hattori Botanical Laboratory* 68：1-192。

1990 年，桑涛[698]（1966—）赴美国俄亥俄州位于 Athens 的俄亥俄大学任助教，1991 年于俄亥俄州立大学植物学系攻读博士研究生，1995 年获博士学位；1995—1996 年哈佛大学博士后；1996 年起任密西根州立大学植物生物学系助理教授、副教授、教授；2010 年 12 月入选国家千人计划；现任中国科学院植物研究所北方资源植物重点实验室主任、系统与进化植物学国家重点实验室研究员。

1990, Tao SANG (1966-), teaching assistant, Department of Botany, Ohio University, Athens, Ohio, studied at Ohio State University, Columbus, Ohio, United States of America, 1991, received Ph. D. degree[699] in 1995; postdoctoral research fellow of Harvard University, 1995-1996; assistant, associate and full professor, Michigan State University, since 1996; Thousand Talents Plan of China since December 2010; now director of Key Laboratory of Plant Resources, Institute of Botany, Chinese Academy of Sciences, and Researcher of State Key Laboratory of Systematic and Evolutionary Botany, Institute of Botany, Chinese Academy of Sciences[700].

698 四川（重庆）人，1982—1986 年复旦大学生物系，获学士学位；1987—1989 年复旦大学生物系，获硕士学位（自中国科学院植物研究所网址：http://klpr.ibcas.ac.cn/news/150，2017 年 12 月进入，以及博士论文简历介绍）。

699 Ph.D. dissertation: Phylogeny and Biogeography of *Paeonia* (Paeoniaceae); Advisor: Tod F. Stuessy.

700 Both from his Ph.D. dissertation and his professional website at the Institute of Botany, Chinese Academy of Science: http://klpr.ibcas.ac.cn/news/150, accessed at December 2017.

1990, Stephen A. Spongberg, ***A Reunion of Trees***—*The discovery of Exotic Plants and Their Introduction into North American and European Landscape*, 270 p; Harvard University Press, Cambridge, Massachusetts.[701]

1990年，Stephen A. Spongberg，**树的重聚**——异域植物的发现及其引入北美和欧洲园林的历程，270页；哈佛大学出版社，剑桥，麻州。

1990年，王文采，当代四被子植物分类系统简介（一、二），植物学通报7（2）：1-17，7（3）：1-18。

1990, Wen-Tsai WANG, An introduction to four important current systems of classification of the angiosperms (1 and 2), *Chinese Bulletin of Botany* 7(2): 1-17, 7(3): 1-18.

1990年，郑儒永（真菌英文名词）、魏江春（地衣）、胡鸿钧（藻类）、余永年（真菌拉丁学名）、吴鹏程（苔藓）、邢公侠（蕨类）、刘波（真菌英文俗名）编，**孢子植物名词及名称**，961页；科学出版社，北京。

1990, Ju-Yung CHENG (fungi English names), Jiang-Chun WEI (lichens), Hong-Jun HU (algae), Yung-Nien YU (fungi Latin names), Pan-Cheng WU (bryophytes), Kung-Hsia SHING (ferns) and Bo LIU (fungi English names), ***A Glossary of Terms and Names of Cryptogamia***, 961 p; Science Press, Beijing.

1990.10+，**中国蕨协简报** 1：12，1990；2：28，1991；3：12，1994；4：10，1995；**中国花卉协会蕨类植物分会简讯** 5：24，2002；6：28，2002；7：20，2003；8：22，2003；9：28，2004；10：28，2004；11：41，2005；12：32，2006；13：24，2008；14：36，2009；15：36，2009；16：33，2010。***Sinopteris* 中国蕨**——中国花卉协会蕨类植物分会、中国野生植物保护协会蕨类植物保育委员会通讯 17：40，2011；18：44，2012；19：38，2013；20：38，2014；21：36，2015；22：64，2016。

1990.10+, ***Newsletter of the Chinese Fern Society***,[702] No. 1: 12, 1990; 2: 28, 1991; 3: 12, 1994; 4: 10, 1995; 5: 24, 2002; 6: 28, 2002; 7: 20, 2003; 8: 22, 2003; 9: 28, 2004; 10: 28, 2004; 11: 41, 2005; ***Newsletter of Fern Committee of China Flower Association*** 12: 32, 2006; 13: 24, 2008; 14: 36, 2009; 15: 36, 2009; 16: 33, 2010; ***Sinopteris*** 17: 40, 2011; 18: 44, 2012; 19: 38, 2013, 20: 38, 2014; 21: 36, 2015; 22: 64, 2016.

1990年10月29日至1991年6月15日，中国科学院昆明植物研究所独龙江考察队赴贡山县独龙江流域越冬考察，黄锦岭、李恒（领队）、杨建昆，采集标本7 075号，1万多份（CAS、KUN）。

1990.10.29-1991.06.15, Dulongjiang Expedition from Kunming Institute of Botany, Chinese Academy

701 This book provides an account of how Asian plants, including many from China, were introduced and spread around the world.

702 English title from No. 6, 2002.

of Sciences, expedition overwinters in Dulongjiang (N'Mai River) valley. Jin-Ling HUANG, Heng LI (leader) and Jian-Kun YANG, collected 7,075 numbers and more than 10,000 plant specimens (CAS, KUN).

1990+, ***Acta Bryolichenologica Asiatica***, Studies on bryophytes of Southeast Asia, vol. 1(1-2), 1990, 2(1-2), 1990, 3, 2010, 4, 2011, 5, 2014, 6, 2017, 7, 2017, established by Ming-Jou LAI, Taiwan, through the Asian Bryological and Lichenological Club, published by Finnish-Chinese Botanical Foundation, Lohja, Finland.

1990+，**亚洲苔藓与地衣学报**，vol. 1（1-2），1990，2（1-2），1990，3，2010，4，2011，5，2014，6，2017，7，2017；台湾赖明洲通过亚洲苔藓与地衣俱乐部建立，位于芬兰洛赫亚的芬中植物基金会出版发行。[703]

1990、1992、1993、1996、1999、2001、2003、2006 年，马金双（Ⅰ-Ⅲ）、刘全儒、马金双（Ⅳ-Ⅷ），我国地方植物志出版情况简介（Ⅰ-Ⅷ），广西植物 10（3）：268-269，1990；12（2）：190，1992；13（2）：192，1993；16（4）：338，1996；19（4）：308，1999；21（4）：381-382，2001；23（1）：48，2003；26（1）：13，2006。

1990, 1992, 1993, 1996, 1999, 2001, 2003, 2006, Jin-Shuang MA (I-Ⅲ), Quan-Ru LIU and Jin-Shuang MA (Ⅳ-VIII), Introduction to local floras of China (I-VIII), *Guihaia* 10(3): 268-269, 1990; 12(2): 190, 1992; 13(2): 192, 1993; 16(4): 338, 1996; 19(4): 308, 1999; 21(4): 381-382, 2001; 23(1): 48, 2003; 26(1): 13, 2006.

1990 and 1996, Christopher D. K. Cook, ***Aquatic Plant Book***, 228 p; SPB Academic Publishing, the Hague; ***Aquatic Plant Book***, Ed. 2, 228 p, 1996; SPB Academic Publishing, Amsterdam. Translated into Chinese in 1993.

1993 年，王徽勤、游浚、王建波译，陈家宽、郭友好校；**世界水生植物**，306 页；武汉大学出版社，武汉。

703 赖明洲逝世前将版权移给芬中基金会。

1991

1991.05-09 and 1995.04-07, Chi-Ming HU, South China Institute of Botany, Chinese Academy of Sciences, studied Primulaceae and Myrsinaceae for ***Flore du Cambodge, du Laos et du Vietnam*** at Museum National d'Histoire Naturelle in Paris, France[704].

1991 年 5 月至 9 月，1995 年 4 月至 7 月，中国科学院华南植物研究所胡启明赴法国巴黎自然历史博物馆显花植物部，参加**柬埔寨、老挝和越南植物志**报春花科和紫金牛科的编研。

1991.07.29, Agreement to cooperate on *Bryoflora of China*, English edition, signed by Chinese and foreign bryologists, five scientists each from within and outside China: Marshall R. Crosby, Missouri Botanical Garden, St. Louis, the United States of America, Chien GAO, Institute of Applied Ecology, Chinese Academy of Sciences, Shenyang, Ren-Liang HU, East China Normal University, Shanghai, Zennoske Iwatsuki, Hiroshima University, Hiroshima, Japan, Timo J. Koponen, University of Helsinki, Helsinki, Finland, Shin-Chiang LEE, Kunming Institute of Botany, Chinese Academy of Sciences, Kunming, Zhi-Hua LI, Sun Yat-Sen University, Guangzhou, Benito C. TAN, Farlow Herbarium of Harvard University, Cambridge, the United States of America, Dale H. Vitt (1944-), University of Alberta, Edmonton, Canada, Pan-Cheng WU, Institute of Botany, Chinese Academy of Sciences, Beijing, China. [705]

1991 年 7 月 29 日，中外苔藓学家签署英文版中国苔藓植物志合作协议。中方代表 5 名：中国科学院沈阳应用生态研究所高谦、上海华东师范大学胡人亮、中国科学院昆明植物研究所黎兴江、中山大学李植华、中国科学院植物研究所吴鹏程；外方代表 5 名：芬兰赫尔辛基大学 Timo J. Koponen、日本广岛大学岩月善之助、加拿大阿尔伯塔大学 Dale H. Vitt（1944— ）、美国密苏里植物植物园 Marshall R. Crosby 和美国哈佛大学隐花植物标本室 Benito C. TAN。

1991.09.03-09.22, Sino-Finland bryological expedition to Sichuan: Hongyuan, Jiuzhaigou, Songpan. Participants Timo J. Koponen (H), Jian-Xin LUO and Pan-Cheng WU (PE) collected 2,000 numbers of specimens (H, PE).

1991 年 9 月 3 日至 22 日，中芬苔藓联合考察四川的红原、九寨沟和松潘，芬兰赫尔辛基大学 Timo J. Koponen，中国科学院植物研究所罗健馨、吴鹏程采集标本 2 000 号（H、PE）。

704 Chi-Ming HU, 1992, *Flore du Cambodge, du Laos et du Vietnam*, 26: 115-144, Primulaceae; Museum National d'Histoire Naturelle, Paris; Chi-Ming HU and Jules E. Vidal, 2004, *Flore du Cambodge, du Laos et du Vietnam*, 32: 1-228, Myrsinaceae; Museum National d'Histoire Naturelle, Paris.

705 Benito C. TAN and Pan-Cheng WU, 1991, The Bryoflora of China project (English edition), *The Bryological Times* 64: 9-10.

1991.10.10-11.02, North American China Plant Expedition Consortium initial feasibility expedition to Beijing, Heilongjiang, Jiangsu, Jilin and Shaanxi. Participants: Peter Bristol, Holden Arboretum, Kirtland, Ohio, Lawrence Lee, US National Arboretum, Washington, D.C., Paul Meyer, Morris Arboretum, University of Pennsylvania, Philadelphia, Pennsylvania.

1991 年 10 月 10 日至 11 月 2 日，由美国俄亥俄州 Kirtland 市霍顿树木园 Peter Bristol、华盛顿特区国家树木园 Lawrence Lee 和宾夕法尼亚州费城宾夕法尼亚大学莫里斯树木园 Paul Meyer 组成的北美中国植物考察联盟首次到北京、黑龙江、吉林、陕西和江苏等地采集。

1991 年 10 月 19 日至 24 日，第二届系统与进化植物学青年研讨会在湖北武汉中国科学院武汉植物研究所召开，全国近百人出席。[706]

1991.10.19-10.24, Second Youth Symposium of Systematic and Evolutionary Botany held at Wuhan Institute of Botany, Chinese Academy of Sciences, Wuhan, Hubei; nearly 100 attendees.

1991.10.25-11.20, Chinese-American bryological expedition to northern Guizhou: Chishui, Fanjing Shan, Huangguoshu, Kuankuoshui and Leigong Shan. Marshall R. Crosby (MO), Xiao-Lan HE (PE), Shu-Qiao HUANG (Guizhou Bureau of Forestry), Qi-Wei LIN (Guizhou Bureau of Environmental Protection), Benito C. TAN (FH) and Pan-Cheng WU (PE) collected more than 1,500 moss specimens (FH, MO, PE and a local herbarium in Guizhou).[707]

1991 年 10 月 25 日至 11 月 20 日，中美苔藓联合考察贵州北部的赤水、梵净山、黄果树、宽阔水和雷公山，美国密苏里植物园 Marshall R. Crosby，哈佛大学隐花植物标本室 Benito C. TAN，中国科学院植物研究所何小兰、吴鹏程，贵州省林业局黄署桥与环保局林齐维，采集 1 500 余号标本（FH、MO、PE 和贵州当地一个标本馆[708]）。

1991 年 11 月，中国科学院植物研究所研究员洪德元[709]当选为中国科学院学部委员；2001 年当选发展中国家科学院院士。

1991.11, De-Yuan HONG, Curator, Institute of Botany, Chinese Academy of Sciences, elected academician of Chinese Academy of Sciences, and academician of The World Academy Sciences for the advancement of science in developing countries in 2001.

1991, John Illingworth and Jane Routh, ***Reginad Farrer—Dalesman, planthunter, gardener***, 102 p;

706 会讯，1992，长江后浪推前浪，植物分类学报，30（2）：192。

707 Benito C. TAN, Qi-Wei LIN, Marshall R. Crosby and Pan-Cheng WU, 1994, A report on the 1991 Sino-American bryological expedition to Guizhou province, China: New and noteworthy additions of Chinese moss taxa, *The Bryologist* 97(2): 127-137.

708 未在国际植物标本馆数据库中注册。

709 安徽绩溪人，1962 年毕业于复旦大学，考入中国科学院植物研究所研究生，1966 年毕业留所工作至今。

Occasional paper No 19, Centre for North-West Regional Studies, University of Lancaster, Lancaster.

1991 年，John Illingworth、Jane Routh，**Reginad Farrer—— 谷地人、植物猎人、园艺师**，102 页；Occasional paper No 19, Centre for North-West Regional Studies, University of Lancaster, Lancaster。

1991 年，刘昭民编著，刘棠瑞订正，**中华生物学史**，496 页；台湾商务印书馆，台北。

1991, Chao-Min LIU, Tang-Shui LIU (proofreader), '***Chinese History of Biology***', 496 p; The Commercial Press, Ltd., Taipei [Taibei].

1991 年，马金双，我国植物标本馆代号介绍，广西植物 11（3）：283-285。

1991, Jin-Shuang MA, Introduction to abbreviations of Chinese herbaria, *Guihaia* 11(3): 283-285.

1991 年，蔡希陶纪念文集编辑委员会，**蔡希陶纪念文集**，201 页；云南科技出版社，昆明。

1991, Editorial Committee of Memorial Works of Hse-Tao TSAI, '***Memorial Works of Hse-Tao TSAI***', 201 p; Yunnan Science and Technology Press, Kunming.

1991 年，王雨宁著，**绿之魂** —— 中国著名植物学家蔡希陶，101 页；科学普及出版社，北京。

1991, Yu-Ning WANG, '***Lü Zhi Hun***—*Famous Chinese Botanist Hse-Tao TSAI*', 101 p; Popular Science Press, Beijing.

1991 年，吴兆洪、秦仁昌著，**中国蕨类植物科属志**，630 页；科学出版社，北京。

1991, Shiew-Hung WU and Ren-Chang CHING, ***Fern Families and Genera of China***, ***Familiae Generaque Pteridophytorum Sinicorum***, 630 p; Science Press, Beijing.

1991 年，中国科学院新疆生物土壤沙漠研究所，**新疆植物学研究文集**，230 页；科学出版社，北京。

1991, Xinjiang Institute of Biology, Pedology and Desert, Chinese Academy of Sciences, ***The Collection of Botanical Papers in Xinjiang,*** 230 p; Science Press, Beijing.

1991—2017，广西科学院广西植物研究所、广西壮族自治区中国科学院广西植物研究所编著，李树刚主编，**广西植物志**，6 卷本；第 1 卷：976 页，1991；第 2 卷：947 页，2005；第 3 卷：1 024 页，2011；第 4 卷：1 082 页，2017；第 5 卷：1 073 页；2016；第 6 卷：474 页，蕨类植物门，2013；广西科学技术出版社，南宁。

1991-2017, Guangxi Institute of Botany, Academia Guangxiana, Guangxi Institute of Botany, Guangxi Zhuang Autonomous Region and the Chinese Academy of Sciences, Shu-Kang LEE (editor in chief), ***Flora of Guangxi***, vols. 1-6; 1: 976, p, 1991; 2: 947 p, 2005; 3: 1,024 p, 2011; 4: 1,082p, 2017; 5: 1,073 p, 2016; 6:

474 p, Pteridophyta, 2013; Guangxi Science and Technology Publishing House, Nanning.

1991、1993 年，吴征镒，中国种子植物属的分布区类型，云南植物研究 增刊 IV，1–139 页，1991；“中国种子植物属的分布区类型”的增订和勘误，141–178 页，1993。

1991 and 1993, Zheng-Yi WU[710], Areal-types of Chinese genera of seed plants, *Acta Botanica Yunnanica* Supplement IV, 1-139 p, 1991; Revised and Errata to 'Areal-types of Chinese genera of seed plants', 141-178 p, 1993.

1991 年 5 月至 1999 年 9 月，西北植物研究所更名为陕西省中国科学院西北植物研究所并实行双重领导；1991—1993 年白守信、1993—1999 年李璋（1938—）任所长。

1991.05-1999.09, Northwest Institute of Botany renamed Northwest Institute of Botany, Shaanxi Province and Chinese Academy of Sciences, with dual leadership under Shaanxi Province and Chinese Academy of Sciences; with directors Shou-Xin BAI, 1991-1993, and Zhang LI (1938-), 1993-1999.

710 Pinyin spelling of Cheng-Yih WU.

1992

1992.06.29-06.30, Joint editorial committee meeting of Flora of China held at Kunming Institute of Botany, Chinese Academy of Sciences, Kunming, China. Attendees: Ihsan A. Al-Shehbaz (MO), Bo-Jian BAO (1959-, PE), Bruce M. Bartholomew (CAS), David E. Boufford (A/GH), Paul Pui-Hay BUT (CUHK), Chia-Jui CHEN (PE), Shou-Liang CHEN (NAS), Sing-Chi CHEN (PE), Shu-Kun CHEN (KUN), Hong-Bin CUI (PE), Lun-Kai DAI (PE), Michael G. Gilbert (1943-, BM/MO), Hong-Ya GU (PEY), Ching-Chiu HUANG (IBSC), Xiang-Sheng JI (KUN), De-Zhu LI (KUN), Hsi-Wen LI (KUN), Nancy R. Morin (MO), Jin Murata (1952-, TI), Peter H. Raven (MO), Orbelia R. Robinson (CAS), Hong-Bin SHI (1963-, KUN), Laurence E. Skog (US), William TAI (MO), Cheng-Yih WU (KUN) and Shu-Ren ZHANG (1962-, PE).

1992 年 6 月 29 日至 30 日，英文版中国植物志编辑委员会联席会议在中国科学院昆明植物研究所举行，中国科学院植物研究所包伯坚（1959—）、陈家瑞、陈心启、崔鸿宾、戴伦凯、张树仁（1962—），加州科学院 Bruce M. Bartholomew、Orbelia R. Robinson，哈佛大学 David E. Boufford，香港中文大学毕培曦，江苏省植物研究所 南京中山植物园陈守良，中国科学院昆明植物研究所陈书坤、姬翔生、李德铢、李锡文、师红斌（1963—）和吴征镒，密苏里植物园 Ihsan A. Al-Shehbaz，Michael G. Gilbert（1943—）、Nancy R. Morin、Peter H. Raven、戴威廉，北京大学顾红雅，中国科学院华南植物研究所黄成就，东京大学邑田仁（1952—），史密森学会的 Laurence E. Skog 出席。

1992.07.04, Chinese-American bryologists signed agreement on *Bryoflora of China.* Representatives included Peter H. Raven, director of Missouri Botanical Garden and Donald H. Pfister (1945-), director, Harvard University Herbaria, the United States of America and Cheng-Kui TSENG, Editor in Chief, and Jiang-Chun WEI, Vice Editor in Chief of the editorial committee of the Cryptogamic Flora of China. Joint editorial committee composed of five representatives each from China and the United States of America, with Pan-Cheng WU (PE) and Marshall R. Crosby (MO) as co-presidents.

1992 年 7 月 4 日，中美苔藓植物学家签署合作英文版中国苔藓植物志协议，美方代表为密苏里植物园园长 Peter H. Raven 和哈佛大学植物标本馆馆长 Donald H. Pfister（1945—），中方代表为中国孢子植物志主编曾呈奎和副主编魏江春。联合编辑委员会由中美双方各 5 人组成，中美双方的主席分别为中国科学院植物研究所吴鹏程和密苏里植物园 Marshall R. Crosby。

1992 年 8 月，周呈维编，**高等植物常用名词术语解释**，138 页；重庆大学出版社，重庆。

1992.08, Cheng-Wei ZHOU, '***Explanation of Common Terminology of Higher Plants***', 138 p; Chongqing University Press, Chongqing.

1992 年 8 月 26 日至 28 日，第一届中国苔藓植物学会议在北京中国科学院植物研究所举行，来自国内外24家单位的51位代表出席；会议决定出版隐花植物生物学以纪念陈邦杰对苔藓学研究的贡献。

1992.08.26-08.28, First Chinese Bryology Symposium held at Institute of Botany, Chinese Academy of Sciences, Beijing; 51 attendees from 24 national and international institutions. *CHENIA, a journal for bryology and other spore producing organisms*, was initiated to commemorate Professor Pan-Chieh CHEN's lifetime of contributions to the study of bryology.

1992年10月，全国植物标本馆会议在北京中国科学院植物研究所植物标本馆举行，全国48个单位共63位代表与会；会议决定编写中国植物标本馆索引、中国植物采集史和中国植物分类学历史。遗憾的是只有中国植物标本馆索引[711]和云南省的部分采集历史出版。[712]

1992.10, Symposium on national herbaria of China held at the herbarium, Institute of Botany, Chinese Academy of Sciences, Beijing; 63 attendees from 48 institutions nationwide. It was decided that a national herbarium index, collection history and taxonomic history should be compiled. Regrettably, only *Index Herbariorum Sinicorum*[713] and part of the history of collecting in Yunnan[714] have been published.

1992年，傅德志，裸子植物一新科 —— 竹柏科，植物分类学报30（6）：515-528。

1992, De-Zhi FU, Nageiaceae—A new gymnosperm family, *Acta Phytotaxonomica Sinica* 30(6): 515-528.

1992、1996年，傅立国主编，**中国植物红皮书 —— 稀有濒危植物**，第1册，736页，1992；科学出版社，北京；傅立国主编，**中国稀有濒危植物**，中文繁体版，1：253页，2：253页，3：257页，1996；淑馨出版社，台北。

1992, 1996, Li-Kuo FU (editor in chief), ***China Plant Red Data Book****—Rare and Endangered Plants*, 1: 736 p, 1992; Science Press, Beijing; Li-Kuo FU (editor in chief), '***Chinese Rare and Endangered Plants***', Traditional Chinese Edition, 1: 253 p, 2: 253 p, 3: 257 p, 1996; Hsuhsin Press, Taipei [Taibei].

1992年，马金双，九十年代植物志的编研动态与展望（一、二），生物学通报11：2-4，12：4-5。

1992, Jin-Shuang MA, 'Compiled trend and prospect of the floras in 1990s (I and II)'. *Bulletin of Biology* 11: 2-4 and 12: 4-5.

711 详细参见：马金双，2011，东亚高等植物分类学文献概览的附录3，《中国植物标本馆索引》编后记（附首届全国植物标本馆研讨会参加人员名单），324-325页；高等教育出版社，北京。

712 包士英、毛品一、苑淑秀著，1998，云南植物采集史略1919—1950，中英文双语版，214页；中国科学技术出版社，北京。

713 Li-Kuo FU (editor in chief), 1993, *Index Herbariorum Sinicorum*, bilingual edition (English and Chinese), 458 p; China Science and Technology Press, Beijing.

714 Shih-Ying BAO, Pin-Yi MAO and Shu-Xiu YUAN, 1998, *A Brief History of Plant Collection in Yunnan 1919-1950*, bilingual edition (English and Chinese), 214 p; China Science and Technology Press, Beijing.

1992，野田光藏[715]，**満州植物誌の思い出 —— 中国動乱の中にて**（**对满洲植物志的回忆 —— 身处中国动乱之中**），278 页；风间书房，东京。

1992, Mitsuzo Noda, '***Memories of Flora of Manchuria—****At Upheaval China*', 278 p; Kazama Shobo, Tokyo.

1992 年，吴鹏程，苔藓植物研究的新进展，植物分类学报 30（2）：183-192。

1992, Pan-Cheng WU, Recent advances in bryological research, *Acta Phytotaxonomica Sinica* 30(2): 183-192.

1992 年，吴兆洪、朱家柟、杨纯瑜编著，**中国现代及化石蕨类植物科属辞典**，200 页；中国科学技术出版社，北京。

1992, Shiew-Hung WU, Jia-Nan ZHU and Chun-Yu YANG, ***A Dictionary of the Extant and Fossil Families and Genera of Chinese Ferns***, 200 p; China Science and Technology Press, Beijing.

1992, Zhao-Ran XU and Ḍan H. Nicolson, Don't abbreviate Chinese names, *Taxon* 41(3): 499-504.

1992 年，许兆然、Dan H. Nicolson，不要缩写中国人名，*Taxon* 41（3）：499-504。

1992 年，于兆英、李学禹、狄维忠主编，**西北地区现代植物分类学研究** 1：151 页；科学技术文献出版社，北京。

1992, Zhao-Ying YU, Xue-Yu LI and Wei-Zhong DI (editors in chief), ***Advance in Plant Taxonomy in Northwest China***, 1: 151 p; Science and Technical Documentation Press, Beijing.

---

715 野田光藏（1909—1995）于 1927 年（昭和 2 年）3 月 8 日从日本乘船经韩国釜山、朝鲜沙里垸抵达中国东北东安（丹东），再转赴奉天（沈阳），入满洲教育专门学校学习；受大贺一郎教授影响开始在东北各地考察并研究陆地植物；1934—1937 年入北海道大学师从山田幸男从事东北空白的藻类研究，1937—1940 年任教于公主岭农业学校，1940—1945 年任职新京第二中学；这期间继续大规模考察与研究中国东北的海洋和陆地植物；1945 年日本投降后继续留在东北，先后任职于国立长春大学农学院、沈阳农学院、东北农学院，同时还参加了 20 世纪 50 年代初期刘慎谔领导的东北植物资源普查，并发表过大量的有关植物学文章（包括参加 1955 年刘慎谔主编的东北木本植物图志的编写）。1953 年（昭和 28 年）7 月 5 日离开沈阳，经塘沽码头离境，由舞鹤登陆返回日本。1954 年任新泻大学理学院助理教授，次年晋升为新泻大学理学院教授。他离开中国时其东北（满洲）植物志手稿在沈阳被中国政府没收；1955 年（昭和 30 年）12 月访问日本的中国科学院院长郭沫若归还其手稿。1956 年（昭和 31 年）以东北（满洲）植物志手稿 11 篇中的内容之一、东北和朝鲜藻类为题，获得母校北海道大学博士学位，同时萌生出版东北（满洲）植物志手稿的想法。经过 15 年的整理，1971 年（昭和 46 年）中国东北（满洲）植物志正式出版，同时送给日本天皇、中国毛泽东主席和郭沫若院长。1974 年（昭和 49 年）退休后继续从事日本海的海藻研究；1982 年（昭和 57 年），野田光藏受当年东北农学院的弟子、黑龙江省自然资源研究所朱有昌（1924—）研究员的邀请到中国访问，特别是走访了东北三省并在黑龙江省自然资源研究所做了 3 天的报告，来自东北多家单位 80 余人参加。详细参见：本刊记者，1982，日本国野田教授到我所讲学，自然资源研究 4：71。

1992年，植物学名词审定委员会，**植物学名词**，191页，1991；科学出版社，北京。

1992, Determination Committee of Names of Botany, '***Names of Botany***', 191 p, 1991; Science Press, Beijing.

1992[716]、1997年，陈汉斌主编（第1卷），陈汉斌、郑亦津、李法曾主编（第2卷），**山东植物志**，2卷本；第1卷：1 210页，1992；第2卷：1 518页，1997；青岛出版社，青岛。[717]

1992 and 1997, Han-Bin CHEN (editors in chief, vol. 1) and Han-Bin CHEN, Yi-Jin ZHENG and Fa-Zeng LI (editors in chief, vol. 2), '***Flora of Shandong***', vols. 1-2; vol. 1: 1,210 p, 1992; 2: 1,518 p, 1997; Qingdao Publishing House, Qingdao.

1992+，江苏省植物研究所/江苏省中国科学院植物研究所、江苏省植物学会、中国环境科学会植园保护分会，**植物资源与环境**，vols. 1+，1992+；季刊，2000年第9卷起中文名称更为**植物资源与环境学报**，英文不变。本刊实为南京中山植物园论文集的续刊；1992—2002年贺善安、2003—2009年夏冰（1960—）、2010年至今庄娱乐（1957—）任主编。

1992+, Jiangsu Institute of Botany / Institute of Botany, Jiangsu Province and Chinese Academy of Sciences, Jiangsu Society of Botany and Branch of Botanical Gardens Conservation, Chinese Society of Environmental Sciences, ***Journal of Plant Resources and Environment***, vols. 1+, quarterly; Editors in Chief: Shan-An HE, 1992-2002, Bing XIA (1960-), 2003-2009, and Yu-Le ZHUANG (1957-) since 2010.

1992年9月至1993年6月，中国科学院昆明植物研究所孙航、周浙昆、俞宏渊（1963—）赴藏东南墨脱等地做越冬考察，采集标本约7 100号2万余份（KUN）。[718]

1992.09-1993.06, Hang SUN, Zhe-Kun ZHOU and Hong-Yuan YU (1963-), Kunming Institute of Botany, Chinese Academy of Sciences, explored Mêdog County, southeastern Xizang over the winter; collected 7,100 numbers and about 20,000 plant specimens (KUN).[719]

1992—2004年，山西植物志编辑委员会编，刘天慰主编（1-3卷），刘天慰、岳建英主编（4-5卷），**山西植物志**，5卷本；第1卷：702页，1992；第2卷：575页，1998；第3卷：655页，2000；第4卷：670页，2004；第5卷：532页，2004；中国科学技术出版社，北京。

1992-2004, Editorial Committee of Shanxi Flora, Tian-Wei LIU (editor in chief, vols. 1-3), Tian-Wei LIU and Jian-Ying YUE (editors in chief, vols. 4-5), ***Flora Shanxiensis***, vols. 1-5; 1: 702 p, 1992, 2: 575 p,

---

716 版本为1990年12月，但是印刷是1992年5月。

717 李法曾、樊首金，2004，山东植物志补遗（一），广西植物24（2）：122-124；李法曾、张学杰，2006，山东植物志补遗（二），广西植物26（6）：581-582。

718 孙航、周浙昆，2002，雅鲁藏布江大峡弯河谷地区种子植物，425页；云南科技出版社，昆明。

719 Hang SUN and Zhe-Kun ZHOU, 2002, *Seed Plants of the Big Bend Gorge of Yalu Tsangpo in SE Tibet, E Himalayas*, 425 p; Yunnan Science and Technology Press, Kunming.

1998, 3: 655 p, 2000, 4: 670 p, 2004, 5: 532 p, 2004; China Science and Technology Press, Beijing.

1993

1993 年 2 月，洪德元，分类学的新进展和我们的对策，生命科学 5（1）：6-9。

1993.02, De-Yuan HONG, 'New progress of taxonomy and our strategy', *Life Science* 5(1): 6-9.

1993.05.19-06.29, Sino-Foreign botanical expedition to northwest Yunnan: Dali, Dêqên [Deqin], Weixi, Xiaguan, Zhongdian; Bjorn Alden (Goteborg, Sweden), J. Crinan M. Alexander (1944-, E), David G. Long (1948-, E), Ronald J. D. McBeath (E), Henry J. Noltie (1957-, E), Mark F. Watson (1964-, E), Kai-Yun GUAN (KUN), Yong FEI (KUN) and Shao-Wen YU (KUN); collected 1,820 numbers of vascular plants (E, KUN) and 1,100 numbers of bryophytes (E, KUN)[720].

1993 年 5 月 19 日至 6 月 29 日，中外植物联合考察滇西北的下关、大理、中甸、德钦和维西，瑞典 Bjorn Alden（Goteborg），爱丁堡植物园 J. Crinan M. Alexander（1944 —）、David G. Long（1948 —）、Ronald J. D. McBeath、Henry J. Noltie（1957 —）、Mark F. Watson（1964 —），中国科学院昆明植物研究所管开云、费勇、俞绍文，采集维管束植物 1 820 号（E、KUN），苔藓植物 1 100 号（E、KUN）。

1993.07-08, Chinese-Foreign botanical expedition to Qinghai: Darley [Dari], Gadê [Gande], Maqên [Maqin]. Bruce M. Bartholomew (CAS), De-Shan DENG (HNWP), Michael G. Gilbert (BM/MO), Ting-Nong HO (1938-2001, leader, HNWP), Jian-Quan LIU (1969-, HNWP), Shang-Wu LIU (HNWP), Xue-Feng LU (1965-2019, HNWP) and Min PENG (1958-, HNWP) collected 1,054 numbers of seed plants (BM, CAS, HNWP).

1993 年 7 月至 8 月，中外植物联合考察青海的达日、甘德、玛沁，美国加州科学院 Bruce M. Bartholomew，中国科学院西北高原生物研究所邓德山、何廷农（1938—2001，领队）、刘建全（1969—）、刘尚武、卢学峰（1965—2019）、彭敏（1958—），英国自然历史博物馆、密苏里植物园 Michael G. Gilbert，采集种子植物标本 1 054 号（BM、CAS、HNWP）。

1993.08, Li-Kuo FU (editor in chief), ***Index Herbariorum Sinicorum***, English and Chinese edition, 458 p; China Science and Technology Press, Beijing.

---

720 Henry J. Noltie, David G. Long, Ronald J. D. McBeath, Mark F. Watson, Bjorn Alden and J. Crinan M. Alexander, 1994, *Report of the Kunming Edinburgh, Goteborg (KEG) Expedition to NW Yunnan* 1993, 89 p; RBGE, Edinburgh.

1993 年 8 月，傅立国主编，**中国植物标本馆索引**[721]，中英文双语版，458 页；中国科学技术出版社，北京。

1993.08.05-08.09, Chinese-American botanical expedition to Inner Mongolia: Xilinggol League, Abagnar Banner. David E. Boufford (A/GH), Shu-Yin SONG (PE), Emily W. Wood (A), Tsun-Shen YING (PE), collected 218 numbers of vascular plants (David E. Boufford et al. nos. 25843-26052, A, CAS, MO, PE, TI).

1993 年 8 月 5 日至 9 日，中美植物联合考察内蒙古锡林郭勒盟阿巴嘎旗，哈佛大学 David E. Boufford 和 Emily W. Wood，中国科学院植物研究所宋澍因和应俊生，采集维管束植物 218 号（A、CAS、MO、PE、TI）。

1993.08.28-09.03, Fifteenth International Botanical Congress held in Tokyo (Yokohama), Japan, 132 attendees from China.[722]

1993 年 8 月 28 日至 9 月 3 日，第十五届国际植物学大会于日本东京（横滨）举行，中国 132 位代表出席。

1993.08.25-09.28, North America-China Plant Expedition Consortium to Heilongjiang. Participants: Kris Bachtell, Morton Arboretum, Lisle, Illinois, Peter Bristol, Holden Arboretum, Kirtland, Ohio, Paul Meyer, Morris Arboretum, University of Pennsylvania, Philadelphia, Pennsylvania, and Shi-Xin GAO, Tie-Shan JIN and Jun LIU, Heilongjiang Academy of Forestry.

1993 年 8 月 25 日至 9 月 28 日，由美国伊利诺伊州 Lisle 市莫顿树木园 Kris Bachtell，俄亥俄州 Kirtland 市霍顿树木园 Peter Bristol 和宾夕法尼亚州费城宾夕法尼亚大学莫里斯树木园 Paul Meyer 组成的北美中国植物考察联盟到黑龙江等地采集；中方合作成员为黑龙江省林业科学院的高士新、金铁山和刘君。

1993.09.04-09.05, Joint editorial committee meeting of Flora of China held at Nikko Botanical Garden, Nikko, Japan. Attendees: Ihsan A. Al-Shehbaz (MO), Bruce M. Bartholomew (CAS), David E. Boufford (A/GH), Paul Pui-Hay BUT (CUHK), Chia-Rui CHEN (PE), Shou-Liang CHEN (NAS), Sing-Chi CHEN (PE), Chang-Hung CHOU (HAST), Marshall Crosby (BM/MO), Hong-Bin CUI (PE), Lun-Kai DAI (PE), Michael G. Gilbert (BM/MO), Hong-Ya GU (PEY), Jian-Guo HAN (Chinese Academy of Sciences), Chi-Ming HU (IBSC), Ren-Liang HU (HSNU), David S. Ingram (E), Zennoske Iwatsuki (HIRO), Shin-Chiang LEE (KUN), Hsi-Wen LI (KUN), Zhi-Hua LI (SYS), Nancy R. Morin (MO), Jin Murata (TI), Peter H. Raven

721 本书是1992年10月全国植物标本馆会议与会的48个单位63位全体代表商讨的计划编写工作之一；详细参见：马金双，2011，东亚高等植物分类学文献概览的附录 3，《中国植物标本馆索引》编后记（附首届全国植物标本馆研讨会参加人员名单），324-325 页；高等教育出版社，北京。

722 XV International Botanical Congress, 1994, Participants of the XV International Botanical Congress 1993, *Proceedings of XV International Botanical Congress*, 84 page, Tokyo.

(MO), Laurence E. Skog (US), William TAI (MO), Pan-Cheng WU (PE), Cheng-Yih WU (KUN) and Zhen-Dai XIA (PE). 1993, Ching-Chiu HUANG (IBSC) left the editorial committee, and Chi-Ming HU (IBSC) joined the editorial committee.

1993年9月4日至5日，英文版中国植物志编辑委员会联席会议在日本日光植物园举行，密苏里植物园 Ihsan A. Al-Shehbaz、Marshall Crosby、Michael G. Gilbert、Nancy R. Morin、Peter H. Raven、戴威廉，加州科学院 Bruce M. Bartholomew，哈佛大学 David E. Boufford，香港中文大学毕培曦，中国科学院植物研究所陈家瑞、陈心启、崔鸿宾、戴伦凯、吴鹏程、夏振岱，江苏省中国科学院植物研究所 南京中山植物园陈守良，"中央"研究院植物研究所周昌弘，北京大学顾红雅，中国科学院韩建国，中国科学院华南植物研究所胡启明，华东师范大学胡人亮，爱丁堡植物园 David S. Ingram，广岛大学岩月善之助，东京大学邑田仁，中国科学院昆明植物研究所黎兴江、李锡文、吴征镒，中山大学李植华，史密森学会的 Laurence E. Skog 等出席。1993年，中国科学院华南植物研究所黄成就离开编辑委员会，胡启明加入编辑委员会。

Joint editorial committee meeting of *Flora of China*, Nikko, Japan, 4-5 September 1993 (Photo provided by Wu Zhengyi Science Foundation)
1993年9月4日至5日，英文版中国植物志编辑委员会联席会议于日本日光（相片提供者：吴征镒科学基金会）

1993月10月13日至16日，中国植物学会第十一届会员代表大会暨六十周年学术年会于北京召开，全国各省（市、区及港、台）500多位代表出席了大会。理事长为张新时。[723]

1993.10.13-10.16, Eleventh National Congress and 60th anniversary meeting of Botanical Society

723 中国植物学会编，1993，中国植物学会六十周年年会学术报告及论文摘要汇编，518页；中国科学技术出版社，北京。

Participants at 60th anniversary meeting of Botanical Society of China, Beijing, 13-16 October 1993 (Photo provided by Botanical Society of China)
1993年10月13日至16日，中国植物学会六十周年年会于北京（相片提供者：中国植物学会）

of China held in Beijing, with more than 500 attendees from every provinces; Hsin-Shi CHANG elected president.

1993 年 11 月，中国科学院植物研究所王文采[724]研究员当选为中国科学院院士[725]。

1993.11, Wen-Tsai WANG, curator, Institute of Botany, Chinese Academy of Sciences, Beijing, elected academician of Chinese Academy of Sciences.

1993 年，曹子余，十年来世界植物标本馆发展概况，广西植物 13（3）：289-293。

1993, Zi-Yu CAO, 'Outline of development of world herbaria in the past ten years', *Guihaia* 13(3): 289-293.

1993, Roy W. Briggs, ***Chinese Wilson****—A Life of Ernest H. Wilson 1876-1930*. 154 p; HMSO[726], London.

1993 年，Roy W. Briggs，**中国的威尔逊** —— 威尔逊的一生 1876—1930，154 页；Her Majesty's Stationary Office，伦敦。

1993 年，陈德懋著，**中国植物分类学史**，356 页；华中师范大学出版社，武汉。

1993, De-Mao CHEN, '***Chinese History of Plant Taxonomy***', 356 p; Central China Normal University Press, Wuhan.

1993, Sing-Chi CHEN, Jiao-Lan LI, Xiang-Yun ZHU and Zhi-Yun ZHANG, ***Bibliography of Chinese Systematic Botany (1949-1990)***, 810 p; Guangdong Science and Technology Press, Guangzhou.

1993 年，陈心启、李娇兰、朱相云、张志耘，**中国植物系统学文献要览（1949—1990）**，810 页；广东科技出版社，广州。

1993, Yue-Ie HSING and Chang-Hung CHOU, ***Recent Advances in Botany***, 350 p; Institute of Botany, Academia Sinica, Taipei [Taibei].

1993 年，邢禹依、周昌弘，**植物学学术研讨会论文集**，350 页；"中央" 研究院植物研究所，台北。

1993, Hsuan KENG, De-Yuan HONG and Chia-Jui CHEN, ***Orders and Families of Seed Plants of***

724 山东掖县人，1949 年毕业于北京师范大学生物系，留系任教半年后经胡先骕介绍于 1950 年 3 月调到中国科学院植物分类研究所工作至今。

725 1955—1993 年 10 月为学部委员，1993 年 11 月改为院士。参见：王扬宗，2014 年 5 月 23 日，学部委员改称院士的曲折过程，中国科学报 第 19 版。

726 Her Majesty's Stationary Office.

Some of the authors of *Flora Reipublicae Popularis Sinicae* at 60th anniversary of Botanical Society of China, Beijing, 13–16 October 1993 (left-right): Jin-Shuang MA, Yu-Chuan MA, You-Hao GUO, Li-Kuo FU, Yue-Jia ZHANG, ?, Kai-Yung LANG, Ting-Nung HO, ?, Jie-Mei XU, Nai-Ran CUI, Kuo-Fang WU, Shang-Wu LIU, Pao-Chun KAO, Shao-Xing CHEN, Shu-Kun CHEN, Ting-Zhi XU, Kuan-Mien SHEN, Yong-Tian CHANG, Sing-Chi CHEN, Cheng-Yih WU, Lang-Rang XU (Photo provided by Wu Zhengyi Science Foundation)

1993 年 10 月 13 日至 16 日，中国植物志部分作者于北京参加中国植物学会 60 周年合影（从左至右）：马金双、马毓泉、郭友好、傅立国、张耀甲、？、郎楷永、何廷农、？、许介眉、崔乃然、吴国芳、刘尚武、高宝纯、陈绍煋、陈书坤、徐廷志、沈观冕、张永田、陈心启、吴征镒、徐郎然（相片提供者：吴征镒科学基金会）

China, 444 p; World Scientific Publishing Co., Pte. Ltd., Singapore.

1993年，耿煊、洪德元、陈家瑞，**中国种子植物目科志**，444页；World Scientific Publishing Co., Pte. Ltd.，新加坡。

1993年，林邦娟、张力，中国苔藓植物新文献Ⅱ，隐花植物生物学1：133-134。

1993, Pang-Juan LIN and Li ZHANG, Current Chinese bryological literature II, *Chenia* 1: 133-134.

1993年，刘胜祥主编，**中国被子植物科的花图式——设计、注释与校正**，197页；湖北科学技术出版社，武汉。

1993, Sheng-Xiang LIU (editor in chief), '***Design, Notes and Correction of Flower Diagrams of Chinese Families of Angiosperms***', 197 p; Hubei Science and Technology Press, Wuhan.

1993年，内蒙古植物志编辑委员会、内蒙古植物学会编，**内蒙古植物检索表**，293页；内蒙古植物志编辑委员会、内蒙古植物学会，呼和浩特（内部印制）。

1993, Compiling Commission of Flora Intramongolica, Botanical Society of Inner Mongolia, ***Key to the Plants of Inner Mongolia***, 293 p; Editorial Committee of Flora of Inner Mongolia, Botanical Society of Inner Mongolia, Hohhot (internal publication).

1993年，吴鹏程，中国苔藓植物研究的回顾与展望，隐花植物生物学1：1-6。

1993, Pan-Cheng WU, A review and the prospect of bryological research in China, *Chenia* 1:1-6.

1993年，旭文、王振淮、晓戈著，**蔡希陶传略**，358页；国际文化出版公司，北京。

1993, Wen XU, Zhen-Huai WANG and Ge XIAO, '***Biography of Hse-Tao TSAI***', 358 p; International Culture Press, Beijing.

1993, Tsun-Shen YING, Yu-Long ZHANG and David E. Boufford, ***The Endemic Genera of Seeds Plants of China***, 824 p; Science Press, Beijing.

1994年，应俊生、张玉龙著[727]，**中国种子植物特有属**，699页；科学出版社，北京。

1993年，张美珍、赖明洲等编著，**华东五省一市植物名录**，491页；上海科学普及出版社，上海。

1993, Mei-Zhen ZHANG, Ming-Jou LAI et al., '***Checklist of Plants From Five Provinces and One City in Eastern China***', 491 p; Shanghai Popular Science Press, Shanghai.

727 中文版和英文版出版不同步，作者也不完全相同。

1993 年，郑重，**湖北植物大全**，677 页；武汉大学出版社，武汉。

1993, Zhong ZHENG, ***Hubei Plants Complete***, 677 p; Wuhan University Press, Wuhan.

1993 年，中国蕨类植物科属的系统排列和历史来源获国家自然科学一等奖，完成人为中国科学院植物研究所秦仁昌。

1993, Chinese Fern Families and Genera: Systematic Arrangement and Historical Origin, finished by Ren-Chang CHING, Institute of Botany, Chinese Academy of Sciences, Beijing, posthumously awarded National Natural Science First Prize.

证书

第六次(1993)国家自然科学奖

项目名称：中国蕨类植物科属系统排列和历史来源
主要研究者：秦仁昌（中国科学院植物研究所）
奖励等级：一等

中华人民共和国
国家科学技术委员会主任 宋健
1994年1月

证书号：Z931011

Ren-Chang CHING posthumously awarded National Natural Science First Prize，1993 (Photo provided by Qi LIN)
1993 年，秦仁昌获国家自然科学一等奖（相片提供者：林祁）

1993+，江西植物志编辑委员会编著，林英主编，**江西植物志**，5 卷本；第 1 卷：541 页，1993；第 2 卷：1 112 页，2004；第 3 卷[728] 第 1 册：410 页，2014，第 3 卷第 2 册：503 页，2014；江西科学技术出版社，南昌，中国科学技术出版社，北京。

1993+, Editorial Committee for Flora of Jiangxi, Ying LIN (editor in chief), ***Flora of Jiangxi***, vols.1-5; 1: 541 p, 1993; 2: 1,112 p, 2004; 3[729](1): 410 p, 2014, 3(2): 503 p, 2014; Jiangxi Science and Technology Press, Nanchang, China Science and Technology Press, Beijing.

1993+，中国科学院华南植物研究所 / 华南植物园、广东省植物学会，**热带亚热带植物学报**，vols. 1+；季刊，2004 年改为双月刊。本刊 1992—1993 年共试刊 2 期，刊名为**华南植物学报**，1993 年仅出版第 1 卷第 1 期，1994 年始为第 2 卷；1992—2002 年吴德邻、2002—2007 年彭少麟、2007 年至今黄宏文任主编。

1993+, South China Institute of Botany / South China Botanical Garden, Chinese Academy of Sciences, and Botanical Society of Guangdong, Guangzhou, ***Journal of Tropical and Subtropical Botany***, vol. 1+; quarterly 1994, bimonthly 2004; two trial issues in 1992 and in 1993 named ***Botanical Journal of South***

728 主编：赖书绅。

729 Editor in Chief, Shu-Shen LAI.

China. Only one issue of volume one published in 1993, and volume two from 1994; Editor in Chief: Te-Lin WU, 1992-2002, Shao-Lin PENG, 2002-2007, Hong-Wen HUANG, since 2007.

1993+，中国植物学会，**隐花植物生物学**，1+[730]，1993，2，1994，3-4，1997，5，1998，6，1999，7，2000，8，2005，9，2007，10，2011，11，2013，12，2016，13，2018；中文版，1998 年第 5 卷始为英文版。

1993+, Botanical Society of China, ***Chenia***, *Contributions to Cryptogamic Biology*[731], Chinese Botanical Society[732]; 1+, 1993; 2, 1994; 3-4, 1997; 5, 1998; 6, 1999; 7, 2000; 8, 2005; 9, 2007; 10, 2011; 11, 2013; 12, 2016; 13, 2018; Chinese, English edition since volume 5, in 1998.

1993+，中国科学院生物多样性委员会、中国植物学会、中国科学院植物研究所、动物研究所和微生物研究所，**生物多样性**，1+，1993，季刊，2003 年改为双月刊；1993—1998 年钱迎倩、1999—2003 年韩兴国、2004—2008 年汪小全、2009 年至今马克平任主编。

1993+, The Biodiversity Committee of the Chinese Academy of Sciences, Botanical Society of China, Institute of Botany, Institute of Zoology and Institute of Microbiology, Chinese Academy of Sciences, ***Biodiversity Science*** [733], vol. 1+, 1993, quarterly, and bimonthly since 2003; Editors in Chief: Ying-Chien CHIEN, 1993-1998, Xing-Guo HAN, 1999-2003, Xiao-Quan WANG, 2004-2008, and Ke-Ping MA, since 2009.

1993、1994 年，中国科学院青藏高原综合科学考察队编，王文采主编，**横断山区维管植物**，上册：1-1 364 页，1993；下册：1 365-2 608 页，1 994；科学出版社，北京。

1993 and 1994, The Comprehensive Scientific Expedition to the Qinghai-Xizang Plateau, Academia Sinica, Wen-Tsai WANG (editor in chief), ***Vascular Plants of the Hengduan Mountains***, 1: 1-1,364 p, 1993; 2: 1,365-2,608 p, 1994; Science Press, Beijing.

1993—1995 年，中国植物志第六届编辑委员会成立，主编：中国科学院昆明植物研究所吴征镒，副主编：中国科学院植物研究所崔鸿宾；委员 23 人：中国科学院昆明植物研究所陈书坤、李锡文，江苏省中国科学院植物研究所 南京中山植物园陈守良，中国科学院华南植物研究所陈德昭、胡启明、林有润，中国科学院植物研究所陈艺林、陈心启、傅国勋（1932—）、戴伦凯、傅立国、李安仁、夏振岱，中国科学院西北高原生物研究所郭本兆、刘尚武，东北林业大学黄普华，复旦大学胡嘉琪，中国科学院成都生物研究所孔宪需，广西植物研究所李树刚，福建师范大学林来官，浙江自然博物馆韦直，

730 编辑 Benito C. TAN 的中文名为贝尼托 陈。

731 The title for the first issue is *Contributions to the Cryptogamic Biology*.

732 i.e. Botanical Society of China.

733 First issue as *Chinese Biodiversity*, 1993.

中国科学院西北植物研究所徐朗然，科学出版社曾建飞。

1993-1995, Sixth Editorial Committee of Flora Reipublicae Popularis Sinicae established, Cheng-Yih WU (KUN) as Editor in Chief and Hung-Pin TSUI (PE) as Vice Editor in Chief, 23 editorial members: Shu-Kun CHEN (KUN), Shou-Liang CHEN (NAS), Te-Chao CHEN (IBSC), Yi-Ling CHEN (PE), Sing-Chi CHEN (PE), Guo-Xun FU (1932-, PE), Lun-Kai DAI (PE), Pung-Chao KUO (HNWP), Li-Kuo FU (PE), Pu-Hwa HUANG (NEFI), Chia-Chi HU (FUS), Chi-Ming HU (IBSC), Hsien-Shiu KUNG (CDBI), An-Jen LI (PE), Shu-Kang LEE (IBK), Hsi-Wen LI (KUN), Yuou-Ruen LING (IBSC), Lai-Kuan LING (FNU), Shang-Wu LIU (HNWP), Zhi WEI (ZM), Zhen-Dai XIA (PE), Lang-Rang XU (WUK) and Jian-Fei ZENG (Science Press).

1993—2011 年，新疆植物志编辑委员会编，杨昌友（编辑，第 1 卷），毛祖美（编辑，第 2 卷），沈观冕（主编，第 3 卷），米吉提•胡达拜尔地、潘晓玲（主编，第 4 卷），安争夕（编辑，第 5 卷），崔乃然（编辑，第 6 卷），**新疆植物志**，6 卷本；第 1 卷：337 页，1993；第 2 卷第 1 册：394 页，1994；第 2 卷第 2 册：425 页，1995；第 3 卷：710 页，2011；第 4 卷：573 页，2004；第 5 卷：534 页，1999；第 6 卷：669 页，1996；新疆科技卫生出版社、新疆科学技术出版社，乌鲁木齐。

1993-2011, Commissione Redactorum Florae Xinjiangensis, Chang-You YANG (editor, vol. 1), Zu-Mei MAO (editor, vol. 2), Kuan-Mien SHEN (editor in chief, vol. 3), Mijit Hudaberdi and Xiao-Ling PAN (editors in chief, vol. 4), Zheng-Xi AN (editor, vol. 5), Nai-Ran CUI (editor, vol. 6), ***Flora Xinjiangensis***, vols. 1-6; 1: 337 p, 1993; 2(1): 394 p, 1994; 2(2): 425 p, 1995; 3: 710 p, 2011; 4: 573 p, 2004; 5: 534 p, 1999; 6: 669 p, 1996; Xinjiang Science and Technology and Hygiene Publishing House, Xinjiang Science and Technology Publishing House, Urumqi.

1994

1994.05.19-06.08, Chinese-American botanical expedition to Southwest Henan: Neixiang (Baotianman) David E. Boufford (A/GH), Bao-Dong LIU (Baotianman Nature Reserve), Zheng-Yong WANG (Baotianman Nature Reserve), Cheng-Jun ZHANG (Baotianman Nature Reserve), Tsun-Shen Ying (PE) and Xian-Chun ZHANG (PE, 1964-), collected 484 numbers of vascular plants (David E. Boufford et al. nos. 26055-26539, A, CAS, MO, PE, TI).

1994 年 5 月 19 日至 6 月 8 日，中美植物联合考察河南西南部内乡（宝天曼），哈佛大学 David E. Boufford，宝天曼自然保护区刘宝东、王正勇、张成军，中国科学院植物研究所应俊生和张宪春（1964 —），采集维管束植物 484 号（A、CAS、MO、PE、TI）。

1994 年 6 月 8 日，中国科学院选出首批 14 位外籍院士，世界著名植物学家、密苏里植物园主任 Peter H. Raven 博士位列其中。

1994.06.08, Dr. Peter H. Raven, director of Missouri Botanical Garden, world-famous botanist, appointed one of first 14 foreign academicians of Chinese Academy of Sciences.

1994.07.28-07.29, Joint editorial committee meeting of Flora of China held at Institute of Botany, Jiangsu Province and Chinese Academy of Sciences, Nanjing Botanical Garden Memorial Sun Yat-Sen, Nanjing, China. Attendees: Ihsan A. Al-Shehbaz (MO), Bruce M. Bartholomew (CAS), David E. Boufford (A/GH), Anthony R. Brach (1963-, A/GH), Chia-Jui CHEN (PE), Shou-Liang CHEN (NAS), Sing-Chi CHEN (PE), Chang-Hung CHOU (HAST), Lun-Kai DAI (PE), Michael G. Gilbert (MO), Lin HE (NAS), Shan-An HE (NAS), Chi-Ming HU (IBSC), Hsi-Wen LI (KUN), Peter H. Raven (MO), Orbelia R. Robinson (CAS), Laurence E. Skog (US), William TAI (MO), Mark F. Watson (E), Cheng-Yih WU (KUN) and Da-Bao ZHU (National Natural Science Foundation of China). Lun-Kai DAI (PE) and William TAI (MO) named co-directors of the project; Michael G. Gilbert (MO) named representative in Europe. Editorial assistants: Ihsan A. Al-Shehbaz (MO), Bo-Jian BAO (PE), Anthony R. Brach (A/GH), Bryan E. Dutton (A/GH), Orbelia R. Robinson (CAS) and Shu-Ren ZHANG (PE). 1994, Ching-I PENG (HAST) and Mark F. Watson (E) joined the editorial committee.

1994 年 07 月 28 日至 29 日，英文版中国植物志编辑委员会联席会议在江苏省中国科学院植物研究所 南京中山植物园举行，密苏里植物园 Ihsan A. Al-Shehbaz、Michael G. Gilbert、Peter H. Raven、戴威廉，加州科学院 Bruce M. Bartholomew、Orbelia R. Robinson，哈佛大学 David E. Boufford、Anthony R. Brach（1963— ），中国科学院植物研究所陈家瑞、陈心启、戴伦凯，江苏省中国科学院植物研究所 南京中山植物园陈守良、贺林、贺善安，中国科学院华南植物研究所胡启明，中国科学院昆明植物研究所李锡文、吴征镒，史密森学会的 Laurence E. Skog，爱丁堡植物园 Mark F. Watson，“中央”

Joint editorial committee meeting of *Flora of China*, Nanjing, 28–29 July 1994 (Photo provided by Wu Zhengyi Science Foundation)
1994 年 7 月 28 日至 29 日，英文版中国植物志编辑委员会联席会议于南京（相片提供者：吴征镒科学基金会）

研究院植物研究所周昌弘和国家自然科学基金委员会朱大保出席；中国科学院植物研究所戴伦凯和密苏里植物园戴威廉共同担任项目主任，密苏里植物园 Michael G. Gilbert 为欧洲代表，编辑助理为：密苏里植物园 Ihsan A. Al-Shehbaz、中国科学院植物研究所包伯坚和张树仁、哈佛大学 Anthony R. Brach 和 Bryan E. Dutton、加州科学院 Orbelia R. Robinson。1994 年，"中央" 研究院植物研究所彭镜毅和爱丁堡植物园 Mark F. Watson 加入编辑委员会。

1994.09.13-10.03, North American China Plant Expedition Consortium to Beijing. Participants: Kris Bachtell, Morton Arboretum, Lisle, Illinois, Charles Tubesing, Holden Arboretum, Kirtland, Ohio, Edward Garvey, US National Arboretum, Washington, D.C., Rick Lewandowski, Morris Arboretum, University of Pennsylvania, Philadelphia, Pennsylvania, Ming-Wang LIU, Beijing Botanical Garden, Institute of Botany, Chinese Academy of Sciences, Beijing.

1994 年 9 月 13 日至 10 月 3 日，由美国伊利诺伊州 Lisle 市莫顿树木园 Kris Bachtell、俄亥俄州 Kirtland 市霍顿树木园 Charles Tubesing、华盛顿特区国家树木园 Edward Garvey 和宾夕法尼亚州费城宾夕法尼亚大学莫里斯树木园 Rick Lewandowski 组成的北美中国植物考察联盟到北京采集，中方为中国科学院植物研究所北京植物园刘明旺。

1994.09.06-10.11, North American China Plant Expedition Consortium to Hubei: Wudong Shan. Participants: Peter Del Tredici (1945-), Arnold Arboretum, Harvard University, Jamaica Plain, Massachusetts, R. William Thomas, Longwood Gardens, Kennett Square, Pennsylvania, Kevin Conrad, US National Arboretum, Washington, D.C., Paul Meyer, Morris Arboretum, University of Pennsylvania, Philadelphia, Pennsylvania, Ri-Ming HAO (1959-) and Cai-Liang MAO, Nanjing Botanical Garden Memorial Sun Yat-Sen[734].

1994 年 9 月 6 日至 10 月 11 日，由美国麻州 Jamaica Plain 市哈佛大学阿诺德树木园 Peter Del Tredici（1945—）、宾夕法尼亚州 Kennett Square 市长木植物园 R. William Thomas、华盛顿特区国家树木园 Kevin Conrad 和宾夕法尼亚州费城宾夕法尼亚大学莫里斯树木园 Paul Meyer 组成的北美中国植物考察联盟到湖北武当山采集；中方为南京中山植物园郝日明（1959—）、毛才良。

1994 年 10 月，中国植物学会编，**中国植物学史**，376 页；科学出版社，北京。

1994.10，Botanical Society of China, '***Botanical History of China***', 376 p; Science Press, Beijing.

1994 年 12 月 12 日至 16 日，第三届系统与进化植物学青年研讨会在广州中国科学院华南植物研究所召开，全国 14 个省（市、区）87 人出席。

1994.12.12-12.16, Third Youth Symposium of Systematic and Evolutionary Botany held at South China

---

734 Peter Del Tredici, Paul Meyer, Hao Riming, Mao Cailiang, Kevin Conrad and R. William Thomas, 1995, Plant collecting on Wudang Shan, *Arnoldia* 55(1): 12-20.

Third Youth Symposium of Systematic and Evolutionary Botany, Guangzhou, 12-16 December 1994 (Photo provided by Cheng-Xin FU)
1994 年 12 月 12 日至 16 日，第三届系统与进化植物学青年研讨会于广州（相片提供者：傅承新）

Institute of Botany, Chinese Academy of Sciences, Guangzhou, 87 attendees from 14 provinces.

1994 年，张宏达，再论华夏植物区系的起源，中山大学学报（自然科学版）33（2）：1-9。

1994, Hung-Ta CHANG, A review on the origin of the Cathaysian flora, *Acta Scientiarum Naturalium Universitatis Sunyatseni* 33(2): 1-9.

1994 年，张宏达，地球植物区系分区提纲，中山大学学报（自然科学版）33（3）：73-80。

1994, Hung-Ta CHANG, An outline on the regionalisation of the global flora, *Acta Scientiarum Naturalium Universitatis Sunyatseni* 33(3): 73-80.

1994, Werner R. Greuter, Fred R. Barrie, Herve M. Burdet, William G. Chaloner, Vincent Demoulin, David L. Hawksworth, Peter M. Jørgensen, Dan H. Nicolson, Paul C. Silva, P. Trehane, and John McNeill, ***International Code of Botanical Nomenclature* (Tokyo Code)**, adopted by the Fifteenth International Botanical Congress, Tokyo, August-September 1993. *Regnum Vegetabile* 131: 1-389 p. Translated into Chinese in 2000.

2000 年，黄增泉译，**植物命名指南**，364 页；“行政院”农业委员会，台北。

1994 年，林齐维、吴鹏程，贵州苔藓植物研究（一）——研究简史（1896—1949 年），隐花植物生物学 2：75-88。

1994, Qi-Wei LIN and Pan-Cheng WU, A study of bryophytes of Guizhou (I) — An outline of Guizhou bryology (1896-1949), *Chenia* 2: 75-88.

1994 年，林邦娟、张力，中国苔藓植物新文献 III，隐花植物生物学 2：121-127。

1994, Pang-Juan LIN and Li ZHANG, Current Chinese bryological literature III, *Chenia* 2: 121-127.

1994 年，林邦娟、张力、吴鹏程、李植华，海南岛苔藓植物研究概述[735]，隐花植物生物学 2：47-73。

1994, Pang-Juan LIN, Li ZHANG, Pan-Cheng WU and Zhi-Hua LI, A survey of bryological research activities in Hainan, China[736], *Chenia* 2: 47-73.

1994 年，吴德邻主编，**海南及广东沿海岛屿植物名录**，334 页；科学出版社，北京。

1994, Te-Lin WU (editor in chief), ***A Checklist of Flowering Plants of Islands and Reefs of Hainan***

735 文章的第 5 部分（第 58-73 页）为海南岛苔藓植物名录。

736 Part five of this paper (pages 58-73) is a Checklist of the bryophytes of Hainan Island.

and Guangdong Province, 334 p; Science Press, Beijing.

1994 年，周鸿、吴玉著，**绿色的开拓者 —— 中国著名植物学家吴征镒**，169 页；科学普及出版社，北京。

1994, Hong ZHOU and Yu WU, '***Lü Se De Kai Tuo Zhe***—*Famous Chinese Botanist Cheng-Yih WU*', 169 p; Popular Science Press, Beijing.

1994 年，朱长山、杨好伟编，**河南种子植物检索表**，560 页；兰州大学出版社，兰州。

1994, Chang-Shan ZHU and Hao-Wei YANG, ***Claves Familiarum Generum Specierumque Spermatophytorum Henanensis***, 560 p; Lanzhou University Press, Lanzhou.

1994 年，朱石麟、马乃训、傅懋毅编，**中国竹类植物图志**，244 页；中国林业出版社，北京。

1994, Shi-Lin ZHU, Nai-Xun MA and Mao-YI FU, '***Illustration of Chinese Bamboo***', 244 p; China Forestry Publishing House, Beijing.

1994 and 2001, James G. Harris and Melinda W. Harris, ***Plant Identification Terminology—An Illustrated Glossary***, 197 p, 1994; ***Plant Identification Terminology—An Illustrated Glossary***, ed. 2, 206 p, 2001; Spring Lake Publishing, Spring Lake, Payson, Utah. Translated into Chinese in 2001.

2001 年，王宇飞等译，王文采审校，**图解植物学词典**，302 页，图 1-1 927；科学出版社，北京。[737]

1994、1996 年，姚荣鼐编，**台湾维管束植物植种名录**，191 页，1994；台湾大学农学院试验林管理处，台北；姚荣鼐编，**台湾维管束植物植种名录**，272 页，1996；台湾大学农学院实验林管理处，台湾南投。

1994, 1996, Nathan Yung-Nai YAO, ***Species List of Vascular Plants of Taiwan***, 191 p, 1994; The Experimental Management of Agriculture College, Taiwan University, Taipei; Nathan Yung-Nai YAO, ***Species list of Vascular Plants of Taiwan***, 272 p, 1996; The Experimental Management of Agriculture College, Taiwan University, Nantou, Taiwan.

1994 年，江苏省植物研究所 南京中山植物园归江苏省和中国科学院双重领导并更名为江苏省中国科学院植物研究所 南京中山植物园，1994—1998 年贺善安、1998—2008 年夏冰、2008—2017 年庄娱乐、2017 年至今薛建辉（1962—）任主任。

1994, Jiangsu Institute of Botany, Nanjing Botanical Garden Memorial Sun Yat-Sen under dual leadership of Jiangsu Province and Chinese Academy of Sciences, renamed Institute of Botany, Jiangsu Province and Chinese Academy of Sciences, Nanjing Botanical Garden Memorial Sun Yat-Sen; with directors Shan-An HE, 1994-1998, Bing XIA, 1998-2008, Yu-Le ZHUANG, 2008-2017, Jian-Hui XUE

737 译自第 2 版。

Institute of Botany, Jiangsu Province and Chinese Academy of Sciences, Nanjing Botanical Garden Memor al Sun Yat-Sen: 1. 80th anniversariys (1929-2009); 2. Herbarium (NAS); 3. Information Building; 4. Public Education Buiding (Photo provided by Mei LI)

江苏省中国科学院植物研究所 南京中山植物园：1. 八十周年纪念（1929—2009）；2. 植物标本馆（NAS）；3. 信息大楼；4. 科普教育楼（相片提供者：李梅）

(1962-), since 2017.

1994+，中国科学院中国孢子植物志编辑委员会编，**中国苔藓志**，12 卷本：第 1 卷：368 页，1994；第 2 卷：293 页，1996；第 3 卷：157 页，2000；第 4 卷：263 页，2006；第 5 卷：493 页，2011；第 6 卷：290 页，2002；第 7 卷：288 页，2005；第 8 卷：482 页，2004；第 9 卷：323 页，2003；第 10 卷：464 页，2008；第 11、12 卷：XXX；科学出版社，北京。[738]

1994+, Consilio Florarum Cryptogamarum Sinicarum Academiae Sinicae Edita, ***Flora Bryophytorum Sinicorum***, vols. 1-12; 1: 368 p, 1994; 2: 293 p, 1996; 3: 157 p, 2000; 4: 263, 2006; 5: 493, 2011; 6: 290 p, 2002; 7: 288 p, 2005; 8: 482 p, 2004; 9: 323 p, 2003; 10: 464 p, 2008; 11 and 12: XXX; Science Press, Beijing.[739]

1994、1999、2007 年，靳淑英编，**中国高等植物模式标本汇编**[740]，716 页，1994；靳淑英编，**中国高等植物模式标本汇编（补编）**，264 页，1999；靳淑英编，**中国高等植物模式标本汇编（补编二）**，205 页，2007；科学出版社、中国林业出版社，北京。[741]

1994, 1999, 2007, Shu-Ying JIN, ***A Catalogue of Type Specimens (Cormophyta) in the Herbaria of China***, 716 p, 1994; ***A Catalogue of Type Specimens (Cormophyta) in the Herbaria of China (the Supplement)***, 264 p, 1999; ***A Catalogue of Type Specimens (Cormophyta) in the Herbaria of China (the Supplement II)***, 205 p, 2007; Science Press, China Forestry Publishing House, Beijing.

1994-2013 and 1998-2013, Zheng-Yi WU and Peter H. Raven (eds., 1994-2001) and Zheng-Yi WU, Peter H. Raven and De-Yuan HONG (eds., 2001-2013)[742], ***Flora of China***, vols. 1-25, 1994-2013, ***Flora of China Illustrations***, 2-25[743], 1998-2013; Science Press, Beijing and Missouri Botanical Garden Press, St. Louis, Missouri.

1994—2013 年、1998—2013 年，吴征镒、Peter H. Raven（1994—2001，主编），吴征镒、Peter H.

738 吴鹏程、贾渝、王幼芳，2008，《中国苔藓志》藓类系统简介，中国植物学会七十五周年年会论文摘要汇编（1933—2008），93-94 页；兰州大学出版社，兰州。

739 Ning-Ning YU, Qing-Hua WANG, Yu JIA, Mei-Zhi WANG and Peng-Cheng WU, 2013, Introduction of the moss system of *Flora Bryophytorum Sinicorum* vols. 1-8, *Chenia* 11: 252-256.

740 靳叔英，1988，中国高等植物新分类群文献及模式标本汇编（一），植物研究（增刊）：117，曾报道过类似 1994 年的内容，但所包含的具体数字不同。

741 该工作涉及一些物种名称的合格发表问题，详细参见：Wen-Bin YU, Hong WANG and De-Zhu LI, 2011, Names of Chinese seed plants validly published in *A Catalogue of Type Specimens (Cormophyta) in the Herbaria of China and it's two supplements*, *Taxon* 60(4): 1168-1172。

742 De-Yuan HONG succeeds Zheng-Yi WU as vice co-chair of Flora of China project in 2001. For details, see *History of the Flora of China*, *Flora of China*, volume 1, 1-11 p, 2013.

743 There was no part one of the illustrations.

Editorial committee meeting of *Flora of China* project, Beijing, 1994. First volume (vol. 17, including Lamiaceae, Solanaceae and Verbenaceae) published (Photo provided by Wu Zhengyi Science Foundation)

1994 年英文版中国植物志编辑委员会于北京举行首发式，第一本（第 17 卷，包括唇形科、茄科和马鞭草科）出版（相片提供者：吴征镒科学基金会）

Raven、洪德元（2001—2013，主编）[744]，**中国植物志**，1–25 卷，1994—2013；**中国植物志 图册**，2–25 卷[745]，1998—2013；科学出版社，北京，密苏里植物园出版社，圣路易斯，密苏里。

744 2001 年洪德元接替吴征镒以副主编主持英文版中国植物志中方工作。

745 此图册没有第 1 卷。

1995

1995.03.31-04.17, North American China Plant Expedition Consortium to Shaanxi. Participants: Edward Garvey, US National Arboretum, Washington, D.C., Rick Lewandowski, Morris Arboretum, University of Pennsylvania, Philadelphia, Pennsylvania, Tie-Cheng CUI (1956-), Xi'an Botanical Garden.

1995 年 3 月 31 日至 4 月 17 日，由美国华盛顿特区国家树木园 Edward Garvey 和宾夕法尼亚州费城宾夕法尼亚大学莫里斯树木园 Rick Lewandowski 组成的北美中国植物考察联盟到陕西采集，中方为西安植物园崔铁成（1956— ）。

1995.04.08-04.09. Joint editorial committee meeting of Flora of China held at Royal Botanic Garden, Edinburgh, Scotland. Attendees: Ihsan A. Al-Shehbaz (MO), Bruce M. Bartholomew (CAS), David E. Boufford (A/GH), Anthony R. Brach (A/GH), Paul Pui-Hay BUT (CUHK), David F. Chamberlain (E), Chia-

Joint editorial committee meeting of *Flora of China*, Edinburgh, 8-9 April 1995 (Photo provided by Royal Botanic Garden, Edinburgh)
1995 年 4 月 8 日至 9 日，英文版中国植物志编辑委员会联席会议于爱丁堡植物园（相片提供者：爱丁堡植物园）

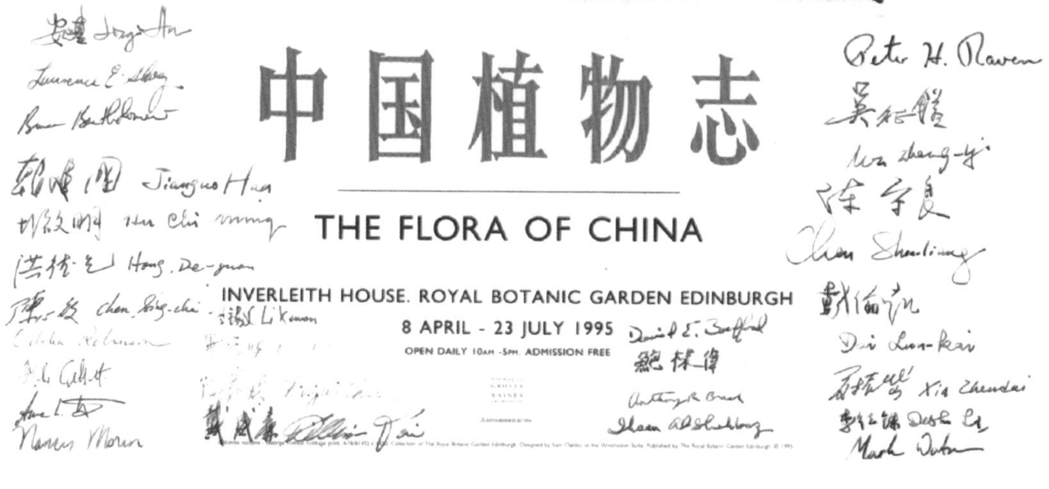

Poster signed by members of Editorial Committee, *Flora of China*, Edinburgh, 8–9 April 1995 (Photo provided by Royal Botanic Garden, Edinburgh)

1995 年 4 月 8 日至 9 日，英文版中国植物志编辑委员会联席会议于爱丁堡的集体签字壁报（相片提供者：爱丁堡植物园）

Jui CHEN (PE), Shou-Liang CHEN (NAS), Sing-Chi CHEN (PE), Derek Clayton (E), Lun-Kai DAI (PE), Michael G. Gilbert (MO), De-Yuan HONG (PE), David S. Ingram (E), Helen Jones (E), De-Zhu LI (KUN), Hsi-Wen LI (KUN), Robert Mill (1950-, E), Nancy R. Morin (MO), Peter H. Raven (MO), Orbelia R. Robinson (CAS), Laurence E. Skog (US), William TAI (MO), Mark F. Watson (E), Anna L. Weitzman (1958-, US) and Cheng-Yih WU (KUN). Ihsan A. Al-Shehbaz (MO) replaced Nancy R. Morin (MO) and and De-Yuan HONG (PE) replaced Hong-Bin CUI (PE) as members of editorial committee.

1995 年 4 月 8 日至 9 日，英文版中国植物志编辑委员会联席会议在苏格兰爱丁堡植物园举行，密苏里植物园 Ihsan A. Al-Shehbaz、Michael G. Gilbert、Nancy R. Morin、Peter H. Raven、戴威廉，加州科学院 Bruce M. Bartholomew、Orbelia R. Robinson，哈佛大学 David E. Boufford、Anthony R. Brach，香港中文大学毕培曦，爱丁堡植物园 David F. Chamberlain、Derek Clayton、David S. Ingram、Helen Jones、Robert Mill（1950—）、Mark F. Watson，中国科学院植物研究所陈家瑞、陈心启、戴伦凯、洪德元，江苏省中国科学院植物研究所 南京中山植物园陈守良，中国科学院昆明植物研究所李德铢、李锡文和吴征镒，史密森学会的 Laurence E. Skog 和 Anna L. Weitzman（1958—）出席；密苏里植物园 Ihsan A. Al-Shehbaz 和中国科学院植物研究所洪德元取代密苏里植物园 Nancy R. Morin 和中国科学院植物研究所崔鸿宾加入编辑委员会。

1995.06.19-07.07, Chinese-American botanical expedition to Qinghai: Chindu [Chengduo], Madoi [Maduo], Nangqên [Nangqian], Yushu. David E. Boufford (A/GH), Michael J. Donoghue (1952-, A), Benito C. TAN (FH), Xue-Feng LU (HNWP), Tsun-Shen YING (PE), collected 535 numbers of seed plants and 500 numbers of bryophytes (A, FH, HNWP, PE).

1995 年 6 月 19 日至 7 月 7 日，中美植物联合考察青海的称多、玛多、囊谦、玉树，哈佛大学 David E. Boufford、Michael J. Donoghue（1952—）、Benito C. TAN，中国科学院西北高原生物研究所卢学峰，中国科学院植物研究所应俊生，采集种子植物标本 535 号和苔藓标本 500 号（A、FH、HNWP、PE）。

1995 年 8 月 1 日，陕西省中国科学院西北植物研究所编，**陕西省中国科学院西北植物研究所建所三十周年（1965—1995）**，210 页；陕西省中国科学院西北植物研究所，杨陵（内部印制）。

1995.08.01, Northwest Institute of Botany, Shaanxi Province and Chinese Academy of Sciences, '***30th anniversary (1965-1995) of Northwest Institute of Botany, Shaanxi Province and Chinese Academy of Sciences***', 210 p; Northwest Institute of Botany, Shaanxi Province and Chinese Academy of Sciences, Yangling (internal publication).

1995 年 8 月 25 日至 28 日，第二届全国苔藓植物学学术研讨会在济南山东师范大学召开，来自全国 15 个高校和研究单位近 40 人参加了大会。

Chinese-American botanical expedition to Qinghai Province, 19 June–7 July 1995 (Photo provided by David E. Boufford)
1995年6月19日至7月7日，中美植物学家考察青海省（相片提供者：鲍棣伟）

1995.08.25-08.28, Second National Bryophyte Symposium held at Shandong Normal University, Jinan, nearly 40 attendees from 15 institutions.

1995 年，张宏达文集编辑组编，**张宏达文集**，770 页；中山大学出版社，广州。

1995, Editorial Committee of Collected Works of Prof. Hung-Ta CHANG, '***Collected Works of Hung-Ta CHANG***', 770 p; Sun Yat-Sen University Press, Guangzhou.

1995 年，李树刚，广西现代植物分类学研究的发展，广西植物 15（3）：256-267。

1995, Shu-Kang LEE, The development of study of modern plant taxonomy of Guangxi, *Guihaia* 15(3): 256-267.

1995 年，李法曾、姚敦义主编，**山东植物研究**，369 页；北京科学技术出版社，北京。

1995, Fa-Zeng LI and Dun-Yi YAO (editors in chief), '***Research on Shandong Plants***', 369 p; Beijing Science and Technology Press, Beijing.

1995 年，天津自然博物馆，刘家宜编，**天津植物名录**[746]，229 页；天津教育出版社，天津。

1995, Tianjin Natural History Museum, Jia-Yi LIU, ***Enumeration of Plants in Tianjin***, 229 p; Tianjin Education Press, Tianjin.

1995 年，马毓泉著，**马毓泉文集**[747]，323 页；内蒙古人民出版社，呼和浩特。

1995, Yu-Chuan MA, ***Collected Works of Ma Yu-Chuan***, 323 p; Inner Mongolia People's

Yu-Chuan MA (1916-2008) (Photo provided by Ping MA)
马毓泉（1916—2008）（相片提供者：马平）

746 油印本曾于 1977 年印刷，计 159 页。

747 马毓泉（1916—2008），江苏苏州人；1935 年考入国立北平师范大学生物系，1936 年转入北京大学生物系。不久抗战军兴，投笔从戎，1938 年春考入第 15 期黄埔军校，受军官训练 15 个月，毕业后分配到 71 军 36 师参谋处服役，曾担任少尉、中尉附员、上尉参谋等；1943 年，马毓泉所在部队在云南大理休整，在老师张景钺督促之下，马毓泉退伍到昆明进入国立西南联合大学生物系三年级继续学习；1945 年毕业，留校任教。1947 年马毓泉随北京大学复员北平，并在该校就读研究生，导师为张肇骞；硕士论文 1951 年发表：中国龙胆科一新属——扁蕾属，植物分类学报 1（1）：5-19，图版 1-5。1957 年春，北京大学支援边疆创建内蒙古大学，副校长李继侗率领 10 余人前往，马毓泉从此在呼和浩特工作直至终老。著名龙胆科专家。详细参见：胡宗刚、夏振岱，2016，中国植物志编纂史，259 页；上海交通大学出版社，上海。

Press, Hohhot.

1995, Benito C. TAN, Jian-Cheng ZHAO and Ren-Liang HU, An updated checklist of mosses of Xinjiang, China, *Arctoa* 4: 1–14.

1995，Benito C. TAN、赵建成、胡人亮，中国新疆藓类植物新名录，*Arctoa* 4：1–14。

1995 年，王文采主编，**武陵山地区维管植物检索表**，626 页；科学出版社，北京。

1995，Wen-Tsai WANG (editor in chief), ***Keys to the Vascular Plants of the Wuling Mountains***, 626 p; Science Press, Beijing.

1995 and 1996, May Ling SO and Rui-Liang ZHU (vol. 1) and May Ling SO (vol. 2), ***Mosses and Liverworts of Hong Kong***, vols. 1: 162 p, 1995; 2: 130 p, 1996; Heavenly People Depot, Hong Kong.

1995、1996 年，苏美灵、朱瑞良（第 1 卷），苏美灵（第 2 卷），**香港苔藓植物志**，1：162 页，1995；2：130 页，1996；Heavenly People Depot，香港。

1995、1996 年，张大为、胡德熙、胡德焜合编，**胡先骕文存**，上卷：744 页，1995；江西高校出版社，南昌；下卷：913 页，1996；中正大学校友会，南昌（内部印制）。

1995 and 1996, Da-Wei ZHANG, De-Xi HU, De-Kun HU (eds.), '***Documents and Writings of Hsen-Hsu HU***', I: 744 p, 1995; Jiangxi University Corporation Press, Nanchang; II: 913 p, 1996; Chung Cheng University Alumni Association, Nanchang (internal publication).

1995、2000、2003 年，赖明洲，**苔藓植物研究手册**，169 页，1995；第 2 版，169 页，2000；第 3 版，169 页，2003；台湾大学农学院实验林管理处，南投。

1995, 2000, 2003, Ming-Jou LAI, '***Bryological Research Handbook***', 169 p, 1995; Ed. 2, 169 p, 2000; Ed. 3, 169 p, 2003; Experimental Management of Agriculture College, Taiwan University, Nantou.

1996

1996.07.16–07.18, Chinese-American botanical expedition to Beijing: Mentougou. David E. Boufford (A/GH), Mei-Duo SURI (Beijing Museum of Natural History), Shao-Yong YANG (PE), Xian-Chun ZHANG (PE), collected 138 numbers of vascular plants (David E. Boufford et al. nos. 27008–27145, A, CAS, PE, TI).

1996年7月16日至18日，中美植物联合考察北京的门头沟，哈佛大学 David E. Boufford，北京自然博物馆梅朵粟黎，中国科学院植物研究所杨少永、张宪春；采集维管束植物138号（A、CAS、PE、TI）。

1996年7月19日至21日，第四届系统与进化植物学青年研讨会及第二届分类学原理讨论会在昆明中国科学院昆明植物研究所召开，全国36个单位66人出席。[748]

1996.07.19–07.21, Fourth Youth Symposium of Systematic and Evolutionary Botany and Second National Symposium on Taxonomic Principles held at Kunming Institute of Botany, Chinese Academy of Sciences, Kunming, 66 attendees from 36 institutions.

1996.07.23–07.24, Joint editorial committee meeting of Flora of China held at Kunming Institute of Botany, Chinese Academy of Sciences, Kunming, China. Attendees: Ihsan A. Al-Shehbaz (MO), Chia-Jui CHEN (PE), Shou-Liang CHEN (NAS), Shu-Kun CHEN (KUN), Lun-Kai DAI (PE), Michael G. Gilbert (MO), Hong-Ya GU (PEY), Xiang-Sheng JI (KUN), De-Zhu LI (KUN), Hsi-Wen LI (KUN), Chun-Chao LÜ (KUN), Nancy R. Morin (MO), Peter H. Raven (MO), Orbelia R. Robinson (CAS), Laurence E. Skog (US), Han-Dong SUN (KUN), William TAI (MO), Mark F. Watson (E) and Cheng-Yih WU (KUN). William TAI (MO) left the editorial committee and Ihsan A. Al-Shehbaz (MO) replaced him as co-director of the project; Anna L. Weitzman (US) and Guang-Hua ZHU (MO) became editorial assistants.

1996年7月23日至24日，英文版中国植物志编辑委员会联席会议在昆明植物研究所举行，密苏里植物园 Ihsan A. Al-Shehbaz、Michael G. Gilbert、Nancy R. Morin、Peter H. Raven、戴威廉，中国科学院植物研究所陈家瑞、戴伦凯，中国科学院昆明植物研究所陈书坤、姬翔生、李德铢、李锡文、吕春朝、孙汉董、吴征镒，江苏省中国科学院植物研究所 南京中山植物园陈守良，北京大学顾红雅，加州科学院 Orbelia R. Robinson，史密森学会的 Laurence E. Skog，爱丁堡植物园 Mark F. Watson 出席。

748 中国科学院昆明植物研究所、云南省植物学会，1996，全国第四届系统与进化植物学青年学术研讨会及第二届分类学原理讨论会论文摘要汇编，71页；中国科学院昆明植物研究所、云南省植物学会，昆明（内部印制）。

密苏里植物园戴威廉离开编辑委员会，密苏里植物园 Ihsan A. Al-Shehbaz 继任为项目联合主任，史密森学会的 Anna L. Weitzman 和密苏里植物园朱光华任编辑助理。

1996.08-09, Chinese-Foreign botanical expedition to Qinghai: Chindu [Chengduo], Nangqên [Nangqian], Yushu. Bruce M. Bartholomew (CAS), Shi-Long CHEN (1967-, HNWP), De-Shan DENG (HNWP), Michael G. Gilbert (BM/MO), Ting-Nong HO (leader, HNWP), Jian-Quan LIU (HNWP), Shang-Wu LIU (HNWP) and Xue-Feng LU (HNWP); collected 1,350 numbers of vascular plants (A, CAS, HNWP).

1996 年 8 月至 9 月，中外植物联合考察青海的称多、囊谦、玉树，加州科学院 Bruce M. Bartholomew，英国自然历史博物馆、密苏里植物园 Michael G. Gilbert，中国科学院西北高原生物研究所陈世龙（1967—）、邓德山、何廷农（领队）、刘建全、刘尚武、卢学峰；采集维管植物标本 1 350 号（A、CAS、HNWP）。

1996.08.30-10.18, North American China Plant Expedition Consortium to Shaanxi and Gansu. Participants: James Ault, Longwood Gardens, Kennett Square, Pennsylvania, Kevin Conrad, US National Arboretum, Washington, D.C., Kunso Kim, Morton Arboretum, Lisle, Illinois, Rick Lewandowski, Morris Arboretum, University of Pennsylvania, Philadelphia, Pennsylvania, Tie-Cheng CUI, Xi'an Botanical Garden.

1996 年 8 月 30 日至 10 月 18 日，由美国宾夕法尼亚州 Kennett Square 市长木植物园 James Ault、华盛顿特区国家树木园 Kevin Conrad、伊利诺伊州 Lisle 市莫顿树木园 Kunso Kim 和宾夕法尼亚州费城宾夕法尼亚大学莫里斯树木园 Rick Lewandowski 组成的北美中国植物考察联盟到陕西和甘肃采集，中方为西安植物园崔铁成。

1996.10.07-10.30, Chinese-British botanical expedition to Yunnan: Gaoligong Shan, on way from Dali, Baoshan, Lushui and Fugong. J. Crinan M. Alexander (E), Zhi-Ling DAO (1965-, KUN), Heng LI (leader, KUN), Mark Newman (E), Philip Thomas (E), Shi-Xiong YANG (KUN) and Andrea Qwong (Anthropologist, Harvard University), collected 1,033 numbers of plant specimens (E, KUN).

1996 年 10 月 7 日至 30 日，中英植物联合考察云南的高黎贡山，并沿途对大理、保山、泸水、福贡进行采集，中国科学院昆明植物研究所刀志灵（1965—）、李恒（领队）、杨世雄，爱丁堡植物园 J. Crinan M. Alexander、Mark Newman、Philip Thomas，哈佛大学人类学者 Andrea Qwong，采集植物标本 1 033 号（E、KUN）。[749]

1996 年，陈焕镛纪念文集编辑委员会编，**陈焕镛纪念文集**，350 页；中国科学院华南植物研究所，

749 李恒、郭辉军、刀志灵编著，2000，高黎贡山植物，1 344 页；科学出版社，北京。

广州（内部印制）。[750]

1996, Editorial Committee of Memorial Works of Woon-Young CHUN, '***Memorial Works of Woon-Young CHUN***', 350 p; South China Institute of Botany, Chinese Academy of Sciences, Guangzhou (internal

Main gate and Herbarium (IBK), Guangxi Institute of Botany, Guilin, 2018 (Photo provided by Sai-Chun TANG)
2018 年的广西植物研究所的大门与植物标本馆（IBK）（相片提供者：唐赛春）

750 原书没有具体出版时间，本书根据该书副主编胡启明提供的信息加入。

publication).

1996 年，广西植物研究所归属广西壮族自治区和中国科学院双重领导，并更名为广西壮族自治区中国科学院广西植物研究所，1996—2010 年李锋、2010—2012 年文永新（1953—）、2012—2016 年何成新（1965—）、2017 至今李典鹏（1968—）任所长。

1996, Guangxi Institute of Botany under dual leadership of Guangxi Zhuang Autonomous Region and Chinese Academy of Sciences, renamed Guangxi Institute of Botany, Guangxi Zhuang Autonomous Region and Chinese Academy of Sciences; with directors Feng LI, 1996-2010, Yong-Xin WEN (1953-), 2010-2012, Cheng-Xin HE (1965-), 2012-2016, Dian-Peng LI (1968-), since 2017.

1996 年，江西省庐山植物园归属江西省和中国科学院双重领导，并更名为江西省中国科学院庐山植物园，1996—1999 年王永高（1942—）、1999—2001 年胡星卫（1957—）、2001—2006 年郑翔（1961—）、2006—2010 年张青松（1965—）、2010 年至今吴宜亚（1969—）任园长（或负责人）。

1996, Jiangxi Lushan Botanical Garden under dual leadership of Jiangxi Province and Chinese Academy of Sciences, renamed Lushan Botanical Garden, Jiangxi Province and Chinese Academy of Sciences. Directors (or in charge): Yong-Gao WANG (1942-), 1996-1999, Xing-Wei HU (1957-), 1999-2001, Xiang ZHENG (1961-), 2001-2006, Qing-Song ZHANG (1965-), 2006-2010, Yi-Ya WU (1969-), since 2010.

1996, Lucien A. Lauener, David K. Ferguson (ed.), ***The Introduction of Chinese Plants into Europe***, 269 p; SPB Academic Publishing, Amsterdam.

1996 年，Lucien A. Lauener、David K. Ferguson 编辑，**中国植物引入欧洲的介绍**，269 页；SPB Academic Publishing，阿姆斯特丹。

1996，Paul L. Redfearn, Jr, Benito C. TAN and Si HE, A newly updated and annotated checklist of Chinese mosses, *The Journal of the Hattori Botanical Laboratory* 79: 163-357.

1996，Paul L. Redfearn，Jr、Benito C. TAN、何思，中国藓类最新名录及注释，*The Journal of the Hattori Botanical Laboratory* 79：163-357。

1996 年，邢福武、吴德邻主编，**南沙群岛及其邻近岛屿植物志**，375 页；海洋出版社，北京。

1996, Fu-Wu XING and Te-Lin WU (editors in chief), ***Flora of Nansha Islands and Their Neighbouring Islands***, 375 p; China Ocean Press, Beijing.

1996 年，赵运林、潘晓玲编著，**湘黔桂交界地区植物名录**，396 页；湖南科学技术出版社，长沙。

1996, Yun-Lin ZHAO and Xiao-Ling PAN, '***Checklist of Plants from the boundaries areas of Hunan,***

Jiangxi Lushan Botanical Garden, 2017: 1. Aerial view; 2. San-Yi-Xiang; 3. Library; 4. Plantings (Photo provided by Zong-Gang HU)
2017 年的江西省庐山植物园：1. 鸟瞰图；2. 三逸乡；3. 图书馆；4. 园林（相片提供者：胡宗刚）

Guizhou and Guangxi', 396 p; Hunan Science and Technology Press, Changsha.

1996 年，中国科学院植物研究所编，王宗训主编，**新编拉汉英植物名称**，1 166 页；航空工业出版社，北京。

1996, Institute of Botany, Chinese Academy of Sciences, Zong-Xun WANG (editor in chief), '***New Latin-Chinese-English Plant Names***', 1,166 p; Aviation Industry Press, Beijing.

1996—1999 年，中国科学院西北高原生物研究所、青海植物志编辑委员会编，刘尚武主编，**青海植物志**，4 卷本；第 1 卷：544 页，1997；第 2 卷：463 页，1999；第 3 卷：547 页，1996；第 4 卷：353 页，1999；青海人民出版社，西宁。

1996–1999, Editorial Committee of Flora Qinghaiica, Northwest Institute of Plateau Biology, Chinese Academy of Sciences, Shang-Wu LIU (editor in chief), ***Flora Qinghaiica***, vols. 1–4; 1: 544 p, 1997; 2: 463 p, 1999; 3: 547 p, 1996; 4: 353 p, 1999; Qinghai People's Publishing House, Xining.

1996—2004 年，中国植物志第七届编辑委员会成立，主编为中国科学院昆明植物研究所吴征镒，副主编为中国科学院植物研究所陈心启；委员 16 人，分别为中国科学院昆明植物研究所陈书坤、李锡文，中国科学院植物研究所陈艺林、戴伦凯、傅国勋、傅立国、李安仁、夏振岱，东北林业大学黄普华，中国科学院华南植物研究所（园）胡启明、林有润，中国科学院成都生物研究所孔宪需，中国科学院西北高原生物研究所刘尚武，浙江自然博物馆韦直，西北农林科技大学徐朗然，科学出版社曾建飞；顾问 7 人，分别为中国科学院华南植物园陈德昭、黄成就，江苏省中国科学院植物研究所 南京中山植物园陈守良，广西壮族自治区、中国科学院广西植物研究所李树刚，福建师范大学林来官，北京医学院任波涛（1919—2003），中国科学院植物研究所汤彦承；1997 年 5 月 7 日，增设特聘青年编辑委员 4 人，分别为中国科学院植物研究所傅德志、包伯坚、张宪春，中国科学院昆明植物研究所李德铢。

1996–2004, Seventh Editorial Committee of Flora Reipublicae Popularis Sinicae established, Cheng-Yih WU (KUN) as Editor in Chief and Sing-Chi CHEN (PE) as Vice Editor in Chief; 16 editorial members: Shu-Kun CHEN (KUN), Yi-Ling CHEN (PE), Lun-Kai DAI (PE), Guo-Xun FU (PE), Li-Kuo FU (PE), Chi-Ming HU (IBSC), Pu-Hwa HUANG (NEFI), Hsien-Shiu KUNG (CDBI), An-Jen LI (PE), Hsi-Wen LI (KUN), Yuou-Ruen LING (IBSC), Shang-Wu LIU (HNWP), Zhi WEI (ZM), Zhen-Dai XIA (PE), Lang-Rang XU (WUK), Jian-Fei ZENG (Science Press), plus seven council members: Shou-Liang CHEN (NAS), Te-Chao CHEN (IBSC), Ching-Chieu HUANG (IBSC), Shu-Kang LEE (IBK), Lai-Kuan LING (FNU), Po-Tao REN (1919–2003, PEM), Yen-Cheng TANG (PE). Four special young members, Bo-Jian BAO (PE), De-Zhi FU (PE), De-Zhu LI (KUN) and Xian-Chun ZHANG (PE), were added on May 7, 1997.

1997

1997 年 1 月，西双版纳热带植物园脱离中国科学院昆明植物研究所独立，隶属于中国科学院；1997—2001 年许再富、2001—2005 年刘宏茂（1961—2006）、2005 年至今陈进（1965—）任主任。

1997.01, Xishuangbanna Tropical Botanical Garden, under Chinese Academy of Sciences, becomes independent of Kunming Institute of Botany, with directors Zai-Fu XU, 1997-2001; Hong-Mao LIU (1961-

Aerial view of Xishuangbanna Tropical Botanical Garden, 2017 (Photo provided by Xishuangbanna Tropical Botanical Garden)
2017 年的西双版纳热带植物园鸟瞰图（相片提供者：西双版纳热带植物园）

Xishuangbanna Tropical Botanical Garden: 1. Herbarium (HITBC); 2. Library; 3. Main Gate; 4. Suspension Bridge (Photo provided by Xishuangbanna Tropical Botanical Garden)
西双版纳热带植物园：1. 植物标本馆 (HITBC)；2. 图书馆；3. 大门；4. 吊桥（相片提供者：西双版纳热带植物园）

2006), 2001-2005; and Jin CHEN (1965-), since 2005.

1997.02-1998.02, Chi-Ming HU, South China Institute of Botany, Chinese Academy of Sciences, traveled to the Rijksherbarium, Leiden University, The Netherlands, to study *Ardisia* (Myrsinaceae) for ***Flora Malesiana***.[751]

1997 年 2 月至 1998 年 2 月，中国科学院华南植物研究所胡启明赴荷兰莱顿大学国立标本馆，编研**马来西亚植物志**紫金牛属（紫金牛科）。

1997.05.26-05.30, International Symposium, Bryology in the 21st Century, held at the Institute of Botany, Chinese Academy of Sciences, Beijing; 108 attendees from 20 countries, including 55 from outside of China. Symposium followed by a one week field excursion to Jiuzhaigou, Sichuan.

1997 年 5 月 26 日至 30 日，“21 世纪的苔藓植物学”国际研讨会在北京中国科学院植物研究所召开，来自 20 个国家和地区的 108 位代表参加了会议，其中外宾 55 人。会后到四川九寨沟野外考察 1 周。

1997.08.25-09.27, North American China Plant Expedition Consortium to Jilin. Participants: Kris Bachtell, Morton Arboretum, Lisle, Illinois, Peter Del Tredici, Arnold Arboretum, Harvard University, Jamaica Plain, Massachusetts, Jeffrey Lynch, Longwood Gardens, Kennett Square, Pennsylvania, Paul Meyer, Morris Arboretum, University of Pennsylvania, Philadelphia, Pennsylvania, Charles Tubesing, Holden Arboretum, Kirtland, Ohio, with Wei CAO (1962-, IFP), Xian-Li WANG (IFP), and Ning SHENG, Nanjing Botanical Garden Memorial Sun Yat-Sen.

1997 年 8 月 25 日至 9 月 27 日，由美国伊利诺伊州 Lisle 市莫顿树木园 Kris Bachtell，麻州 Jamaica Plain 市哈佛大学阿诺德树木园 Peter Del Tredici，宾夕法尼亚州 Kennett Square 市长木植物园 Jeffrey Lynch，宾夕法尼亚州费城宾夕法尼亚大学莫里斯树木园 Paul Meyer 和俄亥俄州 Kirtland 市霍顿树木园 Charles Tubesing 组成的北美中国植物考察联盟到吉林采集，中方成员为中国科学院沈阳应用生态研究所王献礼和曹伟（1962—），以及南京中山植物园盛宁。

1997.08-09, Sino-British botanical expedition to Qinghai: Huzhu, Maqên [Maqin], Tongde, Tongren, Zêkog [Zeku]; De-Shan DENG (HNWP), David G. Long (E), Xue-Feng LU (leader, HNWP) et al. collected 1,300 numbers of seed plants and 800 numbers of bryophytes (E, HNWP).

1997 年 8 月至 9 月，中英植物联合考察青海的互助、玛沁、同德、同仁、泽库，中国科学院西北高原生物研究所邓德山、卢学峰（领队），爱丁堡植物园 David G. Long 等，采集种子植物标本 1 300

751 Chi-Ming HU, 1999, New synonyms and combinations in Asiatic *Ardisia* (Myrsinaceae), *Blumea* 44: 391-406; and Chi-Ming HU, 2002, New species of *Ardisia* (Myrsinaceae) from Malesia, *Blumea* 47: 493-512.

Participants at International Symposium, Bryology in the 21st Century, Beijing, 26-30 May 1997 (Photo provided by Tong CAO)
1997 年 5 月 26 日至 30 日，21 世纪的苔藓植物学国际研讨会于北京（相片提供者：曹同）

号和苔藓植物标本 800 号（E、HNWP）。

1997.08.13-09.08, Chinese-American botanical expedition to Sichuan: Daofu, Kangding, Lixian, Luding, Luhuo, Barkam [Maerkang], Zamtang [Rangtang] and Sêrtar [Seda]. David E. Boufford (A/GH), Michael J. Donoghue (A), David S. Hibbett (1962-, FH), Yu JIA (1965-, PE), Richard H. Ree (1973-, A/F), Zheng WANG (1969-, HMAS), Kai-Pu YIN (1943-, CDBI), collected 745 numbers of vascular plants (David E. Boufford et al. nos. 27163-28004, A, CAS, E, KUN, MO, NY, P, TI); 2,010 numbers of bryophytes (nos. J01725-J03674, FH, PE); 526 numbers of fungi (nos. DSH97-013-DSH97-341; WZ2005-WZ2239, FH, HMAS).

1997 年 8 月 13 日至 9 月 8 日，中美植物联合考察：四川的道孚、康定、理县、泸定、炉霍、马尔康、壤塘、色达，哈佛大学 David E. Boufford、Michael J. Donoghue、David S. Hibbett（1962—），哈佛大学、芝加哥自然博物馆 Richard H. Ree（1973—），中国科学院植物研究所贾渝（1965—），中国科学院微生物研究所王征（1969—），中国科学院成都生物研究所印开蒲（1943—），采集维管植物 745 号（A、CAS、E、KUN、MO、NY、P、TI），苔藓 2 010 号（FH、PE），真菌 526 号（FH、HMAS）。

1997.09.04-10.07, Sino-Foreign botanical expedition to Yunnan: Gaoligong Shan, on the way via Fugong, Lushui, Xiangyun, Yongping and Yunlong. Zhi-Ling DAO (KUN), Heng LI (leader, KUN), Greg Pedano (MO), Andrea Qwong (Anthropologist, Harvard University), Greg Ruckert (horticulturist, Australia), Philip Thomas (E), Nicholas J. Turland (1966-, MO) and Jia-Rong XUE (KUN) collected 1,673 numbers of plant specimens (E, KUN, MO).

1997 年 9 月 4 日至 10 月 7 日，中外植物联合考察云南的高黎贡山，沿途对福贡、泸水、祥云、永平、云龙进行采集，中国科学院昆明植物研究所刀志灵、李恒（领队）、薛嘉绒，哈佛大学人类学者 Andrea Qwong，澳大利亚园艺学者 Greg Ruckert，密苏里植物园 Greg Pedano 和 Nicholas J. Turland（1966—），爱丁堡植物园 Philip Thomas，采集植物标本 1 673 号（E、KUN、MO）。

1997 年 11 月 12 日至 13 日，海峡两岸植物多样性与保育学术研讨会于台中台湾自然科学博物馆举行，两岸共有 80 多人参加。[752]

1997.11.12-11.13, Cross-Strait Symposium on floristic diversity and conservation held at Taiwan Museum of Natural Science, Taichung, Taiwan, more than 80 attendees from Taiwan and mainland.

1997.11.15-11.16, Joint editorial committee meeting of Flora of China held at Taiwan Museum of Natural Sciences, Taichung, Taiwan, China. Attendees: Ihsan A. Al-Shehbaz (MO), Bruce M. Bartholomew (CAS), David E. Boufford (A/GH), Anthony R. Brach (A/GH), Paul Pui-Hay BUT (CUHK), Chia-Jui CHEN (PE), Shou-Liang CHEN (NAS), Sing-Chi CHEN (PE), Shau-Ting CHIU (1961-, TNM), Lun-Kai DAI (PE), Michael G. Gilbert (MO), De-Yuan HONG (PE), Chi-Ming HU (IBSC), Shiu-Ying HU (CUHK), Hong-Ya GU (PEY), Hsi-Wen LI (KUN), Ching-I PENG (HAST), Peter H. Raven (MO), Orbelia R. Robinson (CAS),

752 邱少婷、彭镜毅，1998；海峡两岸植物多样性与保育学术研讨会论文集，327 页；自然科学博物馆，台中。

Chinese-American botanical expedition to western Sichuan, 13 August-8 September 1997: 1 (left-right): Michael J. Donoghue, David S. Hibbett; 2. Richard H. Ree; 3. Yu JIA; 4. Zheng WANG (Photo provided by David E. Boufford)

1997 年 8 月 13 日至 9 月 8 日，中美植物学家考察四川：1（从左至右）：Michael J. Donoghue 和 David S. Hibbett；2. Richard H. Ree；3. 贾渝；4. 王征（相片提供者：鲍棣伟）

Mark F. Watson (E), Cheng-Yih WU (KUN), Zhen-Dai XIA (PE), Tsung-Yu YANG (TNM) and Hsin-Fu YEN (TNM). 1998, Nicholas J. Turland (MO) joined as an editorial assistant. 1999, David S. Ingram (E) left the editorial committee; Simon J. Owens (1947–, K) and Guang-Hua ZHU (MO) joined the editorial committee; Bryan E. Dutton (A/GH) and Anna L. Weitzman (US), editorial assistants, left the project. Robert DeFilipps (1939–2004, US) joined as editorial assistant, Chang-Hung CHOU (HAST) and Laurence E. Skog (US) left the editorial committee, 2000. Joel Jeremie (1944–, P) and W. John Kress (1951–, US) joined the editorial committee. 2001, Qin-Er YANG (PE) joined the editorial committee and replaced Lun-Kai DAI (PE) as co-director of the project; Guang-Hua ZHU (MO) replaced Ihsan A. Al-Shehbaz (MO) as co-director of the project; Lun-Kai DAI (PE) and Ihsan A. Al-Shehbaz (MO) remained members of the editorial committee. William A. McNamara of Quarryhill Botanical Garden joined the editorial committee as an associate member.

1997 年 11 月 15 日至 16 日，英文版中国植物志编辑委员会联席会议在台湾自然科学博物馆举行，密苏里植物园 Ihsan A. Al-Shehbaz、Michael G. Gilbert、Peter H. Raven，加州科学院 Bruce M. Bartholomew、Orbelia R. Robinson，哈佛大学 David E. Boufford、Anthony R. Brach，香港中文大学毕培曦、胡秀英，中国科学院植物研究所陈家瑞、陈心启、戴伦凯、洪德元、夏振岱，江苏省中国科学院植物研究所 南京中山植物园陈守良，中国科学院华南植物研究所胡启明，北京大学顾红雅，中国科学院昆明植物研究所李锡文、吴征镒，“中央”研究院植物研究所彭镜毅，台湾自然科学博物馆邱少婷（1961— ）、严新富、杨宗愈，爱丁堡植物园 Mark F. Watson 出席。1998 年，密苏里植物园 Nicholas J. Turland 加入编辑助理。1999 年，爱丁堡植物园 David S. Ingram 离开编辑委员会，英国皇家植物园邱园 Simon J. Owens（1947— ）和密苏里植物园朱光华加入编辑委员会；哈佛大学 Bryan E. Dutton、史密森学会的 Anna L. Weitzman 离开编辑委员会；史密森学会的 Robert DeFilipps（1939—2004）加入编辑助理。2000 年“中央”研究院植物研究所周昌弘、史密森学会的 Laurence E. Skog 离开编辑委员会，法国巴黎自然历史博物馆 Joel Jeremie（1944— ）和史密森学会的 W. John Kress（1951— ）加入编辑委员会。2001 年中国科学院植物研究所杨亲二取代戴伦凯为联合项目主任并参加编辑委员会；密苏里植物园朱光华取代 Ihsan A. Al-Shehbaz 为项目联合主任；中国科学院植物研究所戴伦凯和密苏里植物园 Ihsan A. Al-Shehbaz 仍留在编辑委员会。Quarryhill Botanical Garden 的 William A. McNamara 作为附属成员参加编辑委员会。

1997 年，白学良主编，**内蒙古苔藓植物志**，541 页；内蒙古大学出版社，呼和浩特。

1997, Xue-Liang BAI (editor in chief), ***Flora Bryophytarum Intramongolicarum***, 541 p; Inner Mongolia University Press, Hohhot.

1997, Tzen-Yuh CHIANG and Tsai-Wen HSU, ***The Literature of Taxonomic Studies on Mosses of China and the Adjacent Regions***, 122 p; Taiwan Endemic Species Research Institute, Nantou.

1997 年，蒋镇宇、许再文，**中国及临近地区藓类植物文献**，122 页；台湾省特有生物研究保育中心，

Joint editorial committee meeting of *Flora of China*, Taichung, 15–16 November 1997 (Photo provided by Wu Zhengyi Science Foundation)
1997 年 11 月 15 日至 16 日，英文版中国植物志编辑委员会联席会议于台中（相片提供者：吴征镒科学基金会）

南投。

1997 年，廖日京，**植物拉丁语**，293 页；台湾大学农学院森林系，台北。

1997, Jih-Ching LIAO, ***Lingua Latina Botanica***, 293 p; Department of Forestry, College of Agriculture, Taiwan University, Taipei [Taibei].

1997 年，满都拉主编，**蒙拉汉英植物名称**（内蒙古栽培植物名称），242 页；内蒙古人民出版社，呼和浩特。

1997, Mandola (editor in chief), ***Mongolo-Latino-Sinico-Anglicum Nominum Plantarum*** (*Nominum Sativum Plantarum Intramongolicum*), *Plant Name in Mongol-Latin-Chinese and English*, 242 p; Inner Mongolia People's Press, Hohhot.

1997, Brenda McLean, ***A Pioneering Plantsman A. K. Bulley and the Great Plant Hunters***, 184 p; The Stationery Office, London.

1997 年，Brenda McLean，**园艺爱好者先驱 *A. K. Bulley* 和伟大的植物猎人们**[753]，184 页；The Stationery Office，伦敦。

1997, Sinikka S. Piippo, Xiao-Lan HE and Timo J. Koponen, Hepatics from northwestern Sichuan, China, with a checklist of Sichuan hepatics, *Annales Botanici Fennici* 34(1): 51-63.

1997 年，Sinikka S. Piippo、何小兰、Timo J. Koponen，中国四川西北部苔类植物，以及四川苔类植物名录，*Annales Botanici Fennici* 34（1）：51-63。

1997, Peng-Cheng RAO, Johannes Enroth, Sinikka S. Piippo and Timo J. Koponen, The bryophytes of Hunan Province, China: an annotated checklist, *Hikobia* 12: 181-203.

1997 年，饶鹏程，Johannes Enroth，Sinikka S. Piippo，Timo J. Koponen，中国湖南苔藓植物名录，*Hikobia* 12：181-203。

1997 年，吴永华著，**被遗忘的日籍台湾植物学者**，台湾历史馆 4，474 页；晨星出版社，台中。

1997, Yung-Hua WU, '***Forgotten Japanese Taiwanese Botanists***', Taiwan History Room 4, 474 p; Morning Star Publisher Inc., Taichung [Taizhong].

1997, Li ZHANG and Pang-Juan LIN, A checklist of bryophytes from Hong Kong, *The Journal of the*

753 英国利物浦棉花商 Arthur K. Bulley（1861—1942），著名园艺家，曾多次资助 George Forrest、Frank Kingdon Ward、Reginald Farrer 等人赴中国西南以及喜马拉雅地区大规模采集并引种高山植物。

Hattori Botanical Laboratory 81: 307-326.

1997 年，张力、林邦娟，香港苔藓植物名录，*The Journal of the Hattori Botanical Laboratory* 81：307-326。

1997—2002 年，杨远波、刘和义等，**台湾维管束植物简志**，6 卷本：第 1 卷：255 页[754]，1997；第 2 卷：352 页，1997；第 3 卷：389 页，1998；第 4 卷：432 页，1999；第 5 卷：457 页，2002；第 6 卷：665 页[755]，2002；"行政院"农业委员会，台北。[756]

1997-2002, Yuen-Po YANG, Ho-Yih LIU et al., ***Manual of Taiwan Vascular Plants***[757], vols. 1-6; 1: 255 p, 1997; 2: 352 p, 1997; 3: 389 p, 1998; 4: 432 p, 1999; 5: 457 p, 2002; 6: 665 p, 2002; Council of Agriculture, Executive Yuan, Taipei [Taibei].

754 第 1 卷作者为郭城孟（1948— ），台湾大学植物学教授。1971 年中兴大学理学学士，1974 年台湾大学理学硕士，1984 年瑞士苏黎世大学系统研究所博士（http://ge.tnua.edu.tw/ge-mna/teachers.html，2017 年进入）；指导教授：Kramer, Karl Ulrich (1928-1994)。

755 第 6 卷为检索表和名录以及索引等。

756 网络版（http://subject.forest.gov.tw/species/vascular/index.htm，2018 年进入）。

757 There was no English title for the first volume.

1998

1998年4月，中国科学院中国孢子植物志编辑委员会第四届编辑委员会改组，主编为曾呈奎，常务副主编为魏江春，副主编为余永年、吴鹏程、毕列爵[758]（1917—2004），编辑委员会29人。

1998.04, Fourth Consilio Florarum Cryptogamarum Sinicarum, Academiae Sinicae Edita changed, Cheng-Kui TSENG as Editor in Chief, Jiang-Chun WEI as Standing Vice Editor in Chief, Yung-Nien YU, Pan-Cheng WU and Lie-Jue BI (1917-2004) as Vice editors in Chief; 29 editorial committee members.

1998年7月7日，中国科学院新疆生物土壤沙漠研究所和中国科学院新疆地理研究所联合重组，成立中国科学院新疆生态与地理研究所。1998—2002年宋郁东（1943—）、2002—2006年张小雷（1963—）、2006—2016年陈曦（1964—）、2017年至今雷加强（1961—）任所长。

1998.07.07, Institute of Biology, Pedology and Desert and Institute of Geology and Geography, Chinese Academy of Sciences, combined as Xinjiang Institute of Ecology and Geography, Chinese Academy of Sciences; with directors Yu-Dong SONG (1943-), 1998-2002; Xiao-Lei ZHANG (1963-), 2002-2006; Xi CHEN (1964-), 2006-2016; Jia-Qiang LEI (1961-), since 2017.

1998.07.01-07.31, Chinese-American botanical expedition to Sichuan: Daocheng, Xiangcheng, Yunnan: Dali, Lijiang, Zhongdian. Bruce M. Bartholomew (CAS), David E. Boufford (A/GH), Wen-Yun CHEN (KUN), Michael J. Donoghue (A), Yu JIA (PE), Richard H. Ree (A/F), Hang SUN (KUN), Shu-Kung WU (KUN), Zheng WANG (HMAS) and Zhu-Liang YANG (1965-, KUN), collected 1,337 numbers of vascular plants (A, CAS, E, F, KUN, MO, NY, P, TI), 900 numbers of bryophytes (FH, PE) and 543 numbers of fungi (FH, HMAS).

1998年7月1日至31日，中美植物联合考察四川的稻城、乡城，云南的大理、丽江、中甸，哈佛大学 David E. Boufford、Michael J. Donoghue，加州科学院 Bruce M. Bartholomew，哈佛大学、芝加哥自然博物馆 Richard H. Ree，中国科学院昆明植物研究所陈文允、孙航、武素功、杨祝良（1965—），中国科学院植物研究所贾渝，中国科学院微生物研究所王征，采集维管植物1 337号（A、CAS、E、F、KUN、MO、NY、P、TI），苔藓900号（FH、PE），真菌543号（FH、HMAS）。

1998.10.05-10.22, North American China Plant Expedition Consortium to Anhui, Guangxi and Jiangxi. Participants: Rick Lewandowski, Morris Arboretum, University of Pennsylvania, Philadelphia, Pennsylvania, Edward Garvey, US National Arboretum, Washington D.C., with Wei-Lin LI and Qing WANG, Nanjing

758 湖北浠水人，1942年毕业于国立西南联合大学生物系，先后在云南省医药研究所、北平研究院植物学研究所、武昌华中大学、华中师范学院、开封师范学院、新乡师范学院、武汉师范学院（今湖北大学）等校生物系工作，先后任助理研究员、讲师、副教授、教授及系主任；著名藻类学家。

Xinjiang Institute of Ecology and Geography, Chinese Academy of Sciences: 1. Herbarium (XJBI), 2003; 2. Participants at annual meeting of Botanical Society of Xinjiang, 2013; 3. Training in plant taxonomy, 2006 (Photo provided by Bo-Rong PAN)

中国科学院新疆生态与地理研究所：1. 2003 年植物标本馆(XJBI)；2. 新疆植物学会 2013 年年会；3. 2006 年植物分类学培训班(相片提供者：潘伯荣)

Chinese-American botanical expedition to western Sichuan, 1-31 July 1998: 1 (left-right): Michael J. Donoghue, Hang SUN, and local people; 2. Su-Kung WU (left) and Yu JIA (right) (Photo provided by David E. Boufford)

1998 年 7 月 1 日至 31 日，中美植物学家考察四川：1（从左至右）：Michael J. Donoghue、孙航、当地群众；2. 武素功（左）和贾渝（右）（相片提供者：鲍棣伟）

Chinese-American botanical expedition to western Sichuan, 1–31 July 1998: 1. Zheng WANG; 2. Zhu-Liang YANG (Photo provided by David E. Boufford)

1998年7月1日至31日，中美植物学家考察四川：1. 王征；2. 杨祝良（相片提供者：鲍棣伟）

Botanical Garden Memorial Sun Yat-Sen.

1998年10月5日至22日，由美国宾夕法尼亚州费城宾夕法尼亚大学莫里斯树木园 Rick Lewandowski 和华盛顿特区国家树木园 Edward Garvey 组成的北美中国植物考察联盟到安徽、广西和江西采集，中方成员为南京中山植物园李维林、汪庆。

1998.10.11–11.08, Chinese-American botanical expedition to Yunnan: Baoshan, Gaoligong Shan, Longling, Tengchong, Lushui. Bruce M. Bartholomew (CAS), Zhi-Ling DAO (KUN), Heng LI and Chun-Lin LONG (leader, KUN) collected 1,544 numbers of vascular plants (CAS, E, GH, KUN).

1998年10月11日至11月8日，中美植物联合考察云南的高黎贡山、保山、龙陵、腾冲、泸水，加州科学院 Bruce M. Bartholomew，中国科学院昆明植物研究所刀志灵、李恒、龙春林（领队），采集维管植物标本 1 544 号（CAS、E、GH、KUN）。

1998年10月，中国林业科学研究院吴中伦文集编辑委员会编，**吴中伦文集**，773页；中国科学技术出版社，北京。

1998.10, Editorial Committee of Selected Works of Wu Zhonglun, ***Selected Works of Wu Zhonglun*** *(Chung-Lun WU* or *Chung-Lwen WU)*, 773 p; China Science and Technology Press, Beijing.

1998年12月4日至7日，中国植物学会第十二届会员代表大会暨六十五周年学术年会在深圳召开，会议代表共450人；第五届系统与进化植物学青年研讨会也同期举行。理事长为匡廷云(1934—)。[759]

1998.12.04-12.07, Twelfth National Congress and 65th anniversary meeting of Botanical Society of China, and Fifth Youth Symposium of Systematic and Evolutionary Botany held in Shenzhen, Guangdong; more than 450 attendees; Ting-Yun KUANG (1934-) elected president.

1998年12月，包士英、毛品一、苑淑秀编写，**云南植物采集史略1919—1950**，中英文双语版，214页；中国科学技术出版社，北京。[760]

1998.12, Shih-Ying BAO, Pin-Yi MAO and Shu-Xiu YUAN, ***A Brief History of Plant Collection in Yunnan 1919-1950***, Chinese and English edition, 214 p; China Science and Technology Press, Beijing.

1998年，陈之端、冯旻编译，**植物系统学进展**，177页；科学出版社，北京。

1998, Zhi-Duan CHEN and Min FENG, '***Progress of Plant Systematics***', 177 p; Science Press, Beijing.

1998, Yan-Ming FANG, Johannes Enroth, Timo J. Koponen and Sinikka S. Piippo, The bryophytes of Jiangxi province, China: An annotated checklist, *Hikobia* 12: 343-363.

1998年，方炎明，Johannes Enroth，Timo J. Koponen，Sinikka S. Piippo，中国江西省苔藓植物名录，*Hikobia* 12：343-363。

1998年，徐炳声，中国植物分类学中的物种问题，植物分类学报36（5）：470-480。

1998, Ping-Sheng HSU, The species problem in plant taxonomy in China, *Acta Phytotaxonomica Sinica* 36(5): 470-480.

1998, Jin-Shuang MA and Quan-Ru LIU, The present situation and prospects of plant taxonomy in China, *Taxon* 47(1): 67-74.

1998年，马金双、刘全儒，中国植物分类学的现状与展望，*Taxon* 47（1）：67-74。

1998年，缪汝槐编著，**植物地理学**，381页；中山大学出版社，广州。

1998, Ru-Huai MIAO, '***Plant Geography***', 381 p; Sun Yat-Sen University Press, Guangzhou.

759 中国植物学会编，1998，中国植物学会六十五周年年会学术报告及论文摘要汇编，638页；中国林业出版社，北京。
760 马金双，2001，《云南植物采集史略》读后感，云南植物研究23（3）：350-351。

1998, Toby Musgrave, Chris Gardner and Will Musgrave, ***The Plant Hunters****—Two hundred years of adventure and discovery around the world*, 224 p; Ward Lock, London. Translated into Chinese in 2005, 2006 and 2014.

2005、2006、2014 年，杨春丽、袁瑀译，**植物猎人** —— 探索发掘世界上两百年历程的植物，200 页，2005；希望出版社，太原；杨春丽、袁瑀译，**植物猎人** —— 探索发掘世界上两百年历程的植物，中文繁体版，276 页，2006；高谈文化事业，台北；杨春丽、袁瑀译，**改变世界的探险家 —— 植物猎人**，中文繁体版，288 页，2014；华滋出版，台北。

1998, Sinikka S. Piippo, Xiao-Lan HE, Timo J. Koponen, Paul L. Redfearn. Jr. and Shin-Chiang LEE, Hepaticae from Yunnan, China, with a checklist of Yunnan Hepaticae and Anthocerotae, *The Journal of the Hattori Botanical Laboratory* 84: 135-158.

1998 年，Sinikka S. Piippo、何小兰、Timo J. Koponen、Paul L. Redfearn，Jr.、黎兴江，中国云南的苔类植物，附云南苔类植物与角苔类植物名录，*The Journal of the Hattori Botanical Laboratory* 84：135-158。

1998, Alan T. Whittemore, Rui-Liang ZHU, Ren-Liang HU and Jian-Cheng ZHAO, A Checklist of the Liverworts of Xinjiang, China, *The Bryologist* 101(3): 439-443.

1998 年，Alan T. Whittemore、朱瑞良、胡人亮、赵建成，中国新疆苔类植物名录，*The Bryologist* 101（3）：439-443.

1998 年，吴鹏程，苔藓植物的分类及其系统排列，**苔藓植物生物学**，9-22 页；科学出版社，北京。

1998, Peng-Cheng WU[761], Taxonomy and systematic arrangement of the primary taxa of Bryophytes, ***Bryological Biology***, 9-22 p; Science Press, Beijing.

1998 年，吴玉虎主编，**青海植物名录**，411 页；青海人民出版社，西宁。

1998, Yu-Hu WU (editor in chief), ***Index Florae Qinghaiensis***, 411 p; Qinghai People's Publishing House, Xining.

1998 年，吴征镒、汤彦承、路安民、陈之端，试论木兰植物门的一级分类 —— 一个被子植物八纲系统的新方案，植物分类学报 36（5）：385-402。

1998, Zheng-Yi WU, Yen-Cheng TANG, An-Ming LU and Zhi-Duan CHEN, On primary subdivisions of the Magnoliophyta—Towards a new scheme for an eight-class system of classification of the angiosperms, *Acta Phytotaxonomica Sinica* 36(5): 385-402.

761 Pinyin spelling of Pan-Cheng WU.

1998, Zheng-Yi WU et al. (ed.), A Comprehensive Study of Magnoliidae sensu lato, with special consideration on the possibility and necessity for proposing a new Polyphyletic-polychronic-polytopic system of Angiosperms, ***Floristic Characteristics and Diversity of East Asian Plants***, 269–334 p; China Higher Education Press, Beijing and Springer-Verlag, Berlin.

1998 年，吴征镒等，综论广义木兰亚纲 —— 并探讨建立一个“多系、多期、多域”的可能性和必要性，**东亚植物区系特征和多样性**，269–334 页；中国高等教育出版社，北京，Springer-Verlag，柏林。

1998, Ao-Luo ZHANG and Su-Gong WU (editors in chief), ***Floristic Characteristics and Diversity of East Asian Plants***, Proceedings of the First International Symposium on Floristic Characteristics and Diversity of East Asian Plants, 25–27 July 1996, 486 p; China Higher Education Press, Beijing, Springer-Verlag, Berlin.

1998 年，张敖罗、武素功主编，**东亚植物区系特征和多样性**，首届东亚国际植物区系与多样性研讨会论文集，1996 年 7 月 25 日至 27 日，486 页；高等教育出版社，北京；Springer-Verlag，柏林。

1998 年，中国科学院昆明植物研究所编，**原本山川 极命草木** —— 中国科学院昆明植物研究所建所六十周年纪念文集，238 页；中国科学院昆明植物研究所，昆明（内部印制）。

1998, Kunming Institute of Botany, Chinese Academy of Sciences, '***Yuan Ben Shan Chuan Ji Ming Cao Mu***—*Memorial Works on the 60th anniversary of Kunming Institute of Botany, Chinese Academy of Sciences*', 238 p; Kunming Institute of Botany, Chinese Academy of Sciences, Kunming (internal publication).

1998 年，中国科学院植物研究所编，**中国科学院植物研究所建所七十周年纪念文集**，276 页；中国科学院植物研究所，北京（内部印制）。

1998, Institute of Botany, Chinese Academy of Sciences, '***Memorial Works on the 70th anniversary of Institute of Botany, Chinese Academy of Sciences***', 276 p; Institute of Botany, Chinese Academy of Sciences, Beijing (internal publication).

1998, Rui-Liang ZHU, May Ling SO and Li-Xin YE, A synopsis of the Hepatic flora of Zhejiang, China, *The Journal of the Hattori Botanical Laboratory* 84: 159–174.

1998 年，朱瑞良、苏美灵、叶立新，中国浙江苔类植物纪要，*The Journal of the Hattori Botanical Laboratory* 84：159–174。

1998, Shan-An HE (editor in chief), ***Rare and Precious Plants of China***, English and Chinese edition, 184 p; Shanghai Scientific and Technical Publishers, Shanghai.

1998 年，贺善安主编，**中国珍稀植物**，中英文双语版，184 页；上海科学技术出版社，上海。

1998 年，赵遵田、曹同主编，**山东苔藓植物志**，339 页；山东科学技术出版社，济南。

1998, Zun-Tian ZHAO and Tong CAO (editors in chief), ***Flora Bryophytorum Shandongicorum***, 339 p; Shandong Science and Technology Press, Jinan.

1998, Richard H. Ree, David E. Boufford and Susan L. Kelley, **Biodiversity of the Hengduan Mountains** and adjacent areas of south-central China website launched at http://hengduan.huh.harvard.edu/fieldnotes.

1998 年，Richard H. Ree、David E. Boufford、Susan L. Kelley，**横断山生物多样性** http://hengduan.huh.harvard.edu/fieldnotes 网站开通。

1998-1999, *Metasequoia glyptostroboides* after fifty years, Special Issue of *Arnoldia*, 58(4)/59(1): 1-84, Part I, A brief history, 11 papers re-published, and Part II, An update, 6 original papers published.

1998—1999 年，水杉引种五十年，*Arnoldia* 专刊 58（4）/59（1）：1-84；第 1 部分为简史，11 篇旧文重新发表，第 2 部分为进展，6 篇文章首次发表。

1998—2007 年，李承森主编，**植物科学进展**，7 卷本；高等教育出版社，北京[762]。

1998-2007, Cheng-Sen LI (editor in chief), ***Advances in Plant Sciences***, vols. 1-7; Higher Education Press, Beijing.

762 其中第 3 和第 4 卷与 Springer-Verlag，Heidelberg 联合出版。

1999

1999 年 6 月 30 日，吴永华著，**台湾植物探险**——19 世纪西方人在台湾采集植物的故事；台湾历史馆 11，302 页；晨星出版社，台中。

1999.06.30, Yung-Hua WU, ***Plant Hunting in Formosa***—*A History of Botanical Exploration in Formosa in the nineteenth Century*, Taiwan History Room 11, 302 p; Morning Star Publisher Inc., Taichung [Taizhong].

1999.08.01-08.07, Sixteenth International Botanical Congress held in St. Louis, Missouri, United States of America, 133 attendees from China.[763]

1999 年 8 月 1 日至 7 日，第十六届国际植物学大会于美国密苏里州圣路易斯举行，中国代表 133 人参加。

1999 年 9 月 11 日，位于陕西杨陵的陕西省中国科学院西北植物研究所合并入新成立的西北农林科技大学。原西北植物研究所标本馆（WUK）隶属于生命科学学院植物科学研究所。

1999.09.11, Northwest Institute of Botany, Shaanxi Province and Chinese Academy of Sciences, Yangling, Shaanxi, merged into new formed Northwest A&F University. The former herbarium (WUK) is within the Institute of Botany, College of Life Sciences.

1999.10.01-10.20, North American China Plant Expedition Consortium to Sichuan. Participants: Shawn Belt and Edward Garvey, US National Arboretum, Washington, D.C., Jeffrey Stites, Longwood Gardens, Kennett Square, Pennsylvania, with Qing WANG, Nanjing Botanical Garden Memorial Sun Yat-Sen.

1999 年 10 月 1 日至 20 日，由美国华盛顿特区国家树木园 Shawn Belt 和 Edward Garvey、宾夕法尼亚州 Kennett Square 市长木植物园 Jeffrey Stites 组成的北美中国植物考察联盟到四川采集，中方成员为南京中山植物园汪庆。

1999 年，陈忠毅主编，**中国科学院华南植物研究所建所七十周年纪念文集 1929—1999**，160 页；中国科学院华南植物研究所，广州（内部印制）。

1999, Zhong-Yi CHEN (editor in chief), '***Memorial Works on the 70th anniversary of South China Institute of Botany, Chinese Academy of Sciences, 1929–1999***', 160 p; South China Institute of Botany,

763 XVI International Botanical Congress, 1999, Participants of the XVI International Botanical Congress 1999, *Proceedings of the XVI International Botanical Congress*, St. Louis, 153 pages.

Chinese Academy of Sciences, Guangzhou (internal publication).

1999, Yu JIA and Peng-Cheng WU, Current Chinese bryological literature IV, *Chenia* 6: 127–135.

1999 年，贾渝、吴鹏程，中国苔藓植物新文献 IV，隐花植物生物学 6：127-135。

1999 年，上海科学院编著，徐炳声主编，**上海植物志**，2 卷本；上册：953 页；下册：596 页；上海科学技术文献出版社，上海。

1999, Science and Technology Academy of Shanghai, Ping-Sheng HSU (editor in chief), ***The Plants of Shanghai***, vols. 1: 953 p; 2: 596 p; Shanghai Scientific and Technological Literature Press, Shanghai.

1999 年，路安民主编，**种子植物科属地理**，664 页；科学出版社，北京。

1999, An-Ming LU (editor in chief), ***The Geography of Spermatophytic Families and Genera***, 664 p; Science Press, Beijing.

1999 年，张宪春、邢公侠编，**纪念秦仁昌论文集 —— **纪念秦仁昌一百周年诞辰蕨类植物学研究论文集，503 页；中国林业出版社，北京。

1999, Xian-Chun ZHANG and Kung-Hsia SHING, ***Ching Memorial Volume****—a collection of pteridological papers published to commemorate the centenary of the birth of Professor Ren-Chang CHING*, 503 p; China Forestry Publishing House, Beijing.

1999, 2000, Dian-Xiang ZHANG, Richard M. K. Saunders and Chi-Ming HU, *Corsiopsis chinensis* gen. et sp. nov. (Corsiaceae): first record of the family in Asia, *Systematic Botany* 24(3): 311–314, 1999; Dian-Xiang ZHANG, Addition to the Flora Reipublicae Popularis Sinicae: the family Corsiaceae, *Acta Phytotaxonomica Sinica* 38(6): 578–581, 2000.

1999、2000 年，张奠湘、Richard M. K. Saunders、胡启明，中华白玉簪新属和新种 —— 亚洲首次记载白玉簪科，*Systematic Botany* 24（3）: 311-314，1999；张奠湘，《中国植物志》增补：白玉簪科，植物分类学报 38（6）：578-581，2000。

1999, 2004, 2010, 2012, Gurcharan Singh, ***Plant Systematics***, 258 p, 1999; Science Publishers, Enfield, NH; ***Plant Systematics – Theory and Practice***, 370 p, 1999; Oxford and IBH, New Delhi; ***Plant Systematics – An Integrated Approach***, Ed. 2, 562 p, 2004; Science Publishers, Enfield, NH; ***Plant Systematics – Theory and Practice***, completely revised and enlarged 2nd edition, 518 p, 2004; Oxford and IBH, New Delhi; ***Plant Systematics – An Integrated Approach***, Ed. 3, 742 p, 2010; Science Publishers, Enfield, NH, 756 p, 2012; Science Publishers, Enfield, NH. Translated into Chinese in 2008.

2008 年，刘全儒、郭延平、于明译，**植物系统分类学 —— **综合理论及方法，436 页；化学工业出版社，北京。[764]

1999-2011, Editorial Committee of Moss Flora of China, Pan-Cheng WU and Marshall R. Crosby (editors in chief), ***Moss Flora of China***, vols. 1-8; 1: 273 p, 1999; 2: 283 p, 2001; 3: 141 p, 2003; 4: 211 p, 2007; 5: 423 p, 2011; 6: 221 p, 2002; 7: 258 p, 2008; 8: 385 p, 2005; Science Press, Beijing and Missouri Botanical Garden Press, St. Louis, Missouri.

1999—2011 年，中国藓类植物志编辑委员会，吴鹏程、Marshall R. Crosby 主编，**中国藓类植物志**，8 卷本；第 1 卷：273 页，1999；第 2 卷：283 页，2001；第 3 卷：141 页，2003；第 4 卷：211 页，2007；第 5 卷：423 页，2011；第 6 卷：221 页，2002；第 7 卷：258 页，2008；第 8 卷：385 页，2005；科学出版社，北京，密苏里植物园出版社，圣路易斯，密苏里。

1999—2013 年，2012 年，傅立国等主编，**中国高等植物**，14 卷本；第 1 卷：1 013 页，2012；第 2 卷：825 页，2008；第 3 卷：757 页，2000；第 4 卷：745 页，2000；第 5 卷：775 页，2003；第 6 卷：833 页，2003；第 7 卷：929 页，2001；第 8 卷：748 页，2001；第 9 卷：627 页，1999；第 10 卷：719 页，2004；第 11 卷：826 页，2005；第 12 卷：1 227 页，2009；第 13 卷：806 页，2002；第 14 卷：721 页，2013；青岛出版社，青岛；傅立国等主编，**中国高等植物**，修订版，14 卷本，2012[765]；青岛出版社，青岛。

1999-2013, 2012, Li-Kuo FU et al. (editors in chief), ***Higher Plants of China***, vols. 1-14; 1: 1,013 p, 2012; 2: 825 p, 2008; 3: 757 p, 2000; 4: 745 p, 2000; 5: 775 p, 2003; 6: 833 p, 2003; 7: 929 p, 2001; 8: 748 p, 2001; 9: 627 p, 1999; 10: 719 p, 2004; 11: 826 p, 2005; 12: 1,227 p, 2009; 13: 806 p, 2002; 14: 721 p, 2013; Qingdao Publishing House, Qingdao; Revised ed., vols. 1-14, 2012; Qingdao Publishing House, Qingdao.

1999, 2002, 2008, 2015, Walter S. Judd, Christopher S. Campbell, Elizabeth A. Kellogg, Peter F. Stevens, Michael J. Donoghue, ***PLANT SYSTEMATICS***: *A Phylogenetic Approach*, 464 p, 1999; 2nd edition, 576 p, 2002; 3rd edition, 620 p, 2008; 4th edition, 667 p, 2015; Sinauer Associate Sunderland. Translated into Chinese from third edition in 2012.

2012，李德铢等译，**植物系统学**，613 页；高等教育出版社，北京。

---

764 译自第 2004 年版。这本书目前有三版，每版又分为国际版（International edition）和印度版（Indian edition），而且每版的出版社、出版与印刷时间以及副标题、页码等都不完全相同（部分页码信息网络获得，仅供参考）。

765 2017 年 9 月 27 日作者从林祁（中国科学院植物研究所标本馆）处得知：①虽然第 1 版的第 14 卷 2013 年才出版，但修订版确实是 2012 年出版的；②第 2 版全书 14 卷共用一个刊号，而第 1 版每卷的刊号都是独立的，因此，两版共 15 个刊号；③林祁在第 7 卷主编和副主编同时署名的原因是，当时作为傅立国的博士后不能任主编，后来从华南正式调入植物所，可以任主编了，但出版时忘了从副主编栏删除；④第 2 版将前 13 卷的索引收在第 14 卷，处于第 1 版第 14 卷的两个科则改放在适当位置；因此，前后两版科属排列上略有不同。

2000

2000 年 2 月，郑武灿，**台湾植物图鉴**，上册：1–1 010 页、1–1 839 页；下册：1 011–1 837 页、1 839–1 987 页；编译馆，台北。[766]

2000.02, Wu-Tsang CHENG, '***The Illustrated Flora of Taiwan***', 1: 1-1,010 p and 1-1,839 p; 2: 1,011-1,837 p and 1,839-1,987 p; Translation House, Taipei [Taibei].

2000.06-07, Sino-British botanical expedition to Qinghai: Banma, Darley [Dari], Jigzhi [Jiuzhi], Maqên [Maqin]; Gansu: Minxian, Wenxian, Wudu; Sichuan: Aba, Jiuzhaigou, Songpan, Zoigê (Norgay). Participants: Shi-Long CHEN (HNWP), Christopher Grey-Wilson[767] (1944-, K), Ting-Nong HO (leader, HNWP), David G. Long (E), Ronald J. D. McBeath (E), Shang-Wu LIU (HNWP), Xue-Feng LU (HNWP), also Rosemary Steele and Elizabeth J. Strangman.

2000 年 6 月至 7 月，中英植物联合考察青海的班玛、达日、久治、玛沁，甘肃的岷县、文县、武都，四川的阿坝、九寨沟、松潘、诺尔盖，成员有中国科学院西北高原生物研究所陈世龙、何廷农（领队）、刘尚武、卢学峰，爱丁堡植物园 David G. Long 和 Ronald J. D. McBeath，英国皇家植物园邱园 Christopher Grey-Wilson（1944— ），以及 Elizabeth J. Strangman 和 Rosemary Steele。

2000.06.25-07.20: Chinese-American botanical expedition to Yunnan: Gaoligong Shan, Fugong, Lushui and Yongping. Bruce M. Bartholomew (CAS), Zhi-Ling DAO (KUN), Peter Fritsch (CAS), Heng LI (leader, KUN), Rong LI (1975-, KUN), Philip Thomas (E) and Zhong-Lang WANG (KUN) collected 1,357 numbers of vascular plants (CAS, E, GH, KUN).

2000 年 6 月 25 日至 7 月 20 日，中美植物联合考察云南的高黎贡山、福贡、泸水、永平，加州科学院 Bruce M. Bartholomew、Peter Fritsch，爱丁堡植物园 Philip Thomas，中国科学院昆明植物研究所刀志灵、李恒（领队）、李嵘（1975— ）、王仲朗，采集 1 357 号维管植物标本（CAS、E、GH、KUN）。

2000.07.05-08.18, Chinese-American botanical expedition to Yunnan: Dali, Lijiang, Zhongdian

766 民国八十九年。

767 Christopher Grey-Wilson was principal scientific officer at the Royal Botanic Gardens, Kew, research botanist, and editor of *Curtis' Botanical Magazine* for many years, editor of the *Alpine Gardener* (1990-2011) and also a freelance horticultural writer and photographer. He has participated in a number of major botanical expeditions in pursuit of the mountain floras of Europe and Asia, particularly of the Himalayan region and southwestern China. In 2008, he was awarded a Victoria Medal of Honour (VMH) by the Royal Horticultural Society for services to horticulture.

Chinese-American botanical expedition to Xizang, 5 July–18 August 2000: 1. Susan L. Kelley collecting specimens on slope, Bangda, Xizang; 2. Dry valley of the Nu Jiang (Salween River) (Photo provided by David E. Boufford)

2000 年 7 月 5 日至 8 月 18 日，中美植物学家在西藏野外：1. Susan L. Kelley 于邦达山坡上采集标本；2. 怒江（萨尔温江）的干旱河谷（相片提供者：鲍棣伟）

[Xianggelila] and Xizang: Baqên [Baqing], Baxoi [Basu], Biru, Bomê [Bomi], Chamdo [Qamdo, Changdu], Damxung [Dangxiong], Dêngqên [Dingqing], Gongbo'gyamda, Nyingchi [Linzhi], Maizhokunggar, Markham [Mangkang], Zogang [Zuogong]. David E. Boufford (A/GH), Yu JIA (PE), Susan L. Kelley (1954-, A), Brian A. Perry (1971-, FH), Richard H. Ree (A/F) and Shu-Kung WU (KUN) collected 928 numbers of vascular plants (A, CAS, E, F, KUN, MO, TI), 1,081 numbers of bryophytes (nos. J05060-J06141, FH, PE) and 175 numbers of fungi (nos. BAP209-BAP384, FH, HMAS).

2000年7月5日至8月18日，中美植物联合考察云南的大理、丽江、中甸，西藏的巴青、

Chinese-American botanical expedition to Xizang, 5 July-18 August 2000: 1. Ranwu Lake, Basu; 2. Bomi, National highway (G318) Shanghai to Zhangmu via Lhasa, the longest road (5476km) in China (Photo provided by David E. Boufford)

2000年7月5日至8月18日，中美植物学家在西藏野外：1. 西藏八宿县然乌湖；2. 上海经拉萨至樟木，中国最长的公路（5 476千米）国道（318）之波密（相片提供者：鲍棣伟）

八宿、比如、波密、昌都、当雄、丁青、工布江达、林芝、墨竹工卡、芒康、左贡；哈佛大学 David E. Boufford、Susan L. Kelley（1954—）和 Brian A. Perry（1971—），哈佛大学、芝加哥自然博物馆 Richard H. Ree，中国科学院昆明植物研究所武素功，中国科学院植物研究所贾渝，采集维管植物 928 号（A、CAS、E、F、KUN、MO、TI），苔藓 1 081 号（FH、PE），真菌 175 号（FH、HMAS）。

2000 年 8 月 26 日至 28 日，第三届全国苔藓植物学学术研讨会在贵阳贵州大学召开，来自全国 16 个高校和研究单位的 35 名代表参加了大会。

2000.08.26-08.28, Third National Bryophyte Symposium held at Guizhou University, Guiyang; 35 attendees from 16 institutions.

Participants at 3rd National Bryophyte Symposium, Guizhou, 26-28 August 2000 (Photo provided by Wei SHA)
2000 年 8 月 26 日至 28 日，第三届全国苔藓植物学学术研讨会于贵阳（相片提供者：沙伟）

2000 年 10 月，洪德元，Biosystematics—— 它的内含和中文翻译问题，植物分类学报 38（5）：490-496。

2000.10, De-Yuan HONG, Biosystematics—it's content and translation in Chinese, *Acta Phytotaxonomica Sinica* 38(5): 490-496.

2000 年 11 月 12 日至 15 日，第六届系统与进化植物学青年研讨会在南京江苏省中国科学院植物研究所 南京中山植物园和杭州浙江大学举行，110 人参加。

Participants at 6th Youth Symposium of Systematic and Evolutionary Botany, Nanjing, 12–15 November 2000 (Photo provided by Cheng-Xin FU)
2000 年 11 月 12 日至 15 日，第六届全国系统与进化植物学青年学术研讨会于南京（相片提供者：傅承新）

2000.11.12–11.15, Sixth Youth Symposium of Systematic and Evolutionary Botany held at Institute of Botany, Jiangsu Province and Chinese Academy of Sciences, Nanjing Botanical Garden Memorial Sun Yat-Sen, Nanjing, and Zhejiang University, Hangzhou; 110 attendees.

2000年，张宏达，种子植物新系统，植物学通报17：152–160。

2000，Hung-Ta CHANG, New system of seed plants (Spermatophyta), *Chinese Bulletin of Botany* 17: 152–160.

2000, Werner R. Greuter, John McNeill, Fred R. Barrie, Herve M. Burdet, Vincent Demoulin, Tarciso S. Filgueiras, Dan H. Nicolson, Paul C. Silva, Judith E. Skog, P. Trehane, Nicholas J. Turland and David L. Hawksworth (editors and compilers), ***International Code of Botanical Nomenclature* (St. Louis Code)**, adopted by the Sixteenth International Botanical Congress, St. Louis, Missouri, July–August, 1999; *Regnum Vegetabile* 138: 1–474 p. Translated into Chinese in 2001 and 2002.

2001、2002年，朱光华译，**国际植物命名法规**，412页，2001；科学出版社，北京；密苏里植物园出版社，圣路易斯；黄增泉译，**植物命名指南**，修订版，329页，2002；“行政院”农业委员会，台北。

2000年，蒋镇宇、牟善杰、许再文、陈建志著，**台湾苔类植物彩色图鉴**，399页；“行政院”农业委员会，台北。

2000, Tzen-Yuh CHIANG, Shann-Jye Moore, Tsai-Wen HSU, Chien-Chi CHEN, ***The Mosses of Taiwan***, 399 p; Council of Agriculture, Executive Yuan, Taipei [Taibei].

2000年，和匠宇、和铹宇著，**孤独之旅**——植物学家、人类学家约瑟夫·洛克和他在云南的探险经历，349页；云南教育出版社，昆明。

2000, Jiang-Yu HE and Lao-Yu HE, ‘***Journey Alone****—Joseph F. Rock, botanist, anthropologist and his adventures in Yunnan*’, 349 p; Yunnan Education Publishing House, Kunming.

2000年，米吉提·胡达拜尔地、徐建国主编，**新疆高等植物检索表**，788页；新疆大学出版社，乌鲁木齐。

2000, Mijit Hudaberdi and Jian-Guo XU (editors in chief), ***Claves Plantarum Xinjiangensis***, 788 p; Xinjiang University Press, Urumqi.

2000年，良振、成志著，**走向绿野**——蔡希陶传，525页；云南教育出版社，昆明。

2000, Zhen LIANG and Zhi CHENG, ‘***Zou Xiang Lü Ye****—A biography of Hse-Tao TSAI*’, 525 p;

Yunnan Education Publishing House, Kunming.

2000 年，林善雄编著，**台湾藓类植物彩色图鉴**，431 页；“行政院”农业委员会，台北。

2000, Shan-Hsiung LIN, ***The Liverwort Flora of Taiwan***, 431 p; the Council of Agriculture, The Executive Yuan, Taipei [Taibei].

2000, Chun-Liang PENG, Johannes Enroth, Timo J. Koponen and Sinikka S. Piippo, The bryophytes of Hubei province, China: an annotated checklist, *Hikobia* 13: 195–211.

2000 年，彭春良、Johannes Enroth、Timo J. Koponen、Sinikka S. Piippo，中国湖北苔藓名录，*Hikobia* 13：195–211。

2000, Ian M. Turner, A bibliography of materials pertaining to the marine and littoral vascular-plant floras and vegetations of the south China Sea, *The Raffles Bulletin of Zoology, Supplement* 8: 117–124.

2000 年，Ian M. Turner，中国南海海洋和沿海维管植物志和植被文献目录，*The Raffles Bulletin of Zoology, Supplement* 8：117–124。

2000 年，张志英、李继瓒、陈彦生著，**陕西种子植物名录**，128 页；陕西旅游出版社，西安。

2000, Zhi-Ying ZHANG, Ji-Zuan LI and Yan-Sheng CHEN, ‘***Checklist of Seed Plants from Shaanxi***’, 128 p; Shaanxi Tour Press, Xi’an.

2000 年，中国科学院青藏高原综合科学考察队，吴鹏程主编，**横断山区苔藓志**，742 页；科学出版社，北京。

2000, The Comprehensive Scientific Expedition to the Qinghai-Xizang Plateau, Academia Sinica, Pan-Cheng WU (editor in chief), ***Bryoflora of Hengduan Mts (Southwest China)***, 742 p; Science Press, Beijing.

2000 年，周厚高主编，**广西蕨类植物概览**，148 页；气象出版社，北京。

2000, Hou-Gao ZHOU (editor in chief), ***The Preliminary Study on Pteridophyte Flora from Guangxi, China***, 148 p; China Meteorological Press, Beijing.

2000+，湖南植物志编辑委员会，李丙贵主编，**湖南植物志**，7 卷本；第 1 卷：509 页，2004；第 2 卷：887 页，2000；第 3 卷：1 008 页，2010；湖南科学技术出版社，长沙。

2000+, Editorial Board of Flora of Hunan, Bing-Gui LI (editor in chief), ***Flora of Hunan***, vols. 1–7; 1: 509 p, 2004; 2: 887 p; 2000; 3: 1,008 p; 2010; Hunan Science and Technology Press, Changsha.

2000–2010, ***Type Specimens in the Herbarium of the Taiwan Forestry Research Institute***, Wen-Liang CHIOU, Kouh-Cheng YANG, Yuen-Po YANG, Kou-Shi HSU and Shun-Ying CHEN, 1: 92 p, 2000; Jer-Haur LI, Chien-Wen CHEN, Kuoh-Cheng YANG, Shih-Wen CHUNG and Wen-Liang CHIOU, 2: 230 p, 2004; 3: 219 p, 2005; 4: 213 p, 2006; 5: 204 p, 2007; and 6: 186 p, 2008; Jer-Haur LI, Chien-Wen CHEN, Kuoh-Cheng YANG, Wen-Liang CHIOU and Shih-Wen CHUNG, 7: 176 p, 2010; TFRI Extension Series No. 118, 165, 169, 172, 176, 181, 195; Taiwan Forestry Research Institute, Taipei [Taibei].

2000—2010 年，**台湾林业试验所标本馆之模式标本**，邱文良、杨国祯、杨远波、徐国士、陈舜英，第 1 卷：92 页，2000；李哲豪、陈建文、杨国祯、钟诗文、邱文良，第 2 卷：230 页，2004；第 3 卷：219 页，2005；第 4 卷：213 页，2006；第 5 卷：204 页，2007；第 6 卷：186 页，2008；李哲豪、陈建文、杨国祯、邱文良、钟诗文，第 7 卷：176 页，2010；台湾林业试验所林业丛刊第 118、165、169、172、176、181、195 号；台湾林业试验所，台北。

2000.03, Jin-Shuang MA, www.metasequoia.org (www.metasequoia.net, since August 2019), website for dawn redwood, ***Metasequoia* (Cupressaceae) and plant taxonomy** launched.

2000 年 3 月，马金双，**水杉（柏科）与植物分类学**网站 www.metasequoia.org（自 2019 年 8 月 www.metasequoia.net) 在线。

2001

2001.05.26-05.29, International Symposium on Chinese Pteridophytes held at Institute of Botany, Chinese Academy of Sciences, Beijing; 56 attendees, including 27 from China.

2001 年 5 月 26 日 至 29 日，中国蕨类植物国际研讨会在北京中国科学院植物研究所召开，56 人出席，包括中国学者 27 人。[768]

2001, Tzen-Yuh CHIANG, Tsai-Wen HSU, Shann-Jye Moore and Benito C. TAN, ***An Updated Checklist of Taiwan Mosses***, 36 p; Chinese Society of Biology, Taiwan.

2001 年，蒋镇宇、许再文、牟善杰、Benito C. TAN，**台湾藓类植物新目录**，36 页；中国生物学会，台湾。

2001, Susan M. Rossi-Wilcox, ***Chinese-English Glossary of Botanical Terms***, 270 p; Harvard University Herbaria, Cambridge, Massachusetts.

2001 年，Susan M. Rossi-Wilcox，**中英植物学术语词汇**，270 页；哈佛大学植物标本馆，剑桥，麻州。

2001 +，深圳仙湖植物园 / 深圳市中国科学院仙湖植物园，**仙湖**，vols. 1 +；本刊于 2001—2002 年试刊 2 年，刊名为**深圳仙湖植物园通讯**；2001—2002 年主编陈潭清；2003—2010 年主编李勇，执行主编张寿洲；2011—2012 年，主编朱伟华，执行主编张力；2012—2015 年，主编王晓明，执行主编张力；2016 年至今，主编张国宏，执行主编张力。

2001+, Shenzhen Fairylake Botanical Garden / Fairylake Botanical Garden, Shenzhen, and Chinese Academy of Sciences, ***Journal of Fairylake Botanical Garden*** 1+; as ***Newsletter of Fairylake Botanical Garden*** between 2001 and 2002; Editors in Chief: Tan-Qing CHEN, 2001-2002, Yong LI, with Shou-Zhou ZHANG as acting editor in chief, 2003-2010, Wei-Hua ZHU, with Li ZHANG as acting editor in chief, 2011-2012, Xiao-Ming WANG, with Li ZHANG as acting editor in chief, 2013-2015, Guo-Hong ZHANG, with Li ZHANG as acting editor in chief, since 2016.

2001 年，王培善、王筱英编著，**贵州蕨类植物志**，727 页；贵州科学技术出版社，贵阳。

2001, Pei-Shan WANG and Xiao-Ying WANG, ***Pteridophyte Flora of Guizhou***, 727 p; Guizhou Science and Technology Publishing House, Guiyang.

---

768 张宪春，2002，中国蕨类植物国际研讨会在北京香山举行，中国花卉协会蕨类植物分会简讯 5：18-19。

2001 年，杨亲二，过分依赖 SCI 正在损害我国的传统分类学研究 —— 从《Nature》上的两封信说开去，植物分类学报 39（3）：283-288。

2001, Qin-Er YANG, Over-reliance of SCI damages the research of traditional taxonomy in China—some thoughts after reading two letters in *Nature*, *Acta Phytotaxonomica Sinica* 39(3): 283-288.

2001、2007 年，焦瑜、李承森著，**中国云南蕨类植物**，238 页，2001；科学出版社，北京；**中国云南蕨类植物新编**，328 页，2007；科学出版社，北京。

2001 and 2007, Yu JIAO and Cheng-Sen LI, ***Yunnan Ferns of China***, 238 p, 2001; Science Press, Beijing; ***Yunnan Ferns of China Supplement***, 328 p, 2007; Science Press, Beijing.

2002

2002 年 9 月 24 日至 27 日，林祁、张春光、覃海宁，**全国生物标本馆技术研讨会论文集**[769]，341 页；中国科学技术出版社，北京。

2002.09.24-09.27, Qi LIN, Chun-Guang ZHANG and Hai-Ning QIN, ***Proceedings of Technical Symposium of Biological Museums***, 24-27 September 2000 (Xiangshan, Beijing), 341 p; China Science and Technology Press, Beijing.

2002.09.09-09.30, North American China Plant Expedition Consortium to Shaanxi. Participants: Anthony Aiello, Morris Arboretum, University of Pennsylvania, Philadelphia, Pennsylvania, Kris Bachtell, Morton Arboretum, Lisle, Illinois, Carole Bordelon, US National Arboretum, Washington, D.C., Peter Bristol, Holden Arboretum, Kirtland, Ohio, Yu-Dan TANG, Beijing Botanical Garden, Institute of Botany, Chinese Academy of Sciences, Beijing.

2002 年 9 月 9 日至 30 日，由美国宾夕法尼亚州费城宾夕法尼亚大学莫里斯树木园 Anthony Aiello、伊利诺伊州 Lisle 市莫顿树木园 Kris Bachtell、华盛顿特区国家树木园 Carole Bordelon 和俄亥俄州 Kirtland 市霍顿树木园 Peter Bristol 组成的北美中国植物考察联盟到陕西采集，中方成员为中国科学院植物研究所北京植物园唐玉丹。

2002.09.22-10.15, Chinese and Foreign botanical expedition to western Yunnan: Gaoligong Shan and Lushui. Bruce M. Bartholomew (CAS), Zhi-Ling DAO (KUN), Peter Fritsch (CAS), Yun-Heng JI (1972-), David Knott (E), Heng LI (leader, KUN), Rong LI (KUN), Ben-Xi LIU (KUN) and Mark F. Watson (E) collected 1,267 numbers of plant specimens (CAS, E, GH, KUN).

2002 年 9 月 22 日至 10 月 15 日，中外植物联合考察云南的高黎贡山和泸水，美国加州科学院 Bruce M. Bartholomew、Peter Fritsch，中国科学院昆明植物研究所刀志灵、纪运恒（1972— ）、李恒（领队）、李嵘、刘本玺，爱丁堡植物园 David Knott、Mark F. Watson，采集植物标本 1 267 号（CAS、E、GH、KUN）。

2002 年 11 月 12 日至 15 日，第七届系统与进化植物学青年研讨会在广东广州中山大学举行，全国 28 个省（市、区）65 个单位 180 多位代表出席。[770]

2002.11.12-11.15, Seventh Youth Symposium of Systematic and Evolutionary Botany held at Sun Yat-

769 国内 100 余家单位的 148 位代表参加会议。

770 第七届全国系统与进化植物学青年学术研讨会论文摘要集，139 页，2002；中国植物学会，广东广州。

Sen University, Guangzhou; more than 180 attendees from 65 institutions in 28 provinces.

2002.12.05-12.06, Chinese-American botanical expedition to Guangdong: Fengkai. David E. Boufford (A/GH) and Su-Hua SHI (1956-, SYS) collected 78 numbers of vascular plants (David E. Boufford and Su-Hua SHI nos. 30324-30402, A, CAS, PE, TI).

2002 年 12 月 5 日至 6 日，中美植物联合考察广东的封开，哈佛大学 David E. Boufford 和中山大学施苏华（1956—），采集维管束植物 78 号（A、CAS、PE、TI）。

2002, Irish Garden Plant Society, ***An Irish Plant Collector in China—Augustine Henry***, 42 p; Irish Garden Plant Society in association with Bord Glas, Dublin.

2002 年，Irish Garden Plant Society，**Augustine Henry—— 爱尔兰植物采集者在中国**，42 页；Irish Garden Plant Society in association with Bord Glas，都柏林。

2002, Yu JIA and Peng-Cheng WU, Current Chinese bryological literature V, *Chenia* 7: 193-195.

2002 年，贾渝、吴鹏程，中国苔藓植物新文献 V，隐花植物生物学 7：193-195。

2002, David B. Lellinger, ***A Modern Multilingual Glossary for Taxonomic Pteridology***, 263 p; American Fern Society, Washington, D.C.. Translated into Chinese in 2007.

2007 年，向建英、武素功译，马启盛校，**现代英中对照蕨类植物分类学词汇**，222 页；云南科技出版社，昆明。

2002, Jin-Shuang MA and Yun-Ping HUANG, Options and errors in citing Chinese personal names, *Taxon* 51(3): 521-522.

2002 年，马金双、黄运平，引证中国人名的选择与错误，*Taxon* 51（3）：521-522。

2002 年，祁承经、喻勋林编，**湖南种子植物总览**，615 页；湖南科学技术出版社，长沙。

2002, Cheng-Jing QI and Xun-Lin YU, ***A Survey of Hunan Seed Plants***, 615 p; Hunan Science and Technology Press, Changsha.

2002, Nicola Shulman, ***A Rage for Rock Gardening****—The Story of Reginald Farrer, Gardener, Writer and Plant Collector*, 118 p; David R. Godine Publisher, Boston.

2002 年，Nicola Shulman，**对岩石园艺的狂热激情** —— 园丁、作家、植物采集家 Reginald Farrer 的故事，118 页；David R. Godine Publisher，波士顿。

2002 年，汪庆，Courtois 在中国东部沿海的采集活动 [771]，隐花植物生物学 7：45-49。

2002, Qing WANG, Courtois' collecting activities in east China[772], *Chenia*, 7: 45-49.

2002, Yu-Huan WU, Chien GAO, & Benito C. TAN, The bryophytes of Gansu Province, China: A new annotated checklist, *Arctoa* 11: 11-22.

2002 年，吴玉环、高谦、Benito C. TAN，中国甘肃苔藓植物 —— 一个注释的新名录，*Arctoa* 11：11-22。

2002 年，吴征镒、路安民、汤彦承、陈之端、李德铢，被子植物的一个"多系—多期—多域"新分类系统总览，植物分类学报 40（4）：289-322。

2002, Zheng-Yi WU, An-Ming LU, Yen-Cheng TANG, Zhi-Duan CHEN and De-Zhu LI, Synopsis of a new "polyphyletic-polychronic-polytopic" system of the angiosperms, *Acta Phytotaxonomica Sinica* 40(4): 289-322.

2002, Hartmut Walravens, ***Joseph Franz Rock (1884-1962)—Berichte, Briefe und Dokumente des Botanikers, Sinologen und Nakhi-Forschers, mit einem Schriftenverzeichnis*** *[Joseph Franz Rock (1884-1962)—Reports, letters and documents of the Botanist, Sinologist and Naxi Scholar, with a list of publications]*, 452 p; Franz Steiner Verlag, Stuttgart.

2002 年，Hartmut Walravens，**植物学家、汉学家和纳西学者 Joseph Franz Rock（1884—1962）的报告、通信和文件以及论著目录**，452 页；Franz Steiner Verlag，Stuttgart。

771 戴丽娟，2013，从徐家汇博物院到震旦博物院 —— 法国耶稣会士在近代中国的自然史研究活动，"中央"研究院历史语言研究所集刊，第 84 本，第 2 分册，第 329-385 页；"中央"研究院，台北。

772 French Jesuit missionary and naturalist. Pére Frédéric Courtois arrived in China in 1901. He collected plants in the eastern part of the country from 1903 until his death in 1928. Born in Mayenne, France, Courtois was also a keen ornithologist. In China he organized numerous expeditions in Jiangsu, collecting some 40,000 plants from various locations including Huang Shan and other mountains in Anhui. Many of his collections were sent to the Muséum national d'Histoire naturelle in Paris. Courtois served as director of the Musée Heude (Heude Museum) in Zikawei (Xujiahui), Shanghai [(from https://plants.jstor.org/stable/history/10.5555/al.ap.person.bm000375502; accessed 5 July 2018)]. See also Tracey L. D. LU, 2014, *Museums in China—Power, politics and identities*, 235 p; Routledge, Abingdon, Oxon, UK.

2003

2003 年 8 月，高谦、赖明洲主编，**中国苔藓植物图鉴**，1 313 页；南天书局，台北。

2003.08, Chien GAO and Ming-Jou LAI (editors in chief), ***Illustrations of Bryophytes of China***, 1,313 p; Southern Media Publishing Co., Taipei [Taibei].

2003.08.19-09.10, Chinese-American botanical expedition to Yunnan: Bao Shan, Gaoligong Shan, Longling, Tengchong. Bruce M. Bartholomew (CAS), Zhi-Ling DAO (KUN), Zhu-Tan JIANG (1980-, KUN), Heng LI (leader, KUN), Rong LI (KUN), Xiao-Chun SHI (Gaoligongshan National Conservation Protection Agency) and Li-Hua ZHOU (1963-, CAS) collected 2,018 numbers of plant specimens (CAS, E, GH, KUN).

2003 年 8 月 19 日至 9 月 10 日，中美植物联合考察云南的保山、高黎贡山、龙陵、腾冲，加州科学院 Bruce M. Bartholomew 和周丽华（1963—），中国科学院昆明植物研究所李恒（领队）、刀志灵、蒋柱檀（1980—）、李嵘，高黎贡山国家级自然保护区施晓春，采集植物标本 2 018 号（CAS、E、GH、KUN）。

2003 年 10 月 10 日至 13 日，中国植物学会第十三届会员代表大会暨七十周年学术年会在四川省成都市召开，与会代表 600 多人。理事长：韩兴国。[773]

2003.10.10-10.13, Thirteenth National Congress and 70th anniversary meeting of Botanical Society of China held in Chengdu, Sichuan, more than 600 attendees; Xing-Guo HAN elected president.

2003 年 10 月，中国科学院华南植物研究所更名为中国科学院华南植物园；2003—2006 年陈勇（1957—）、2006—2015 年黄宏文、2015 年至今任海（1970—）任主任。

2003.10, South China Institute of Botany, Chinese Academy of Sciences, renamed South China Botanical Garden, Chinese Academy of Sciences, with directors Yong CHEN (1957-), 2003-2006; Hong-Wen HUANG, 2006-2015; Hai REN (1970-), since 2015.

2003 年 11 月，中国科学院武汉植物研究所更名为中国科学院武汉植物园；2003—2009 年黄宏文、2009—2014 年李绍华、2014—2015 年罗志强（1962—，副主任[774]）、2015 年至今张全发（1965—）任主任。

773 中国植物学会编，2003，中国植物学会七十周年年会论文摘要汇编（1933—2003），583 页；高等教育出版社，北京。
774 主持工作。

Participants at 13th National Congress and 70th anniversary meeting of Botanical Society of China, Chengdu, 10–13 October 2003 (Photo provided by Botanical Society of China)
2003 年 10 月 10 日至 13 日，中国植物学会第十三届会员代表大会暨七十周年年会于成都（相片提供者：中国植物学会）

Participants at 13th National Congress and 70th anniversary meeting of Botanical Society of China, Chengdu, 10–13 October 2003 (Photo provided by Botanical Society of China)

2003 年 10 月 10 日至 13 日，中国植物学会第十三届会员代表大会暨七十周年年会于成都（相片提供者：中国植物学会）

Today's South China Botanical Garden, Chinese Academy of Sciences: 1. Members of Plant Taxonomy Department, 10 January 2005; 2–3. Herbarium (IBSC, 2. inside, and 3. outside); 4. Entrance; 5. Conservation (Photo provided by Xiang-Xu HUANG and Rui-Jiang WANG)

今日中国科学院华南植物园：1. 2005 年 1 月 10 日植物分类室合影；2–3. 植物标本馆（2. 内部，3. 外部）；4. 植物园大门；5. 温室（相片提供者：黄向旭，王瑞江）

Sixty anniversary meeting (1956-2016) of Wuhan Botanical Garden, Chinese Academy of Sciences (Photo provided by Wuhan Botanical Garden)

中国科学院武汉植物园 60 周年纪念（1956—2016）（相片提供者：武汉植物园）

2003.11, Wuhan Institute of Botany, Chinese Academy of Sciences, renamed Wuhan Botanical Garden, Chinese Academy of Sciences, with directors Hong-Wen HUANG, 2003-2009; Shao-Hua LI, 2009-2014; Zhi-Qiang LUO (1962-, Vice Director), 2014-2015; Quan-Fa ZHANG (1965-), since 2015.

2003 年，黄增泉、萧锦隆，**台湾维管束植物名录**，254 页；南天书局，台北。

2003, Tseng-Chieng HUANG and Ching-Long HSIAO, ***Checklist of Taiwan Vascular Plants***, 254 p; Southern Media Publishing Co., Taipei [Taibei].

2003, Meng-Cheng JI & Benito C. TAN, A new checklist of mosses of Jiangxi Province, China. *Hikobia* 14: 87-106.

2003 年，季梦成、Benito C. TAN，江西藓类植物新名录，*Hikobia* 14：87-106。

2003 年，贾渝、吴鹏程，中国几个类群的研究进展Ⅲ：中国苔藓植物学研究（1993—2003），植物学报 45（增刊）：27-34。

2003, Yu JIA and Peng-Cheng WU, Advances in some plant groups in China III: Progress of the bryology in China (1993-2003), *Acta Botanica Sinica* 45 (Supplement): 27-34.

2003 年，李添进、周锦超、吴兆洪编著，**香港植物志 —— 蕨类植物门**，469 页；嘉道理农场暨植物园，香港。

2003，Wicky Tim-Chun LEE, Lawrence Kam-Chiu CHAU and Shiew-Hung WU, ***Flora of Hong Kong - Pteridophyta***, Flora Conservation Publication No 1, 469 p, Kadoorie Farm and Botanic Garden, Hong Kong.

2003 年，马其云编著，**中国蕨类植物和种子植物名称总汇**，1 561 页；青岛出版社，青岛。

2003, Qi-Yun MA, '***Names Compilation of Chinese Ferns and Seed Plants***', 1,561 p; Qingdao Publishing House, Qingdao.

2003 年，汤彦承、路安民，被子植物非国产科汉名的初步拟订，植物分类学报 41（3）：285-304。

2003, Yen-Cheng TANG and An-Ming LU, A tentative designation of Chinese names for the non-Chinese angiospermous families, *Acta Phytotaxonomica Sinica* 41(3): 285-304.

2003, Rui-Liang ZHU and May Ling SO, Liverworts and hornworts of Shangsi County of Guangxi (Kwangsi), with an updated checklist of the hepatic flora of Guangxi province of China, *Cryptogamie Bryologie* 24(4): 319-334.

2003 年，朱瑞良、苏美灵，广西上思县苔类和角苔类植物，及中国广西最新苔类植物名录，*Cryptogamie Bryologie* 24（4）：319-334。

2003 年，吴征镒、周浙昆、李德铢、彭华、孙航，世界种子植物科的分布区类型系统，云南植物研究 25（3）：245-257。

2003, Zheng-Yi WU, Zhe-Kun ZHOU, De-Zhu LI, Hua PENG and Hang SUN, The Areal-types of the world families of seed plants, *Acta Botanica Yunnanica* 25(3): 245-257.

2003 年，吴征镒，"世界种子植物科的分布区类型系统"的修订，云南植物研究 25（5）：535-538、543。

2003, Zheng-Yi WU, Revision on the Areal-types of the world families of seed plants, *Acta Botanica Yunnanica* 25(5): 535-538 and 543.

2003 年，吴征镒、路安民、汤彦承、陈之端、李德铢著，**中国被子植物科属综论**，1 209 页；科学出版社，北京。

2003, Zheng-Yi WU, An-Ming LU, Yen-Cheng TANG, Zhi-Duan CHEN and De-Zhu LI, ***The Families***

and Genera of Angiosperms in China—*A Comprehensive Analysis*, 1,209 p; Science Press, Beijing.

2003, Hartmut Walravens, ***Joseph F. Rock, Expedition zum Amnye Machhen in Südwest-China im Jahre 1926—im Spiegel von Tagebüchem und Briefen*** (*Joseph F. Rock: Expedition to Amnye Machen in southwest-China in 1926—in the mirror of diaries and letters*), 237 p; Harrassowitz, Wiesbaden.

2003 年，Hartmut Walravens，**Joseph F. Rock 1926 年对中国西南阿尼玛卿山的考察 —— 日记与信件之见**，237 页；Harrassowitz，Wiesbaden。

2003, Tom Christopher, ***In the Land of the Blue Poppies—****the collected plant hunting writings of Frank Kingdon Ward*, 243 p; Modern Library, New York.

2003 年，Tom Christopher，**绿绒蒿的故乡** ——Frank Kingdon Ward 关于采集植物探险之旅的著作，243 页；Modern Library，纽约。

2003 年，吴永华著，**台湾森林探险** —— 日治时期西方人在台采集植物的故事，台湾历史馆 28，205 页；晨星出版社，台中。

2003, Yung-Hua WU, ***Plant Hunting in Taiwan—****A History of Botanical Exploration in Taiwan in the Japanese colonial Taiwan (1895-1945)*, Taiwan History Room 28, 205 p; Morning Star Publisher Inc., Taichung [Taizhong].

2003 年，胡秀英等，**秀苑撷英**（胡秀英教授论文集），349 页；商务印书馆，香港。

2003, Shiu-Ying HU et al., ‘***Xiu Yuan Xie Ying—****Collected Works of Professor Shiu-Ying HU*’, 349 p; The Commercial Press, Hong Kong.

2003 年，贾渝、吴鹏程、汪楣芝、何思，藻苔纲，一个独特的苔藓植物类群，植物分类学报 41（4）：350-361。

2003, Yu JIA, Peng-Cheng WU, Mei-Zhi WANG and Si HE, *Takakiopsida*, a unique taxon of bryophytes, *Acta Phytotaxonomica Sinica* 41(4): 350-361.

2003 年，姜玉平，北平研究院植物学研究所的二十年，中国科技史料 24（1）：34-46。

2003, Yu-Ping JIANG, Twenty Years of the Institute of Botany of National Academy of Peiping, *China Historical Materials of Science and Technology* 24(1): 34-46.

2004

2004.04.21-05.10, Chinese and Foreign botanical expedition to Yunnan: Fugong and Gaoligong Shan. Kate Armstrong (E), Bruce M. Bartholomew (CAS), Zhi-Ling DAO (KUN), Peter Fritsch (CAS), Yun-Heng JI (KUN), Zhu-Tan JIANG (KUN), Heng LI (leader, KUN), Rong LI (KUN) and Li-Hua ZHOU (CAS) collected 1,676 numbers of plant specimens (A, CAS, E, GH, KUN).

2004 年 4 月 21 日至 5 月 10 日，中外植物联合考察云南的高黎贡山以及福贡，美国加州科学院 Bruce M. Bartholomew、Peter Fritsch 和周丽华，苏格兰爱丁堡植物园 Kate Armstrong，中国科学院昆明植物研究所刀志灵、纪运恒、蒋柱檀、李恒（领队）、李嵘，采集植物标本 1 676 号（A、CAS、E、GH、KUN）。

Chinese-Foreign botanical expedition to Yunnan, May 2004 (Photo provided by Heng LI / Rong LI)
2004 年 5 月中外植物联合考察云南（相片提供者：李恒 / 李嵘）

2004.06.28-06.29. Joint editorial committee meeting of Flora of China held at Institute of Botany, Chinese Academy of Sciences, Beijing. Attendees: Ihsan A. Al-Shehbaz (MO), Bruce M. Bartholomew (CAS), David E. Boufford (A/GH), Anthony R. Brach (A/GH), Shi-Long CHEN (HNWP), A. Michele Funston (MO), Michael G. Gilbert (MO), De-Yuan HONG (PE), Chi-Ming HU (IBSC), W. John Kress (US), De-Zhu

LI (KUN), Simon J. Owens (K), Ching-I PENG (HAST), Peter H. Raven (MO), Nicholas J. Turland (MO), Mark F. Watson (E), Bing XIA (NAS), Qin-Er YANG (PE), Xian-Chun ZHANG (PE) and Guang-Hua ZHU (MO). Paul Pui-Hay BUT (CUHK), Shou-Liang CHEN (NAS), Sing-Chi CHEN (PE), Lun-Kai DAI (PE) and Hsi-Wen LI (KUN) left the editorial committee in 2004. Anthony R. Brach (A/GH), Shi-Long CHEN (HNWP), Michael G. Gilbert (MO), De-Zhu LI (KUN), Nicholas J. Turland (MO), Bing XIA (NAS) and Xian-Chun ZHANG (PE) joined the editorial committee; Xiao-Lin GONG (PE) replaced Chia-Jui CHEN (PE) and Hong-Ya GU (PEY) as secretary.

2004 年 6 月 28 日至 29 日，英文版中国植物志编辑委员会联席会议在北京中国科学院植物研究所举行，密苏里植物园 Ihsan A. Al-Shehbaz、A. Michele Funston、Michael G. Gilbert、Peter H. Raven、Nicholas J. Turland、朱光华，加州科学院 Bruce M. Bartholomew，哈佛大学 David E. Boufford、Anthony R. Brach，中国科学院西北高原生物研究所陈世龙，中国科学院华南植物园胡启明，中国科学院植物研究所洪德元、杨亲二、张宪春，史密森学会的 W. John Kress，中国科学院昆明植物研究所李德铢，英国皇家植物园邱园 Simon J. Owens，“中央”研究院植物研究所彭镜毅，爱丁堡植物园 Mark F. Watson，江苏省中国科学院植物研究所 南京中山植物园夏冰出席。2004 年香港中文大学毕培曦，江苏省中国科学院植物研究所 南京中山植物园陈守良，中国科学院植物研究所陈心启、戴伦凯和中国科学院昆明植物研究所李锡文离开编辑委员会，哈佛大学 Anthony R. Brach，中国科学院西北高原生物研究所陈世龙，密苏里植物园 Michael G. Gilbert、Nicholas J. Turland，中国科学院昆明植物研究所李德铢，江苏省中国科学院植物研究所 南京中山植物园夏冰，中国科学院植物研究所张宪春加入编辑委员会；中国科学院植物研究所宫晓林取代陈家瑞和顾红雅为秘书。

2004.07.05-08.25, Chinese-American botanical expedition to western Sichuan: Daocheng, Dêrong [Derong], Litang, Xiangcheng; Xizang: Chamdo [Qamdo, Changdu], Jomda [Jiangda], Riwoqê [Leiwuqi], Markham [Mangkang], Zogang [Zuogong]; Yunnan: Songming, Shangri-La (Xianggelila), David E. Boufford (A/GH), Jia-Hui CHEN (1977-, KUN), Zai-Wei GE (1975-, KUN), Yu JIA (PE), Susan L. Kelley (A), Jun LI (1971-, KUN), Hidetsugu Miwa (1972-, KYO), Richard H. Ree (A/F), Hang SUN (KUN), Ji-Pei YUE (1974-, KUN), Yong-Hong ZHANG (1975-, KUN) and Zhu-Liang YANG (KUN) collected 1,191 numbers of vascular plant (A, CAS, E, F, KUN, MO, P, TI), 1,106 numbers of bryophytes (E, FH, PE) and 806 numbers of fungi (FH, KUN).

2004 年 7 月 5 日至 8 月 25 日，中美植物联合考察四川西部的稻城、得荣、理塘、乡城，西藏的昌都、江达、类乌齐、芒康、左贡，云南的嵩明、香格里拉[775]；哈佛大学 David E. Boufford、Susan L. Kelley，中国科学院昆明植物研究所陈家辉（1977—）、葛再伟（1975—）、李俊（1971—）、孙航、乐霁培（1974—）、张永洪（1975—）、杨祝良，中国科学院植物研究所贾渝，京都大学美和秀胤（1972—）、哈佛大学、芝加哥自然博物馆 Richard H. Ree，采集维管植物 1 191 号（A、CAS、E、F、KUN、MO、P、TI），苔藓 1 106 号（E、FH、PE），真菌 806 号（FH、KUN）。

775 即中甸，2001 年更改为现名。

2004 年 7 月 6 日至 11 日，中国科学院华南植物园和广东省植物学会共同主办的植物学命名法规研习班在华南植物园举行；全国各类高校和研究所 20 多家近 130 人参加。

2004.07.06-07.11, Workshop on botanical nomenclature, co-organized by the South China Botanical Garden, Chinese Academy of Sciences, and Botanical Society of Guangdong, was held at the South China Botanical Garden, Chinese Academy of Sciences; nearly 130 attendees from more than 20 universities, colleges and research institutions.

2004 年 7 月 21 日至 25 日，第八届系统与进化植物学青年研讨会在上海的复旦大学和合肥的安徽大学举行，约 100 人出席。

2004.07.21-07.25, Eighth Youth Symposium of Systematic and Evolutionary Botany held at Fudan University, Shanghai, and Anhui University, Hefei, 100 attendees.

2004.10.26-11.14, Chinese and Foreign botanical expedition to Yunnan: Dulongjiang valley and surrounding areas. J. Crinan M. Alexander (E), Kerstin Bach (Germany), Bruce M. Bartholomew (CAS),

Chinese-Foreign botanical expedition to Yunnan, October 2004 – Lunch along the way (Photo provided by Heng LI / Rong LI)
2004 年 10 月中外植物联合考察云南，午饭于路上（相片提供者：李恒 / 李嵘）

Participants at 8th Youth Symposium of Systematic and Evolutionary Botany, Shanghai (photo) and Hefei, 21-25 July 2004 (Photo provided by Cheng-Xin FU)
2004 年 7 月 21 日至 25 日，第八届全国系统与进化植物学青年学术研讨会于上海（合影）和合肥（相片提供者：傅承新）

Martyn Dickson (E), Zhi-Ling DAO (KUN), Zhu-Tan JIANG (KUN), Heng LI (leader, KUN), Rong LI (KUN), David G. Long (E), Xue-Mei ZHANG (1981-, KUN) and Li-Hua ZHOU (CAS) collected 2,096 numbers of plant specimens (CAS, E, GH, KUN).

2004 年 10 月 26 日至 11 月 14 日，中外植物联合考察云南的贡山独龙江流域及其周边地区，中国科学院昆明植物研究所刀志灵、蒋柱檀、李恒（领队）、李嵘、张雪梅（1981—），美国加州科学院 Bruce M. Bartholomew 和周丽华，苏格兰爱丁堡植物园 J. Crinan M. Alexander、David G. Long、Martyn Dickson，德国的 Kerstin Bach，采集植物标本 2 096 号（CAS、E、GH、KUN）。

2004.12.13-12.17, International Workshop on Conservation of Biodiversity of Chinese Bryophytes held at Shanghai Normal University, Shanghai, 22 attendees, including 6 invited from abroad.[776]

2004 年 12 月 13 日至 17 日，中国苔藓植物多样性保护国际研讨会在上海的上海师范大学举行，国内 16 位代表出席，国际代表 6 位。

Participants at International Workshop on Conservation of Biodiversity of Chinese Bryophytes, Shanghai, 13-17 December 2004 (Photo provided by Tong CAO)
2004 年 12 月 13 日至 17 日，中国苔藓植物多样性保护国际研讨会于上海（相片提供者：曹同）

2004 年 12 月，张宪春，中国蕨类植物系统分类学研究百年之回顾与前瞻，中国花卉协会蕨类植

776 Benito C. TAN, Tong CAO, Rui-Liang ZHU, 2005, Successful workshop on first official red list of endangered Chinese bryophytes, *The Bryological Times* 115: 13-15.

物分会简讯 10：1-9。

2004.12, Xian-Chun ZHANG, One Hundred Years of Systematic and Taxonomical Studies on Chinese Pteridophytes, *Newsletter of the Chinese Fern Society* 10: 1-9.

2004 年，张宏达、黄云晖、缪汝槐、叶创兴、廖文波、金建华著，**种子植物系统学**，699 页；科学出版社，北京。

2004, Hung-Ta CHANG, Yun-Hui HUANG, Ru-Huai MIAO, Chuang-Xing YE, Wen-Bo LIAO and Jian-Hua JIN, ***Systematics of Spermatophyta***, 699 p; Science Press, Beijing.

2004, Fa-Ti FAN, ***British Naturalists in Qing China—****Science, Empire and Cultural Encounter*, 238 p; Harvard University Press, Cambridge, Massachusetts.[777] Translated into Chinese in 2011 and 2018.

2011、2018 年，袁剑译，**清代在华的英国博物学家** —— 科学、帝国与文化遭遇，283 页，2011；中国人民大学出版社，北京；袁剑译，**知识帝国** —— 清代在华的英国博物学家，276 页，2018；中国人民大学出版社，北京。[778]

2004, De-Zhi FU, Yong YANG and Guang-Hua ZHU, A new scheme of classification of living gymnosperms at family level. *Kew Bulletin* 59(1): 111-116.

2004 年，傅德志、杨永、朱光华，现存裸子植物的一个科级新类系统，*Kew Bulletin* 59（1）：111-116。

2004 年，李法曾主编，**山东植物精要**，600 页；科学出版社，北京。

2004, Fa-Zeng LI (editor in chief), '***Essentials of Shandong Plants***', 600 p; Science Press, Beijing.

2004, Shan-Hsiung LIN, The history and present status of bryophyte herbaria in Taiwan, *'National' Science Museum Monographs* 24: 63-78.

2004 年，林善雄，台湾藓苔类植物标本馆的历史与现状，*'National' Science Museum Monographs* 24：63-78。

2004 年，刘家宜主编，**天津植物志**，994 页；天津科学技术出版社，天津。

2004, Jia-Yi LIU (editor in chief), ***Flora of Tianjin***, 994 p; Tianjin Science and Technology Press, Tianjin.

777 祝平一，2005，评介，新史学 16（3）：181-186。

778 袁剑，2018，分类、博物学与中国空间，读书 5：131-138。

2004, Brenda McLean, ***George Forrest Plant Hunter***, 239 p; Antique Collectors' Club, Woodbridge, Suffolk.

2004 年，Brenda McLean，**植物猎人 George Forrest**，239 页；Antique Collectors' Club，Woodbridge，Suffolk。

2004 年，汤彦承、路安民，《中国植物志》和《中国被子植物科属综论》所涉及“科”界定之比较，云南植物研究 26（2）：129–138。

2004, Yen-Cheng TANG and An-Ming LU, A comparison of family circumscription between Flora Reipublicae Popularis Sinicae [*Flora Reipublicae Popularis Sinicae*] and FGAC [*Families and Genera of Angiosperms in China*], *Acta Botanica Yunnanica* 26(2): 129–138.

2004，Sung WANG and Yan XIE (editors in chief), ***China Species Red List***, Chinese and English Editions, vol. 1: Red List, 468+224 p; Higher Education Press, Beijing.

2004 年，汪松、解炎主编，**中国物种红色名录**，中英双语版，第 1 卷：红色名录，468+224 页；高等教育出版社，北京。

2004 年，王印政、覃海宁、傅德志，中国植物采集简史，中国植物志 第 1 卷：第 658–732 页；科学出版社，北京。

2004, Yin-Zheng WANG, Hai-Ning QIN and De-Zhi FU, '*General History of Plant Collection of China*', *Flora Reipublicae Popularis Sinicae* 1: 658–732 p; Science Press, Beijing.

2004 年，邢福武主编，**澳门植物名录**，239 页；中国科学院华南植物园、澳门特别行政区民政总署园林绿化部，广州、澳门。

2004, Fu-Wu XING (editor in chief), ***Catálogo de Plantas de Macau—Check List of Macao Plants***, 239 p; South China Botanical Garden, Chinese Academy of Sciences and Department of Gardens and Green Areas, Civic and Municipal Affairs Bureau of Macao Special Administrative Region, Guangzhou and Macao.

2004 年，杨永、傅德志、王祺，被子植物花的起源 —— 假说与证据，西北植物学报 24（12）：2 366–2 380。

2004, Yong YANG, De-Zhi FU and Qi WANG, Origin of flower: hypotheses and evidence, *Acta Botanica Boreali-Occidentalia Sinica* 24(12): 2,366–2,380.

2004 年，中国科学院中国植物志编辑委员会，中国植物志，第 1 卷，1 044 页；科学出版社，北京。本卷为中国植物志 80 卷的最后 1 卷，详细参见：http://frps.iplant.cn。

2004, Editorial Committee of Flora Reipublicae Popularis Sinicae, *Flora Reipublicae Popularis Sinicae* 1: 1,044 p; Science Press, Beijing. This was the last of the 80 volumes of *Flora Reipublicae Popularis Sinicae* to be published. For details, please see: http://frps.iplant.cn.[779]

Flora Reipublicae Popularis Sinicae, final volume (Volume1) of 80 volumes published, 2004. Front Row (left-right): Li-Kuo FU, De-Shui NIU, Jie WANG, Feng-Qin TONG, Cheng-Yih WU, Da-Bao ZHU, Yi-Ling CHEN, Sing-Chi CHEN, Yan WANG; Middle Row: Zhen-Dai XIA, You-Xing LIN, An-Jen LI, Wen-Shu GAO, Lun-Kai DAI, Song-Jun LIANG, Jian-Fei ZENG, Han-Bi YANG, Tsue-Chih KU, Shu-Qin CAI; Back Row: Guo-Xun FU, Cheng-Sen LI, Zhong-Ren WANG, Chia-Jui CHEN, Zhan-Huo TSI, Ying-Jie ZHU, Xian-Chun ZHANG, Bo-Jian BAO, Lei SHI, Shu-Ren ZHANG (Photo provided by Institute of Botany)

2004 年，中国植物志 80 卷的最后一卷（第 1 卷）出版。前排（左起）：傅立国、牛德水、王捷、佟凤勤、吴征镒、朱大保、陈艺林、陈心启、王燕；中排：夏振岱、林尤兴、李安仁、高文淑、戴伦凯、梁松筠、曾建飞、杨汉碧、谷翠芝、蔡淑琴；后排：傅国勋、李承森、王中仁、陈家瑞、吉占和、朱颖杰、张宪春、包伯坚、石雷、张树仁（相片提供者：植物研究所）

779 Qin-Er YANG, Guang-Hua ZHU, De-Yuan HONG, Zheng-Yi WU and Peter H. Raven, 2005, World's largest flora completed, *Science* volume 309 Issue 5744, 2163 p; Jin-Shuang MA and Steve Clemants, 2006, A history and overview of the *Flora Reipublicae Popularis Sinicae* (FRPS, Flora of China, Chinese edition, 1959–2004), *Taxon* 55(2): 451–460.

2005

2005 年 2 月 25 日至 27 日，深圳市仙湖植物园[780]和植物分类学报在深圳联合举办中国植物系统学百年回顾学术交流会，暨庆祝胡秀英博士九十七华诞，全国 110 人参加。植物分类学报 2005 年第 43 卷第 5 期第 389–488 页刊登植物分类学报主编杨亲二、深圳市仙湖植物园主任李勇（1967— ）、著名植物分类学家吴征镒、王文采等数篇相关论文。

2005.02.25-02.27, Symposium on centennial review of plant systematics in China on the occasion of Dr. Shiu-Ying HU's 97th birthday held in Shenzhen at the Fairylakc Botanical Garden, Shenzhen, Guangdong; 110 attendees. Special issue 43(5): 389–488, 2005, of *Acta Phytotaxonomica Sinica* included related papers presented by Dr. Qin-Er YANG, chief editor of *Acta Phytotaxonomica Sinica*, Dr. Yong LI (1967-) , director of Shenzhen Fairylake Botanical Garden, Prof. Zheng-Yi WU and Prof. Wen-Tsai WANG et al.

2005.05.11-06.03, Chinese and Foreign botanical expedition to Yunnan (Gaoligong Shan): Baoshan, Lushui, Longling, Tengchong. Zhi-Ling DAO (KUN), Martyn Dickson (E), Peter Fritsch (CAS), Fu GAO (KUN), Yun-Heng JI (KUN), Heng LI (leader, KUN), Ben-Xi LIU (1976–, KUN) and Li-Hua ZHOU (CAS) collected 2,819 numbers of plant specimens (CAS, E, GH, KUN).

2005 年 5 月 11 日至 6 月 3 日，中外植物联合考察云南（高黎贡山）的保山、泸水、龙陵、腾冲，苏格兰爱丁堡植物园 Martyn Dickson，中国科学院昆明植物研究所刀志灵、高富、纪运恒、李恒（领队）、刘本玺（1976— ），美国加州科学院 Peter Fritsch 和周丽华，采集 2 819 号（CAS、E、GH、KUN）。

2005.07.17-07.23, Seventeenth International Botanical Congress held in Vienna, Austria, 37 attendees from China.[781]

2005 年 7 月 17 日至 23 日，第十七届国际植物学大会于奥地利维也纳举行，中国代表 37 人参加。

2005.07.07-08.28: Chinese-American botanical expedition to Sichuan: Ganzi (Garzê), Jiulong, Kangding, Luhuo, Barkam [Maerkang], Mianning, Zamtang [Rangtang], Sêrtar [Seda], Shimian, Shiqu, Xinlong and Yajiang; Yunnan: Anning and Songming; Qinghai: Banma; Xizang (Tibet): Jomda [Jiangda], David E. Boufford (A/GH), Jia-Hui CHEN (KUN), Kazumi Fujikawa ((1970–, MBK), Zai-Wei GE (KUN), Yu JIA (PE), Susan L. Kelley (A), Richard H. Ree (A/F), Hang SUN (KUN), Ji-Pei YUE (KUN), Da-Cai ZHANG (1977–, KUN), Jian-Wen ZHANG (1980–, KUN) and Yong-Hong ZHANG (KUN), collected 2,394

780 该园 1982 年规划，1983 年建设，1988 年正式对外开放，2008 年更名为深圳市中国科学院仙湖植物园；1982—1988 年叶锡洪、1989—2001 年陈谭清、2001—2011 年李勇、2011 年朱伟华、2012—2015 年王晓明、2015—2017 年张国宏、2018 年至今杨义标任主任。引自深圳仙湖植物园手册（中英文版电子版，2018）。

781 XVII International Botanical Congress, 2005, Participants List of the XVII International Botanical Congress, Vienna.

Participants at Symposium, centennial review of plant systematics in China on the occasion of Dr. Shiu-Ying HU's 97th birthday, Shenzhen, 25-27 February 2005 (Photo provided by Shou-Zhou ZHANG)
2005 年 2 月 25 日至 27 日，中国植物系统学百年回顾学术交流会暨庆祝胡秀英博士九十七华诞于深圳（相片提供者：张寿洲）

1

2

Chinese-American botanical expedition to Sichuan, Qinghai, Yunnan and Xizang, 7 July–28 August 2005: 1. Lamagetcu Nature Reserve—Meal time at campsite (left-right): Kazumi Fujikawa, Zai-Wei GE, Jian-Wen ZHANG, Jia-Hui CHEN, and guides; 2. Luhuo to Gyangze, Sichuan—pressing specimens (left-right): Kazumi Fujikawa, Jian-Wen ZHANG, Hang SUN; 3. Xizang-Yanjing to Markam, truck stuck in mud and blocking the way (Photo provided by David E. Boufford)

2005 年 7 月 7 日至 8 月 28 日：中美植物学家联合考察四川、青海、云南、西藏：1. 喇嘛格头自然保护区午餐营地（从左至右）：藤川和美、葛再伟、张建文、陈家辉、当地导游；2. 四川炉霍至岗责路上压制标本（从左至右）：藤川和美、张建文、孙航；3. 西藏盐井至马尔康，卡车陷入泥中堵住了路（相片提供者：鲍棣伟）

numbers of vascular plants (David E. Boufford et al. nos. 32600-34994, A, CAS, F, KUN, MBK, MO, P, PE, TI), 797 numbers of bryophytes (FH, PE) and 118 numbers of fungi (FH, KUN).

2005 年 7 月 7 日至 8 月 28 日，中美植物联合考察四川的甘孜、九龙、康定、炉霍、马尔康、冕宁、壤塘、色达、石棉、石渠、新龙、雅江，青海的班玛，云南的安宁、崇明，西藏的江达，哈佛大学 David E. Boufford、Susan L. Kelley，中国科学院昆明植物研究所陈家辉、葛再伟、孙航、乐霁培、张大才（1977—)、张建文（1980—)、张永洪，日本牧野植物标本馆藤川和美（1970—)，中国科学院植物研究所贾渝，哈佛大学、芝加哥自然博物馆 Richard H. Ree，采集维管植物 2 394 号（A、CAS、F、KUN、MBK、MO、P、PE、TI），苔藓 797 号（FH、PE），真菌 118 号（FH、KUN）。

2005 年 8 月 15 日至 19 日，全国植物进化生物学研讨会在安徽芜湖安徽师范大学召开，全国与会者 100 多位。

2005.08.15-08.19, National Symposium of Evolutionary Botanical Biology held at Anhui Normal University, Wuhu, Anhui, more than 100 attendees from throughout China.

2005.08.02-08.24, Chinese and Foreign botanical expedition to Yunnan: Fugong and Lushui. Simon Anthony (E), Bruce M. Bartholomew (CAS), Zhi-Ling DAO (KUN), Yun-Heng JI (KUN), Heng LI (leader, KUN), Ben-Xi LIU (KUN), Jin Hyub Paik (E), An-Jun TANG (KUN) and Li-Hua ZHOU (CAS) collected 3,123 numbers of vascular plants (CAS, E, GH, KUN).

2005 年 8 月 2 日至 24 日，中外植物联合考察云南的福贡和泸水，苏格兰爱丁堡植物园 Simon Anthony 和 Jin Hyub Paik，中国科学院昆明植物研究所刀志灵、纪运恒、李恒（领队）、刘本玺、唐安军，加州科学院 Bruce M. Bartholomew 和周丽华，采集维管植物 3 123 号（CAS、E、GH、KUN）。

2005.09.14-10.12, North American China Plant Expedition Consortium to Gansu. Participants: Anthony Aiello, Morris Arboretum, University of Pennsylvania, Philadelphia, Pennsylvania, Kris Bachtell, Morton Arboretum, Lisle, Illinois, Martin Scanlon, US National Arboretum, Washington, D.C., Kang WANG (1971-), Beijing Botanical Garden and Xue-Gang SUN (1960-), Forestry College of Gansu Agricultural University.

2005 年 9 月 14 日至 10 月 12 日，由美国宾夕法尼亚州费城宾夕法尼亚大学莫里斯树木园 Anthony Aiello、伊利诺伊州 Lisle 市莫顿树木园 Kris Bachtell 和华盛顿特区国家树木园 Martin Scanlon 组成的北美中国植物考察联盟到甘肃采集，中方成员为北京植物园王康（1971— ）和甘肃农业大学林学院孙学刚（1960— ）。

2005.10.26-10.28, International Bryological Symposium for Professor Pan-Chieh CHEN's Centennial Birthday held at Nanjing Normal University, Nanjing, Jiangsu; nearly 100 attendees from China, Hungary,

Japan, Russia and the United States of America.[782]

2005 年 10 月 26 日 —28 日，中国苔藓植物学奠基人陈邦杰先生百年诞辰国际学术研讨会于江苏南京的南京师范大学召开，来自中国、匈牙利、日本、俄罗斯、美国的 100 余名代表与会。[783]

2005 年，广西壮族自治区、中国科学院广西植物研究所，**广西自治区、中国科学院广西植物研究所发展七十年（1935—2005）**，242 页；广西壮族自治区、中国科学院广西植物研究所，桂林，广西（内部印制）。

2005, Guangxi Institute of Botany, Guangxi Zhuang Autonomous Region and Chinese Academy of Sciences, '***70 Years (1935–2005) of Development of Guangxi Institute of Botany, Guangxi Autonomous Region and Chinese Academy of Sciences***', 242 p; Guangxi Institute of Botany, Guangxi Zhuang Autonomous Region and Chinese Academy of Sciences, Guilin, Guangxi (internal publication).

2005 年，胡宗刚著，**静生生物调查所史稿**，250 页；山东教育出版社，济南。

2005, Zong-Gang HU, ***Historical Manuscript of Fan Memorial Institute of Biology***, 250 p; Shandong Education Press, Jinan.

2005 年，胡宗刚著，**不该遗忘的胡先骕**，200 页；长江文艺出版社，武汉。

2005, Zong-Gang HU, '***An Undeserved Forgotten Hsen-Hsu HU***', 200 p; Changjiang Literature and Art Press, Wuhan.

2005 年，黄普华、孙洪志编，**汉英拉植物分类群描述常见词汇**，299 页；东北林业大学出版社，哈尔滨。

2005, Pu-Hwa HUANG and Hong-Zhi SUN, ***A Common Glossary of Description to Plant Taxa in Chinese-English-Latin***, 299 p; Northeast Forestry University Press, Harbin.

2005，Yan LIU, Tong CAO and Shui-Liang GUO, The mosses of Zhejiang province, China: an annotated checklist, *Arctoa* 14: 95–133.

2005 年，刘艳、曹同、郭水良，中国浙江省藓类植物名录，*Arctoa* 14：95–133。

---

782 Editorial Committee of International Bryological Symposium for Prof. Pan-Chieh CHEN's Centennial Birthday, *Proceedings of International Bryological Symposium for Prof. Pan-Chieh CHEN's Centennial Birthday*, 78 p; Editorial Committee of International Bryological Symposium for Prof. Pan-Chieh CHEN's Centennial Birthday, Nanjing (issued internally).

783 中国苔藓植物学奠基人陈邦杰先生百年诞辰国际学术研讨会论文集编辑委员会，中国苔藓植物学奠基人陈邦杰先生百年诞辰国际学术研讨会论文集，78 页；中国苔藓植物学奠基人陈邦杰先生百年诞辰国际学术研讨会论文集编辑委员会，南京（内部印制）。吴鹏程，2006，中国苔藓植物学的开拓者 —— 陈邦杰教授，仙湖 1：36–37。

2005, Shu-Gang LU and Tsung-Yu YANG, The checklist of Taiwanese pteridophytes following Ching's system. *Taiwania* 50(2): 137-165.

2005 年，陆树刚、杨宗愈，按秦仁昌系统排列的台湾蕨类植物，*Taiwania* 50（2）：137-165。

2005 年，罗桂环著，**近代西方识华生物史**，434 页；山东教育出版社，济南。

2005, Gui-Huan LUO, ***History of Western Botanical and Zoological Studies in China***, 434 p; Shandong Education Press, Jinan.

2005, Jin-Shuang MA and Kerry Barringer, Dr. Hsen-Hsu HU (1894-1968)—a founder of modern plant taxonomy in China, *Taxon* 54(2): 559-566.

2005 年，马金双、Kerry Barringer，中国当代植物分类学的奠基人 —— 胡先骕博士（1894—1968），*Taxon* 54（2）：559-566。

2005 年，彭镇华主编，**中国长江三峡植物大全**，上卷：1-886 页；下卷：887-1 771 页；科学出版社，北京。

2005, Zhen-Hua PENG (editor in chief), ***Encyclopedia of Plants in Three Gorges of Yangtze River of China***, 1: 1-886 p, 2: 887-1,771 p; Science Press, Beijing.

2005 年，汤彦承、路安民，浅评当今植物系统学中争论的三个问题 —— 并系类群、谱系法规和系统发育种概念，植物分类学报 43（5）：403-419。

2005, Yen-Cheng TANG and An-Ming LU, Paraphyletic group, PhyloCode and Phylogenetic Species—the current debate and a preliminary commentary, *Acta Phytotaxonomica Sinica* 43(5): 403-419.

2005 年，邢福武主编，**中国的珍稀植物**，278 页；湖南教育出版社，长沙。

2005, Fu-Wu XING (editor in chief), ***Rare Plants of China***, 278 p; Hunan Education Publishing House, Changsha.

2005 年，王文采，关于我国植物系统学研究的一些感想和建议，植物分类学报 43（5）：398-402。

2005, Wen-Tsai WANG, Miscellaneous notes on plant systematics in China, *Acta Phytotaxonomica Sinica* 43(5): 398-402.

2005 年，吴家睿，论 Systems Biology 的译名问题，科学 57（4）：25-27。

2005, Jia-Rui WU, 'On the translation of Systems Biology', *Science (China)* 57(4): 25-27.

2005, Peng-Cheng WU, Paul Pui-Hay BUT and Mei-Zhi WANG, Hepatic flora of Hong Kong, *Chenia* 8: 127-131.

2005 年，吴鹏程、毕培曦、汪楣芝，香港苔类名录，隐花植物生物学 8：127-131。

2005 年，吴征镒、孙航、周浙昆、彭华、李德铢，中国植物区系中的特有性及其起源和分化，云南植物研究 27（6）：577-604。

2005, Zheng-Yi WU, Hang SUN, Zhe-Kun ZHOU, Hua PENG and De-Zhu LI, Origin and differentiation of endemism in the flora of China, *Acta Botanica Yunnanica* 27(6): 577-604.

2005 年，叶华谷、邢福武主编，**广东植物名录**，500 页；世界图书出版广东公司，广州。

2005, Hua-Gu YE and Fu-Wu XING (editors in chief), ***Checklist of Guangdong Plants***, 500 p; World Publishing Guangdong Corporation, Guangzhou.

2005 年，赵建成、王振杰、李琳主编，**河北高等植物名录**，132 页；科学出版社，北京。

2005, Jian-Cheng ZHAO, Zhen-Jie WANG and Lin LI (editors in chief), ***Higher Plant Catalogue of Hebei Province, China***, 132 p; Science Press, Beijing.

2005 年，郑朝宗主编，**浙江种子植物检索鉴定手册**，538 页；浙江科学技术出版社，杭州。

2005, Chao-Zong ZHENG (editor in chief), '***Identification Handbook of Seed Plants of Zhejiang***', 538 p; Zhejiang Science and Technology Publishing House, Hangzhou.

2005 年，周桂玲主编，**新疆高等植物科属检索表**[784]，124 页；新疆大学出版社，乌鲁木齐。

2005, Gui-Ling ZHOU (editor in chief), '***Keys to Families and Genera of Vascular Plants from Xinjiang***', 124 p; Xinjiang University Press, Urumqi.

2005+，甘肃植物志编辑委员会编，廉永善、孙坤主编，**甘肃植物志**，8 卷本；第 2 卷：607 页，2005；甘肃科学技术出版社，兰州。

2005+, Editorial Committee of Flora of Gansu, Yong-Shan LIAN (Yung-Shan LIAN) and Kun SUN (editors in chief), ***Flora of Gansu***, vols. 1-8, 2: 607 p, 2005; Gansu Science and Technology Press, Lanzhou.

2005、2010、2015 年，沈显生编著，**植物学拉丁文**，162 页，2005；中国科学技术大学出版社，合肥；

784 本书的起点类群为石松类和蕨类植物，不包括苔藓植物。

沈显生编著，**植物学拉丁文**，第 2 版，219 页，2010；中国科学技术大学出版社，合肥；沈显生、蔡斯编著，**植物学拉丁文**，第 3 版，232 页，2015；中国科学技术大学出版社，合肥。

2005, 2010, 2015, Xian-Sheng SHEN, ***Lingua Latinae Botanicae***, 162 p, 2005; University of Science and Technology of China Press, Hefei; Xian-Sheng SHEN, ***Lingua Latinae Botanicae***, ed. 2, 219 p, 2010; University of Science and Technology of China Press, Hefei; Xian-Sheng SHEN and Antonio Ziosi[785], ***Lingua Latinae Botanicae*** (*Lingua Latina Botanica*), ed. 3, 232 p, 2015; University of Science and Technology of China Press, Hefei.

2005—2007 年，邢福武主编，**澳门植物志**[786]，3 卷本；第 1 卷：328 页，2005；第 2 卷：405 页，2006；第 3 卷：314 页，2007；澳门特别行政区民政总署园林绿化部、中国科学院华南植物园，澳门、广州。澳门植物志总索引，61 页。

2005-2007, Fu-Wu XING (editor in chief), ***Flora de Macau* (*Flora of Macao*)**, vols. 1-3; 1: 328 p, 2005; 2: 405 p, 2006; 3: 314 p, 2007; Index, 61 p; Department of Gardens and Green Areas, Civic and Municipal Affairs Bureau of Macao Special Administrative Region, Macao, South China Botanical Garden, Chinese Academy of Sciences, Guangzhou.

785 Antonio Ziosi 博士在意大利 Universita Di Bologna 讲授拉丁文 (https://www.unibo.it/sitoweb/antonio.ziosi/en)。第 3 版随书附有他为本书配制的拉丁文字母名称的发音以及部分植物分类群学名的经典拉丁文配音的光盘。详细参见该书第 3 版前言。

786 马金双、刘全儒，2009，书评：《香港植物志》和《澳门植物志》，广西植物 29（4）：568。

2006

2006 年 1 月，洪菊生主编，**吴中伦云南考察日记**（1934 年 6 月 29 日至 1935 年 3 月 31 日），236 页；中国林业出版社，北京。

2006. 01, Ju-Sheng HONG (editor in chief), ***Reviewing Diary on Yunnan by Wu Zhonglun*** *(Chung-Lun WU's diaries on exploration in Yunnan, 29 June 1934-31 March 1935)*, 236 p; China Forestry Publishing House, Beijing.

2006.04.28-04.29, Joint editorial committee meeting of Flora of China held at the National Museum of Natural History, Smithsonian Institution, Washington, D.C., United States of America. Attendees: Ihsan A. Al-Shehbaz (MO), Bruce M. Bartholomew (CAS), David E. Boufford (A/GH), Anthony R. Brach (A/GH), Shi-Long CHEN (HNWP), Yan-Feng FU (Chinese Academy of Sciences), A. Michele Funston (MO), Michael G. Gilbert (MO), Xiao-Lin GONG (PE), De-Yuan HONG (PE), Chi-Ming HU (IBSC), W. John Kress (US), De-Zhu LI (KUN), William A. McNamara (Quarryhill Botanical Garden), Simon J. Owens (K), Ching-I PENG (HAST), Peter H. Raven (MO), Tetsuo Koyama (MBK), Nicholas J. Turland (MO), Mark F. Watson (E), Jun WEN (US), Qin-Er YANG (PE), Li-Bing ZHANG (1966-, MO) and Xian-Chun ZHANG (PE). Nicholas J. Turland (MO) replaced deceased Guang-Hua ZHU (MO) as co-director of the project. A. Michele Funston (MO) joined the editorial committee. A. Michele Funston (MO) and Simon J. Owens (K) left the editorial committee, and David A. Simpson (1955-, K) and Li-Bing ZHANG (MO) joined the editorial committee in 2007.

2006 年 4 月 28 日至 29 日，英文版中国植物志编辑委员会联席会议在华盛顿特区史密森学会举行，密苏里植物园 Ihsan A. Al-Shehbaz、A. Michele Funston、Michael G. Gilbert、Peter H. Raven、Nicholas J. Turland、张丽兵[787]（1966— ），加州科学院 Bruce M. Bartholomew，哈佛大学 David E. Boufford、Anthony R. Brach，中国科学院西北高原生物研究所陈世龙，中国科学院傅燕凤，中国科学院植物研究所宫晓林、洪德元、杨亲二、张宪春，中国科学院华南植物园胡启明，史密森学会的 W. John Kress、文军，中国科学院昆明植物研究所李德铢，Quarryhill Botanical Garden 的 William A. McNamara，英国皇家植物园邱园 Simon J. Owens，"中央"研究院植物研究所彭镜毅，牧野植物园小山铁夫，爱

787 四川人，1983 年 9 月至 1987 年 6 月兰州大学生物系本科毕业获学士学位，1987 年 9 月至 1990 年 7 月中国科学院成都生物研究所研究生毕业获硕士学位，1990 年 8 月至 1992 年 1 月任中国科学院成都生物研究所实习研究员，1992 年 1 月至 1995 年 1 月任助理研究员，1995 年 1 月至 1997 年 1 月任副研究员，1997 年 2 月至 2002 年 2 月于德国美茵茨大学（Johannes Gutenberg-Universität, Mainz）研究生毕业获博士学位；Ph.D. Dissertation: Phylogeny, biogeography and systematics of *Soldanella* L. and *Primula* L. sect. *Auricula* Duby (Primulaceae) based on molecular and morphological evidence, supervised by Joachim W. Kadereit (1956-); 2002—2006 年分别在密苏里大学、科罗拉多州立大学和杨百翰大学从事博士后研究，2006 年 10 月至 2011 年 5 月任密苏里植物园助理研究员，2011 年 6 月至今任副研究员；其中 2013 年任英文版中国植物志项目主任直至项目完成，2016 年任密苏里植物园 *Annals of the Missouri Botanical Garden* 主编。

丁堡植物园 Mark F. Watson 出席。密苏里植物园 Nicholas J. Turland 取代密苏里植物园已故朱光华为项目联合主任，密苏里植物园 A. Michele Funston 加入编辑委员会。2007 年密苏里植物园 A. Michele Funston 和英国皇家植物园邱园 Simon J. Owens 离开编辑委员会，英国皇家植物园邱园 David A. Simpson（1955—）和密苏里植物园张丽兵加入编辑委员会。

2006.05.14-06.05, Chinese and Foreign botanical expedition to Yunnan: Tengchong. Bruce M. Bartholomew (CAS), Zhi-Ling DAO (KUN), Yun-Heng JI (KUN), Xiao-Hua JIN (1975-, KUN), Heng LI (leader, KUN), Neil S. McCheyne (E) and Li-Hua ZHOU (CAS) collected 2,344 numbers of plant specimens (CAS, E, GH, KUN).

2006 年 5 月 14 日至 6 月 5 日，中外植物联合考察云南的腾冲，中国科学院昆明植物研究所刀志灵、纪运恒、金效华（1975—）、李恒（领队），美国加州科学院 Bruce M. Bartholomew 和周丽华，苏格兰爱丁堡植物园 Neil S. McCheyne，采集植物标本 2 344 号（CAS、E、GH、KUN）。

2006 年 6 月 9 日至 12 日，2006 年全国蕨类植物学术研讨会“21 世纪的中国蕨类植物学”在云南香格里拉高山植物园举行，来自 21 个省（市、区）及海外学者 81 人参加会议[788]。

2006.06.09-06.12, 2006 National Fern Symposium, Chinese Pteridophytes of the 21st Century, held at Shangri-La Alpine Botanical Garden, Shangri-La, Yunnan, 81 attendees from 21 provinces and abroad.

2006.07.14-08.31, Chinese-American botanical expedition to Sichuan: Baiyu, Batang, Daocheng, Dêgê [Dege], Ganzi (Garzê), Kangding, Luhuo, Litang, Xinlong and Yajiang; Yunnan: Shangri-La (Xianggelila) (Haba Xueshan). Bruce M. Bartholomew (CAS), David E. Boufford (A/GH), Jia-Hui Chen (KUN), Zai-Wei GE (KUN), Susan L. Kelley (A), Richard H. Ree (A/F), Hang SUN (KUN), Bo XU (1983-, YNUB), Zhu-Liang YANG (KUN), Ji-Pei YUE (KUN), Liang-Liang YUE (1982-, KUN), Da-Cai ZHANG (KUN), Jian-Wen ZHANG (KUN), Yong-Hong ZHANG (KUN) and Wei-Dong ZHU (1982-, KUN) collected 2,424 numbers of vascular plants (David E. Boufford et al. nos. 34997-37421, A, CAS, F, KUN, MO, PE, TI) and 556 numbers of fungi (FH, KUN).

2006 年 7 月 14 日至 8 月 31 日，中美植物联合考察四川的白玉、巴塘、稻城、德格、甘孜、康定、炉霍、理塘、新龙、雅江，云南的香格里拉（哈巴雪山），哈佛大学 David E. Boufford、Susan L. Kelley，加州科学院 Bruce M. Bartholomew，哈佛大学、芝加哥自然博物馆 Richard H. Ree，中国科学院昆明植物研究所陈家辉、葛再伟、孙航、乐霁培、岳亮亮（1982—）、杨祝良、张大才、张建文、张永洪、朱卫东（1982—），云南师范大学徐波（1983—），采集维管植物标本 2 424 号（A、CAS、F、KUN、MO、PE、TI），真菌 556 号（FH、KUN）。

788 中国花卉协会蕨类植物分会，2006，2006 年全国蕨类植物学术研讨会论文摘要，95 页；中国花卉协会蕨类植物分会，云南香格里拉。

Participants at National Fern Symposium, Shangri-La, 2006 (Photo provided by Xian-Chun ZHANG)
2006 年全国蕨类植物学术研讨会于香格里拉（相片提供者：张宪春）

Chinese-American botanical expedition to Sichuan and Yunnan, 14 July-31 August 2006 (left-right): Richard H. Ree, Ji-Pei YUE, Liang-Liang YUE, Bo XU, Jia-Hui CHEN, Da-Cai ZHANG, Hong HUI, Hui-Xin LI, Wei-Dong ZHU, Ying-Hui LI, Zai-Wei GE, David E. Boufford, Hang SUN, photo by Susan Kelley (Photo provided by http://hengduan.huh.harvard.edu/fieldnotes/expeditions/2006%20expedition%20team.JPG/view, accessed December 2017)

2006 年 7 月 14 日至 8 月 31 日，中美植物学家考察四川和云南（从左至右）：Richard H. Ree、乐霁培、岳亮亮、徐波、陈家辉、张大才、惠宏、李会新、朱卫东、李英辉、葛再伟、David E. Boufford、孙航，摄影者：Susan L. Kelley（相片提供者：横断山生物多样性网站）

2006.08.11-09.02, Chinese and Foreign botanical expedition to Yunnan: Gongshan, Dulongjiang valley and surrounding areas. Catherine Buch (CAS), Simon Crutchey (E), Zhi-Ling DAO (KUN), Peter Fritsch (CAS), Guang-Wan HU (1974-, KUN), Yun-Heng JI (KUN), Xiao-Hua JIN (KUN), Heng LI (leader, KUN) and Yi-Tao LIU (1955-, KUN) collected 3,197 numbers of vascular plants (CAS, E, GH, KUN).

2006 年 8 月 11 日至 9 月 2 日，中外植物联合考察云南的贡山独龙江流域及周边地区，中国科学院昆明植物研究所刀志灵、胡光万（1974— ）、纪运恒、金效华、李恒（领队）、刘怡涛（1955— ），美国加州科学院 Peter Fritsch 和 Catherine Buch，苏格兰爱丁堡植物园 Simon Crutchey，采集维管植物 3 197 号（CAS、E、GH、KUN）。

2006 年 8 月 20 日至 23 日，全国系统与进化植物学研讨会暨第九届系统与进化植物学青年研讨

Participants at National and Ninth Youth Symposium of Systematic and Evolutionary Botany, Xi'an, 20–23 August 2006 (Photo provided by Liang-Qian LI)
2006 年 8 月 20 日至 23 日，全国系统与进化植物学研讨会暨第九届系统与进化植物学青年研讨会于西安（相片提供者：李良千）

会于陕西西安西北大学举行，全国 43 个单位的 160 名代表出席。

2006.08.20–08.23, National and Ninth Youth Symposium of Systematic and Evolutionary Botany held at Northwest University, Xi'an, Shaanxi, 160 attendees from 43 institutions.

2006.09.02–09.17, Chinese-American botanical expedition to Xizang: Dagzê [Dazi], Gongbo'gyamda, Gyirong [Jilong], Lhasa, Nyingchi (Linzhi), Maizhokunggar, Nyalam [Nielamu], Tingri [Dingri]; and Yunnan: Gongshan. Deborah Bell (US), Vicki A. Funk (1941–2019, US), Jun-Feng LIANG (1971–, KUN), Ying MENG (1973–, KUN), Greg Mueller (F), Ze-Long NIE (1973–, KUN), Richard H. Ree (A/F), Hang SUN (KUN), Jun WEN (US), Shu-Kung WU (KUN), Yangjin ZHUOGA and KELSANG (XZ), Ji-Pei YUE (KUN) and Zhe-Kun ZHOU (KUN) collected 922 numbers of vascular plants (MacArthur Tibet Expedition nos. 1–922, F, KUN, US).

2006 年 9 月 2 日至 17 日，中美植物联合考察西藏的达孜、工布江达、吉隆、拉萨、林芝、墨竹工卡、聂拉木、定日，云南的贡山，史密森学会的 Deborah Bell、Vicki A. Funk（1941—2019）和文军，哈佛大学、芝加哥自然博物馆 Richard H. Ree，芝加哥自然博物馆 Greg Mueller，中国科学院昆明植物研究所梁俊峰（1971—）、孟颖（1973—）、聂泽龙（1973—）、孙航、武素功、乐霁培、周浙昆，西藏高原生物研究所央金卓嘎和格桑，采集维管束植物 922 号（F、KUN、US）。

2006 年，中国科学院武汉植物园，**中国科学院武汉植物园五十周年史料集（1956—2006）**，224 页；中国科学院武汉植物园，武汉（内部印制）。

2006, Wuhan Botanical Garden, Chinese Academy of Sciences, '***Fifty Anniversary (1956-2066) Collection of Historical Materials of Wuhan Botanical Garden, Chinese Academy of Sciences***', 224 p; Wuhan Botanical Garden, Chinese Academy of Sciences, Wuhan (internal publication).

2006 年，朱宗元，十七世纪至二十世纪中叶西方引种中国园林和经济植物史记，仙湖 1：2–12。

2006, Zong-Yuan ZHU, The history of introduction of landscape plants and economic plants from China to the West since 17th century, *Journal of Fairylake Botanical Garden* 1: 2–12.

2006, Tong CAO, Rui-Liang ZHU, Benito C. TAN, Shui-Liang GUO, Chien GAO, Peng-Cheng WU and Xing-Jiang LI, A report of the first national red list of Chinese endangered bryophytes. *The Journal of the Hattori Botanical Laboratory* 99: 275–295.

2006 年，曹同、朱瑞良、Benito C. TAN、郭水良、高谦、吴鹏程、黎兴江，中国首批国家级濒危苔藓植物红色名录报告，*The Journal of the Hattori Botanical Laboratory* 99：275–295。

2006 年，曹同、朱瑞良、郭水良、左本荣、于晶，中国首批濒危苔藓植物红色名录简报，植物研究 26（6）：756–762。

2006, Tong CAO, Rui-Liang ZHU, Shui-Liang GUO, Ben-Rong ZUO and Jing YU, A brief report of the first red list of endangered bryophytes in China, *Bulletin of Botanical Research* 26(6): 756-762.

2006, Tsung-Yu YANG, Type specimens of Taiwanese plants named by Dr. C. J. Maximowicz and housed at the herbarium, Komarov Botanical Institute of the Russian Academy of Sciences, St. Petersburg, Russia (LE), *'National' Museum of Natural Science, Special Publication Number* 10: 1-90 p.

2006 年，杨宗愈，典藏于俄罗斯圣彼得堡科马洛夫植物研究所植物标本馆（LE）Maximowicz 博士所命名的台湾植物模式标本，“国立”自然科学博物馆专刊 10：1-90 页。

2006 年，中国科学院中国植物志编辑委员会编，**中国植物志　中名和拉丁名总索引 1959—2004**，1 155 页；科学出版社，北京。

2006, Editorial Committee of Flora Reipublicae Popularis Sinicae, ***Indeces Nominum Sinensium et Latinorum Generales—Flora Reipublicae Popularis Sinicae 1959-2004,*** 1,155 p; Science Press, Beijing.

2006, Zhao-Hui ZHANG and Jia-Kuan CHEN, Marchantiophyta and Anthocerophyta in Guizhou province, P. R. China, *Journal of Bryology* 28: 170-176.

2006 年，张朝晖、陈家宽，中国贵州省苔纲和角苔纲，*Journal of Bryology* 28：170-176。

2006, Jin-Shuang MA, Recent progress in systematics in China, *Progress in Botany* 67: 361-382.

2006 年，马金双，中国系统学进展，*Progress in Botany* 67：361-382。

2006 年，叶华谷、彭少麟主编，**广东植物多样性编目**，657 页；世界图书出版公司，广州。

2006, Hua-Gu YE and Shao-Lin PENG (editors in chief), ***Plant Diversity Inventory of Guangdong***, 657 p; World Publishing Corporation, Guangzhou.

2006 年，吴征镒、周浙昆、孙航、李德铢、彭华，**种子植物分布区类型及其起源和分化**，566 页；云南科技出版社，昆明。

2006, Zheng-Yi WU, Zhe-Kun ZHOU, Hang SUN, De-Zhu LI and Hua PENG, ***The Areal-types of Seed Plants and Their Origin and Differentiation***, 566 p; Yunnan Scientific and Technological Press, Kunming.

2006，Jim Goodman, ***Joseph F. Rock and His Shangri-La***, 196 p; Caravan Press, Hong Kong.

2006 年，Jim Goodman，**Joseph F. Rock 和他的香格里拉**，196 页；Caravan Press，香港。

2006, Hartmut Walravens, ***Joseph Franz Rock Briefwechsel mit E. H. Walker 1938-1961*** (*Joseph*

Franz Rock correspondence with E. H. Walker—1938-1961), 328 p; Österreichische Akademie der Wissenschaften, Wien.

2006 年, Hartmut Walravens, **Joseph Franz Rock 与 E. H. Walker 1938—1961 年间的通信**, 328 页; 奥地利科学院，维也纳。

2006 年，吴永华著，**台湾特有植物发现史** —— 西元 1854—2003 年台湾特有维管束植物研究，台湾历史馆 32，797 页；晨星出版社，台中。

2006, Yung-Hua WU, ***The History of the Discovery of Endemic Plants in Taiwan***—*Study on the Endemic Vascular Plants in Taiwan during 1854-2003*, Taiwan History Room 32, 797 p; Morning Star Publisher Inc., Taichung [Taizhong].

2006 年，吴征镒，**吴征镒文集**，954 页；科学出版社，北京。

2006, Zheng-Yi WU, '***Collected Works of Zheng-Yi WU***', 954 p; Science Press, Beijing.

2006 年，郑重编著，**绿色之音**[789]，630 页；天马出版有限公司，香港。

2006, Zhong ZHENG, ***The Music of Green***, 630 p; Tianma Publisher, Hong Kong.

2006, John McNeill, Fred R. Barrie, Herve M. Burdet, Vincent Demoulin, David L. Hawksworth, Karol Marhold, Dan H. Nicolson, Jefferson Prado, Paul C. Silva, Judith E. Skog, John H. Wiersema and Nicholas J. Turland (eds.), ***International Code of Botanical Nomenclature* (Vienna Code)**, adopted by the Seventeenth International Botanical Congress, Vienna, Austria, July 2005, *Regnum Vegetabile* 146: 1-568 p; Gantner Verlag, Ruggell, Liechtenstein. Translated into Chinese in 2007.

2007 年，张丽兵译，**国际植物命名法规**，295 页，2007；科学出版社，北京、密苏里植物园，圣路易斯；黄增泉译，**植物命名指南**，繁体摘译版，279 页，2007；“行政院”农业委员会林务局，台北。

2006—，中国科学院植物研究所[790]，**中国数字植物标本馆**（CVH，http://www.cvh.ac.cn/）上线。

789 本书是作者“从事区域植物学基础研究工作和有关科研业务活动的论著和作品的选集”。

790 共建单位：中国科学院昆明植物研究所（KUN）、中国科学院华南植物园（IBSC）、中国科学院成都生物研究所（CDBI）、中国科学院武汉植物园（HIB）、中国科学院西双版纳热带植物园（HITBC）、中国科学院沈阳应用生态研究所（IFP）、中国科学院西北高原生物研究所（HNWP）、中国科学院新疆生态与地理研究所（XJBI）、广西壮族自治区中国科学院广西植物研究所（IBK）、西北农林科技大学生命科学学院植物科学研究所（WUK）、江苏省中国科学院植物研究所 南京中山植物园（NAS）、江西省中国科学院庐山植物园（LBG）、广西中医药研究院（GXMI）、贵州省生物研究所（HGAS）、湖南科技大学（HUST）、杭州师范大学（HTC）、杭州植物园（HHBG）、河南师范大学（HENU）、南京大学（N）、中山大学（SYS）、四川大学（SZ）、东北林业大学（NEFI）等。

中国最大的植物分类数据库网站，提供中国植物标本馆的植物标本数据且多数具有图片，进入中国植物志、彩色图库以及相关的其他数据。

2006-, Institute of Botany, Chinese Academy of Sciences, **Chinese Virtual Herbarium** (CVH) http://www.cvh.ac.cn/ launched. This is the largest online resources for plant taxonomy in China. The Chinese Virtual Herbarium (CVH) provides a consolidated database of plant specimens, many with images, from the herbaria of China, access to the flora of China, color photos of plants, as well as related data on Chinese plants.

2006，赵一之，**鄂尔多斯高原植物**，300 页；内蒙古大学出版社，呼和浩特。

2006, Yi-Zhi ZHAO, ***Vascular Plants of Plateau Ordos,*** 300p; Inner Mongolia University Press, Hohhot.

2007

2007.05.06-05.21, Chinese-American botanical expedition to Gansu: Wenxian. David E. Boufford (A/GH), Yu JIA (PE), Qing TIAN (GAUF) and Zhi-Yun ZHANG (1950-, PE) collected 428 numbers of vascular plants (David E. Boufford et al. nos. 37422-37850, A, CAS, MO, PE, TI) and 261 numbers of bryophytes (FH, MO, PE).

2007 年 5 月 6 日至 21 日，中美植物联合考察甘肃的文县，哈佛大学 David E. Boufford，中国科学院植物研究所贾渝和张志耘（1950—），甘肃农业大学田青，采集维管植物标本 428 号（A、CAS、MO、PE、TI），苔藓植物标本 261 号（FH、MO、PE）。

2007 年 5 月，中国科学院中国孢子植物志编辑委员会第五届编辑委员会成立，主编魏江春，副主编夏邦美、胡征宇（1957—）、庄文颖（1948—）、吴鹏程，编辑委员会 23 人。

2007.05, Fifth Consilio Florarum Cryptogamarum Sinicarum, Academiae Sinicae Edita organized, Jiang-Chun WEI as Editor in Chief, Bang-Mei XIA, Zheng-Yu HU (1957-), Wen-Ying ZHUANG (1948-) and Peng-Cheng WU as Vice Editors in Chief; 23 editorial committee members.

2007.07.15-08.31, Chinese-American botanical expedition to Gansu: Maqu; Qinghai: Jigzhi [Jiuzhi]; and Sichuan: Aba, Danba, Daofu, Heishui, Hongyuan, Jinchuan, Kangding, Barkam [Maerkang], Zamtang [Rangtang], Zoigê (Norgay), Sêrtar [Seda], Songpan and Xiaojin. David E. Boufford (A/GH), Kazumi Fujikawa (MBK), Zai-Wei GE (KUN), Yu JIA (PE), Susan L. Kelley (A), Richard H. Ree (A/F), Hang SUN (KUN), Bo XU (YNUB), Da-Cai ZHANG (KUN), Jian-Wen ZHANG (KUN), Ti-Chao ZHANG (KUN) and Wei-Dong ZHU (KUN) collected 2,643 numbers of vascular plants (David E. Boufford et al. nos. 37851-40494, A, CAS, F, KUN, MBK, MO, PE, TI), 652 numbers of bryophytes (A, FH, KUN) and 527 numbers of fungi (FH, KUN).

2007 年 7 月 15 日至 8 月 31 日，中美植物联合考察甘肃的玛曲，青海的久治，四川的阿坝、丹巴、道孚、黑水、红原、金川、康定、马尔康、壤塘、诺尔盖、色达、松潘、小金，哈佛大学 David E. Boufford、Susan L. Kelley，日本牧野植物标本馆藤川和美，中国科学院昆明植物研究所葛再伟、孙航、张大才、张建文、张体操、朱卫东，中国科学院植物研究所贾渝，哈佛大学、芝加哥自然博物馆 Richard H. Ree，云南师范大学徐波，采集维管植物标本 2 643 号（A，CAS、F、KUN、MBK、MO、PE、TI），苔藓标本 652 号（A、FH、KUN），真菌标本 527 号（FH、KUN）。

2007.07.22-07.28, World Conference of Bryology held at University of Malaya, Kuala Lumpur,

Chinese-American botanical expedition to south Gansu (Wenxian), 6-21 May 2007: 1. Zhi-Yun ZHANG and David E. Boufford; 2. Group photo in local conservation station (Left: Yu JIA, Middle: David E. Boufford, Right: Zhi-Yun ZHANG, other are local staff members) (Photo provided by Zhi-Yun ZHANG)

2007 年 5 月 6 日至 21 日，中美植物联合考察甘肃（文县）：1. 张志耘和 David E. Boufford；2. 当地保护区合影（左：贾渝，中间：David E. Boufford，右：张志耘；其余为当地工作人员）（相片提供者：张志耘）

Chinese-American botanical expedition to Gansu, Qinghai and Sichuan, 15 July-31 August 2007; front row (left-right): Richard H.Ree, Yu JIA, Susan Kelley, Kazumi Fujikawa, back row: Shun-Jun HUO, Hong HUI, Jian-Wen ZHANG, David E. Boufford, Bo XU, Hui-Xin LI, Local guide, Zai-Wei GE (Photo provided by David E. Boufford, http://hengduan.huh.harvard.edu/fieldnotes/expeditions/DSC05585.JPG/image_view_fullscreen, accessed December 2017)

2007 年 7 月 15 日至 8 月 31 日，中美植物联合考察甘肃、青海、四川；前排（从左至右）：Richard H. Ree、贾渝、Susan L. Kelley、藤川和美；后排：霍顺军、惠宏、张建文、David E. Boufford、徐波、李辉新、当地导游、葛再伟（相片提供者：鲍棣伟，来自横断山生物多样性网站）

Malaysia; attendees included Tong CAO, Ying CHANG (1970-), Wei SHA (1963-), Rui-Ping SHI (1982-), Wen YE (1982-), Li ZHANG (1967-), Yuan-Ming ZHANG (1972-) and Rui-Liang ZHU (1963-) from China.

2007年7月22日至28日，世界苔藓植物学大会在马来西亚吉隆坡马来亚大学召开，中国代表曹同、常缨（1970—）、沙伟（1963—）、师瑞萍（1982—）、叶文（1982—）、张力（1967—）、张元明（1972—）、朱瑞良（1963—）参加。

2007 年 8 月 10 日至 15 日，全国国际植物命名法规最新版（维也纳法规）讲习班在中国科学院昆明植物研究所举行，来自全国科研机构、大专院校的 150 多人参加。

2007.08.10-08.15, Workshop for the most recent edition of the *International Code of Botanical*

Participants at World Conference of Bryology, Kuala Lumpur, Malaysia, 22–28 July 2007 (Upper: Group photo; Lower: Field photo) (Photo provided by Wen YE)

2007 年 7 月 22 日至 28 日，世界苔藓植物学大会于马来西亚吉隆坡（上：全体合影；下：野外合影）（相片提供者：叶文）

Participants at workshop for the newest edition of International Code of Botamical Nomenclature (ICBN), Kunming, 10–15 August 2007 (Photo provided by Jing-Hua WANG)
2007 年 8 月 10 日至 15 日，最新版国际植物命名法规讲习班于昆明（相片提供者：王静华）

Nomenclature (Vienna Code) was held at the Kunming Institute of Botany, Chinese Academy of Sciences; attended by more than 150 participants from national institutions and universities.

2007.11.12-11.18, Fourth Symposium on Asian Pteridology held at Central Mindanao University, Mindanao, the Philippines; more than 170 attendees from throughout the world, including Yan-Fen CHANG (1980-), Wen-Liang CHIOU (1953-), Shi-Yong DONG (1970-), Wicky Tim-Chun LEE (1968-), Chun-Xiang LI (1969-), Yea-Chen LIU (1973-) and Yi-Shan ZHAO (1976-) from China.[791]

2007 年 11 月 12 日至 18 日，第四届亚洲蕨类植物学大会在菲律宾棉兰老中央大学举行，来自世界各地的 170 多位出席，中国学者包括常艳芬（1980—）、邱文良（1953—）、董仕勇（1970—）、李添进（1968—）、李春香（1969—）、刘以诚（1973—）、赵怡珊（1976—）。

2007—，*Sinopteris* 中国蕨 - 花协蕨类分会网址（www.fern.ac.cn）开通。

2007-, *Newsletter of Fern Committee of China Flower Association* website launched (www.fern.ac.cn).

2007 年，王锦秀、汤彦承，略论植物中文名称的统一，中国生物多样性保护与研究进展Ⅶ：135-148；附文：中国种子植物中（汉）名命名法规刍议，中国生物多样性保护与研究进展Ⅶ：149-152。

2007, Jin-Xiu WANG and Yen-Cheng TANG, Brief discussion on the unity of Chinese names for plants, *Advances in Biodiversity Conservation and Research in China* VII: 135-148; Appendix: Primary discussion on the nomenclatural rules for Chinese names of Chinese seed plants, VII: 149-152.

2007 年，张丽兵、杨亲二、Nicholas J. Turland、John McNeill，新版国际植物命名法规（维也纳法规）中的主要变化，植物分类学报 45（2）：251-255。

2007, Li-Bing ZHANG, Qin-Er YANG, Nicholas J. Turland and John McNeill, Main changes in the new edition of the International Code of Botanical Nomenclature—The "Vienna Code", *Acta Phytotaxonomica Sinica* 45(2): 251-255.

2007 年，张丽兵、Paul C. Silva、John McNeill、Nicholas J. Turland，国际植物命名法规中的术语介绍，植物分类学报 45（4）：593-598。

2007, Li-Bing ZHANG, Paul C. Silva, John McNeill, Nicholas J. Turland, On the terms in the International Code of Botanical Nomenclature, *Acta Phytotaxonomica Sinica* 45(4): 593-598.

---

791 Christopher R. Fraser-Jenkins, 2008, The 4th Symposium on Asian Pteridology in the Philippines, *Fiddlehead Forum* 35(4): 25-31.

2007 年，刘夙，《中国植物志》植物中文普通名的订正和读音的统一，中国生物多样性保护与研究进展Ⅶ：153-214。

2007, Su LIU, The correction of the Chinese common names in *Flora Reipublicae Popularis Sinicae* and the correspondence of their pronunciations, *Advances in Biodiversity Conservation and Research in China* VII: 153-214.

2007, Cun-Gen CHEN and Fischer Anton (general editors), Herrmann Walter, Ping-Hou YANG, Yan-Sheng CHEN, Shuo-Xin ZHANG, Zai-Min JIANG and Harald Forther (co-editors), ***Flora of the Loess Plateau in Central China****—a field guide*, 336 p; IHW-Verlag, Berchtesgaden.

2007 年，陈存根、Fischer Anton（总编辑），Herrmann Walter、杨平厚、陈彦生、张硕新、姜在民、Harald Forther（共同编辑），**华中黄土高原植物志** —— 野外指导，336 页；IHW-Verlag，Berchtesgaden。

2007, Hartmut Walravens, ***Joseph Franz Rock (1884-1962) Tagebuch der Reise von Chieng Mai nach Yunnan 1921-1922*** [*Joseph Franz Rock (1884-1962) Diary of the journey from Chieng Mai to Yunnan in 1921-1922*], 580 p; Österreichische Akademie der Wissenschaften, Wien.

2007 年，Hartmut Walravens，**1921—1922 年 Joseph Franz Rock（1884—1962）从清迈至云南的旅行日记**，580 页；奥地利科学院，维也纳。

2007, Jane Kilpatrick, ***Gifts from the Gardens of China****—The Introduction of Traditional Chinese Garden Plants to Britain 1698-1862*, 288 p; Frances Lincoln Ltd., London. Translated into Chinese in 2011.

2011 年，俞蘅译，**异域盛放** —— 倾靡欧洲的中国植物，287 页；南方日报出版社，广州。

2007, International Bryological Symposium on the occasion of Prof. Pan-Chieh Chen's centennial birthday, *Chenia* 9: 1-398.

2007 年，陈邦杰诞辰一百周年国际苔藓学术研讨会，隐花植物生物学 9：1-398。

2007 年，成晓、焦瑜著，**中国云南野生蕨类植物彩色图鉴**，314 页；云南科技出版社，昆明。

2007, Xiao CHENG and Yu JIAO, ***Native Ferns and Fern Allies of Yunnan China in Colour***, 314 p; Yunnan Science and Technology Press, Kunming.

2007 年，刘全儒、于明、马金双，中国地方植物志评述，广西植物 27（6）：844-849。

2007, Quan-Ru LIU, Ming YU and Jin-Shuang MA, Review on the Chinese local floras, *Guihaia* 27(6): 844-849.

2007, Peng-Cheng WU, Yu JIA and Mei-Zhi WANG, Retrospection of a century of Chinese bryology and the prospect, *Chenia* 9: 51-60.

2007 年，吴鹏程、贾渝、汪楣芝，中国苔藓植物学百年回顾与展望，隐花植物生物学 9：51-60。

2007、2011 年，冯双，**中山大学生命科学学院（生物学系）编年史 1924—2007**，281 页，2007；中山大学出版社，广州；冯双，**中山大学生命科学学院（生物学系）编年史 1924—2011**，346 页，2011；中山大学出版社，广州。

2007 and 2011, Shuang FENG, '***Chronicle of School of Life Sciences (Department of Biology), Sun Yat-Sen University 1924-2007***', 281 p, 2007, Sun Yat-Sen University Press, Guangzhou; Shuang FENG, '***Chronicle of School of Life Sciences (Department of Biology), Sun Yat-Sen University 1924-2011***', 346 p, 2011; Sun Yat-Sen University Press, Guangzhou.

2007-2011, Hong Kong Herbarium, Agriculture, Fisheries and Conservation Department and South China Botanical Garden, Chinese Academy of Sciences, Chi-Ming HU and Te-Lin WU (editors in chief), ***Flora of Hong Kong***, vols. 1-4; 1: 329 p, 2007; 2: 331p, 2008; 3: 352 p, 2009; 4: 379 p, 2011; Agriculture, Fisheries and Conservation Department, Government of the Hong Kong Special Administrative Region, Hong Kong. For the Chinese edition, please see 2015+.

2007—2011 年，渔农自然护理署香港植物标本室、中国科学院华南植物园，胡启明、吴德邻主编，**香港植物志**[792]，4 卷本；第 1 卷：329 页，2007；第 2 卷：331 页，2008；第 3 卷：352 页，2009；第 4 卷：379 页，2011；香港特别行政区政府渔农自然护理署，香港。另外，中文版参见 2015+ 条目。

792 马金双、刘全儒，2009，书评：《香港植物志》和《澳门植物志》，广西植物 29（4）：568。

2008

2008.02.28-03.02, Asian Bryophyte Conservation Workshop held at Singapore Botanic Gardens, Singapore; 16 attendees, including Tong CAO, Li ZHANG and Rui-Liang ZHU, from China.[793]

2008 年 2 月 28 日至 3 月 2 日，亚洲苔藓保护会议在新加坡的新加坡植物园举行，16 位代表出席，包括中国的曹同、张力和朱瑞良。

2008 年 2 月，**自然标本馆**上线（http://www.cfh.ac.cn/default.html），已拥有近 900 万张彩色图片（截至 2017 年 7 月）。

2008.02, '**Chinese Field Herbaria**' online (http://www.cfh.ac.cn/default.html), with nearly nine million color photos (to July 2017).

2008 年 3 月，中国科学院植物研究所植物标本馆的**中国植物图像库**上线（http://ppbc.iplant.cn），现有 27 263 分类群的图片（2017 年 12 月进入）。

2008.03, **The Plant Photo Bank of China** (PPBC), Herbarium of Institute of Botany, Chinese Academy of Sciences, online (www.ppbc.iplant.cn), with photos of 27,263 taxa (accessed December 2017).

2008.04.02-04.03[794], Joint editorial committee meeting of Flora of China held at Zhejiang University, Hangzhou, China. Attendees: Ihsan A. Al-Shehbaz (MO), Bruce M. Bartholomew (CAS), Stephen Blackmore (1952-, E), David E. Boufford (A/GH), Anthony R. Brach (A/GH), Shi-Long CHEN (HNWP), Cheng-Xin FU (1954-, HZU), Michael G. Gilbert (MO), Xiao-Lin GONG (PE), De-Yuan HONG (PE), Liang HONG (1959-, PE), Chi-Ming HU (IBSC), De-Zhu LI (KUN), William A. McNamara (Quarryhill Botanical Garden), Ching-I PENG (HAST), Lisa J. Pepper (MO), Peter H. Raven (MO), Nicholas J. Turland (MO), Mark F. Watson (E), Alexandra H. Wortley (E), Bing XIA (NAS), Qin-Er YANG (PE), Li-Bing ZHANG (MO) and Xian-Chun ZHANG (PE).

2008 年 4 月 2 日至 3 日，英文版中国植物志编辑委员会联席会议在杭州浙江大学举行，密苏里植物园 Ihsan A. Al-Shehbaz、Michael G. Gilbert、Peter H. Raven、Lisa J. Pepper、Nicholas J. Turland、张丽兵，加州科学院 Bruce M. Bartholomew，爱丁堡植物园 Stephen Blackmore（1952— ）、Mark F. Watson、Alexandra H. Wortley，哈佛大学 David E. Boufford、Anthony R. Brach，中国科学院西北高原

793 Thomas Hallingbck, 2009, Report from the Asian Bryophyte Conservation Workshop held in Singapore, February 28-March 2, 2008, *The Bryological Times* 128: 20-25.

794 Original agenda of Editorial Meeting of *Flora of China* from Zhejiang University provided by Yun-Peng ZHAO, who assisted at the meeting, shows that it was not April 28-29 as stated in page 17, volume one, *Flora of China*, 2013.

Joint editorial committee meeting of *Flora of China*, Zhejiang University, Hangzhou, 2-3 April 2008 (Photo provided by Yun-Peng ZHAO)

2008 年 4 月 2 日至 3 日，英文版中国植物志编辑委员会联席会议于杭州浙江大学举行（相片提供者：赵云鹏）

生物研究所陈世龙，浙江大学傅承新（1954— ），中国科学院植物研究所宫晓林、洪德元、洪亮（1959— ）、杨亲二、张宪春，中国科学院华南植物园胡启明，中国科学院昆明植物研究所李德铢，Quarryhill Botanical Garden 的 William A. McNamara，“中央”研究院植物研究所彭镜毅，江苏省中国科学院植物研究所 南京中山植物园夏冰出席。

2008 年 7 月 12 日至 15 日，中国植物学会第十四届会员代表大会暨七十五周年学术年会在甘肃省兰州市召开，全国代表 800 多人参加[795]。理事长为洪德元。

2008.07.12-07.15, Fourteenth National Congress and 75th anniversary meeting of Botanical Society of China held in Lanzhou, Gansu; more than 800 attendees; De-Yuan HONG elected president.

2008.09-10, Sino-UK Upper Dulong River botanical expedition to Northwest Yunnan; Zhi-Ling DAO (leader, KUN), Michael Wickenden (Cally Gardens), Michael Lear (Lear Associates), Yu-Hua LI (KUN), made 175 seed collections, 150 herbarium samples, among 350 collections[796].

2008 年 9 月至 10 月，中英植物联合考察滇西北的独龙江上游，中国科学院昆明植物研究所刀志灵（领队）、李玉华，以及英方 Cally Gardens 的 Michael Wickenden 和 Lear Associates 的 Michael Lear，采集植物 350 份，其中种子 175 份，标本 150 份。

2008.09.16-10.08, North American China Plant Expedition Consortium to Shaanxi. Participants: Anthony Aiello, Morris Arboretum, University of Pennsylvania, Philadelphia, Pennsylvania, Kris Bachtell, Morton Arboretum, Lisle, Illinois, Chris Carley, U.S. National Arboretum, Washington, D.C., Kang WANG, Beijing Botanical Garden.

2008 年 9 月 16 日至 10 月 8 日，由美国宾夕法尼亚州费城宾夕法尼亚大学莫里斯树木园 Anthony Aiello、伊利诺伊州 Lisle 市莫顿树木园 Kris Bachtell 和华盛顿特区国家树木园 Chris Carley 组成的北美中国植物考察联盟到陕西采集，中方成员为北京植物园王康。

2008 年 10 月，孙英宝、马履一、覃海宁，中国植物科学画小史，植物分类学报 46（5）：772-784。

2008.10, Ying-Bao SUN, Lü-Yi MA and Hai-Ning QIN, A brief history of botanical scientific illustration in China, *Journal of Systematics and Evolution* 46(5): 772-784.

2008 年 12 月 17 日至 19 日，2008 中国蕨类植物学术研讨会在广东深圳市中国科学院仙湖植物

795 中国植物学会编，2008，中国植物学会七十五周年年会论文摘要汇编（1933—2008），432 页；兰州大学出版社，兰州。

796 Michael Wickenden, Cally Gardens, Gatehouse of Fleet, Castle Douglas, Scotland, and Michael Lear, Lear Associates, England, 2010, *Exploring the upper Dulong river, the KWL Expedition to northwest Yunnan, September – October 2008*, 79 p; authors.

Fourteenth National Congress and 75th anniversary meeting of Botanical Society of China, Lanzhou, 12-15 July 2008 (Photo provided by Botanical Society of China)
2008 年 7 月 12 日至 15 日，中国植物学会第十四届会员代表大会暨七十五周年学术年会于兰州（相片提供者：中国植物学会）

Participants at National Fern Symposium, Shenzhen, 2008 (Photo provided by Xian-Chun ZHANG)
2008 年，中国蕨类植物学术研讨会于深圳（相片提供者：张宪春）

园举行，来自 21 个省（市、区）市及海外的 100 余人参加。[797]

2008.12.17-12.19, 2008 National Fern Symposium held at Fairylake Botanical Garden, Shenzhen and Chinese Academy of Sciences, Shenzhen, Guangdong, about 100 attendees from 21 provinces and foreign countries.

2008 年 12 月，杨远波、廖俊奎、唐默诗、杨智凯，**台湾种子植物要览**，278 页；“行政院”农业委员会林务局，台北。

2008.12, Yuen-Po YANG, Chun-Kuei LIAO, Mo-Shih TANG and Chih-Kai YANG, ***A Synopsis of Taiwan Seed Plants***, 278 p; Forestry Bureau, Council of Agriculture, Executive Yuan, Taipei [Taibei].

2008 年 12 月，刘红梅、王丽、张宪春、曾辉，石松类和蕨类植物研究进展 —— 兼论国产类群的科级分类系统，植物分类学报 46(6): 808-829。

2008.12, Hong-Mei LIU, Li WANG, Xian-Chun ZHANG and Hui ZENG, Advances in the studies of lycophytes and monilophytes with reference to systematic arrangement of families distributed in China, *Journal of Systematics and Evolution* 46(6): 808-829.

2008 年，中国科学院成都生物研究所，**岁月足迹**（1958—2008），140 页；中国科学院成都生物研究所，成都（内部印制）。

2008, Chengdu Institute of Biology, Chinese Academy of Sciences, '***Sui Yue Zu Ji - The Footprint of Time (1958-2008)***', 140 p; Chengdu Institute of Biology, Chinese Academy of Sciences, Chengdu (internal publication).

2008 年，王文采，植物分类学的历史回顾与展望，生物学通报 43（6）：1-3。

2008, Wen-Tsai WANG, 'Historical review and prospect of plant taxonomy', *Bulletin of Biology* 43(6): 1-3.

2008 年，吴玉虎著，**青藏高原维管植物及其生态地理分布**，1 369 页；科学出版社，北京。

2008，Yu-Hu WU, ***The Vascular Plants and Their Eco-Geographical Distribution of the Qinghai-Tibetan Plateau***, 1,369 p; Science Press, Beijing.

2008, De-Zhu LI, Floristics and plant biogeography in China, *Journal of Integrative Plant Biology* 50(7): 771-777.

2008 年，李德铢，中国植物区系和植物地理学，植物学报 50（7）：771-777。

797 中国花卉协会蕨类植物分会，2008 中国蕨类植物学术研讨会论文摘要集，149 页；中国花卉协会蕨类植物分会，深圳。

2008 年，郑一帆、郑瑾华编著，**植物拉丁学名及其读音**，126 页；广东科技出版社，广州。

2008, Yi-Fan ZHENG and Jin-Hua ZHENG, '***Botanical Latin Name and Its Pronunciation***', 126 p; Guangdong Science and Technology Press, Guangzhou.

2008 年，胡宗刚著，**胡先骕先生年谱长编**，688 页；江西教育出版社，南昌[798]。

2008, Zong-Gang HU, '***Compilation of Chronicle of Mr. Hsen-Hsu HU***', 688 p; Jiangxi Education Publishing House, Nanchang.

2008 年，郑万钧专集编辑委员会，张守攻主编，**郑万钧专集** —— 郑万钧林业学术思想研究，684 页；科学出版社，北京[799]。

2008，Editorial Committee of Collected Works on Zheng Wanjun, Shou-Gong ZHANG (editor in chief), ***Collected Works on Zheng Wanjun*** *(Wan-Chun CHENG)*, 684 p; Science Press, Beijing.

2008 年，冯耀宗编著，**大青树下** —— 跟随老师蔡希陶的三十年，100 页；云南科技出版社，昆明。

2008, Yao-Zong FENG, '***Da Qing Shu Xia****—Thirty Years Followed Teacher Hse-Tao TSAI*', 100 p; Yunnan Science and Technology Press, Kunming.

2008 年，吴征镒，**百兼杂感随忆**，594 页；科学出版社，北京。

2008, Zheng-Yi WU, '***Miscellany with Memory of Baijian***'[800], 594 p; Science Press, Beijing.

2008 年，中国科学院植物研究所志编纂委员会编，**中国科学院植物研究所所志**，970 页；高等教育出版社，北京。

2008, Editorial Committee of Records of Institute of Botany, Chinese Academy of Sciences, ***Records of Institute of Botany, Chinese Academy of Sciences***, 970 p; Higher Education Press, Beijing.

2008 年，中国科学院昆明植物研究所简史编纂委员会编，**中国科学院昆明植物研究所简史 1938—2008**，332 页；中国科学院昆明植物研究所简史编纂委员会，昆明（内部印制）。

2008, Editorial Committee of the Brief History of Kunming Institute of Botany, Chinese Academy of Sciences, '***The Brief History of Kunming Institute of Botany, Chinese Academy of Sciences, 1938-2008***', 332 p; Editorial Committee of Brief History of Kunming Institute of Botany, Chinese Academy of Sciences,

798 马金双，2008，新书介绍：胡先骕先生年谱长篇，植物分类学报 46（5）：793-794。

799 马金双，2009，书评：郑万钧专集，云南植物研究 31（3）：386-388。

800 Baijian is an courtesy name of Cheng-Yih WU.

Kunming (internal publication).

2008 年，中国科学院昆明植物研究所，**放飞梦想 收获希望 —— 中国科学院昆明植物研究所建所七十周年纪念文集 1938—2008**，128+ XI页；中国科学院昆明植物研究所，昆明（内部印制）。

2008, Kunming Institute of Botany, Chinese Academy of Sciences, '***Fang Fei Meng Xiang Shou Huo Xi Wang****—Memorial Works of the 70th anniversary of Kunming Institute of Botany, Chinese Academy of Sciences, 1938-2008*', 128+XI p; Kunming Institute of Botany, Chinese Academy of Sciences, Kunming (internal publication).

2008 年，许再富著，**沧桑葫芦岛 —— 中国科学院西双版纳热带植物园 50 年回顾**，110 页；云南科技出版社，昆明。

2008, Zai-Fu XU, '***Cang Sang Hu Lu Dao*** *(Gourd Island)—Fifty Year's Review of Xishuangbanna Tropical Botanical Garden, Chinese Academy of Sciences*', 110 p; Yunnan Science and Technology Press, Kunming.

2008 年，易富科主编，**中国东北湿地野生维管束植物**，上册：1–728 页，下册：729–1 268 页；科学出版社，北京。

2008, Fu-Ke YI (editor in chief), ***Wetland Wild Vascular Plants in Northeastern China***, 1: 1-728 p, 2: 729-1,268 p; Science Press, Beijing.

2008-2015, Alexander V. Galanin (editor in chief), ***Flora of Dahuria*** *(Vascular Plants)*[801], vols. 1-6; 1: 184 p, 2008; 2: 279 p, 2009; 3: 280 p, 2011; 4: 259 p, 2014; 5: 285 p, 2015; 6: 228 p, 2015; Botanical Garden-Institute, Far Eastern Branch, Russian Academy of Sciences, Vladivostok.

2008—2015 年，Alexander V. Galanin 主编，**达乌里植物志**（维管束植物），6 卷本：第 1 卷：184 页，2008；第 2 卷：279 页，2009；第 3 卷：280 页，2011；第 4 卷：259 页，2014；第 5 卷：285 页，2015；第 6 卷：228 页，2015；俄罗斯科学院远东分院植物园—植物研究所，海参崴。

801 地理范围包括俄罗斯达乌里地区和蒙古东北以及中国内蒙古大兴安岭西坡。目前只印刷了前 3 卷纸制版，其余仅为网络版；详细参见：http://ukhtoma.ru/geobotany/dahuria_01.html（accessed 6 October, 2016）。

2009

2009.07.03, Jiao LI, China searches for an 11th-hour lifesaver for a dying discipline, *Science* 325: 31.

2009 年 7 月 3 日，李娇，中国在为一个消亡中的学科寻找最后的救星，*Science* 325: 31。

2009.07.16–08.08, Chinese-American botanical expedition to Sichuan: Batang, Daocheng, Dêgê [Dege], Xiangcheng; Yunnan: Dêqên [Deqin], Shangri-La (Xianggelila); Xizang (Tibet): Baxoi [Basu], Banbar [Bianba], Chagyab [Zhag'yab, Chaya], Chamdo [Qamdo, Changdu], Jomda [Jiangda], Gonjo [Gongjue],

Chinese-American botanical expedition to Sichuan, Yunnan and Xizang, 16 July–8 August 2009 (left-right): Zhu-Liang YANG, Richard H. Ree, local police, Bang FENG, Ji-Pei YUE, local forester, Bruce M. Bartholomew, David E. Boufford, Deren A. Eaton and Hong HUI in Gongjue, Xizang, photo by Xin-Hui LI (http://hengduan.huh.harvard.edu/fieldnotes/expeditions/41477-41518-Group% 20photo-2.jpg/View, accessed December 2017)

2009 年 7 月 16 日至 8 月 8 日，中美植物学家考察四川、云南和西藏（从左至右）：杨祝良、Richard H. Ree、当地警察、冯邦、乐霁培、当地林业人员、Bruce M. Bartholomew、David E. Boufford、Deren A. Eaton 和惠宏于西藏贡觉；摄影者：李新辉（相片提供者：横断山生物多样性网站）

Lhorong [Luolong], Markham [Mangkang], Riwoqê [Leiwuqi], Zogang [Zuogong]. Bruce M. Bartholomew (CAS), David E. Boufford (A/GH), Deren A. Eaton (F), Bang FENG (1983-, KUN), Xin-Hui LI (1981-, KUN), Richard H. Ree (A/F), Bo XU (YNUB), Zhu-Liang YANG (KUN), Ji-Pei YUE (KUN), Jian-Wen ZHANG (KUN) and Xin-Xin ZHU (1986-, KUN) collected 1,503 numbers of vascular plants (David E. Boufford et al. nos. 40532-42035, A, CAS, KUN, MO, P, PE, TI) and 416 numbers of fungi (KUN).

2009 年 7 月 16 日至 8 月 8 日，中美植物联合考察四川的巴塘、稻城、德格、乡城，云南的德钦、香格里拉，西藏的八宿、边坝、察雅、昌都、江达、贡觉、洛隆、芒康、类乌齐、左贡；加州科学院 Bruce M. Bartholomew，哈佛大学 David E. Boufford，中国科学院昆明植物研究所冯邦（1983—）、李新辉（1981—）、杨祝良、乐霁培、张建文、朱鑫鑫（1986—，KUN），芝加哥自然博物馆 Deren A. Eaton，哈佛大学、芝加哥自然博物馆 Richard H. Ree，云南师范大学徐波，采集维管植物标本 1 503 号（A、CAS、KUN、MO、P、PE、TI），真菌标本 416 号（KUN）。

2009 年 8 月 23 日至 27 日，全国苔藓植物学学术研讨会在黑龙江齐齐哈尔的齐齐哈尔大学召开，来自全国 25 个高校和研究单位的 70 余名代表参加了此次大会。

2009.08.23-08.27, National Bryophyte Symposium held in Qiqihaer University, Qiqihaer, Heilongjiang, nearly 70 attendees from 25 institutions.

2009 年 9 月，杨昌煦、熊济华、钟世理、王海洋、李先源编著，**重庆维管植物检索表**，1 026 页；四川科学技术出版社，成都。

2009.09, Chang-Xun YANG, Ji-Hua XIONG, Shi-Li ZHONG, Hai-Yang WANG and Xian-Yuan LI, ***Keys to the Vascular Plants of Chongqing***, 1,026 p; Sichuan Science and Technology Press, Chengdu.

2009 年 9 月，方精云、王志恒、唐志尧，**中国木本植物分布图集**，4 卷本；第 1 卷：1-624 页；第 2 卷：625-1 264 页；第 3 卷：1 265-1 908 页；第 4 卷：索引，1 909-2 020 页；高等教育出版社，北京。

2011, 09, Jing-Yun FANG, Zhi-Heng WANG and Zhi-Yao TANG, ***Atlas of Woody Plants in China—Distribution and Climate***; vols. I-II; I: 1-984 p, II: 985-1,972 p; Higher Education Press, Beijing, and Springer-Verlag, Berlin.[802]

2009.10.20, ***Phytotaxa***, a peer-reviewed scientific journal for rapid publication on all aspects of systematic botany, founded online (https://biotaxa.org/Phytotaxa/index)[803], with Zhi-Qiang ZHANG (1963-) as executive editor. Until the end of December 2017, 4,156 papers in 331 volumes had been

802 该书中英文出版不同步且卷数与页码不同。另参见：Jin-Shuang MA, 2011, Reviews: Atlas of woody plants in China: Distribution and Climate, *Taxon* 60(2): 621-622.

803 Maarten J.M. Christenhusz, Mark W. Chase, Michael F. Fay, Thorsten Lumbsch, Alex Monro, Maria Vorontsova, Zhi-Qiang Zhang, 2009, A new international journal for rapid publication of botanical taxonomy, *Phytotaxa* 1(1): 1-2.

Participants at National Bryophyte Symposium, Qiqihaer, 23-27 August 2009 (Photo provided by Tong CAO)
2009 年 8 月 23 日至 27 日，全国苔藓植物学学术研讨会于齐齐哈尔（相片提供者：曹同）

published. Among them, at least 441 papers, about 10.6 percent of the total, involved Chinese plants. Those publications represented 23.2% of the 1,900 papers on Chinese plants published during the same period (2009 to 2017) in all other journal combined.

2009 年 10 月 20 日，***Phytotaxa***，经过同行评议的快速发表的系统植物学学术期刊在线成立（https://biotaxa.org/Phytotaxa/index），张智强[804]（1963—）任执行编辑。至 2017 年底共发表 331 卷 4 156 篇文章。其中，有关中国植物的文章至少 441 篇，约占文章总数的 10.6%。这些文章占同期（2009—2017）有关中国植物系统学文章总数 1900 篇的 23.2%。[805]

2009 年 11 月，任海主编，**风雨兼程八十载**——中国科学院华南植物园八十周年纪念画册（1929—2009），189 页；华中科技大学出版社，武汉。

2009,11, Hai REN (editor in chief), '***Feng Yu Jian Cheng Ba Shi Zai***' ——*South China Botanical Garden, Chinese Academy of Sciences: 80 Years of Elaboration and Development (1929-2009),* 189 p; Huazhong University of Science and Technology Press, Wuhan.

2009 年 12 月，杨远波、廖俊奎、唐默诗、杨智凯、叶秋妤，**台湾种子植物科属志**，231 页；"行政院"农业委员会林务局，台北。

2009.12, Yuen-Po YANG, Chun-Kuei LIAO, Mo-Shih TANG, Chih-Kai YANG and Chiou-Yu YEH, ***Family and Genus Flora of Taiwan Seed Plants***, 231 p; Forestry Bureau, Council of Agriculture, Executive Yuan, Taipei [Taibei].

2009 年，陈进，**五十年回眸**——中国科学院西双版纳热带植物园 1959—2009，145 页；中国科学院西双版纳热带植物园，勐仑，云南（内部印制）。

2009, Jin CHEN, '***Fifty Year's Review***—*Xishuangbanna Tropical Botanical Garden, Chinese Academy of Sciences, 1959-2009*', 145 p; Xishuangbanna Tropical Botanical Garden, Chinese Academy of Sciences, Menglun, Yunnan (internal publication).

---

804 上海人，1981—1985 年复旦大学生物系本科，动物学学士，1985—1987 年复旦大学生物系硕士（导师忻介六教授），1987 年夏直升转博，攻读博士学位，1988 年入美国康奈尔大学昆虫系攻读博士学位，1993 年获昆虫学博士学位，1993—1994 年美国俄勒冈州立大学博士后，1994—1995 年受聘于国际昆虫研究所（英国自然历史博物馆）任分类学家，1995—1999 年任国际昆虫研究所终身分类学家，1999 年至今任新西兰皇家研究院土地环境保护研究所终身研究员；其中 1996 年创立国际 *Systematic and Applied Acarology* 系统及应用蜱螨学（SCI 收录期刊）并任主编，2000 年至今当选动物分类学报编委，创立并编辑英文系列丛书 *Fauna of China* 和 *Fauna of China Synopsis*，2001 年创立国际杂志 *Zootaxa*（SCI 收录杂志）并任主编，2008 年创立国际杂志 *Zoosymposia* 并任主编，2009 年创立国际杂志 *Phytotaxa*（SCI 收录杂志）并任执行编辑。

805 Cheng DU & Jin-Shuang MA, 2019, *China Plant Names Index, 2000-2009*, 606 p; Cheng DU & Jin-Shuang MA, 2019, *China Plant Names Index, 2010-2017*, 603 p; Science Press, Beijing.

2009, Pan-Cheng WU and Paul Pui-Hay BUT, ***Hepatic Flora of Hong Kong***, 192 p; Northeast Forestry University Press, Harbin.

2009 年，吴鹏程、毕培曦，**香港苔类植物**，192 页；东北林业大学出版社，哈尔滨。

2009 年，赵家荣、刘艳玲主编，**水生植物图鉴**，320 页；华中科技大学出版社，武汉。

2009, Jia-Rong ZHAO and Yan-Ling LIU (editors in chief), ***Aquatic Plant***, 320 p; Huazhong University of Science and Technology Press, Wuhan.

2009 年，刘永英、张为民、赵建成、李琳、王育水，河南省藓类植物名录，河南科学 27（8）：935–946。

2009, Yong-Ying LIU, Wei-Min ZHANG, Jian-Cheng ZHAO, Lin LI and Yu-Shui WANG, An updated checklist of mosses from Henan province, *Henan Science* 27(8): 935–946.

2009, David Sox, ***Père David 1826–1900—Early Nature Explorer in China***, 38 p; Sessions Book Trust, York, UK.

2009 年，David Sox，**中国的早期博物考察者 David 神父 1826—1900**，38 页；Sessions Book Trust，约克，英国。

2009, Mark Flanagan and Tony Kirkham, ***Wilson's China—a Century On***, 256 p; Kew Publishing, London.

2009 年，Mark Flanagan、Tony Kirkham，**威尔逊的中国 —— 一个世纪后**，256 页；Kew Publishing，伦敦。

2009, Judith M. Taylor, ***The Global Migrations of Ornamental Plants****—How the World got into Your Garden*, 312 p; Missouri Botanical Garden Press, St. Louis, Missouri.

2009 年，Judith M. Taylor，**观赏植物的全球迁移** —— 世界是如何进入你的花园，312 页；密苏里植物园出版社，圣路易斯，密苏里。

2009 年，王文采口述，胡宗刚访问整理，**王文采口述自传**，20 世纪中国科学口述史，239 页；湖南教育出版社，长沙。

2009, Wen-Cai WANG[806] (dictated), Zong-Gang HU (transcriber), ***Wang Wencai: An Oral Biography*** (*Wen-Tsai Wang's dictated autobiography); An Oral History of Chinese Science in the 20 Century*, 239 p; Hunan Education Publishing House, Changsha.

806 Pinyin spelling of Wen-Tsai WANG.

2009 年，魏平主编，**根深叶茂竞芳菲 —— **中国科学院华南植物园八十周年纪念文集，211 页；广东科技出版社，广州。

2009, Ping WEI (editor in chief), '***Gen Shen Ye Mao Jing Fang Fei***—*Memorial Works on the 80 anniversary of South China Botanical Garden, Chinese Academy of Sciences*', 211 p; Guangdong Science and Technology Press, Guangzhou.

2009 年，江苏省中国科学院植物研究所（南京中山植物园）纪念文集编写委员会编，**江苏省中国科学院植物研究所（南京中山植物园）纪念文集（1929—2009）**，181 页；江苏省中国科学院植物研究所（南京中山植物园），南京（内部印制）。

2009, Editorial Committee of Memorial Works of Institute of Botany, Jiangsu Province and Chinese Academy of Sciences, Nanjing Botanical Garden Memorial Sun Yat-Sen, '***Memorial Works of Institute of Botany, Jiangsu Province and Chinese Academy of Sciences, Nanjing Botanical Garden Memorial Sun Yat-Sen (1929-2009)***, 181 p; Institute of Botany, Jiangsu Province and Chinese Academy of Sciences, Nanjing Botanical Garden Memorial Sun Yat-Sen, Nanjing (internal publication).

2009 年，江苏省中国科学院植物研究所（南京中山植物园）所（园）志编写委员会编，**江苏省中国科学院植物研究所（南京中山植物园）所（园）志（1929—2009）**，143 页；江苏省中国科学院植物研究所（南京中山植物园），南京（内部印制）。

2009, Editorial Committee of Records of Institute of Botany, Jiangsu Province and Chinese Academy of Sciences, Nanjing Botanical Garden Memorial Sun Yat-Sen, '***Records of Institute of Botany, Jiangsu Province and Chinese Academy of Sciences, Nanjing Botanical Garden Memorial Sun Yat-Sen (1929-2009)***, 143 p; Institute of Botany, Jiangsu Province and Chinese Academy of Sciences, Nanjing Botanical Garden Memorial Sun Yat-Sen, Nanjing (internal publication).

2009 年，钟如涛、陈喜英，中国苔藓植物研究现状，林业调查规划 34（5）：43-46。

2009, Ru-Tao ZHONG and Xi-Ying CHEN, Current situation of bryological research in China, *Forest Inventory and Planning* 34(5): 43-46.

2010

2010 年 1 月，中国植物志获国家自然科学一等奖，钱崇澍、陈焕镛、吴征镒、王文采、李锡文、胡启明、陈艺林、陈心启、崔鸿宾、张宏达 10 位代表全国 146 家单位的 312 位作者及 164 位绘图人员获奖（http://frps.iplant.cn/about）。[807]

2010.01, *Flora Reipublicae Popularis Sinicae* awarded National Natural Science First Prize, presented to 312 authors and 164 illustrators from 146 institutions (http://frps.iplant.cn/about). Sung-Shu CHIEN, Woon-Young CHUN, Cheng-Yih WU, Wen-Tsai WANG, Hsi-Wen LI, Chi-Ming HU, Yi-Ling CHEN, Sing-Chi CHEN, Hung-Pin TSUI and Hung-Ta CHANG represented the awardees.

国家自然科学奖

证 书

为表彰国家自然科学奖获得者，特颁发此证书。

项目名称：《中国植物志》的编研

奖励等级：一等

获 奖 者：钱崇澍(中国科学院植物研究所)

中华人民共和国国务院

2009 年 12 月 25 日

证书号：2009-Z-105-1-01-R01

Flora Reipublicae Popularis Sinicae awarded National Natural Science First Prize, 2010 (Photo provided by Qi LIN)

2010 年，中国植物志获国家自然科学奖一等奖（相片提供者：林祁）

807 胡宗刚、夏振岱，2016，中国植物志编纂史，259 页；上海交通大学出版社，上海。

2010 年 2 月，中国科学院沈阳应用生态研究所编著，高谦、吴玉环主编；**中国苔纲和角苔纲植物属志**，636 页；科学出版社，北京。

2010.02, Institute of Applied Ecology, Chinese Academy of Sciences Edita, Chien GAO and Yu-Huan WU (editors in chief), ***Genera Hepaticopsida et Anthocerotopsida Sinicorum***, 636 p; Science Press, Beijing.

2010.05.18-05.19, Joint editorial committee meeting of Flora of China held at the Royal Botanic Garden, Edinburgh, Scotland. Attendees: Ihsan A. Al-Shehbaz (MO), Stephen Blackmore (E), Shi-Long CHEN (HNWP), Michael G. Gilbert (MO), Xiao-Lin GONG (PE), De-Yuan HONG (PE), De-Zhu LI (KUN), Jing LENG (PE), William A. McNamara (Quarryhill Botanical Garden), Simon J. Owens (K), Ching-I PENG (HAST), Peter H. Raven (MO), David A. Simpson (K), Nicholas J. Turland (MO), Mark F. Watson (E), Alexandra H. Wortley (E), Li-Bing ZHANG (MO) and Xian-Chun ZHANG (PE). 2013, Li-Bing ZHANG (MO) replaced Nicholas J. Turland (MO) as co-director of the project, but Nicholas J. Turland (MO) remained on editorial committee.

2010 年 5 月 18 日至 19 日，英文版中国植物志编辑委员会联席会议在苏格兰爱丁堡植物园举行，密苏里植物园 Ihsan A. Al-Shehbaz、Michael G. Gilbert、Peter H. Raven、Nicholas J. Turland、张丽兵，爱丁堡植物园 Stephen Blackmore、Mark F. Watson、Alexandra H. Wortley，中国科学院西北高原生物研究所陈世龙，中国科学院植物研究所宫晓林、洪德元、冷静、张宪春，中国科学院昆明植物研究所李德铢，英国皇家植物园邱园 Simon J. Owens、David A. Simpson，Quarryhill Botanical Garden 的 William A. McNamara，“中央”研究院植物研究所彭镜毅出席。2013 年密苏里植物园张丽兵取代密苏里植物园 Nicholas J. Turland 为联合项目主任，但密苏里植物园 Nicholas J. Turland 仍在编辑委员会。

2010 年 5 月，哈斯巴根主编，**内蒙古种子植物名称手册**，219 页；内蒙古教育出版社，呼和浩特。

2010.05, Khasbagan (editor in chief), '***A Handbook of the Names of Seed Plants of Inner Mongolia***', 219 p; Inner Mongolia Education Press, Hohhot.

2010.07.07-07.09, International Conference, New Frontiers in Plant Systematics and Evolution held in Beijing, 311 attendees from 81 institutions and 14 countries (http://www.ibcas.ac.cn/xueshu/Academic_report/201007/t20100714_2902765.html, accessed 23 October 2017).

2010 年 7 月 7 日至 9 日，系统与进化植物学前沿国际学术研讨会在北京举行，14 个国家 81 个单位 311 人出席。

2010.07.19-08.14, Chinese-American botanical expedition to Sichuan: Daocheng, Jiulong, Kangding, Litang, Muli, Xiangcheng, Yajiang, Yanyuan; Yunnan: Shangri-La (Xianggelila). David E. Boufford (A/GH), Lin-Yang CHEN (KUN), Jin-Long DONG (KUN), James R. Shevock (1950-, CAS) and Ji-Pei YUE (KUN) collected 843 numbers of vascular plants (David E. Boufford et al. nos. 42052-42894, A, CAS, KUN, MO, P,

Chinese-American botanical expedition to Sichuan and Yunnan, 19 July-14 August 2010 (left-right): Jin-Long DONG, Lin-Yang CHEN, Jian-Hua ZHANG, Xin-Hui LI, James R. Shevock, Ying-Hui LI, David E. Boufford, photo by Ji-Pei YUE outside the city of Litang (in background), Sichuan (Photo provided by http://hengduan.huh.harvard.edu/fieldnotes/expeditions/Field%20team%202010%20leaving%20Litang.JPG/view, accessed December 2017)

2010年7月19日至8月14日，中美植物联合考察四川和云南：(从左至右)董金龙、陈林扬、张建华、李新辉、James R. Shevock、李英辉、David E Boufford 于理塘（背景）市郊。摄影者：乐霁培（相片提供者：横断山生物多样性网站）

PE, TI and 699 numbers of bryophytes (James R. Shevock 35652-36350, CAS, FH, KUN, MO).

2010年7月19日至8月14日，中美植物联合考察四川的稻城、九龙、康定、理塘、木里、乡城、雅江、盐源，云南的香格里拉，成员有哈佛大学 David E. Boufford，中国科学院昆明植物研究陈林扬、董金龙、乐霁培，加州科学院 James R. Shevock（1950—），采集维管植物标本 843 号（A、CAS、KUN、MO、P、PE、TI），苔藓植物标本 699 号（CAS、FH、KUN、MO）。

2010年8月23日至27日，全国苔藓植物学学术研讨会在贵阳贵州师范大学召开，来自全国 17 个省（市、区）38 所单位 90 名人员出席。

2010.08.23-08.27, National Bryophyte Symposium held at Guizhou Normal University, Guiyang, nearly 90 attendees from 38 institutions of 17 provinces.

2010.10.15-10.21, International Symposium on Systernatic Lichenology and Bryology held in Zhejiang University, Hangzhou, Zhejiang, more than 80 attendees from 30 institutions in 13 provinces in China, including 10 from abroad.

Participants at National Bryophyte Symposium, Guiyang, 23-27 August 2010 (Photo provided by Tong CAO)
2010 年 8 月 23 日至 27 日，全国苔藓植物学学术研讨会于贵阳（相片提供者：曹同）

2010 年 10 月 15 日至 21 日，地衣与苔藓植物系统学国际研讨会在浙江大学举行。国内外 13 个省（市、区），30 余所高校、科研院所的 80 多位代表参加，包括国际学者 10 余人。

2010.11.15-11.17, Fifth Symposium on Asian Pteridology and Fern Show held at Fairylake Botanical Garden, Shenzhen, and Chinese Academy of Sciences, Shenzhen, Guangdong, 150 attendees from 15 countries in Africa, Asia, Europe and North America.[808]

2010 年 11 月 15 日至 17 日，第五届亚洲蕨类植物学大会在广东深圳市中国科学院仙湖植物园举行，来自亚洲、欧洲、北美和非洲 15 个国家的 150 位代表出席。

2010，白学良等，**贺兰山苔藓植物**，281 页；黄河出版传媒集团宁夏人民出版社，银川。

2010，Xue-Liang BAI et al., '***Bryophytes of Helan Mountain***', 281 p; Yellow River Publishing Media Ningxia People's Press, Yinchuan.

2010 年，印开蒲等著，**百年追寻** —— 见证中国西部环境变迁，582 页；中国大百科全书出版社，北京。

2010, Kai-Pu YIN et al., ***Tracing One Hundred Years of Change****—Illustrating the Environmental Changes in Western China*, 582 p; Encyclopedia of China Publishing House, Beijing.

2010 年，刘启新，**育才尽瘁 事业流芳** —— 纪念单人骅教授百年诞辰，72 页；江苏省中国科学院植物研究所伞形科项目组，南京（内部印制）。

2010, Qi-Xin LIU, '***Yu Cai Jin Cui Shi Ye Liu Fang****— Centennial celebration on Professor Ren-Hwa SHAN's birth anniversary*', 72 p; Apiaceae Group, Institute of Botany, Jiangsu Province and Chinese Academy of Sciences, Nanjing (internal publication).

2010, Special Issue of *Arnoldia* 68(2): 1-76 p, celebrates the upcoming twentieth anniversary of the North America-China Plant Exploration Consortium (NACPEC); six papers pertaining to the NACPEC's past, present and future.

2010 年，*Arnoldia* 68（2）：1-76 页，纪念北美中国植物考察联盟成立二十周年专刊，刊载六篇关于北美中国植物考察联盟的过去、现在及未来的文章。

2010 年，张力主编，**澳门苔藓植物名录**，86 页；澳门特别行政区民政总署园林绿化部、深圳市中国科学院仙湖植物园，澳门、深圳。

2010，Li ZHANG (editor in chief), ***Checklist of Macao Bryophytes***, 86 p; Department of Gardens

808 Xian-Chun ZHANG, 2011, Fifth symposium on Asian pteridology and fern show, *Taxon* 60(1): 299.

Participants at International Symposium of Systematic Lichenology and Bryology, Hangzhou, 15–21 October 2010 (Photo provided by Yu-Huan WU)
2010 年 10 月 15 日至 21 日，地衣与苔藓植物系统学国际研讨会于杭州（相片提供者：吴玉环）

Participants at Fifth Symposium on Asian Pteridology and Fern Show, Shenzhen, 15-17 November 2010 (Photo provided by Hong-Mei LIU)
2010 年 11 月 15 日至 17 日，第五届亚洲蕨类植物学大会于深圳（相片提供者：刘红梅）

and Green Areas, Civic and Municipal Affairs Bureau of Macao Special Administrative Region, Macao and Fairylake Botanical Garden, Shenzhen, and Chinese Academy of Sciences, Shenzhen.

2010 年，张力主编，**澳门苔藓植物志**，361 页；澳门特别行政区民政总署园林绿化部，澳门；深圳市中国科学院仙湖植物园，深圳。

2010, Li ZHANG (editor in chief), ***Flora Briófita de Macau—Bryophyte Flora of Macao***, 361 p; Department of Gardens and Green Areas, Civil and Municipal Affairs Bureau of Macao Special Administrative Region, Macao and Fairylake Botanical Garden, Shenzhen, and Chinese Academy of Sciences, Shenzhen.

2010 年，李宏庆主编，**华东种子植物检索手册**，439 页；华东师范大学出版社，上海。

2010, Hong-Qing LI (editor in chief), '***Identification Handbook of Seed Plants from Eastern China***', 439 p; East China Normal University Press, Shanghai.

2010 年，覃海宁、刘演主编，**广西植物名录**，625 页；科学出版社，北京。

2010, Hai-Ning QIN and Yan LIU (editors in chief), ***A Checklist of Vascular Plants of Guangxi***, 625 p; Science Press, Beijing.

2010 年，吴鹏程、袁生主编，**陈邦杰先生国际学术纪念文集**，342 页；南京师范大学出版社，南京。

2010, Peng-Cheng WU and Sheng YUAN (editors in chief), ***International Academic Memorial Issue for Prof. Pan-Chieh CHEN***, 342 p; Nanjing Normal University Press, Nanjing.

2010 年，刘仁林、张志翔、廖为明，**江西种子植物名录**，365 页；中国林业出版社，北京。

2010, Ren-Lin LIU, Zhi-Xiang ZHANG and Wei-Ming LIAO, '***Checklist of Jiangxi Seed Plants***', 365 p; China Forestry Publishing House, Beijing.

2010 年 10 月 和 2011 年 12 月[809]，傅德志，**世界维管植物**，55 636 页；第 1 卷：科属名称和分布，第 2 卷：分类系统，第 3–50 卷：全球种志初编；青岛出版社，青岛。

2010.10 and 2011.12[810], De-Zhi FU, ***Vascular Plants of the World***, 55,636 p.; vol. 1, Names and Distribution of Families and Genera (A–Z), vol. 2, Classification Systems, vols. 3–50, Species (Aa–Zyzyxia); Qingdao Publishing House, Qingdao.

809 初版：2010 年 10 月，第 2 版：2011 年 12 月。

810 First edition, October 2010; second edition, December 2011.

2011

2011 年 1 月，吴征镒、孙航、周浙昆、李德铢、彭华，**中国种子植物区系地理**，485 页；科学出版社，北京。

2011.01，Zheng-Yi WU, Hang SUN, Zhe-Kun ZHOU, De-Zhu LI and Hua PENG, ***Floristics of Seed Plants from China***, 485 p; Science Press, Beijing.

2011 年 3 月 21 日至 24 日，首届泛喜马拉雅植物志作者会议在北京召开，全国 70 多人参加；洪德元院士带领中外学者开始编研泛喜马拉雅植物志。

2011.03.21-03.24, First meeting of authors for *Flora of Pan-Himalaya* held in Beijing, more than 70 attendees. *Flora of Pan-Himalaya* to be compiled by scholars from China and abroad, led by academician De-Yuan HONG.

2011 年 4 月，冯广平、赵建成、王青主编，**北京植物学史图鉴**，179 页；北京科学技术出版社，北京。

2011.04, Guang-Ping FENG, Jian-Cheng ZHAO and Qing WANG (editors in chief), '***Illustration of Beijing Botanical History***', 179 p; Beijing Science and Technology Press, Beijing.

2011 年 6 月，曹伟主编，**东北生物标本馆维管束植物模式标本考订**，415 页；科学出版社，北京。

2011.06, Wei CAO (editor in chief), ***Textual Criticism of Type Specimens of Vascular Plants Preserved in the Northeast China Herbarium***[811], 415 p; Science Press, Beijing.

2011.07.23-07.30, Eighteenth International Botanical Congress held in Melbourne, Australia, 124 attendees from China. Decision made to hold XIX IBC in Shenzhen, China, in 2017.

2011 年 7 月 23 日至 30 日，第十八届国际植物学大会于澳大利亚墨尔本举行，中国代表 124 人参加[812]。会议决定第十九届国际植物学大会 2017 年在中国深圳举行。

2011 年 8 月 23 日至 25 日，全国苔藓植物学学术研讨会在新疆乌鲁木齐中国科学院新疆生态与地理研究所召开；来自全国 23 个单位的 60 余名代表参加。

2011.08.23-08.25, National Bryophyte Symposium held at Xinjiang Institute of Ecology and Geography, Chinese Academy of Sciences, Urumqi, Xinjiang; nearly 60 attendees from 23 institutions.

---

811 i.e. IFP.

812 夏念和，2011，第十八届国际植物学大会简介，热带亚热带植物学报 19（5）：491-492。

First meeting of participants at *Flora of Pan-Himalaya*, Beijing, 21–24 March 2011 (Photo provided by Min LI)
2011 年 3 月 21 日至 24 日，首届泛喜马拉雅植物志会议于北京（相片提供者：李敏）

Participants at National Bryophyte Symposium, Urumqi, 23–25 August 2011 (Photo provided by Tong CAO)
2011 年 8 月 23 日至 25 日，全国苔藓植物学学术研讨会于乌鲁木齐（相片提供者：曹同）

2011 年 9 月，罗桂环、李昂，哈佛大学阿诺德树木园对我国植物学早期发展的影响，北京林业大学学报（社会科学版）10（3）：1-8。

2011.09, Gui-Huan LUO and Ang LI, Impacts of Arnold Arboretum in Harvard University to early development of botany in China, *Journal of Beijing Forestry University (Social Science)* 10(3): 1-8.

2011 年 10 月 25 日至 27 日，全国系统与进化植物学暨第十届青年学术研讨会在云南昆明举行，全国 70 多个单位 300 多人参加。[813]

2011.10.25-10.27, National and Tenth Youth Symposium of Systematic and Evolutionary Botany held in Kunming, Yunnan, over 300 attendees from more than 70 institutions.

2011 年，臧穆、黎兴江主编，**中国隐花（孢子）植物科属辞典**，990 页；高等教育出版社，北京。

2011, Mu ZANG and Xing-Jiang LI (editors in chief), ***Dictionary of the Families and Genera of Chinese Cryptogamic (Spore) Plants***, 990 p; Higher Education Press, Beijing.

2011, Ji-Hong HUANG, Jian-Hua CHEN, Jun-Sheng YING and Ke-Ping MA, Features and distribution patterns of Chinese endemic seed plant species, *Journal of Systematics and Evolution* 49(2): 81-94.

2011 年，黄继红、陈建华、应俊生、马克平，中国特有种子植物的特征与分布类型，*Journal of Systematics and Evolution* 49（2）：81-94。

2011, Maarten J. M. Christenhusz, Xian-Chun ZHANG and Harald Schneider, A linear sequence of extant families and genera of lycophytes and ferns, *Phytotaxa* 19: 7-54, including Diplaziopsidaceae X.C.Zhang & Christenh. and Rhachidosoraceae X.C.Zhang, two new families.

2011 年，Maarten J. M. Christenhusz、张宪春、Harald Schneider，现代石松类和蕨类植物科属的线性分类系统，*Phytotaxa* 19：7-54，包括 2 个新科 —— 肠蕨科和轴果蕨科。

2011, Denise M. Glover, Stevan Harrell, Charles F. McKhann and Margaret B. Swain, ***Explorers and Scientists in China's Borderlands 1880-1950***, 300 p; University of Washington Press, Seattle, Washington.

2011 年，Denise M. Glover、Stevan Harrell、Charles F. McKhann、Margaret B. Swain，**1880—1950 年间中国边境之地的考察者与科学家**，300 页；华盛顿大学出版社，西雅图，华盛顿。

813 全国系统与进化植物学暨第十届青年学术研讨会，2011 年全国系统与进化植物学暨第十届青年学术研讨会论文集，147 页；云南省植物学会，昆明（内部印制）。

National and 10th Youth Symposium of Systematic and Evolutionary Botany, Kunming, 2011 (Photo provided by Kunming Institute of Botany)
2011 年，系统与进化植物学暨第十届青年学术研讨会于昆明（相片提供者：昆明植物所）

2011 年，马金双著，**东亚高等植物分类学文献概览**，505 页；高等教育出版社，北京[814]。

2011, Jin-Shuang MA, ***The Outline of Taxonomic Literature of Eastern Asian Higher Plants***, 505 p; Higher Education Press, Beijing[815].

2011, Erik Mueggler, ***The Paper Road***—*Archive and Experience in the Botanical Exploration of West China and Tibet*, 361 p; University of California Press, Berkeley.

2011 年，Erik Mueggler，**纸路** —— 华西与西藏的植物学考察的档案与经历[816]，361 页；加州大学出版社，伯克利。

2011, Seamus O'Brien, ***In the Footsteps of Augustine Henry and his Chinese plant collectors***, 367 p; Garden Art Press, Suffolk.

2011 年，Seamus O'Brien，**沿着 Augustine Henry 和他的中国采集者的脚印**，367 页；Garden Art Press，Suffolk。[817]

2011 年，胡宗刚著，**北平研究院植物学研究所史略**，212 页；上海交通大学出版社，上海。

2011, Zong-Gang HU, '***Brief History of the Institute of Botany, National Academy of Peiping***', 212 p; Shanghai Jiao Tong University, Shanghai.

2011 年，应俊生、陈梦玲著，**中国植物地理**，598 页；上海科学技术出版社，上海。

2011, Tsun-Shen YING and Mong-Ling CHEN, ***Plant Geography of China***, 598 p; Shanghai Scientific and Technical Publishers, Shanghai.

2011 年，王战文选编辑委员会，**王战文选**，512 页；科学出版社，北京。

2011, Editorial Committee of Selected Works of Chan WANG, '***Selected Works of Chan WANG***'[818],

---

814 百原新，2012，书评：马金双，2011，东亚高等植物分类学概览，505 页，高等教育出版社，北京，植生史研究 21（1）：20；刘全儒，张宪春，2011，书评：《东亚高等植物分类学文献概览》—— 一本中国分类学者必备的工具书，植物分类与资源学报 33（6）：690–691；马炜梁，2011，书评：读《东亚高等植物分类学文献概览》有感，植物分类与资源学报 33（6）：691–692；夏念和，2012，推荐一本分类学参考书 ——《东亚高等植物分类学文献概览》，热带亚热带植物学报 20（3）：242；胡宗刚，2011 年 12 月 6 日，介绍一部中国植物分类学史著作，科学时报 B4 读书周刊。

815 Li ZHANG, 2012, Review: The outline of taxonomic literature of eastern Asian higher plants, *Taiwania* 57(4): 443–444; Rudi Schimid, 2012, Reviews and notices of publications, *Taxon* 61(1):265–274.

816 李晋，2017，纸路 —— 19 世纪初期西方植物学家在西南中国的实践与体验，西南民族大学学报（人文社会科学版）8：37–50。

817 叶文、马金双，2012，书评：重复的脚印 —— 两个爱尔兰青年相距百年的中国之旅，仙湖 11（3-4）：56–58（2014）。

818 Guo-Fan SHAO, Qi-Jing LIU, Hong QIAN, Ji-Quan CHEN, Jin-Shuang MA and Zheng-Xiang TAN, 2000, Zhan Wang (1911–2000), *Taxon* 49(3): 593–601.

512 p; Science Press, Beijing.

2011 年，张奠湘、李世晋主编，**南岭植物名录**，407 页；科学出版社，北京。

2011, Dian-Xiang ZHANG and Shi-Jin LI (editors in chief), ***A Checklist of Vascular Plants of Nanling Mountains***, 407 p; Science Press, Beijing.

2011, Hartmut Walravens[819], ***Joseph Franz Rock—Phytogeography of Northwest and Southwest China***, 356 p.; Austrian Academy of Sciences Press, Wien.

2011 年，Hartmut Walravens，***Joseph Franz Rock——* 中国西北和西南植物地理学**，356 页；奥地利科学院出版社，维也纳。

2011, Mamtimin Sulayman, A new checklist of bryophytes of Xinjiang, China, *Chenia* 10: 33-53.

2011 年，买买提明·苏来曼，新疆苔藓植物名录，隐花植物生物学 10：33-53。

2011, Peng-Cheng WU, Yu JIA, Lin-Ying PEI, Ning-Ning YU and Li ZHANG, Bryological literatures in China, *Chenia* 10: 139-318.

2011 年，吴鹏程、贾渝、裴林英、于宁宁、张力，中国苔藓植物文献，隐花植物生物学 10：139-318。[820]

2011, Anthony R. Brach and David E. Boufford, Why are we still producing paper floras? [821], *Annals of the Missouri Botanical Garden* 98(3): 297-300.

2011 年，Anthony R. Brach、David E. Boufford，为什么我们还在出版纸制植物志？*Annals of the Missouri Botanical Garden* 98（3）：297-300。

2011 年，董洪进、刘恩德、彭华，中国植物分类编目的过去、现在和将来，植物科学学报 29（6）：755-762。

---

819 Joseph F. Rock wrote his phytogeography as an abstract of his botanical explorations in China in 1952 for the Royal Horticultural Society but because of circumstances beyond his control it remained unpublished. It gives a survey of the vegetation of the areas that Rock visited himself, especially Yunnan and Qinghai and is enhanced by the inclusion of the author's experiences and views. Rock lived for more than twenty years, until 1948, in the Chinese province of Yunnan where besides botanical exploration mainly for Harvard University and the US National Museum, research on the rituals and pictographic manuscripts of the Naxi people were his main occupation. The present work, in German, is based on the typed abstract in the Royal Botanic Garden, Edinburgh and includes many plates of Rock's own photographs. Rock's original captions were added to the plates.

820 此为 1864 年以来的所有文献，且附有学者以及期刊的缩写与全拼及其中文名称。

821 This was based on the *Flora of China*. 本文基于英文版中国植物志。

2011, Hong-Jin DONG, En-De LIU, Hua PENG, Plant cataloguing in China: Past, present and future, *Plant Science Journal* 29(6): 755-762.

2011 年，张晓青、朱瑞良、黄志森、陈允泰、陈文伟，福建苔类和角苔类植物最新名录与区系分析，植物分类与资源学报 33（1）：101-122。

2011, Xiao-Qing ZHANG, Rui-Liang ZHU, Zhi-Sen HUANG, Yun-Tai CHEN and Wen-Wei CHEN, Liverworts and Hornworts of Fujian, China: an updated checklist and bryofloristic accounts, *Plant Diversity and Resources* 33(1): 101-122.

2011, Jian WANG, Ming-Jou LAI and Rui-Liang ZHU, Liverworts and hornworts of Taiwan: an updated checklist and floristic accounts, *Annales Botanici Fennici* 48: 369-395.

2011 年，王健、赖明洲、朱瑞良，台湾苔类和角苔类 —— 更新的名录和区系记载，*Annales Botanici Fennici* 48：369-395。

2011 年，汪梅芝，苔藓植物志第 5 卷，第 9 页，高领藓科（新科），科学出版社，北京。

2011, Mei-Zhi WANG, *Flora Brgophytorum Sinocorum,* vol.5, p.9, Glyphomitriaceae (New family), Science Press, Beijing.

2011、2012 年，傅德志主编，**王文采院士论文集**，上卷（毛茛科）：1 376 页，2011；下卷（苦苣苔科、荨麻科、葡萄科、紫草科、十字花科、大戟科、虎耳草科、山龙眼科、芍药科、金丝桃科、植物区系、术语、杂评、标题论文、附录、新类群目录、大事年表、编后记）：1 244 页，2012；高等教育出版社，北京。[822]

2011 and 2012, De-Zhi FU (editor in chief), ***Paper Collection of W. T. WANG (Wen-Tsai WANG)***, 1 (Ranunculaceae): 1,376 p, 2011; 2 (Gesneriaceae, Urticaceae, Vitaceae, Boraginaceae, Brassicaceae, Euphorbiaceae, Saxifragaceae, Proteaceae, Paeoniaceae, Hypericaceae, Floristics, Terminology, Miscellaneous Comments, Papers in titles, Appendix, List of new taxa, Chronicle, Afterword): 1,244 p, 2012; Higher Education Press, Beijing.

2011、2014 年，黄普华编著，**植物名称研究专集**，484 页，2011；中国林业出版社，北京；黄普华、王洪峰编著，**植物名称研究续集**，352 页，2014；东北林业大学出版社，哈尔滨。

2011, 2014, Pu-Hwa HUANG, '***Collections of Studies on Plant Names***', 484 p, 2011; China Forestry Publishing House, Beijing; Pu-Hwa HUANG and Hong-Feng WANG, '***Sequel Collections of Studies on Plant Names***', 352 p, 2014; Northeast Forestry University Press, Harbin.

822 参见：生命世界 2014 年第 1 期（特别策划栏目）有关王文采先生的 11 篇文章。

2011，朱宗元，梁存柱，李志刚，**贺兰山植物志**，848 页；黄河出版传媒集团阳光出版社，银川。

2011, Zong-Yuan ZHU, Cun-Zhu LIANG, Zhi-Gang LI, ***Flora of Helan Mountain,*** 848 p; Yellow River Publishing Media Sunshine Press, Yinchuan.

2012

2012 年 6 月，中国科学院西北高原生物研究所志编纂委员会，**中国科学院西北高原生物研究所志**[823]，1 102 页；青海人民出版社，西宁。

2012.06, Editorial Committee of Northwest Institute of Plateau Biology, Chinese Academy of Sciences, '***Records of Northwest Institute of Plateau Biology, Chinese Academy of Sciences***', 1,102 p; Qinghai People's Publishing House, Xining.

2012 年 6 月，尚衍重编著，**种子植物名称**，5 卷本；第 1 卷：拉汉英名称 A–D，1–1 991 页；第 2 卷：拉汉英名称 E–O，1 992–3 976 页；第 3 卷：拉汉英名称 P–Z，3 997–5 970 页；第 4 卷：中文名称索引，1–1 128 页；第 5 卷：英日俄名称索引，1–1 477 页；中国林业出版社，北京。

2012.06, Yan-Chong SHANG, ***A Dictionary of Seed Plant Names***, vols. 1–5; 1, Latin, Chinese and English: A–D, 1–1,991 p; 2, Latin, Chinese and English: E–O, 1,992–3,976 p; 3, Latin, Chinese and English: P–Z, 3,977–5,970 p; 4: Chinese Index, 1–1,128 p; 5: English, Japanese and Russian Indices, 1–1,477 p; China Forestry Publishing House, Beijing.

2012 年 8 月 21 日至 25 日，全国苔藓植物学学术研讨会在成都中国科学院成都生物研究所召开，来自 28 个单位的 70 余名代表参加。

2012.08.21–08.25, National Bryophyte Symposium held at Chengdu Institute of Biology, Chinese Academy of Sciences, Chengdu, nearly 70 attendees from 28 institutions.

2012 年 8 月，赵一之著，**内蒙古维管植物分类及其区系生态地理分布**，856 页；内蒙古大学出版社，呼和浩特。

2012.08, Yi-Zhi ZHAO, ***Classification and Its Floristic Ecological Geographic Distributions of Vascular Plants in Inner Mongolia***, 856 p; Inner Mongolia University Press, Hohhot.

2012 年 9 月 20 日，中国科学院华南植物园曾庆文（1963 年 9 月 15 日生）研究员在云南西畴坪寨开展华盖木传粉生物学实验时不幸从 40 多米高的树上坠落遇难。

2012.09.20, Qing-Wen ZENG (born September 15, 1963), curator, South China Botanical Garden, Chinese Academy of Sciences, accidentally died in Pingzhai, Xichou, Yunnan, after falling from an approximately 40 meter tall tree of *Manglietiastrum sinicum* while performing artificial pollination.

---

823 出版日期为 2012 年，内容截止时间为 2006 年。

Participants at National Bryophyte Symposium, Chengdu, 21–25 August 2012 (Photo provided by Tong CAO)
2012 年 8 月 21 日至 25 日，全国苔藓植物学学术研讨会于成都（相片提供者：曹同）

2012 年 10 月，卢琦、王继和、褚建民主编，**中国荒漠植物图鉴**，630 页；中国林业出版社，北京。

2012.10, Qi LU, Ji-He WANG and Jian-Min CHU (editors in chief), ***Desert Plants in China*** '*Icones of Desert Plants of China*', 630 p; China Forestry Publishing House, Beijing.

2012 年 10 月，诚静容先生百年寿辰纪念文集编写组，**中国药用植物学奠基人——诚静容先生百年寿辰纪念文集**，60 页；北京大学药学院，北京（内部印制）。

2012.10, Editorial Group of Memorial Works of Ching-Yung CHENG's Centennial Birthday, '***The Founder of Chinese Medical Botany—Memorial Works of Ching-Yung CHENG's Centennial Birthday***', 60 p; School of Pharmaceutical Sciences, Peking University, Beijing (internal publication).

2012 年 12 月 11 日至 15 日，2012 年中国蕨类植物研讨会在海南海口海南大学举行，全国 55 家单位 75 名人员参加。

2012.12.11-12.15, 2012 National Fern Symposium held at Hainan University, Haikou, Hainan, 75 attendees from 55 institutions.

2012 年，陈耀东、马欣堂、杜玉芬、冯旻、李敏编著，**中国水生植物**，477 页；河南科学技术出版社，郑州。

2012, Yao-Dong CHEN, Xin-Tang MA, Yu-Fen DU, Min FENG and Min LI, ***The Chinese Aquatic Plants***, 477 p; Henan Science and Technology Press, Zhengzhou.

2012 年 12 月，傅德志，**植物科属大辞典**，1 520 页；青岛出版社，青岛。

2012.12, De-Zhi FU, ***The Plant-Dictionary***, 1,520 p; Qingdao Publishing House, Qingdao.

2012 年，骆洋、何延彪、李德铢、王雨华、尹廷双、王红，中国植物志、*Flora of China* 和维管植物新系统中科的比较，植物分类与资源学报 34（3）：231-238。

2012, Yang LUO, Yan-Biao HE, De-Zhu LI, Yu-Hua WANG, Ting-Shuang YI and Hong WANG, A comparison of classifications of families of Chinese vascular plants among *Flora Reipublicae Popularis Sinicae, Flora of China* and the new classifications, *Plant Diversity and Resources* 34(3): 231-238.

2012, Mamtimin Sulayman, New checklist of Xinjiang liverworts, hornworts and mosses, *Journal of Xinjiang University (Natural Science Edition)* 29(3): 259-267.

2012 年，买买提明·苏来曼，新疆苔类、角苔类、藓类植物最新名录，新疆大学学报（自然科学版）29（3）：259-267。

Participants at National Fern Symposium, Haikou, 2012 (Photo provided by Xian-Chun ZHANG)
2012 年中国蕨类植物学术研讨会于海口（相片提供者：张宪春）

2012, Jun-Xia SU, Wei WANG, Li-Bing ZHANG and Zhi-Duan CHEN, Phylogenetic placement of two enigmatic genera, *Borthwickia* and *Stixis*, based on molecular and pollen data, and the description of a new family of Brassicales, Borthwickiaceae, *Taxon* 61(3): 601-611.

2012 年，苏俊霞、王伟、张丽兵、陈之端，利用分子和花粉数据确定节蒴木属和斑果藤属的系统位置 —— 从十字花目描述一新科节蒴木科，*Taxon* 61（3）：601-611。

2012 年，杨永，我国植物模式标本的馆藏量，生物多样性 20（4）：512-516。

2012, Yong YANG, Holdings of type specimens of plants in herbaria of China, *Biodiversity Science* 20(4): 512-516.

2012 年，杨永，全球裸子植物物种编目 —— 现状和问题，生物多样性 20（6）：755-760。

2012, Yong YANG, Catalogue of global gymnosperms: problems and perspectives, *Biodiversity Science* 20(6): 755-760.

2012 年，邢福武、周劲松、王发国、曾庆文、易绮斐、刘东明主编，**海南植物物种多样性编目**，630 页；华中科技大学出版社，武汉。

2012, Fu-Wu XING, Jin-Song ZHOU, Fa-Guo WANG, Qing-Wen ZENG, Qi-Fei YI, Dong-Ming LIU (editors in chief), ***Inventory of Plant Species Diversity of Hainan***, 630 p; Huazhong University of Science and Technology Press, Wuhan.

2012 年，张宪春著，**中国石松类和蕨类植物**，711 页；北京大学出版社，北京。

2012, Xian-Chun ZHANG, ***Lycophytes and Ferns of China***, 711 p; Peking University Press, Beijing.

2012 年，植物分类与资源学报第 34 卷第 6 期，以**新一代植物志：机遇与挑战**为题，刊登 11 篇有关文章。

2012, Eleven papers published under the subject ***iFlora—Opportunities and Challenges for Botanists***, *Plant Diversity and Resources*, 34(6).

2012—2015 年，吴玉虎主编，**昆仑植物志**，4 卷本；第 1 卷：593 页，2014；第 2 卷：776 页，2015；第 3 卷：942 页，2012；第 4 卷：601 页，2013；重庆出版社，重庆。[824]

2012-2015, Yu-Hu WU (editor in chief), ***Flora Kunlunica***, vols. 1-4; 1: 593 p, 2014; 2: 776 p, 2015; 3: 942 p, 2012; 4: 601 p, 2013; Chongqing Publishing House, Chongqing.

824 马金双，2016，书评：昆仑植物志，热带亚热带植物学报 24（2）：242。

Some of the authors of *Flora Kunlunica* (left-right): Yong-Ping YANG, Zhen-Lan WU, Su-Kung WU, Xing-Min ZHOU, Yong-Chang YANG, Yu-Hu WU and Yao FU (Photo provided by Yu-Hu WU)
昆仑植物志部分作者（从左至右）：杨永平、吴珍兰、武素功、周兴民、杨永昌、吴玉虎、傅瑶（相片提供者：吴玉虎）

2012, John McNeill, Fred R. Barrie, William R. Buck, Vincent Demoulin, Werner R. Greuter, David L. Hawksworth, Patrick S. Herendeen, Sandra D. Knapp, Karol Marhold, Jefferson Prado, W. F. Prud'homme van Reine, G. F. Smith, John H. Wiersema, and Nicholas J. Turland (eds.), ***International Code of Nomenclature for algae, fungi and plants*** **(Melbourne Code)**, adopted by the Eighteenth International Botanical Congress, Melbourne, Australia, July 2011; *Regnum Vegetabile* 154: 1-240 p, Koeltz Scientific Books, Königstein.[825] Translated into Chinese in 2013 and 2016.

2013、2016 年，黄增泉、吴明洲、黄星凡著译，**植物命名指南**，331 页，2013；“行政院”农业委员会林务局，台北；张丽兵译，**国际藻类、真菌和植物命名法规（墨尔本法规）**，257 页，2016；密苏里植物园，圣路易斯。[826]

825 John H. Wiersema, John McNeill, Nicholas J. Turland, Fred R. Barrie, William R. Buck, Vincent Demoulin, Werner R. Greuter, David L. Hawksworth, Patrick S. Herendeen, Sandra D. Knapp, Karol Marhold, Jefferson Prado, W. F. Prud'homme van Reine and G. F. Smith (eds.) 2015. International Code of Nomenclature for algae, fungi and plants (Melbourne Code), adopted by the Eighteenth International Botanical Congress, Melbourne, Australia, July 2011; ***Appendices II–VIII.*** *Regnum Vegetabile* 157: 1-492; Koeltz Scientific Books, Königstein.

826 http://www.iapt-taxon.org/files/Melbourne_Code_Chinese.pdf；2016 年 11 月国际植物分类学会（IAPT）网站在线发表。

2013

2013 年 3 月，杨小波主编，**海南植物名录**，579 页；科学出版社，北京。

2013.03, Xiao-Bo YANG (editor in chief), '***Checklist of Hainan Plants***', 579 p; Science Press, Beijing.

2013.05.27-05.29, Chinese-American botanical expedition to Guangdong: Shenzhen (former Bao'an). David E. Boufford (A/GH), Wei GUO (1982-, SYS) and Tian-Tian YUAN (1990-, SYS) collected 69 numbers of vascular plants (David E. Boufford et al. nos. 42945-43014, A, MO, PE, TI).

2013 年 5 月 27 日至 29 日，中美植物联合考察广东深圳[827]，哈佛大学 David E. Boufford、中山大学郭微（1982—）和袁天天（1990—），采集 69 号维管束植物标本（A、MO、PE、TI）。

2013 年 5 月，李剑、张晓红，**此生情怀寄树草**——张宏达传，320 页；中国科学技术出版社，北京，上海交通大学出版社，上海。

2013.05, Jian LI and Xiao-Hong ZHANG, '***Ci Sheng Qing Huai Ji Shu Cao****—A biography of Hung-Ta CHANG*', 320 p; China Science and Technology Press, Beijing and Shanghai Jiao Tong University Press, Shanghai.

2013.06.03-06.07, Chinese-American botanical expedition to Jiangxi: Jinggang Shan. David E. Boufford (A/GH), Wen-Bo LIAO (1963-, SYS), Hui-Min XU (1985-, SYS), Bao-Huan WU (1991-, CANT) and Tian-Tian YUAN (SYS) collected 115 numbers of vascular plants (David E. Boufford et al. nos. 43015-43119, A, CAS, KUN, MO, P, PE, TI).

2013 年 6 月 3 日至 7 日，中美植物联合考察江西的井冈山，哈佛大学 David E. Boufford，中山大学廖文波（1963—）、许会敏（1985—）和袁天天，华南农业大学吴保欢（1991—），采集维管束植物 115 号（A、CAS、KUN、MO、P、PE、TI）。

2013.07.15-07.19, World Conference of Bryology held in Natural History Museum, London, England, attendees from China: Yu-Xia LOU, Tian-Yi MA (1989-), Wei SHA, Zhao-Hui ZHANG (1963-) and Rui-Liang ZHU.

2013 年 7 月 15 日至 19 日，世界苔藓植物学大会在英国伦敦自然历史博物馆召开，中国代表娄玉霞、马天意（1989—）、沙伟、张朝晖（1963—）、朱瑞良参加。

---

827 原宝安，1979 年更为现名。

Participants at World Conference of Bryology, London, 2013 (Photo provided by Wei SHA)
2013 年世界苔藓植物学大会于伦敦（相片提供者：沙伟）

2013.08.19–09.18, Chinese-American botanical expedition to Sichuan: Batang, Daocheng, Dêrong [Derong] and Xiangcheng; Yunnan: Binchuan, Dêqên [Deqin], Eryuan, Heqing, Huize, Ninglang, Qiaojia, Shilin (former Lunan), Shangri-La (Xianggelila). David E. Boufford (A/GH), Yong-Sheng CHEN (KUN), Julian F. Harber (MO), Xin-Hui LI (KUN) and Qia WANG (KUN) collected 616 numbers of vascular plants (David E. Boufford et al. nos. 43125–43741, A, CAS, KUN, MO, P, PE, TI).

2013 年 8 月 19 日至 9 月 18 日，中美植物联合考察四川的巴塘、稻城、得荣、乡城，云南的宾川、德钦、洱源、鹤庆、会泽、宁蒗、巧家、石林 [828]、香格里拉，哈佛大学 David E. Boufford，中国科学院昆明植物研究所陈永生、李新辉和王洽，密苏里植物园 Julian F. Harber，采集维管束植物标本 616 号（A、CAS、KUN、MO、P、PE、TI）。

828 原名路南，1998 年 10 月改为现名。

2013.09.23-09.25, International Symposium of Plant Diversity and Conservation in China to celebrate completion of the *Flora of China*, held at the Institute of Botany, Chinese Academy of Sciences, Beijing, China. The 25 volumes of keys and descriptions and the 24 volumes of illustrations of the *Flora of China* treat 31,362 species of vascular plants in 3,329 genera and 312 families. Of the total, 2,129 species are ferns and lycophytes, 237 are gymnosperms and 28,995 are angiosperms. Just under 50% of the species are endemic to China. The 10 largest families are Asteraceae (2,336 spp.), Poaceae (1,795 spp.), Fabaceae (1,673 spp.), Orchidaceae (1,388 spp.), Rosaceae (950 spp.), Ranunculaceae (921 spp.), Cyperaceae (865 spp.), Ericaceae (826 spp.), Lamiaceae (807 spp.) and Liliaceae (sensu lato 726 spp.). The 25 largest families contain 19,494 species (62% of all species in the flora). The 37 largest genera, each with over 100 species, contain 7,624 species (almost a quarter of all species). The five most species-rich provinces are Yunnan (14,186 spp.), Sichuan (including Chongqing, 9,443 spp.), Guangxi (6,916 spp.), Xizang (6,756 spp.) and Guizhou (5,588 spp.). The five provinces with the highest number of endemic species are Sichuan (including Chongqing, 57.6%), Hubei (46.1%), Yunnan (43.9%), Gansu (42.4%) and Qinghai (41.5%)[829].

2013年9月23日至25日，中国植物多样性与保护国际研讨会在北京中国科学院植物研究所举行，纪念英文版中国植物志全部出版。英文版中国植物志文字25卷、图版24卷，含维管束植物312科3 329属31 362种，其中包括蕨类和石松类植物2 129种、裸子植物237种和被子植物28 995种，特有植物近50%。10个大科是菊科（2 336种）、禾本科（1 795种）、豆科（1 673种）、兰科（1 388种）、蔷薇科（950种）、毛茛科（921种）、莎草科（865种）、杜鹃花科（826种）、唇形科（807种）和百合科(广义,726种)。25个大科含19 494种(约占总数的62%)；超过百种的37个大属含7 624种(约为总数的1/4)。5个最丰富的省份为云南（14 186种）、四川（包括重庆,9 443种）、广西（6 916种）、西藏（6 756种）和贵州（5 588种），而5个特有性最高的省份则为四川（包括重庆，57.6%）、湖北（46.1%）、云南（43.9%）、甘肃（42.4%）和青海（41.5%）。

2013年10月13日至16日，中国植物学会第十五届会员代表大会暨八十周年学术年会在江西省南昌市举行，全国代表800多人参加。理事长为武维华（1956—）。[830]

2013.10.13-10.16, Fifteenth National Congress and 80th anniversary meeting of Botanical Society of China held in Nanchang, Jiangxi; over 800 delegates attended; Wei-Hua WU (1956-) elected president.

2013年10月17日至18日，2013年中国蕨类植物研讨会在上海辰山植物园（中国科学院上海辰山植物科学研究中心）和上海师范大学举行，全国约130名人员参加。

2013.10.17-10.18, 2013 National Fern Symposium held in Shanghai Chenshan Botanical Garden (Shanghai Chenshan Plant Science Research Center, Chinese Academy of Sciences) and Shanghai Normal University, Shanghai, about 130 attendees.

829 For details about these figures, please see *Flora of China* 1: 21-23, 2013.

830 中国植物学会主编，2013，中国植物学会八十周年年会论文摘要汇编，346页；江西高校出版社，南昌。

Participants at International Symposium on Plant Diversity and Conservation in China to celebrate completion of *Flora of China*, Beijing, 2013 (Photo provided by Min LI)
2013 年中国植物多样性与保护国际研讨会于北京庆祝英文版中国植物志完成（相片提供者：李敏）

The authors of treatments of ferns for *Flora of China*, 2011 (left-right): Gang-Min ZHANG, Xue-Ping WEI, Zhao-Rong HE, Zhong-Yang LI, Jia-Xi LI, Xian-Chun ZHANG, Shi-Yong DONG, Quan-Ru LIU, You-Xing LIN, Ran WEI, Li-Juan HE, Kung-Hsia SHING, Jian-Ying XIANG, Li WANG, Yue-Hong YAN, Shu-Kung WU, Fa-Guo WANG, Bin-Hui CHEN, Xiao-Lin GONG, and Jiu-Qiong SUN (Photo provided by Xian-Chun ZHANG)

2011 年，英文版中国植物志蕨类作者（从左至右）：张钢民、魏雪苹、和兆荣、李中阳、刘家熙、张宪春、董仕勇、刘全儒、林尤兴、卫然、何丽娟、邢公侠、向建英、王丽、严岳鸿、武素功、王发国、陈斌惠、宫晓林、孙久琼（相片提供者：张宪春）

2013年10月，吴德邻、张力，**广东苔藓志**[831]，552页；广东科技出版社，广州。

2013.10, Te-Lin WU and Li ZHANG, ***Bryophyte Flora of Guangdong***, 552 p; Guangdong Science and Technology Press, Guangzhou.

2013年11月，严岳鸿、张宪春、马克平编著，**中国蕨类植物多样性与地理分布**，308页；科学出版社，北京。

2013.11, Yue-Hong YAN, Xian-Chun ZHANG and Ke-Ping MA, ***Pteridophytes in China – Diversity and Distribution***, 308 p; Science Press, Beijing.

2013, Wen-Ying WU, You-Fang WANG, Qin ZUO, Wen YE, Benito C. TAN, Zhen-Ji LI, An updated checklist of mosses of Fujian Province, China, *Chenia* 11:144–182.

2013，吴文英、王幼芳、左勤、叶文、Benito C. TAN、李振基，中国福建藓类新名录，隐花植物生物学 11:144–182.

2013年11月，吴玉虎、王玉金主编，**喀喇昆仑山—昆仑山地区植物名录**，456页；青海民族出版社，西宁。

2013.11, Yu-Hu WU and Yu-Jin WANG (editors in chief), ***Index Florae Karakorum-Kunlunensis***, 456 p; Qinghai Ethnic Publishing House, Xining.

2013年11月，秦路平、黄宝康、周秀佳主编，**上海地区高等植物**，234页；第二军医大学出版社，上海。

2013.11, Lu-Ping QIN, Bao-Kang HUANG and Xiu-Jia ZHOU (editors in chief), '***Higher Plants of Shanghai District***', 234 p; Second Military Medical University Press, Shanghai.

2013年，陈丽、董洪进、彭华，云南省高等植物多样性与分布状况，生物多样性 21（3）：359-363。

2013, Li CHEN, Hong-Jin DONG and Hua PENG, Diversity and distribution of higher plants in Yunnan, China, *Biodiversity Science* 21(3): 359–363.

2013年，诚静容论文集编写组，**中国药用植物学奠基人——诚静容论文集**，287+79页；北京大学药学院（内部印制），北京。[832]

2013, Editorial Group of Collected Works of Ching-Yung CHENG, '***The Founder of Chinese Medical Botany—Collected Works of Ching-Yung CHENG***', 287+79 p; School of Pharmaceutical Sciences, Peking

831 本书包括海南的种类。

832 参见：马金双、陈虎彪，2012，缅怀诚静容教授，植物分类与资源学报 34（6）：633-634。

Participants at National Fern Symposium, Shanghai, 2013 (Photo provided by Chun-Xiang LI)
2013 年中国蕨类植物研讨会于上海（相片提供者：李春香）

University, Beijing (internal publication).

2013 年，国家林业局野生动植物保护和自然保护区管理司、中国科学院植物研究所编著，印红主编，**中国珍稀濒危植物图鉴**，378 页；中国林业出版社，北京。

2013, Management Bureau of Wild Animals and Plants Protection and Natural Conservation, National Bureau of Forestry, Institute of Botany, Chinese Academy of Sciences, Hong YIN (editor in chief), ***Rare and Endangered Plants in China***, 378 p; China Forestry Publishing House, Beijing.

2013, Hong-Mei LIU, Ri-Hong JIANG, Jian GUO, Peter Hovenkamp, Leon R. Perrie, Lara Shepherd, Sabine Hennequin and Harald Schneider, Towards a phylogenetic classification of the climbing fern genus *Arthropteris*, *Taxon* 62(4): 688–700, including a new family, Arthropteridaceae.

2013 年，刘红梅、蒋日红、郭捡、Peter Hovenkamp、Leon R. Perrie、Lara Shepherd、Sabine Hennequin、Harald Schneider，攀缘蕨类植物爬树蕨属的系统分类研究，*Taxon* 62（4）：688–700，包括新科——爬树蕨科。

2013, Nicolas J. Turland, ***The Code Decoded*** *– A User's Guide to the International Code of Nomenclature for algae, fungi, and plants*, 169 p; *Regnum Vegetabile* 155: 1–169 p; Koeltz Scientific Books, Königstein. Translated into Chinese in 2014. ***The Code Decoded***, second edition,196 p, 2019; Pensoft Publishers, Sofia.

2014 年，解译法规翻译组译，刘夙审校，**解译法规**——《国际藻类、菌物和植物命名法规》读者指南，190 页；高等教育出版社，北京。Nicolas J. Turland 著，**解译法规**，第二版，196 页，2019；Pensoft Publishers，索菲亚。

2013 年，张宪春、卫然、刘红梅、何丽娟、王丽、张钢民，中国现代石松类和蕨类的系统发育与分类系统，植物学报 48（2）：119–137。

2013, Xian-Chun ZHANG, Ran WEI, Hong-Mei LIU, Li-Juan HE, Li WANG and Gang-Min ZHANG, Phylogeny and classification of the extant lycophytes and ferns from China, *Chinese Bulletin of Botany* 48(2): 119–137.

2013 年，陈艺林、林慰慈主编，**林镕文集**，945 页；科学出版社，北京。

2013, Yi-Ling CHEN and Wei-Ci LIN (editors in chief), ***Collected Works of Ling Yong (Yong LING)***, 945 p; Science Press, Beijing.

2013, De-Yuan HONG and Stephen Blackmore, ***Plants of China—****A companion to the Flora of China*, 472 p; Science Press, Beijing.

2013 年，洪德元、Stephen Blackmore，**中国植物**——英文版中国植物志姊妹篇[833]，472 页；科学出版社，北京。

2013 年，胡宗刚著，**笺草释木六十年——王文采传**，261 页；上海交通大学出版社，上海；中国科学技术出版社，北京。

2013, Zong-Gang HU, '***Jian Cao Shi Mu Liu Shi Nian—A Biography of Wen-Tsai WANG***', 261 p; Shanghai Jiao Tong University Press, Shanghai and China Science and Technology Press, Beijing.

2013 年，胡宗刚著，**华南植物研究所早期史**——中山大学农林植物研究所史事（1928—1954），208 页；上海交通大学出版社，上海。

2013, Zong-Gang HU, '***Early History of South China Institute of Botany****—Historical Events of the Botanical Institute of Sun Yat-Sen University (1928–1954)*', 208 p; Shanghai Jiao Tong University Press, Shanghai.

2013 年，香港渔农自然护理署香港植物标本馆、中国科学院华南植物园编著，夏念和、张国伟主编，**香港植物检索手册**，349 页，2 025 张图片；香港特别行政区政府渔农自然护理署，香港。

2013, Hong Kong Herbarium, Agriculture, Fisheries and Conservation Department and South China Botanical Garden, Chinese Academy of Sciences, Nian-He XIA and Guo-Wei ZHANG (editors in chief), '***Identification Handbook of Hong Kong Plants***', 349 p, 2,025 photos; Agriculture, Fisheries and Conservation Department, Government of the Hong Kong Special Administrative Region, Hong Kong.

2013, Su LIU, Bing LIU and Xiang-Yun ZHU, Corrections of wrongly spelled scientific names in *Flora Reipublicae Popularis Sinicae*, *Journal of Systematics and Evolution* 51(2): 231–234.

2013 年，刘夙、刘冰、朱相云，中国植物志误拼学名的订正，*Journal of Systematics and Evolution* 51（2）：231–234。

2013 年，马金双主编，**上海维管植物名录**，442 页；高等教育出版社，北京。

2013, Jin-Shuang MA (editor in chief), ***The Checklist of Shanghai Vascular Plants***, 442 p; Higher Education Press, Beijing.

2013 年，马金双主编，**中国入侵植物名录**，324 页；高等教育出版社，北京。

2013, Jin-Shuang MA (editor in chief), ***The Checklist of the Chinese Invasive Plants***, 324 p; Higher Education Press, Beijing.

833 2016 年 10 月 10 日星期一，洪德元于北京中国科学院大学雁栖湖第十二届全国生物多样性科学与保护研讨会现场赐教。

2013+，全国开展第 4 次中草药普查。

2013+, Forth National Survey on traditional Chinese medicine begun.

2013+，艾铁民主编，**中国药用植物志**，第 1、2 卷:XXX；韦发南（卷主编），第 3 卷，1 215 页，莲叶桐科—茅膏菜科，2016；陆玲娣（卷主编），第 4 卷，1 142 页，罂粟科—牛栓藤科，2015；朱相云（卷主编），第 5 卷（上、下册），1 816 页，豆科—槭树科，2016；第 6 卷：XXX；李世晋（卷主编），第 7 卷，1 344 页，千屈菜科—报春花科，2018；第 8 卷：XXX；彭华（卷主编），第 9 卷，1 496 页，唇形科—车前科，2017；陈艺林（卷主编），第 10 卷，1 337 页，忍冬科—菊科，2014；张树仁（卷主编），第 11 卷，974 页，泽泻科 — 禾本科，2014；戴伦凯（卷主编），第 12 卷，797 页，棕榈科—兰科，2013。北京大学医学出版社，北京。

2013+, Tie-Min AI (editor in chief), ***Medicinal Flora of China***, vols. 1 & 2: XXX; Fa-Nan WEI (editor), 3: 1,215 p, Hernandiaceae-Droseraceae, 2016; Ling-Di LU (editor), 4: 1,142, Papaveraceae-Connaraceae, 2015; Xiang-Yun ZHU (editor), 5 (part I and II), 1,816 p, Fabaceae-Aceraceae, 2016; 6: XXX; Shi-Jin LI (editor), 7:1,344p, Lythraceae-Primulaceae, 2018; 8: XXX; Hua PENG (editor), 9: 1,496 p, Lamiaceae-Plantaginaceae, 2017; Yi-Ling CHEN (editor), 10: 1,337 p, Caprifoliaceae-Compositae, 2014; Shu-Ren ZHANG (editor), 11: 974 p, Alismataceae-Poaceae, 2014; Lun-Kai DAI (editor), 12: 797 p, Arecaceae-Orchidaceae, 2013; Peking University Medical Press, Beijing.

2013 年 10 月至 2018 年，中国生物物种名录编辑委员会植物卷工作组，**中国生物物种名录**，第 1 卷，**植物**；贾渝、何思，苔藓植物，525 页，2013；严岳鸿、张宪春、周喜乐、孙久琼，蕨类植物，277 页，2016；金效华、杨永，种子植物（Ⅰ，裸子植物、被子植物：莼菜科—兰科），372 页，2015；陈文利、张树仁，种子植物（Ⅱ，被子植物：棕榈科—禾本科），354 页，2018；覃海宁、刘博、何兴金、叶建飞，种子植物（Ⅲ，被子植物：百合科—五桠果科），264 页，2018；朱相云、陈之端、刘博，种子植物（Ⅳ，被子植物：芍药科—远志科），344 页，2015；夏念和、童毅华，种子植物（Ⅴ，被子植物：蔷薇科—叶下珠科），420 页，2018；张志翔、侯元同、廖帅、谢宜飞，种子植物（Ⅵ，被子植物：沟繁缕科—钩枝藤科），343 页，2017；于胜祥、郝刚、金孝锋，种子植物（Ⅶ，被子植物：石竹科—杜鹃花科），329 页，2016；王瑞江、刘演、陈世龙，种子植物（Ⅷ，被子植物：茶茱萸科—胡麻科），332 页，2017；向春雷、刘启新、彭华，种子植物（Ⅸ，被子植物：唇形科—伞形科），313 页，2016；高天刚、张国进，种子植物（Ⅹ，被子植物：桔梗科—忍冬科），300 页，2018；王利松、贾渝、张宪春、覃海宁，第 1 卷，植物总目录（上册：苔藓植物、蕨类植物、裸子植物，1-240 页；中册：爵床科—千屈菜科，241-944 页；下册：木兰科—蒺藜科，945-1 951 页），2018；科学出版社，北京[834]。

2013.10-2018, Editorial Committee of Species Catalogue of China, ***Species Catalogue of China***, Vol. 1, **PLANTS**; Yu JIA and Si HE, Bryophytes, 525 p, 2013; Yue-Hong YAN, Xian-Chun ZHANG, Xi-

834 马克平，2015，中国生物多样性编目取得重要进展，生物多样性 23（2）：137-138。

Le ZHOU and Jiu-Qiong SUN, Pteridophytes, 277 p, 2016; Xiao-Hua JIN and Yong YANG, Spermatophytes (I, Gymnosperms and Angiosperms: Cabombaceae-Orchidaceae), 372 p, 2015; Wen-Li CHEN and Shu-Ren ZHANG, Spermatophytes (II, Angiosperms: Arecaceae-Poaceae), 354 p, 2018; Hai-Ning QIN, Bo LIU, Xing-Jin HE and Jian-Fei YE, Spermatophytes (III, Angiosperms: Liliaceae-Dilleniaceae), 264 p, 2018; Xiang-Yun ZHU, Zhi-Duan CHEN, and Bo LIU, Spermatophytes (IV, Angiosperms: Paeoniaceae-Polygalaceae), 344 p, 2015; Nian-He XIA and Yi-Hua TONG, Spermatophytes (V, Angiosperms: Rosaceae-Phyllanthaceae), 420 p, 2018; Zhi-Xiang ZHANG, Yuan-Tong HOU, Shuai LIAO and Yi-Fei XIE, Spermatophytes (VI, Angiosperms: Elatinaceae-Ancistrocladaceae), 343 p, 2017; Sheng-Xiang YU, Gang HAO and Xiao-Feng JIN, Spermatophytes (VII, Angiosperms: Caryophyllaceae-Ericaceae), 329 p, 2016; Rui-Jiang WANG, Yan LIU and Shi-Long CHEN, Spermatophytes (VIII, Angiosperms: Icacinaceae-Pedaliaceae), 332 p, 2017; Chun-Lei XIANG, Qi-Xin LIU and Hua PENG, Spermatophytes (IX, Angiosperms: Lamiaceae-Apiaceae), 313 p, 2016; Tian-Gang GAO and Guo-Jin ZHANG, Spermatophytes (X, Angiosperms: Campanulaceae-Caprifoliaceae), 300 p, 2018; Li-Song WANG, Yu JIA, Xian-Chun ZHANG and Hai-Ning QIN, Volume 1, Plants, *A Synoptic Checklist* (I: Bryophytes, Pteriophytes, Gymnosperms, 1–240 p, II: Acanthaceae-Lythraceae, 241–944 p, and III: Magnoliaceae-Zygophyllaceae, 945–1,951 p), 2018; Science Press, Beijing.

2013 年，王利松、杨永、张宪春，在线植物志：网络时代分类学的方法和实践，植物学报 48（2）：174-183。

2013, Li-Song WANG, Yong YANG, Xian-Chun ZHANG, Online Flora: Method and Practice of E-taxonomy, *Chinese Bulletin of Botany* 48(2): 174-183.

2014

2014 年 1 月，徐波、李志敏、孙航著，**横断山高山冰缘带种子植物**，413 页；科学出版社，北京。

2014.01, Bo XU, Zhi-Min LI and Hang SUN, ***Seed Plants of the Alpine Subnival Belt from the Hengduan Mountains, SW China***, 413 p; Science Press, Beijing.[835]

2014 年 3 月，田旗主编，**华东植物区系维管束植物多样性编目**，565 页；科学出版社，北京。

2014.03, Qi TIAN (editor in chief), '***Inventory of Vascular Plant Diversity from East China***' 565 p; Science Press, Beijing.

2014 年 4 月，吴征镒述、吕春朝记录整理，**吴征镒自传**，338 页；科学出版社，北京。

2014.04, Zheng-Yi WU (dictated) and Chun-Chao LÜ (recorded and transcribed), '***Autobiography of Zheng-Yi WU***', 338 p; Science Press, Beijing.

2014 年 4 月 21 日至 26 日，国际藻类、菌物和植物命名法规暨植物学拉丁文培训班在深圳市中国科学院仙湖植物园举行，来自全国 30 多个高校、科研机构和植物园的 72 位学员参加。

2014.04.21–04.26, Workshop on the *International Code of Nomenclature for algae, fungi, and plants*, and on Botanical Latin, was held at Fairylake Botanical Garden, Shenzhen and Chinese Academy of Sciences; 72 attendees from more than 30 universities, institutions and botanical gardens nationwide.

2014.05.06–05.15, Chinese-American botanical expedition to Sichuan: Baoxing, Heishui, Lixian, Maoxian, Songpan. David E. Boufford (A/GH), Sharon Christoph and Christopher Davidson (***Flora of the World***, http://floraoftheworld.org/), Yun-Dong GAO (CDBI) and Qiu-Yun XIANG (NCSC) collected 138 numbers of vascular plants (David E. Boufford et al. nos. 43585–43741, A, CAS, CDBI, MO, P, PE, TI).

2014 年 5 月 6 日至 15 日，中美植物联合考察四川的宝兴、黑水、理县、茂县、松潘。哈佛大学 David E. Boufford，**世界植物志**在线 Sharon Christoph 和 Christopher Davidson（http://floraoftheworld.org/），中国科学院成都生物研究所高云东，美国北卡罗来纳州立大学向秋云，采集维管束植物标本 138 号（A、CAS、CDBI、MO、P、PE、TI）。

2014 年 5 月 31 日，**中国植物分类学在线家园**上线：http://www.planttaxonomists.cn/default.php；截至 2017 年 7 月 31 日，631 位注册会员，月访问量超过 3 000。

835 Bo XU, Zhi-Min LI and Hang SUN, 2014, Plant diversity and floristic characters of the alpine subnival belt flora in the Hengduan Mountains, SW China, *Journal of Systematics and Evolution* 52(3): 271–279.

Participants at Workshop on the International Code of Nomenclature for algae, fungi, and plants, and Botanical Latin, Shenzhen, 21–26 April 2014 (Photo provided by Shou-Zhou ZHANG)
2014 年 4 月 21 日至 26 日，国际藻类、菌物和植物命名法规暨植物学拉丁文培训班于深圳（相片提供者：张寿洲）

2014.05.31, Online Home of ***Chinese Plant Taxonomists*** launched; http://www.planttaxonomists.cn/default.php, with 631 members registered by July 31, 2017, and more than 3,000 visits per month.

2014年6月，中国科学院昆明植物研究所编，**吴征镒先生纪念文集**，348页；云南科技出版社，昆明。

2014.06, Kunming Institute of Botany, Chinese Academy of Sciences, ***In Memory of Zheng-Yi WU***, 348 p; Yunnan Science and Technology Press, Kunming.

2014.07-08, Chinese American botanical expdition in Shaanxi: Qinling (Taibai Shan), Wen-Bo LIAO (SYS), Daniel Potter (DAV), Lei WANG (BJTC) and Qian-Yi YIN (SYS) collected more than 500 numbers of plants specimens of Maloideae and Prunoideae (DAV, SYS).

2014年7月至8月，中美植物联合考察陕西秦岭太白山，中山大学廖文波、阴倩怡，加州大学戴维斯分校 Daniel Potter，首都师范大学王蕾，采集苹果亚科、李亚科植物标本500多号（DAV、SYS）。

2014年8月1日至5日，全国苔藓植物学学术研讨会在呼和浩特内蒙古大学召开，来自全国33个高校和研究单位的130余名代表参加了会议。

2014.08.01-08.05, National Bryophyte Symposium held at Inner Mongolia University, Hohhot, nearly 130 attendees from 33 institutions.

2014.08.13-08.21, Chinese-American botanical expedition to Hubei: Shennongjia Forest District and Xing Shan. David E. Boufford (A/GH), Tao DENG (1985-, KUN), Michael J. Donoghue (YU), Ze-Long NIE (KUN), Patrick W. Sweeney (YU) and Dai-Gui ZHANG (JIU) collected 96 numbers of vascular plants (David E. Boufford et al. nos. 43768-43863, A, KUN, MO, P, PE, TI, YU).

2014年8月13日至21日，中美植物联合考察湖北的神农架林区、兴山，哈佛大学 David E. Boufford，中国科学院昆明植物研究所邓涛（1985—）、聂泽龙，耶鲁大学自然历史博物馆 Michael J. Donoghue 和 Patrick W. Sweeney，吉首大学张代贵，采集维管束植物96号（A、KUN、MO、P、PE、TI、YU）。

2014.08.26-08.29, Sixth Asian Fern Symposium held in Bali Botanic Garden, Bali, Indonesia, more than 82 attendees from 13 countries in Asia, Europe and North America, including Cheng-Wei CHEN (1984-), Wen-Liang CHIOU, Tzu-Yun CHIU, Li-Yaung KUO (1986-), Chun-Xiang LI, Shau-Fu LI (1991-), Hong-Mei LIU (1979-), Yea-Chen LIU, Xin-Ping QI (1982-), Hui SHANG (1986-), Ai-Hua WANG (1985-), Fa-Guo WANG (1978-), Jian-Ying XIANG (1976-), Yue-Hong YAN (1974-) and Hao ZHANG (1965-) from China.

Participants at National Bryophyte Symposium, Hohhot, 1-5 August 2014 (Photo provided by Yu-Huan WU)
2014 年 8 月 1 日至 5 日，全国苔藓植物学学术研讨会于呼和浩特（相片提供者：吴玉环）

2014 年 8 月 26 日至 29 日，第六届亚洲蕨类植物学大会在印度尼西亚巴厘植物园举行，来自亚洲、欧洲和北美洲 13 个国家的 82 位学者出席，中国学者包括陈正为（1984—）、邱子芸、邱文良、郭立园（1986—）、李春香、李绍辅（1991—）、刘红梅（1979—）、刘以诚、齐新萍（1982—）、商辉（1986—）、王爱华（1985—）、王发国（1978—）、向建英（1976—）、严岳鸿（1974—）、张豪（1965—）。

2014 年 8 月，吴玉虎、李忠虎主编，**喀喇昆仑山—昆仑山地区植物检索表**，292 页；青海民族出版社，西宁。

2014.08, Yu-Hu WU and Zhong-Hu LI (editors in chief), ***Claves Plantarum Karakorum-Kunlunensis***, 292 p; Qinghai Ethnic Publishing House, Xining.

2014 年 9 月，新疆植物志编辑委员会编著，安峥皙主编，**新疆植物志简本**，1 007 页；新疆人民出版社、新疆科学技术出版社，乌鲁木齐。

2014.09, Commissione Redactorum Florae Xinjiangensis, Zheng-Xi AN (editor in chief), ***Flora Xinjiangensis Simplified Edition***, 1,007 p; Xinjiang People's Publishing Group, Xinjiang Science and Technology Publishing House, Urumqi.

2014.09.25-09.27, The 17th Annual Moss International Conference held at Capital Normal University, Beijing, about 60 attendees, including 10 from abroad.

2014 年 9 月 25 日至 27 日，第十七届国际藓类年会在北京首都师范大学举行，海内外 60 余人参加，包括来自国外 10 个代表。

2014 年 11 月，黄继红、马克平、陈彬著，**中国特有种子植物的多样性及其地理分布**，442 页；高等教育出版社，北京。

2014.11, Ji-Hong HUANG, Ke-Ping MA and Bin CHEN, '***Diversity and Its Geographical Distribution of Seed Plants Endemic to China***', 442 p; Higher Education Press, Beijing.

2014 年 11 月 8 日至 9 日，2014 全国系统与进化植物学研讨会暨第十一届青年学术研讨会在杭州浙江大学举行，全国 28 个省（市、区）112 家单位 480 多人与会。[836]

2014.11.08-11.09, 2014 National and Eleventh Youth Symposium of Systematic and Evolutionary Botany held at Zhejiang University, Hangzhou, over 480 attendees from 112 institutions in 28 provinces.

2014 年 11 月 20 日至 21 日，2014 年中国蕨类植物研讨会在广西南宁广西药用植物园举行，全

836 全国系统与进化植物学研讨会暨第十一届青年学术研讨会，2014 全国系统与进化植物学研讨会暨第十一届青年学术研讨会论文集，137 页；浙江大学，杭州（内部印制）。

Participants at 6th Asian Fern Symposium, Bali, Indonesia, 26-29 August 2014 (upper), and some Chinese delegates (lower) (Photo provided by Hong-Mei LIU)

2014 年 8 月 26 日至 29 日，第六届亚洲蕨类植物学大会于巴厘岛（上），部分中国代表（下）（相片提供者：刘红梅）

Participants at National and Eleventh Youth Symposium of Systematic and Evolutionary Botany, Hangzhou, 8-9 November 2014 (Photo provided by Cheng-Xin FU)
2014 年 11 月 8 日至 9 日，全国系统与进化植物学研讨会暨第十一届青年学术研讨会于杭州（相片提供者：傅承新）

25th reunion of members of the First National Youth Symposium of Systematic and Evolutionary Botany, 2014 (left-right): Xian-Chun ZHANG, Cheng-Xin FU, Qi-Xin LIU, Jia-Kuan CHEN, Jin-Shuang MA, Bing-Yang DING (Photo provided by Jin-Shuang MA)
2014 年首届青年系统与进化植物学研讨会参加者二十五年后再聚（从左至右）：张宪春、傅承新、刘启新、陈家宽、马金双、丁炳扬（相片提供者：马金双）

国 36 家单位 82 人参加。

2014.11.20-11.21, 2014 National Fern Symposium held at Guangxi Medicinal Botanical Garden, Nanning, 82 attendees from 36 institutions.

2014 年，罗桂环著，**中国近代生物学的发展**，417 页；中国科学技术出版社，北京。

2014, Gui-Huan LUO, ***The Development of Modern Biology in China***, 417 p; China Science and Technology Press, Beijing.

2014 年，胡宗刚编，**庐山植物园八十春秋纪念集**，366 页；上海交通大学出版社，上海。

2014, Zong-Gang HU, '***Memorial Works of Eighty Anniversary of Lushan Botanical Garden***', 366 p; Shanghai Jiao Tong University Press, Shanghai.

2014, Christopher Gardner and Başak Gardner, ***Flora of the Silk Road***—*An Illustrated Guide*, 406 p; I. B. Tauris, London, New York.

2019 年，刘夙译，**丝路之花**[837]，404 页；中国工信出版集团，人民邮电出版社，北京。

837 本书所涉地域覆盖从中国至土耳其的上万公里，书中至少有 1/4 的篇幅介绍中国西部的植物。

Participants at National Fern Symposium, Nanning, 2014 (Photo provided by Xian-Chun ZHANG)
2014 年中国蕨类植物研讨会于南宁（相片提供者：张宪春）

2014 年，杜诚、马金双，亚洲植物志编研简述，中国生物多样性保护与研究进展 X：67-80。

2014, Cheng DU and Jin-Shuang MA, Introduction to Asian floras, *Advances in Biodiversity Conservation and Research in China* X: 67-80.

2014 年，胡宗刚著，**西双版纳热带植物园五十年**，370 页；科学出版社，北京。

2014, Zong-Gang HU, '***Fifty Years of Xishuangbanna Tropical Botanical Garden***', 370 p; Science Press, Beijing.

2014 年，贾渝、马欣堂、班勤、李敏、杨改河主编，**大巴山地区高等植物名录**，392 页；科学出版社，北京。

2014, Yu JIA, Xin-Tang MA, Qin BAN, Min LI and Gai-He YANG (editors in chief), ***Higher Plants of the Dabashan Mountains***, 392 p; Science Press, Beijing.

2014 年，马金双主编，**上海维管植物检索表**，411 页；高等教育出版社，北京。

2014, Jin-Shuang MA (editor in chief), ***The Keys of Shanghai Vascular Plants***, 411 p; Higher Education Press, Beijing.

2014 年，马金双主编，**中国外来入侵植物调研报告**[838]，上卷：1-537 页，下卷：542-949 页；高等教育出版社，北京。

2014, Jin-Shuang MA (editor in chief), ***The Survey Reports on Chinese Alien Invasive Plants***, Volume I: 1-537 p, and II: 542-949 p; Higher Education Press, Beijing.

2014 年，赵一之、赵利清著，**内蒙古维管植物检索表**，426 页；科学出版社，北京。

2014, Yi-Zhi ZHAO and Li-Qing ZHAO, ***Key to the Vascular Plants of Inner Mongolia***, 426 p; Science Press, Beijing.

2014, Jane Kilpatrick, ***Fathers of Botany—****The Discovery of Chinese Plants by European Missionaries*, 254 p; Kew Publishing, Royal Botanic Gardens, Kew and University of Chicago Press, London and Chicago.[839]

2014，Jane Kilpatrick，**从事植物学的神父们**——欧洲传教士发现的中国植物，254 页；Kew Publishing，皇家植物园，邱园，伦敦；芝加哥大学出版社，芝加哥。

2014 年，马金双，中国植物分类学的现状与挑战，科学通报 59（6）：510-521。

838 上卷包括东北、华北、西北、华中、华东，下卷包括华南、中南、西南以及学名和中文名索引。

839 Hans W. Lack, 2015, Catholic missionaries and Chinese plants, *Taxon* 64(6): 1362-1363.

2014, Jin-Shuang MA, Current status and challenges of Chinese plant taxonomy, *Chinese Science Bulletin* 59(6): 510-521.

2014 年，马金双，第十九届国际植物学大会——中国植物分类学家的任务，仙湖 13（3-4）：58-60。

2014, Jin-Shuang MA, The XIX IBC—The task of Chinese plant taxonomists, *Journal of Fairylake Botanical Garden* 13(3-4): 58-60.

2014 年，臧穆 黎兴江 论文集编辑组，**臧穆 黎兴江论文集**，1 296 页；论文集编辑组，昆明（内部印制）。

2014, Members of Editorial Group of Mu ZANG and Xingjiang LI Collected Works, ***Collected Works of Academic Papers by Profs. ZANG Mu and LI Xingjiang***, 1,296 p; Members of Editorial Group, Kunming (internal publication).

2014 年，杨亲二，掌叶橐吾地理分布的订正及对我国植物志书中一些明显问题的述评，热带亚热带植物学报 22（2）：107-120。

2014, Qin-Er YANG, Notes on the geographical distribution of *Ligularia przewalskii* and critical comments on some problems apparent in floristic works of China, *Journal of Tropical and Subtropical Botany* 22(2): 107-120.

2014，2018 年，熊源新编著，**贵州苔藓植物志**，第 1 卷：509 页，第 2 卷：686 页 , 2014；第 3 卷：720 页，2018；贵州科技出版社，贵阳。

2014, 2018, Yuan-Xin XIONG, ***Bryophyte Flora of Guizhou China***, vol. 1: 509 p, 2: 686 p, 2014; 3:720p, 2018; Guizhou Science and Technology Publishing House, Guiyang.

2014，白学良（主编），**贺兰山苔藓植物彩图志**，458 页；黄河传媒集团宁夏人民出版社，银川。

2014，Xue-Liang BAI (Editor in Chief), '***Color Flora of Bryophytes of Helan Mountain***' 458 p; Yellow River Publishing Media Ningxia People's Press, Yinchuan.

2014 年，陈健斌，分类都去哪儿了——为分类学正名，向分类学家致敬，菌物研究 12（3）：125-129。

2014, Jian-Bin CHEN, 'Where taxonomy gone – Rectify the taxonomy and salute to the taxonomists', *Journal of Fungal Research* 12(3): 125-129.

2015

2015.01.11-01.15, World Conference of Bryology held at University of Magallanes, Puerto Williams, Chile, attended by Tian-Yi MA, Wei SHA and Li ZHANG from China.

2015年1月11日至15日，世界苔藓植物学大会在智利威廉斯港麦哲伦大学召开，中国代表马天意、沙伟和张力参加。

2015.03.27, The Arnold Arboretum in China: Reflections on a Century of Plant Exploration in China by Harvard University held at the Harvard Center Shanghai, China; presentations by David E. Boufford, Senior Research Scientist, Harvard University Herbaria, Michael S. Dosmann, Curator of Living Collections, the Arnold Arboretum of Harvard University, Guang-You HAO (1980-), Curator, Institute of Applied Ecology, Chinese Academy of Sciences, Putnam Fellow of the Arnold Arboretum of Harvard University, Jin KOU, Chief Director of three episodes of TV series Chinese Wilson for CCTV[840] Channel 9, Lisa Pearson, Head of the Library and Archives, the Arnold Arboretum of Harvard University and Kai-Pu YIN (1943-), Curator, Chengdu Institute of Biology, Chinese Academy of Sciences, Author of *Tracing One Hundred Years of Change-Illustrating the Environmental Changes in Western China*. Symposium hosted by William (Ned) Friedman, Director of the Arnold Arboretum, Arnold Professor of Organismic and Evolutionary Biology, Harvard University; more than 40 attendees from China.

2015年3月27日，阿诺德树木园在中国——回顾哈佛大学对中国植物一个多世纪的考察研讨会在中国上海的哈佛中心举行；报告人：哈佛大学植物标本馆高级科学家 David E. Boufford，阿诺德树木园活植物主管 Michael S. Dosmann，阿诺德树木园 Putnam Fellow 获得者、中国科学院沈阳应用生态研究所研究员郝广友（1980—），央视纪录频道三集电视连续剧中国威尔逊导演寇金，中国科学院成都生物研究所研究员、百年追寻——见证中国西部环境变迁作者印开蒲，阿诺德树木园图书馆与档案负责人 Lisa Pearson；主持人：哈佛大学有机与进化生物学阿诺德教授、阿诺德树木园主任 William (Ned) Friedman；中国 40 多人参加。

2015.06-07, Chinese American botanists explored Xinjiang: Dushanzi, Gongliu, Hutubi, Huocheng, Urho, Xinyuan, Da-Fang CUI (1964-, CANT), Daniel Potter (DAV), Bao-Huan WU (CANT) collected more than 300 numbers of plant specimens of Prunoideae, Maloideae and Spiraeoideae (CANT, DAV).

2015年6月至7月，中美植物联合考察新疆的独山子、巩留、呼图壁、霍城、乌尔禾、新源，华南农业大学崔大方（1964—）、吴保欢，美国加州大学戴维斯 Daniel Potter，采集李亚科、苹果亚科、绣线菊亚科标本 300 多号（CANT、DAV）。

840 i.e. China Central Television.

2015.07.23, The Organizing Committee of the XIX IBC announces the first circular (http://www.ibc2017.cn/index.aspx) in Shenzhen, China, on the occasion of the start of the two year countdown to the XIX IBC, which will be officially opened July 23-29, 2017.

2015 年 7 月 23 日，在 2017 年 7 月 23 日至 29 日第十九届国际植物大会召开倒计时两周年之际，第十九届国际植物学大会组委会在深圳发布大会第一轮通知（http://www.ibc2017.cn/index.aspx）。

2015 年 8 月 21 日至 23 日，全国苔藓植物学学术研讨会在云南昆明中国科学院昆明植物研究所召开，来自全国 38 家单位的 88 人出席。

2015.08.21-08.23, National Bryophyte Symposium held at Kunming Institute of Botany, Chinese Academy of Sciences, Kunming, Yunnan, 88 attendees from 38 institutions.

2015 年 8 月 28 日至 30 日，三集电视连续剧**中国威尔逊**在央视九台每晚 8 时播出。

2015.08.28-08.30, Three episodes of TV series ***Chinese Wilson*** broadcast on CCTV Channel 9 each evening at 8:00 pm.

2015 年 10 月 17 日至 19 日，2015 年中国蕨类植物研讨会在云南昆明中国科学院昆明植物研究所举行，全国 32 个单位 94 名人员参加。

2015.10.17-10.19, 2015 National Fern Symposium held at Kunming Institute of Botany, Chinese Academy of Sciences, Kunming, Yunnan, 94 attendees from 32 institutions.

2015 年 12 月，张宪春、孙久琼编著，**石松类和蕨类名词及名称**，237 页；中国林业出版社，北京。

2015.12, Xian-Chun ZHANG and Jiu-Qiong SUN, ***A Glossary of Terms and Names of Lycopods and Ferns***, 237 p; China Forestry Publishing House, Beijing.

2015 年 12 月，黄普华、薛煜编著，**拉汉植物与真菌分类群描述常见例句**，142 页；东北林业大学出版社，哈尔滨。

2015.12, Pu-Hwa HUANG and Yu XUE, '***Common Examples of Description of Plants and Fungi Taxa in Latin-Chinese***', 142 p; Northeast Forestry University Press, Harbin.

2015 年，广西壮族自治区、中国科学院广西植物研究所，**广西自治区、中国科学院广西植物研究所建所八十年纪念册**（1935—2015），101 页；广西壮族自治区、中国科学院广西植物研究所，桂林，广西（内部印制）。

2015, Guangxi Institute of Botany, Guangxi Zhuang Autonomous Region and Chinese Academy of Sciences, ***The 80th Anniversary of Guangxi Institute of Botany, Guangxi Autonomous Region and Chinese***

Participants at National Bryophyte Symposium, Kunming, 21-23 August 2015 (Photo provided by Yu-Huan WU)
2015 年 8 月 21 日至 23 日，全国苔藓植物学学术研讨会于昆明（相片提供者：吴玉环）

Panticipants at National Fern Symposium, Kunming, 2015 (Photo provided by Xian-Chun ZHANG)

2015 年中国蕨类植物研讨会于昆明（相片提供者：张宪春）

Academy of Sciences, 101 p; Guangxi Institute of Botany, Guangxi Zhuang Autonomous Region and Chinese Academy of Sciences, Guilin, Guangxi (internal publication).

2015 年，杨永，中国裸子植物的多样性与地理分布，生物多样性 23（2）：243-246。

2015, Yong YANG, Diversity and distribution of gymnosperms in China, *Biodiversity Science* 23(2): 243-246.

2015 年，王利松、贾渝、张宪春、覃海宁，中国高等植物多样性，生物多样性 23（2）：217-224。

2015, Li-Song WANG, Yu JIA, Xian-Chun ZHANG and Hai-Ning QIN, Overview of higher plant diversity in China, *Biodiversity Science* 23(2): 217-224.

2015, Li-Bing ZHANG and Michael G. Gilbert, Comparison of classification of vascular plants of China, *Taxon* 64(1): 17-26.

2015 年，张丽兵、Michael G. Gilbert，中国维管植物分类比较，*Taxon* 64（1）：17-26。

2015, Li-Bing ZHANG and Liang ZHANG, Didymochlaenaceae—A new fern family of eupolypods I (Polypodiales), *Taxon* 64(1): 27-38.

2015 年，张丽兵、张良，翼囊蕨科 —— 真水龙骨类 I（水龙骨目）一新科，*Taxon* 64（1）：27-38。

2015 年，杨小波等编著，**海南植物图志**，14 卷本；第 1 卷：557 页，第 2 卷：425 页，第 3 卷：578 页，第 4 卷：400 页，第 5 卷：361 页，第 6 卷：397 页，第 7 卷：379 页，第 8 卷：401 页，第 9 卷：414 页，第 10 卷：539 页，第 11 卷：500 页，第 12 卷：380 页，第 13 卷：502 页，第 14 卷：666 页；科学出版社，北京。

2015, Xiao-Bo YANG et al., '***An Illustrated Flora of Hainan***', vols. 1-14; 1: 557 p, 2: 425 p, 3: 578 p, 4: 400 p, 5: 361 p, 6: 397 p, 7: 379 p, 8: 401 p, 9: 414 p, 10: 539 p; 11: 500 p, 12: 380 p, 13: 502 p, 14: 666 p; Science Press, Beijing.

2015+，渔农自然护理署香港植物标本室、中国科学院华南植物园编著，夏念和主编，**香港植物志**，第 1 卷：458 页，图片 608 幅；渔农自然护理署出版，香港。

2015+, Agriculture, Fisheries and Conservation Department, South China Botanical Garden, Chinese Academy of Sciences, Nian-He XIA (editor in chief), ***Flora of Hong Kong***, vol. 1: 458 p, 608 photos;

Agriculture, Fisheries and Conservation Department, Hong Kong[841].

2015 年，刘冰、叶建飞、刘夙、汪远、杨永、赖阳均、曾刚、林秦文，中国被子植物科属概览——依据 APG Ⅲ系统，生物多样性 23（2）：225-231。

2015, Bing LIU, Jian-Fei YE, Su LIU, Yuan WANG, Yong YANG, Yang-Jun LAI, Gang ZENG and Qin-Wen LIN, Families and genera of Chinese angiosperms: A synoptic classification based on APG III, *Biodiversity Science* 23(2): 225-231.

2015 年，刘夙、刘冰，中国维管植物属中文普通名选定规则新探，生物多样性 23（2）：254-258。

2015, Su LIU and Bing LIU, New comments on the rules of choice of Chinese common names of genera of vascular plants in China, *Biodiversity Science* 23(2): 254-258.

2015 年，武秀之，建立植物中文学名的构想，生物多样性 23（2）：259-263。

2015, Xiu-Zhi WU, A conception of establishing formal Chinese names of plants, *Biodiversity Science* 23(2): 259-263.

2015+, De-Yuan HONG (editor in chief), ***Flora of Pan-Himalaya***, vol. 30: 594 p, 2015 (Ihsan A. Al-Shehbaz: Brassicaceae / Cruciferae), 47: 292 p, 2015 (De-Yuan HONG: Aquifoliaceae, Helwingiaceae, Campanulaceae, Lobeliaceae, Menyanthaceae); 48(2): 340 p, 2015 (You-Sheng CHEN: *Saussurea* of Asteraceae / Compositae); 48(3): 134 p, 2017 (Tian-Gang GAO: Mutisieae, Hyalideae, Pertyeae of Asteraceae / Compositae); Science Press, Beijing and Cambridge University Press, Cambridge, UK.

2015+，洪德元主编，**泛喜马拉雅植物志**，第 30 卷：594 页，2015（Ihsan A. Al-Shehbaz：十字花科）；第 47 卷：292 页，2015（洪德元：冬青科、青荚叶科、桔梗科、半边莲科、睡菜科）；第 48（2）卷：340 页，2015（陈又生：菊科风毛菊属）；第 48（3）卷：134 页，2017（高天刚：菊科须菊木族、粉菊木族、帚菊族）；科学出版社，北京，剑桥大学出版社，剑桥。

2015—2017 年，林祁（第 1–4 卷），林祁、杨志荣（第 5–6 卷），林祁、杨永、杨志荣（第 7 卷），林祁、杨志荣、林云（第 8–11 卷 、第 14 卷），林祁、林云、张小冰、杨志荣（第 12–13 卷）主编，**中国国家植物标本馆（PE）模式标本集**，14 卷本；第 1 卷：545 页，2015，蕨类植物门（1）；第 2 卷：485 页，2015，蕨类植物门（2）；第 3 卷：479 页，2015，蕨类植物门（3）和裸子植物门；第 4 卷：519 页，2015，被子植物门（1）；第 5 卷：528 页，2015，被子植物门（2）；第 6 卷：589 页，2015，被子植物门（3）；第 7 卷：594 页，2017，被子植物门（4）；第 8 卷：562 页，2017，被子

841 This is the Chinese edition of the previous English edition, Chi-Ming HU and Te-Lin WU (editors in chief), *Flora of Hong Kong* (2007-2011).

Type Specimens of China National Herbarium (**PE**), fourteen volumes of photos of type specimens in the Institute of Botany Herbarium, Beijing (Photo provided by Henan Science and Technology Press)

十四卷本的中国科学院植物研究所中国国家植物标本馆模式标本集（相片提供者：河南科学技术出版社）

植物门（5）；第 9 卷：608 页，2017，被子植物门（6）；第 10 卷：578 页，2017，被子植物门（7）；第 11 卷：602 页，2017，被子植物门（8）；第 12 卷：638 页，2017，被子植物门（9）；第 13 卷：595 页，2017，被子植物门（10）；第 14 卷：614 页，2017，被子植物门（11）；河南科学技术出版社，郑州。

2015-2017, Qi LIN (vols. 1-4), Qi LIN and Zhi-Rong YANG (vols. 5-6), Qi LIN, Yong YANG and Zhi-Rong YANG (vol. 7), Qi LIN, Zhi-Rong YANG and Yun LIN (vols. 8-11 and 14), Qi LIN, Yun LIN, Xiao-Bing ZHANG and Zhi-Rong YANG (vols. 12-13, editors in chief), ***Type Specimens in China National Herbarium (PE)***, vols. 1-14: 1: 545, 2015, Pteridophyta (1); 2: 485 p, 2015, Pteridophyta (2); 3: 479 p, 2015, Pteridophyta (3) and Gymnospermae; 4: 519 p, 2015, Angiospermae (1); 5: 528 p, 2015, Angiospermae (2); 6: 589 p, 2015, Angiospermae (3); 7: 594 p, 2017, Angiospermae (4); 8: 562 p, 2017, Angiospermae (5); 9: 608 p, 2017, Angiospermae (6); 10: 578 p, 2017, Angiospermae (7); 11: 602 p, 2017, Angiospermae (8); 12: 638 p, 2017, Angiospermae (9); 13: 595 p, 2017, Angiospermae (10); 14: 614 p, 2017, Angiospermae (11); Henan Science and Technology Press, Zhengzhou.

2015, Ning-Ning YU, Qing-Hua WANG, Yu JIA, Peng-Cheng WU and Li ZHANG, Bryological literatures of China II, *Chenia* 12: 151-178.

2015 年，于宁宁、王庆华、贾渝、吴鹏程、张力，中国苔藓植物文献 Ⅱ，隐花植物生物学 12：151-178。

2016

2016年1月，中国高等植物彩色图鉴编辑委员会主编[842]，**中国高等植物彩色图鉴**，张力、左勤，第1卷，226页，苔藓植物；张宪春、成晓，第2卷，413页，蕨类植物一裸子植物；王文采、刘冰，第3卷，596页，被子植物：木麻黄科一莲叶桐科；于胜祥，第4卷，523页，被子植物：罂粟科一毒鼠子科；刘博、林秦文，第5卷，534页，被子植物：大戟科一山茱萸科；李振宇，第6卷，538页，被子植物：岩梅科一茄科；陈又生，第7卷，669页，被子植物：玄参科 — 菊科；张树仁，第8卷，402页，被子植物：香蒲科一翡若翠科；金效华，第9卷，345页，被子植物：蒟蒻薯科一兰科；总索引，244页。科学出版社，北京。

2016.01, Editorial Committee of Higher Plants of China in Colour (editor in chief [843]), ***Higher Plants of China in Colour***, Li ZHANG and Qin ZUO, vol. 1: 226 p, Bryophytes; Xian-Chun ZHANG and Xiao CHENG, 2: 413 p, Pteridophytes-Gymnosperms; Wen-Tsai WANG and Bing LIU, 3: 596 p, Angiosperms: Casuarinaceae-Hernandiaceae; Sheng-Xiang YU, 4: 523 p, Angiosperms: Papaveraceae-Dichapetalaceae; Bo LIU and Qin-Wen LIN, 5: 534 p, Angiosperms: Euphorbiaceae-Cornaceae; Zhen-Yu LI, 6: 538 p, Angiosperms: Diapensiaceae-Solanaceae; You-Sheng CHEN, 7: 669 p, Angiosperms: Scrophulariaceae-Compositae; Shu-Ren ZHANG, 8: 402 p, Angiosperms: Typhaceae-Velloziaceae; Xiao-Hua JIN, 9: 345 p, Angiosperms: Taccaceae–Orchidaceae; Index for the series: 244 p. Science Press, Beijing.

2016年2月26日至27日，第二届泛喜马拉雅植物志国内作者会议在北京举行，全国82人参加。

2016.02.26-02.27, Second meeting of domestic authors of *Flora of Pan-Himalaya* held in Beijing, 82 attendees.

2016.03, Alistair Watt, ***Robert Fortune—A Plant Hunter in the Orient***, 420 p; Kew Publishing, Kew.

2016年3月，Alistair Watt，**Robert Fortune——一位植物猎人在东方**，420页；Kew Publishing，邱园。

2016年3月13日，我国首个省（市、区）级地方数字植物志——**上海数字植物志**正式上线（http://shflora.ibiodiversity.net/index.html）。

2016.03.13, ***Shanghai Digital Flora***, the first local flora at the province level in China is launched (http://shflora.ibiodiversity.net/index.html).

---

842 编辑委员会主任：王文采，副主任：吴声华、李振宇；作者：费勇等；总策划：吴声华。

843 Chair of the editorial committee: Wen-Tsai WANG, vice chairs: Sheng-Hua WU and Zhen-Yu LI; Authors: Yong FEI et al., and Chief Planner: Sheng-Hua WU.

2016 年 5 月，陈彦生主编[844]，**陕西维管植物名录**，525 页；高等教育出版社，北京。

2016.05, Yan-Sheng CHEN (editor in chief), ***The Checklist of Shaanxi Vascular Plants***, 525 p; Higher Education Press, Beijing.

2016 年 5 月，任昭杰、赵遵田主编，**山东苔藓志**，450 页；青岛出版社，青岛。

2016.05, Zhao-Jie REN and Zun-Tian ZHAO (editors in chief), ***Bryophyte Flora of Shandong***, 450 p; Qingdao Publishing House, Qingdao.

2016 年 5 月，吴鹏程、汪楣芝、贾渝编著，**苔藓名词及名称（新版）**[845]，356 页；中国林业出版社，北京。

2016.05, Peng-Cheng WU, Mei-Zhi WANG and Yu JIA, ***A Glossary of Terms and Names of Bryophytes (New Edition)***, 356 p; China Forestry Publishing House, Beijing.

2016 年 5 月、2017 年 12 月，汪远、马金双主编，**上海植物图鉴 · 草本卷**，335 页，**上海植物图鉴 · 乔灌木卷**，413 页；河南科学技术出版社，郑州。[846]

2016.05 and 2017.12, Yuan WANG and Jin-Shuang MA, ***Illustrated Flora of Shanghai · Herbs***, 335 p, ***Illustrated Flora of Shanghai · Trees and Shrubs***, 413 p; Henan Science and Technology Press, Zhengzhou.

2016 年 6 月，洪德元，关于提高物种划分合理性的意见，生物多样性 24（3）：360-361。

2016.06, De-Yuan HONG, Opinion of raising rationality in species delimitation, *Biodiversity Science* 24(3): 360-361.

2016 年 7 月，谭策铭，**九江森林植物标本馆（JJF）建馆 23 年馆藏植物标本名录**，351 页；九江森林植物研究所，江西九江（内部刊物）。

2016.07, Ce-Ming TAN, '***Catalogue of Plant Specimens of the Herbarium of Jiujiang Forestry Institute (JJF) on the occasion of the twenty-third anniversary of its founding***', 351 p; Jiujiang Forestry Institute, Jiujiang, Jiangxi (internal publication).

2016 年 7 月，周繇编著，**东北珍稀濒危植物彩色图志**，上册：1-752 页，下册：753-1 056 页；

844 副主编之一杜诚（1984—）根据陈彦生（1952—2013）先生生前的手稿编辑并整理、加工后出版。

845 该书附有中国苔藓植物系统表，其中第 265-286 页为陈式系统（1963、1978），而第 287-310 页为贾渝和何思系统（2013，即中国生物物种名录编辑委员会植物卷工作组，中国生物物种名录，第 1 卷，植物；贾渝、何思，苔藓植物，525 页，2013）。

846 上海植物图鉴共分为 3 卷，即草本卷、乔灌木卷和室内观赏卷；室内观赏卷准备中。

东北林业大学出版社，哈尔滨。

2016.07, You ZHOU, '***Rare and Endangered Plants of Northeast China in Colour***', vol. 1: 1-752 p, and 2: 753-1,056 p; Northeast Forestry University Press, Harbin.

2016.07.12-07.13, Centennial Celebration of Zheng-Yi WU's Birth and International Symposium on Biodiversity and Biogeography in east Aaia held at Kunming Institute of Botany, Chinese Academy of Sciences, Kunming, nearly 150 national and international attendees, including Peter H. Raven, Missouri Botanical Garden, Pete Hollingsworth (1968-), Royal Botanic Garden, Edinburgh, Pamela S. Soltis (1957-) and Douglas E. Soltis (1953-), University of Florida, Stephen Blackmore, Botanic Garden Conservation International, David E. Boufford, Harvard University, Jin Murata, University of Tokyo, Jun WEN, Smithsonian Institution. The Wu Zhengyi Science Foundation was officially established.

2016 年 7 月 12 日至 13 日，纪念吴征镒先生百年诞辰暨东亚生物多样性与生物地理国际学术研讨会于昆明中国科学院昆明植物研究所举行，与会代表 150 余人，包括密苏里植物园 Peter H. Raven、爱丁堡植物园 Pete Hollingsworth（1968—）、佛罗里达大学 Pamela S. Soltis（1957—）和 Douglas E. Soltis（1953—）、植物园保护国际 Stephen Blackmore、哈佛大学 David E. Boufford、东京大学邑田仁，以及史密森学会文军等。吴征镒科学基金会正式成立。

Cheng-Yih WU (Cheng-Yi WU, 1916-2013) (Photo provided by Wu Zhengyi Science Foundation)
吴征镒（1916—2013）（相片提供者：吴征镒科学基金会）

2016 年 7 月 25 日至 27 日，全国苔藓植物学学术研讨会在辽宁沈阳中国科学院沈阳应用生态研究所召开，全国 38 家科研院所的近 140 名代表参加了会议。

2016.07.25-07.27, National Bryophyte Symposium held at Institute of Applied Ecology, Chinese Academy of Sciences, Shenyang, Liaoning, nearly 140 attendees from 38 institutions.

Participants at National Bryophyte Symposium, Shenyang, 2016 (Photo provided by Wen YE)
2016 年全国苔藓植物学学术研讨会于沈阳（相片提供者：叶文）

2016 年 8 月，吴永华著，**早田文藏——台湾植物大命名时代**，439 页；台湾大学出版中心，台北。

2016.08, Yung-Hua WU, '***Bunzo Hayata***—*Nomenclature Time of Taiwan Plants*', 439 p; Taiwan University Press, Taipei [Taibei].

2016 年 9 月 10 日至 12 日，2016 全国系统与进化植物学研讨会暨第十二届青年学术研讨会在西宁青海民族大学和拉萨西藏大学举行；全国 52 个单位 200 多人参加了第一阶段西宁的会议，25 人参加了第二阶段拉萨的会议。

2016.09.10-09.12, 2016 National and Twelfth Youth Symposium of Systematic and Evolutionary Botany held at Qinghai Nationalities University, Xining and Tibet University, Lhasa, over 200 attendees from 52 institutions in Xining for the first section and 25 attendees in Lhasa for the second section.

2016 年 9 月 29 日，生物多样性第 24 卷第 9 期，以“生物多样性事业呼唤对物种概念和物种划分标准的深度讨论”为题，刊登 7 篇有关物种概念等文章。

2016.09.29, Seven papers regarding species concepts published in *Biodiversity Science* volume 24 issue 9, entitled Biodiversity undertakings call for extensive discussion on species concept and the criteria for species delimitation.

2016 年 10 月 18 日至 21 日，2016 年中国蕨类植物研讨会在西安西北大学举行，全国 16 个单位的 46 位代表参加。

2016.10.18-10.21, 2016 National Fern Symposium held at Northwest University, Xi'an, 46 attendees from 16 institutions.

2016 年 11 月，中国科学院武汉植物园，**中国科学院武汉植物园史料集（2007—2016）**，284 页；中国科学院武汉植物园，武汉（内部印制）。

2016.11, Wuhan Botanical Garden, Chinese Academy of Sciences, '***A Collection of Historical Materials (2007-2016) of Wuhan Botanical Garden, Chinese Academy of Sciences***', 284 p; Wuhan Botanical Garden, Chinese Academy of Sciences, Wuhan (internal publication).

2016 年 11 月，中国科学院武汉植物园，**中国科学院武汉植物园六十年（1956—2016）**，234 页；中国科学院武汉植物园，武汉（内部印制）。

2016.11, Wuhan Botanical Garden, Chinese Academy of Sciences, '***Sixty Years (1956-2016) of Wuhan Botanical Garden, Chinese Academy of Sciences***', 234 p; Wuhan Botanical Garden, Chinese Academy of Sciences, Wuhan (internal publication).

Participants at National Fern Symposium, Xi'an, 2016 (Photo provided by Xian-Chun ZHANG)
2016 年中国蕨类植物研讨会于西安（相片提供者：张宪春）

2016.12, Chun-Chao LÜ, Chronicle of Wu Zhengyi (Cheng-Yih WU), *Plant Diversity* 38(6): 330-344.[847]

2016 年 12 月，吕春朝，吴征镒年谱，*Plant Diversity* 38（6）：330-344。

2016 年，胡宗刚、夏振岱著，**中国植物志编纂史**，322 页；上海交通大学出版社，上海。[848]

2016, Zong-Gang HU and Zhen-Dai XIA, '***Compiling History of Flora Reipublicae Popularis Sinicae***', 322 p; Shanghai Jiao Tong University Press, Shanghai.

2016 年，**庆祝王文采院士 90 华诞专刊**[849]，广西植物第 36 卷增刊 1：1-244 页。

2016, ***Celebration of Academician Wen-Tsai WANG's 90th Birthday***, Special Issue of *Guihaia* 36 (supplement 1): 1-244.

2016 年，葛斌杰，上海辰山植物标本馆的建设及取得的成绩，仙湖 15（1）：45-55。

2016, Bin-Jie GE, Shanghai Chenshan herbarium and its accomplishments, *Journal of Fairylake Botanical Garden*[850] 15(1): 45-55.

2016 年，李春香、苗馨元，浅议中国高等植物多样性在世界上的排名，生物多样性 24（6）：725-727。

2016, Chun-Xiang LI and Xin-Yuan MIAO, Notes on the rank of China in the world in terms of higher plant diversity, *Biodiversity Science* 24(6): 725-727.

2016, Hong-Mei LIU, Embracing the pteridophyte classification of Ren-Chang CHING using a generic phylogeny of Chinese ferns and lycophytes, *Journal of Systematic and Evolution* 54(4): 307-335.

2016 年，刘红梅，中国蕨类和石松类属级分子系统学支持秦仁昌分类系统，*Journal of Systematic and Evolution* 54（4）：307-335。

2016 年，周喜乐、张宪春、孙久琼、严岳鸿，中国石松类和蕨类植物的多样性与地理分布，生物多样性 24（1）：102-107。

---

847 Dedicated to the 100th anniversary of the birth of Wu Zhengyi, Special issue of *Plant Diversity* 38(6): 259-344, 2016. For details, see Zhe-Kun ZHOU and Hang SUN, 2016, Wu Zhengyi and his contributions to plant taxonomy and phytogeography, *Plant Diversity* 38(6): 259-261.

848 薛伟平，2016年11月7日，一部跨世纪完成的科学巨著——访中国植物志编纂史作者胡宗刚，文汇读书周报，第1 639号，第 1-3 版。

849 刊前有马克平、蒋巧媛、肖培根、艾铁民、傅德志、杨永、谢磊、梁方文、孙英宝撰写的庆贺文，刊后有王文采院士年表。

850 2017 年 7 月 17 日，第十九届国际植物学大会命名法会议首日，深圳市中国科学院仙湖植物园张力博士告知：自 2017 年起仙湖植物园对外正式名称为 Shenzhen Fairy Lake Botanical Garden，而不再是 Shenzhen Fairylake Botanical Garden。

2016, Xi-Le ZHOU, Xian-Chun ZHANG, Jiu-Qiong SUN and Yue-Hong YAN, Diversity and distribution of lycophytes and ferns in China, *Biodiversity Science* 24(1): 102–107.

2016年，赵一之，马文红，赵利清，**贺兰山维管植物检索表**，342页；内蒙古大学出版社，呼和浩特。

2016, Yi-Zhi ZHAO, Wen-Hong MA, Li-Qing ZHAO, ***Key to the Vascular Plants of Helan Mountain***, 342 p; lnner Mongolia University Press, Hohhot.

2016年，陈健斌，分类学发展史概略和坚持分类学的完整概念，菌物研究 14（1）：1-7。

2016, Jian-Bin CHEN, Outline of taxonomic history and insisting on the integrated concept of taxonomy, *Journal of Fungal Research* 14(1): 1-7.

2017

2017年2月，生物多样性第25卷第2期，以中国植物区系地理研究专辑为题刊登12篇相关文章。

2017.02, Twelve papers on Chinese floristic geography in *Biodiversity Science* 25(2), published as a special issue.

2017.02, Jin-Shuang MA, ***A Checklist of Woody Plants from East Asia***, 650 p; Henan Science and Technology Press, Zhengzhou.

2017年2月，马金双，**东亚木本植物名录**，650页；河南科学技术出版社，郑州。

2017.03，大场秀章，**早田文藏**，12+196页；台湾林业试验所，台北（Hideaki Ohba, **'BUNZO HAYATA'**, 12+196 p; Taiwan Forestry Research Institute, Taipei）。

2017年3月，汪佳琳译，**早田文藏**，202页；台湾林业试验所，台北（繁体中文版）。

2017年4月，胡宗刚，**江苏省中国科学院植物研究所 南京中山植物园早期史**，237页；上海交通大学出版社，上海。

2017.04, Zong-Gang HU, '***Early History of Institute of Botany, Jiangsu Province and Chinese Academy of Sciences, Nanjing Botanical Garden Memorial Sun Yat-Sen***', 237 p; Shanghai Jiao Tong University Press, Shanghai.

2017年4月，王伟、张晓霞、陈之端、路安民，被子植物APG分类系统评论，生物多样性25（4）：418–426。

2017.04, Wei WANG, Xiao-Xia ZHANG, Zhi-Duan CHEN and An-Ming LU, Comments on the APG's classification of angiosperms, *Biodiversity Science* 25(4): 418–426.

2017年6月，王瑞江主编，**广东维管植物多样性编目**，372页；广东科技出版社，广州。

2017.06, Rui-Jiang WANG (editor in chief), ***Inventory of Species Diversity of Guangdong Vascular Plants***, 372 p; Guangdong Science and Technology Press, Guangzhou.

2017年6月，于胜祥、许为斌、武建勇、余丽莹、黄云峰，**滇黔桂喀斯特地区种子植物名录**，602页，中国环境出版社，北京。

2017.06, Sheng-Xiang YU, Wei-Bin XU, Jian-Yong WU, Li-Ying YU and Yun-Feng HUANG,

Spermatophytae of Karst Area in Guangxi, Yunnan and Guizhou—A Checklist, 602 p; Chinese Environment Publishing Group, Beijing.

2017年6月，税玉民、陈文红、秦新生，**中国喀斯特地区种子植物名录**，273页，科学出版社，北京。

2017.06, Yu-Min SHUI, Wen-Hong CHEN and Xin-Sheng QIN, ***Checklist of Seed Plants in the Karst Regions in China***, 273 p; Science Press, Beijing.

2017.07.17-07.21, Nomenclature Section, Nineteenth International Botanical Congress held in Shenzhen, China, 155 attendees from 30 countries and regions, including 46 attendees from China[851].

2017年7月17日至21日，第十九届国际植物学大会命名法会议于中国深圳举行，来自30个国家与地区近155位代表出席，包括中国代表46人。

2017.07.23-07.29, Nineteenth International Botanical Congress held in Shenzhen, the first IBC in its more than 100 years history to be held in China, 'Mother of Gardens'. 5,225 international and domestic attended, with 3,713 from the mainland of China, 31 from Chinese Hong Kong and 50 from Chinese Taiwan, the most attendees in the history of the botanical congresses.

2017年7月23日至29日，第十九届国际植物学大会于深圳举行；这是拥有100多年历史的大会首次来到被誉为"园林之母"的中国。77个国家与地区共5 225人与会，包括中国大陆3 713人、香港31人、中国台湾50人[852]，为历史上出席人数最多的植物学盛会。

2017年7月25日，云南吴征镒科学基金会颁发首届吴征镒植物学奖。中国科学院植物研究所洪德元获杰出贡献奖，中国科学院昆明植物研究所高连明（1972— ）和中国科学院植物研究所孔宏智（1973— ）获青年创新奖。

2017.07.25, First Wu Zhengyi Botanical Award issued by the Wu Zhengyi Science Foundation. The Outstanding Contribution Award was presented to De-Yuan HONG, Institute of Botany, Chinese Academy of Sciences. Young Innovation Awards were presented to Lian-Ming GAO (1972-), Kunming Institute of Botany, Chinese Academy of Sciences, and to Hong-Zhi KONG (1973-), Institute of Botany, Chinese Academy of Sciences.

2017年7月26日，2017年中国蕨类植物研讨会暨国际植物学大会蕨类植物卫星会议在广东深圳会展中心举行，国内外约90位代表出席。

2017.07.26, 2017 National Fern Symposium, a satellite meeting of the International Botanical Congress,

851 Nicholas J. Turland, John H. Wiersema, Anna M. Monro, Yun-Fei Deng and Li Zhang, 2017, XIX International Botanical Congress: Report of congress action on nomenclature proposals, *Taxon* 66(5): 1 234-1 245.

852 根据组委会提供的报到数据统计。

Chinese attendees at Nomenclature Section, XIX International Botanical Congress, Shenzhen, 18 July 2017, first time group photo (thirty total): front row (left-right): Bin-Jie GE, Ming ZHANG, Cheng DU, Bo ZHANG, Shuai LIAO, Yi-Hua TONG, Jin-Shuang MA; middle row: Ya-Ning WANG, Su-Zhou ZHANG, Zheng-Duan XIAO, Ting-Jun ZHANG, Xiao-Lan HE, Kuen-Shum PANG, Xin-Sheng QIN, Xuan-Ting SUN, Chao-Qun WANG, Lin BAI; and back row: Shuo SHI, Yong YANG, Ti-Ran HUANG, Tuo YANG, Ho-Yih LIU, Ke WANG, Yun-Fei DENG, Nian-He XIA, Xiang-Yun ZHU, Rui-Liang ZHU, Wen-Jun LI, Pei-Liang LIU, David Liu (Photo provided by Bin-Jie GE)

2017 年 7 月 18 日，第 19 届国际植物学大会命名法分会中国出席者第一次合影（30 人）；前排（从左至右）：葛斌杰、张明、杜诚、张波、廖帅、童毅华、马金双；中排：王亚宁、张苏州、肖正端、张庭筠、何晓兰、彭权森、秦新生、孙煊婷、王超群、白琳；后排：石硕、杨永、黄体冉、杨拓、刘和义、王科、邓云飞、夏念和、朱相云、朱瑞良、李文军、刘培亮、刘大伟（相片提供者：葛斌杰）

Chinese attendees at Nomenclature Section, XIX International Botanical Congress, Shenzhen, 20 July 2017, second time group photo (total twenty nine): front row (left-right): Chao-Qun WANG, Ming ZHANG, Ti-Ran HUANG, Bo LI, Yong YANG, Shuai LIAO, Cheng DU, Shi-Xiu FENG, Bo ZHANG, Wen-Sheng CAI, Bin-Jie GE; middle and back rows (left-right): Hui DENG, Jun-Yan XU, Su-Zhou ZHANG, Xuan-Ting SUN, Zheng-Duan XIAO, Ting-Jun ZHANG, Shuo SHI, Kuen-Shum PANG, Tuo YANG, Li ZHANG, Yun-Fei DENG, Xiang-Yun ZHU, Pei-Liang LIU, Ho-Yih LIU, Rui-Liang ZHU, David LIU, Yea-Chen LIU, Jin-Shuang MA (Photo provided by Bin-Jie GE)

2017 年 7 月 20 日，第 19 届国际植物学大会命名法分会中国出席者第二次合影（29 人）；前排（从左至右）：王超群、张明、黄体冉、李波、杨永、廖帅、杜诚、冯世秀、张波、蔡文圣、葛斌杰；中后排：邓晖、徐隽彦、张苏州、孙煊婷、肖正端、张庭筠、石硕、彭权森、杨拓、张力、邓云飞、朱相云、刘培亮、刘和义、朱瑞良、刘大伟，刘以诚、马金双（相片提供者：葛斌杰）

Participants at Nomenclature Section, XIX International Botanical Congress, Shenzhen, 20 July 2017 (Photo provided by Li ZHANG)
2017 年 7 月 20 日，第 19 届国际植物学会命名法规分会全体合影（相片提供者：张力）

held in Shenzhen Convention and Exhibition Center, Shenzhen, 90 national and international attendees.

2017.07.29, First Shenzhen International Award in Plant Science given to Peter H. Raven, Missouri Botanical Garden, St. Louis, Missouri, during the closing ceremony of the Nineteenth International Botanical Congress.

2017 年 7 月 29 日，第十九届国际植物学大会闭幕式，首届深圳国际植物科学奖授予密苏里植物园的 Peter H. Raven。

2017.07.29, De-Yuan HONG, Institute of Botany, Chinese Academy of Sciences, was awarded the Engler Medal in Gold by the International Association for Plant Taxonomy during the closing ceremony of the Nineteenth International Botanical Congress. This was the first Engler Medal in Gold given to a Chinese botanist.

2017 年 7 月 29 日，中国科学院植物研究所洪德元于第十九届国际植物学大会闭幕式获国际植物分类协会颁发的恩格勒金质奖章。此奖章首次颁发给中国植物学家。

De-Yuan HONG awarded Engler Medal in Gold by the IAPT at XIX International Botanical Congress, Shenzhen, 29 July 2017 (Photo provided by IBC)
2017 年 7 月 29 日，洪德元于第十九届国际植物学大会获国际植物分类协会颁发的恩格勒金质奖章（相片提供者：国际植物学大会）

2017.07 De-Zhi FU, **GLOBAL SPECIES**—Synopsis of Higher Plants in Countries of Africa, Asia, Australia, Europe, North America and South America, SIX CDs; Qingdao Press, Qingdao.

2017 年 7 月，傅德志，**世界高等植物** —— 非洲、亚洲、澳洲、欧洲、北美洲和南美洲各国植物要览[853]，六张光盘；青岛出版社，青岛。

2017 年，熊源新、曹威，贵州植物分类学研究概述，山地农业生物学报 36（1）：1-11。

2017, Yuan-Xin XIONG and Wei CAO, Review of researches of plant taxonomy in Guizhou, *Journal of Mountain Agriculture and Biology* 36(1): 1-11.

853 本套光盘收录全世界 258 个国家（地区）的苔藓、蕨类、裸子和被子植物四大类，总计有 1 033 科、15 409 属和 393 940 种（可靠接受 304 441 种）。

Participants at National Fern Symposium, Shenzhen, 2017 (Photo provided by Xian-Chun ZHANG)
2017 年中国蕨类植物研讨会于深圳（相片提供者：张宪春）

2017 年 7 月，生物多样性第 25 卷第 7 期第 689–795 页**中国高等植物红色名录**专辑刊登 8 篇文章：中国高等植物濒危状况评估、中国高等植物受威胁物种名录[854]、中国被子植物濒危等级的评估、中国裸子植物物种濒危和保育现状、中国石松类和蕨类植物的红色名录评估、中国苔藓植物濒危等级的评估原则和评估结果、野生牡丹的生存状况和保护、中国生物多样性保护的国家意志、科学决策和公众参与 —— 第一份省域物种红色名录研究。

2017.07, '**Red Lists of Higher Plants of China**', Special Issue of *Biodiversity Science* 25(7): 689–795, eight papers published: 1) Evaluating the threat status of higher plants in China; 2) Threatened species list of China's higher plants; 3) Evaluating the endangerment status of China's angiosperms through the red list assessment; 4) Red list assessment and conservation status of gymnosperms from China; 5) Red list assessment of lycophytes and ferns in China; 6) Assessing the threat status of China's bryophytes; 7) Current status of wild tree peony species with special reference to conservation; and 8) The state's will, scientific decision and citizen participation: in memory of the first provincial species red list in China.

2017 年，张丽兵，蕨类植物 PPG Ⅰ系统与中国石松类和蕨类植物分类，生物多样性 25（3）：340–342。

2017, Li-Bing ZHANG, The PPG I classification and pteridophytes of China, *Biodiversity Science* 25(3): 340–342.

2017 年 8 月[855]，第 19 届国际植物学大会组织委员会、深圳市中国科学院仙湖植物园，**芳华修远 ——** 第 19 届国际植物学大会植物艺术画展画集，343 页；江苏凤凰科学技术出版社，南京。

2017.08, The XIX IBC Organizing Committee and Shenzhen Fairy Lake Botanical Garden, Shenzhen, and the Chinese Academy of Sciences, ***Walking the Path to Eternal Fragrance***—*Catalog of the XIX IBC Botanical Art Exhibition*, 343 p; Phoenix Science Press, Nanjing.

854 本工作是基于 2013 年 9 月环境保护部、中国科学院第 54 号公告形式发布的中国生物多样性红色名录——高等植物卷（http://www.zhb.gov.cn/gkml/hbb/bgg/ 201309/t20130912_260061.htm/）修订而成，但原始链接无法进入（2017 年 12 月）。

855 尽管本书出版时间标注为 2017 年 8 月，实际上 2017 年 7 月 24 日植物画展开幕当天已抵达展览现场。

附录
Appendices

附录一　中国年表 1753—2017

Appendix I　Chinese Calendar 1753–2017

1753 年 乾隆十八年[1]
1754 年 乾隆十九年
1755 年 乾隆二十年
1756 年 乾隆二十一年
1757 年 乾隆二十二年
1758 年 乾隆二十三年
1759 年 乾隆二十四年
1760 年 乾隆二十五年
1761 年 乾隆二十六年
1762 年 乾隆二十七年
1763 年 乾隆二十八年
1764 年 乾隆二十九年
1765 年 乾隆三十年
1766 年 乾隆三十一年
1767 年 乾隆三十二年
1768 年 乾隆三十三年
1769 年 乾隆三十四年
1770 年 乾隆三十五年
1771 年 乾隆三十六年
1772 年 乾隆三十七年
1773 年 乾隆三十八年
1774 年 乾隆三十九年

1 Qianlong Emperor Period (18–60).

1775 年 乾隆四十年
1776 年 乾隆四十一年
1777 年 乾隆四十二年
1778 年 乾隆四十三年
1779 年 乾隆四十四年
1780 年 乾隆四十五年
1781 年 乾隆四十六年
1782 年 乾隆四十七年
1783 年 乾隆四十八年
1784 年 乾隆四十九年
1785 年 乾隆五十年
1786 年 乾隆五十一年
1787 年 乾隆五十二年
1788 年 乾隆五十三年
1789 年 乾隆五十四年
1790 年 乾隆五十五年
1791 年 乾隆五十六年
1792 年 乾隆五十七年
1793 年 乾隆五十八年
1794 年 乾隆五十九年
1795 年 乾隆六十年
1796 年 嘉庆元年[2]

2 Jiaqing Emperor Period (1–25).

1797 年 嘉庆二年
1798 年 嘉庆三年
1799 年 嘉庆四年
1800 年 嘉庆五年
1801 年 嘉庆六年
1802 年 嘉庆七年
1803 年 嘉庆八年
1804 年 嘉庆九年
1805 年 嘉庆十年
1806 年 嘉庆十一年
1807 年 嘉庆十二年
1808 年 嘉庆十三年
1809 年 嘉庆十四年
1810 年 嘉庆十五年
1811 年 嘉庆十六年
1812 年 嘉庆十七年
1813 年 嘉庆十八年
1814 年 嘉庆十九年
1815 年 嘉庆二十年
1816 年 嘉庆二十一年
1817 年 嘉庆二十二年
1818 年 嘉庆二十三年
1819 年 嘉庆二十四年
1820 年 嘉庆二十五年
1821 年 道光元年 [3]
1822 年 道光二年
1823 年 道光三年
1824 年 道光四年
1825 年 道光五年
1826 年 道光六年
1827 年 道光七年

3 Daoguang Emperor Period (1-30).

1828 年 道光八年
1829 年 道光九年
1830 年 道光十年
1831 年 道光十一年
1832 年 道光十二年
1833 年 道光十三年
1834 年 道光十四年
1835 年 道光十五年
1836 年 道光十六年
1837 年 道光十七年
1838 年 道光十八年
1839 年 道光十九年
1840 年 道光二十年
1841 年 道光二十一年
1842 年 道光二十二年
1843 年 道光二十三年
1844 年 道光二十四年
1845 年 道光二十五年
1846 年 道光二十六年
1847 年 道光二十七年
1848 年 道光二十八年
1849 年 道光二十九年
1850 年 道光三十年
1851 年 咸丰元年 [4]
1852 年 咸丰二年
1853 年 咸丰三年
1854 年 咸丰四年
1855 年 咸丰五年
1856 年 咸丰六年
1857 年 咸丰七年
1858 年 咸丰八年

4 Xianfeng Emperor Period (1-11).

1859 年 咸丰九年
1860 年 咸丰十年
1861 年 咸丰十一年
1862 年 同治元年 [5]
1863 年 同治二年
1864 年 同治三年
1865 年 同治四年
1866 年 同治五年
1867 年 同治六年
1868 年 同治七年
1869 年 同治八年
1870 年 同治九年
1871 年 同治十年
1872 年 同治十一年
1873 年 同治十二年
1874 年 同治十三年
1875 年 光绪元年 [6]
1876 年 光绪二年
1877 年 光绪三年
1878 年 光绪四年
1879 年 光绪五年
1880 年 光绪六年
1881 年 光绪七年
1882 年 光绪八年
1883 年 光绪九年
1884 年 光绪十年
1885 年 光绪十一年
1886 年 光绪十二年
1887 年 光绪十三年
1888 年 光绪十四年

5 Tongzhi Emperor Period (1-13).
6 Guangxu Emperor Period (1-34).

1889 年 光绪十五年
1890 年 光绪十六年
1891 年 光绪十七年
1892 年 光绪十八年
1893 年 光绪十九年
1894 年 光绪二十年
1895 年 光绪二十一年
1896 年 光绪二十二年
1897 年 光绪二十三年
1898 年 光绪二十四年
1899 年 光绪二十五年
1900 年 光绪二十六年
1901 年 光绪二十七年
1902 年 光绪二十八年
1903 年 光绪二十九年
1904 年 光绪三十年
1905 年 光绪三十一年
1906 年 光绪三十二年
1907 年 光绪三十三年
1908 年 光绪三十四年
1909 年 宣统元年 [7]
1910 年 宣统二年
1911 年 宣统三年
1912 年 民国元年 [8]
1913 年 民国二年
1914 年 民国三年
1915 年 民国四年
1916 年 民国五年
1917 年 民国六年
1918 年 民国七年

7 Xuantong Emperor Period (1-3).
8 Beginning of Republic of China.

1919 年 民国八年
1920 年 民国九年
1921 年 民国十年
1922 年 民国十一年
1923 年 民国十二年
1924 年 民国十三年
1925 年 民国十四年
1926 年 民国十五年
1927 年 民国十六年
1928 年 民国十七年
1929 年 民国十八年
1930 年 民国十九年
1931 年 民国二十年
1932 年 民国二十一年
1933 年 民国二十二年
1934 年 民国二十三年
1935 年 民国二十四年
1936 年 民国二十五年
1937 年 民国二十六年
1938 年 民国二十七年
1939 年 民国二十八年
1940 年 民国二十九年
1941 年 民国三十年
1942 年 民国三十一年
1943 年 民国三十二年
1944 年 民国三十三年
1945 年 民国三十四年
1946 年 民国三十五年
1947 年 民国三十六年
1948 年 民国三十七年
1949 年 民国三十八年[9]

1950 年 庚寅年
1951 年 辛卯年
1952 年 壬辰年
1953 年 癸巳年
1954 年 甲午年
1955 年 乙未年
1956 年 丙申年
1957 年 丁酉年
1958 年 戊戌年
1959 年 己亥年
1960 年 庚子年
1961 年 辛丑年
1962 年 壬寅年
1963 年 癸卯年
1964 年 甲辰年
1965 年 乙巳年
1966 年 丙午年
1967 年 丁未年
1968 年 戊申年
1969 年 己酉年
1970 年 庚戌年
1971 年 辛亥年
1972 年 壬子年
1973 年 癸丑年
1974 年 甲寅年
1975 年 乙卯年
1976 年 丙辰年
1977 年 丁巳年
1978 年 戊午年
1979 年 己未年
1980 年 庚申年
1981 年 辛酉年

9 Beginning of People's Republic of China.

1982 年　壬戌年
1983 年　癸亥年
1984 年　甲子年
1985 年　乙丑年
1986 年　丙寅年
1987 年　丁卯年
1988 年　戊辰年
1989 年　己巳年
1990 年　庚午年
1991 年　辛未年
1992 年　壬申年
1993 年　癸酉年
1994 年　甲戌年
1995 年　乙亥年
1996 年　丙子年
1997 年　丁丑年
1998 年　戊寅年
1999 年　己卯年
2000 年　庚辰年
2001 年　辛巳年
2002 年　壬午年
2003 年　癸未年
2004 年　甲申年
2005 年　乙酉年
2006 年　丙戌年
2007 年　丁亥年
2008 年　戊子年
2009 年　己丑年
2010 年　庚寅年
2011 年　辛卯年
2012 年　壬辰年
2013 年　癸巳年
2014 年　甲午年
2015 年　乙未年
2016 年　丙申年
2017 年　丁酉年

附录二　中外收藏中国植物标本的主要标本馆[1]

Appendix II　Major herbaria with holdings of Chinese specimens both in China and abroad[2]

All of the information is from the website of **Index Herbarium** at www.sweetgum.nybg.org/ih/, accessed December 2017[3].

A: Harvard University Herbaria, **Harvard University**, 22 Divinity Avenue, Cambridge, Massachusetts 02138, U.S.A. (https://huh.harvard.edu).

AU: Herbarium, School of Life Science, **Xiamen University**, Xiamen, Fujian 361005, China (http://life.xmu.edu.cn).

B: Herbarium, **Botanischer Garten und Botanisches Museum Berlin-Dahlem,** Zentraleinrichtung der Freien Universität Berlin, Königin-Luise-Strasse 6-8, D-14195, Berlin, Germany (https://www.bgbm.org).

BM: Herbarium, Department of Botany, **The Natural History Museum**, Cromwell Road, London, SW7 5BD, England, U.K. (http://www.nhm.ac.uk).

BNU: Herbarium, Biology Department, College of Life Sciences, **Beijing Normal University**, Beijing, Beijing 100875, China (http://cls.bnu.edu.cn).

C: Herbarium, Botanical Museum and Library, **University of Copenhagen**, Gothersgade 130, DK - 1123 Copenhagen K, Denmark (www.botanicalmuseum.dk/bot).

CAL: Central National Herbarium, **Botanical Survey of India**, P. O. Botanic Garden, Howrah, Calcutta 711103, West Bengal, India (https://bsi.gov.in/Center/206_7_Central-National-Herbarium.aspx).

1 即本书所涉及的标本馆。

2 i.e. the herbaria involved within this book.

3 With some Chinese herbaria's websites updated.

CANT: Herbarium, Forestry Department, **South China Agricultural University**, Shi-Pai, Guangzhou, Guangdong 510642, China. As of 2017, this record has not been updated in over 15 years[4] (http://www.scau.edu.cn).

CAS: Herbarium, Department of Botany, Institute for Biodiversity Science and Sustainability, **California Academy of Sciences,** 55 Music Concourse Drive, Golden Gate Park, San Francisco, California 94118, U.S.A. (http://www.calacademy.org/research/botany/).

CDBI: Herbarium, **Chengdu Institute of Biology**, Chinese Academy of Sciences, P. O. Box 416, Chengdu, Sichuan 6100641, China (http://www.cib.ac.cn).

CM: Herbarium, **Carnegie Museum of Natural History**, 4,400 Forbes Avenue, Pittsburgh, Pennsylvania, 15213-4080, U.S.A. (http://www.carnegiemnh.org/botany/collection.html).

CQNM: Herbarium, **Chongqing Natural History Museum**, 74 Pipashan Central Street, Chongqing, Chongqing 630013, China. As of 2017, this record has not been updated in over 15 years (https://www.cmnh.org.cn/).

CUHK: Shiu-Ying Hu Herbarium, School of Life Sciences, **The Chinese University of Hong Kong**, Shatin, Hong Kong, China (http://syhuherbarium.sls.cuhk.edu.hk/).

DAV: UC Davis Center for Plant Diversity, **University of California,** Plant Sciences MS#7One Shields Avenue, Davis, California 95616, U.S.A. (http://herbarium.ucdavis.edu).

DBN: Herbarium, **National Botanic Gardens**, Glasnevin, Dublin 9, Ireland (http://botanicgardens.ie).

E: Herbarium, **Royal Botanic Garden**, Edinburgh, EH3 5LR, Scotland, U.K. http://www.rbge.org.uk).

F: Herbarium, Botany Department, **Field Museum of Natural History,** 1,400 South Lake Shore Drive, Chicago, Illinois 60605-2496, U.S.A. (http://fieldmuseum.org/explore/department/botany/collections).

FH: Farlow Reference Library and Herbarium of Cryptogamic Botany, **Harvard University**, 22 Divinity Avenue, Cambridge, Massachusetts 02138, U.S.A. (http://www.huh.harvard.edu).

FI: Herbarium Universitatis Florentinae, Botanical Section, **Natural History Museum**, Via G. La Pira 4, Florence I-50121, Italy (http://parlatore.msn.unifi.it/hci_italy_web.html).

FJFC: Dendrological Herbarium, **Fujian Forestry College**, 182 Xinghua Second Road, Nanping,

4 From Index Herbariorum website (same as below).

Fujian 353001, China. As of 2017, this record has not been updated in over 15 years. Formerly Fujian Forestry College now changed to Fujian Agriculture and Forestry University (http://www.fafu.edu.cn).

FNU: Herbarium, Bioengineering College, **Fujian Normal University,** Cangshan, Fuzhou, Fujian 350007, China. As of 2017, this record has not been updated in over 15 years (http://www.fjnu.edu.cn).

G: Herbarium, **Conservatoire et Jardin botaniques de la Ville de Genève**, Case postale 71, CH-1292 Chambésy/Genève, Switzerland (http://www.ville-ge.ch/cjb/).

GB: Herbarium, Department of Biological and Environmental Sciences, **University of Gothenburg**, Carl Skottsbergs Gata 22 B, Göteborg SE-41319, Sweden (http://bioenv.gu.se/forskning/forskningsresurser/herbarium).

GH: Harvard University Herbaria, **Harvard University**, 22 Divinity Avenue, Cambridge, Massachusetts 02138, U.S.A. (https://huh.harvard.edu).

GNUB: Herbarium, Biology Department, **Guizhou Normal University**, 270 Waihuan East Road, Guiyang, Guizhou 550001, China (http://www.gznu.edu.cn).

H: Herbarium, Botanical Museum, **University of Helsinki**, P. O. Box 7, FIN-00014, Helsinki, Finland (http://www.luomus.fi/english/botany/).

HAST: Herbarium, Biodiversity Research Museum, **Academia Sinica,** 128 Sec. 2, Academia Road, Taipei 11529, China (http://hast.sinica.edu.tw/).

HGAS: Herbarium, Guizhou Institute of Biology, **Guizhou Academy of Sciences,** Xiao He, Guiyang, Guizhou 550009, China. As of 2017, this record has not been updated in over 15 years (http://www.gzpib.org.cn).

HIB: Herbarium, **Wuhan Institute of Botany,** Chinese Academy of Sciences, P.O. Box 74006, Wuhan, Hubei 430074, China (http://www.wbg.cas.cn).

HITBC: Herbarium, **Xishuangbanna Tropical Botanical Garden**, Chinese Academy of Sciences, Menglun Town, Mengla County, Yunnan 666303, China. As of 2017, this record has not been updated in over 15 years. The herbarium code YNTBI (abbreviation of Yunnan Institute of Tropical Botany) was used in various publications for this herbarium prior to publication of *Index Herbariorum Sinicorum* in 1993 (http://www.xtbg.ac.cn).

HK: **Hong Kong Herbarium**, Agriculture, Fisheries, and Conservation Department, 303 Cheung Sha Wan Road, Room 737, Cheung Sha Wan Government Offices, Kowloon, Hong Kong, China (http://www.herbarium.gov.hk/).

HMAS: Herbarium Mycologium, **Institute of Microbiology**, Chinese Academy of Sciences, 1 Beichenxilu, Chaoyang District, Beijing 100101, China (http://www.mycolab.org.cn).

HNWP: Herbarium, Qinghai-Tibet Plateau Museum of Biology, **Northwest Institute of Plateau Biology**, Chinese Academy of Sciences, 23 Xinning Road, Xining, Qinghai 810008, China (http://www.nwipb.cas.cn).

IBK: Herbarium, **Guangxi Institute of Botany**, Guangxi Academy of Sciences, Yanshan, Guilin, Guangxi Zhuangzu 541006, China. As of 2017, this record has not been updated in over 15 years (http://www.gxib.cn).

IBSC: Herbarium, **South China Botanical Garden**, The Chinese Academy of Sciences, Guangzhou, Guangdong 510650, China. As of 2017, this record has not been updated in over 15 years (http://english.scib.cas.cn).

IFP: Herbarium, **Institute of Applied Ecology**, Chinese Academy of Sciences, P.O. Box 417, Shenyang, Liaoning 110016, China. As of 2017, this record has not been updated in over 15 years (http://www.iae.cas.cn).

IMC: Herbarium, Institute of Medicinal Plant Cultivation, **Chongqing Academy of Traditional Chinese Medicine and Pharmacy**, Nanchuan, Chongqing 648408, China. As of 2017, this record has not been updated in over 15 years.

K: Herbarium, **Royal Botanic Gardens**, Kew, Richmond, Surrey TW9 3AB, London, England, U.K. (https://www.kew.org/science/collections).

KUN: Herbarium, **Kunming Institute of Botany**, Chinese Academy of Sciences, 132 Lanhei Road, Heilongtan, Kunming, Yunnan 650204, China (http://www.kib.ac.cn).

L: Nationaal Herbarium Nederland, Botany Section, **Naturalis Biodiversity Center Vondellaan**, 55, Leiden 2332 AA, The Netherlands (https://science.naturalis.nl/en/collection/naturalis-collections/botany/).

LE: Herbarium, **V. L. Komarov Botanical Institute**, Russian Academy of Sciences, Prof. Popov Street 2, Saint Petersburg 197376, Russia (www.binran.ru).

LBG: Herbarium, **Lushan Botanical Garden**, Jiangxi Province and Chinese Academy of Sciences, Lushan, Jiangxi 332900, China; formerly as LUS. As of 2017, this record has not been updated in over 15 years (http://www.lsbg.cn).

LINN: Herbarium, **Linnean Society of London**, Burlington House, Piccadilly, London W1J 0BF, England, U.K. (www.linnean.org).

M: Herbarium, **Botanische Staatssammlung München**, Menzinger Straße 67, München D-80638, Germany (http://www.botanischestaatssammlung.de).

MBK: Herbarium, Department of Botany, **Makino Botanical Garden**, 4200-6 Godaisan, Kochi City, Kochi 781-8125, Japan (http://www.makino.or.jp).

MICH: Herbarium, **University of Michigan**, 3600 Varsity Drive, Ann Arbor, Michigan 48108-2228, U.S.A. (http://herbarium.lsa.umich.edu).

MO: Herbarium, **Missouri Botanical Garden**, P. O. Box 299, Saint Louis, Missouri 63166-0299, U.S.A. (http://www.missouribotanicalgarden.org/).

MSC: Herbarium, Plant Biology Laboratories, **Michigan State University**, 612 Wilson Road, Room 166, East Lansing, Michigan 48824, U.S.A. (http://www.herbarium.msu.edu/).

N: Herbarium, Biology Department, School of Life Science, **Nanjing University**, Nanjing, Jiangsu 210093, China. As of 2017, this record has not been updated in over 15 years (https://life.nju.edu.cn).

NA: Herbarium, **United States National Arboretum**, United States Department of Agriculture, The Agriculture Research Service, 3501 New York Avenue, N.E., Washington, District of Columbia 20002-1958, U.S.A. (http://www.usna.usda.gov/Research/Herbarium/index.html).

NAS: Herbarium, **Institute of Botany**, Jiangsu Province and Chinese Academy of Sciences, Nanjing, Jiangsu 210014, China. As of 2017, this record has not been updated in over 15 years (http://www.cnbg.net).

NF: Dendrological Herbarium, Forest Resources and Environment, **Nanjing Forestry University**, Nanjing, Jiangsu 210037, China. As of 2017, this record has not been updated in over 15 years (http://www.njfu.edu.cn).

NWTC: Herbarium, Institute of Botany, College of Life Sciences, **Northwest Normal University**, Shilidian, Lanzhou, Gansu 730070, China. As of 2017, this record has not been updated in over 15 years (https://www.nwnu.edu.cn/cate.do?dept=0015).

NY: William and Lynda Steere Herbarium, **New York Botanical Garden**, 2900 Southern Blvd., Bronx, New York 10458-5126, U.S.A. (www.nybg.org/).

P: Herbier National de Paris, Département de Systématique et Evolution, Phanérogamie, **Muséum National d'Histoire Naturelle**, 16 rue Buffon, F-75005 Paris, France (http://colhelper.mnhn.fr).

PE: Herbarium, **Institute of Botany**, Chinese Academy of Sciences, 20 Nanxincun, Xiangshan,

Beijing 100093, China; including FM (Fan Memorial Institute of Biology, 1928-1949) (http://pe.ibcas.ac.cn/).

PEY: Herbarium, Biology Department, **Peking University**, Beijing 100871, China. As of 2017, this record has not been updated in over 15 years (http://www.bio.pku.edu.cn).

PR: Herbarium, Department of Botany, **National Museum in Prague**, Cirkusova 1740, Praha 9 - Horni Pocernice CZ-190 00, Czech Republic (http://www.nm.cz/Departments/Natural-History-Museum/?xSET=lang&xLANG=2).

S: Herbarium, **Swedish Museum of Natural History,** Svante Arrhenius väg 3, Stockholm SE-114 18, Sweden (http://www.nrm.se/english/researchandcollections/botany/collections.565_en.html).

SM: Herbarium, **Chongqing Municipal Academy of Chinese Materia Medica,** Chongqing, 400065, China. As of 2017, this record has not been updated in over 15 years. In 1997, Chongqing became a municipality of the Central Government, and the former Sichuan Institute of Chinese Materia Medica was changed to Chongqing Municipal Academy of Chinese Materia Medica (www.cqacmm.com/).

SYS: Herbarium, Biology Department, Botanical Division, **Zhongshan (Sun Yatsen) University,** Guangzhou, Guangdong 510275, China. As of 2017, this record has not been updated in over 15 years (http://www.sysu.edu.cn/2012/cn/index.htm).

SZ: Herbarium, College of Life Sciences, **Sichuan University,** 29 Jiuyanqiao Wangjiang Road, Chengdu, Sichuan 610064, China (http://www.scu.edu.cn).

TAI: Herbarium, College of Life Science, **Taiwan University,** 1 Roosevelt Road, Section 4, Taipei, Taiwan 10617, China (http://tai.ntu.edu.tw/).

TAIF: The Herbarium of Taiwan Forestry Research Institute, Botanical Garden Division, **Taiwan Forestry Research Institute**, 67 San-Yuan St., Taipei 100, Zhong-Zheng Dist., 100 Taiwan, China (http://taif.tfri.gov.tw/en/).

TI: Herbarium, University Museum, **University of Tokyo**, 7-3-1 Hongo, Bunkyo-ku, Tokyo, Tokyo 113-0033, Japan (http://www.bg.s.u-tokyo.ac.jp).

TIE: Herbarium, Botany Department, **Tianjin Natural History Museum**, No. 206, Machangdao, Tianjin, 300074, China (http://taif.tfri.gov.tw/en/).

UC: University Herbarium, **University of California**, 1001 Valley Life Sciences Building #2465, Berkeley, California 94720-2465, U.S.A. (http://ucjeps.berkeley.edu/).

UPS: Herbarium, **Uppsala University**, Botany, Museum of Evolution, Norbyvägen 16, SE-75236,

Uppsala, Sweden (www-hotel.uu.se/evolmuseum/fytotek/).

US: United States National Herbarium, Department of Botany, National Museum of Natural History, MRC-166, **Smithsonian Institution,** P. O. Box 37012, Washington, District of Columbia 20013-7012, U.S.A. (http://botany.si.edu/).

W: Herbarium, Department of Botany, **Natürhistorisches Museum Wien**, Burgring 7, A-1010 Wien, Austria (http://www.nhm-wien.ac.at/en/research/botany).

WU: Herbarium, Faculty Center Botany, Department of Plant Systematics and Evolution, Faculty of Life Sciences, **Universität Wien**, Rennweg 14, A-1030 Wien, Austria (http://herbarium.botanik.univie.ac.at).

WUK: Herbarium, College of Life Sciences, **Northwestern Institute of Botany,** Yangling, Shaanxi 712100, China. As of 2017, this record has not been updated in over 15 years. This institute was combined into Northwest A&F University in 1999 (http://sm.nwafu.edu.cn/index.htm).

XJBI: Herbarium, Xinjiang **Institute of Ecology and Geography**, Chinese Academy of Sciences, 818 South Beijing Road, Urumqi, Xinjiang 830011, China (http://www.egi.cas.cn/jgsz/zcbm/bbg/).

YU: Herbarium, Botany Division, Peabody Museum of Natural History, **Yale University**, 170-210 Whitney Avenue, New Haven, Connecticut 06511, U.S.A. (http://www.peabody.yale.edu/collections/bot/).

附录三　中国植物分类学者名字的新旧拼写对照[1]

Appendix III Comparison of old and new transliterations of names of Chinese plant taxonomists[2]

表 1：中国植物分类学者以名字的汉语拼音为序

Table 1, Chinese plant taxonomists alphabetically by Pinyin of their names

Number	Pinyin Spelling	Old Spelling	Traditional Chinese	Simple Chinese
1	An, Zheng-Xi	An, Cheng-Hsi	安爭夕	安争夕
2	Ao, Liang-De	Aou, Lian-Do	敖良德	敖良德
2	Ao, Liang-De	Aou, Liang-Teh	敖良德	敖良德
3	Ao, Zhi-Wen	Aur, Chih-Wen	敖志文	敖志文
3	Ao, Zhi-Wen	Aur, Chin-Wen	敖志文	敖志文
4	Bai, Cai	Pai, Ts'ai	白　采	白　采
5	Bai, Pei-Yu	Pai, Pei-Yu	白佩瑜	白佩瑜
6	Bai, Yin-Yuan	Pai, Yin-Yuan	白蔭元	白荫元
6	Bai, Yin-Yuan	Pai, Yin-Yüan	白蔭元	白荫元
7	Bi, Lie-Jue	Bi, Lie-Jiu	畢列爵	毕列爵
8	Bi, Pei-Xi	But, Paul Pui-Hay	畢培曦	毕培曦
9	Bi, Zhong-Ben	Pi, Chung-Pen	畢中本	毕中本
10	Bi, Zu-Gao	Pi, T. K.	畢祖高	毕祖高
11	Bu, Yu-Zhi	Bu, Yu-Tze	步毓芝	步毓芝
11	Bu, Yu-Zhi	Pu, Yu-Chih	步毓芝	步毓芝
12	Cai, Jin-Lai	Tsai, Jenn-Lai	蔡進來	蔡进来
12	Cai, Jin-Lai	Tsai, Jinn-Lai	蔡進來	蔡进来
13	Cai, Pan-Sheng	Choi, P. S.	蔡盤生	蔡盘生
13	Cai, Pan-Sheng	Choy, Pun-Sang	蔡盤生	蔡盘生
14	Cai, Shao-Lan	Ts'ai, shao-Lan	蔡少蘭	蔡少兰

1 详细参见：邓玲丽，杜诚，廖帅，马金双，2018，中国植物分类学者姓名拼写的讨论与建议，生物多样性 26（6）：627-635.

2 Ling-Li Deng, Shuai Liao, Cheng Du, Jin-Shuang Ma, David E. Boufford, 2017, Names of Chinese Plant Taxonomists—Order out of Chaos, *Taxon* 66(3): 782-783.

15	Cai, Shu-Hua	Tsai, Su-Hwa	蔡淑華	蔡淑华
16	Cai, Xi-Tao	Tsai, Hse-Tao	蔡希陶	蔡希陶
16	Cai, Xi-Tao	Tsai, Hsi-Tao	蔡希陶	蔡希陶
16	Cai, Xi-Tao	Ts'ai, Hsi-T'ao	蔡希陶	蔡希陶
17	Cai, Zhen-Cong	Tsai, Tseng-Chun	蔡振聰	蔡振聪
18	Cao, Kan	Tsao, Kan	曹　侃	曹　侃
18	Cao, Kan	Ts'ao, K'an	曹　侃	曹　侃
19	Cen, Ming-Shu	Ts'ên, Ming-Shu	岑銘恕	岑铭恕
20	Chang, Xing-Ruo	Chang, Hsing-Jo	暢行若	畅行若
21	Chen, An-Ji	Ch'ên, An-Chi	陳安集	陈安集
22	Chen, Bang-Jie	Chen, Pan-Chieh	陳邦傑	陈邦杰
22	Chen, Bang-Jie	Chen, Pang-Chieh	陳邦傑	陈邦杰
22	Chen, Bang-Jie	Ch'ên, Pang-Chieh	陳邦傑	陈邦杰
23	Chen, Bang-Yu	Chen, Pang-Yu	陳邦餘	陈邦余
24	Chen, Bo-Chuan	Ch'ên, Pê-Ch'uan	陳伯川	陈伯川
24	Chen, Bo-Chuan	Tchen, Te-Tchouan	陳伯川	陈伯川
25	Chen, Chong-Ming	Chen, Chung-Ming	陳重明	陈重明
26	Chen, Chun-Shou	Ch'ên, Ch'ung-Shou	陳椿壽	陈椿寿
26	Chen, Chun-Shou	Ch'ên, Ch'un-Shou	陳椿壽	陈椿寿
27	Chen, De-Mao	Chen, Dao-Men	陳德懋	陈德懋
28	Chen, De-Zhao	Chen, Te-Chao	陳德昭	陈德昭
28	Chen, De-Zhao	Ch'ên, Tê-Chao	陳德昭	陈德昭
29	Chen, Di	Chen, Ti	陳　蒂	陈　蒂
30	Chen, Feng-Huai	Ch'ên, Fêng-Huai	陳封懷	陈封怀
30	Chen, Feng-Huai	Chen, Feng-Hwai	陳封懷	陈封怀
31	Chen, Guan-You	Ch'ên, Kuan-Yu	陳冠友	陈冠友
32	Chen, Gui	Ch'ên, Kuei	陳　貴	陈　贵
33	Chen, Hao-Zi	Chen, Hao-Tzu	陳淏子	陈淏子
34	Chen, Hua-Gui	Chen, Hua-Kwei	陳華癸	陈华癸
35	Chen, Huan-Yong	Ch'ên, Huan-Yung	陳煥鏞	陈焕镛
35	Chen, Huan-Yong	Chun, Woon-Young	陳煥鏞	陈焕镛
35	Chen, Huan-Yong	Chun, Woon-Yung	陳煥鏞	陈焕镛
36	Chen, Jia-Rui	Chen, Chia-Jue	陳家瑞	陈家瑞
36	Chen, Jia-Rui	Chen, Chia-Jui	陳家瑞	陈家瑞
37	Chen, Jie	Chen, Cheih	陳　介	陈　介
38	Chen, Jiong-Song	Chen, Chiung-Sung	陳炯松	陈炯松
38	Chen, Jiong-Song	Ch'ên, Chiung-Sung	陳炯松	陈炯松
39	Chen, Juan-Ren	Ch'ên, Chien-Jên	陳雋人	陈隽人
40	Chen, Jun-Yu	Chen, Chun-Yu	陳俊愉	陈俊愉
40	Chen, Jun-Yu	Chen, Chün-Yü	陳俊愉	陈俊愉
40	Chen, Jun-Yu	Ch'ên, Chün-Yü	陳俊愉	陈俊愉

41	Chen, Li-Qing	Chen, Li-Ching	陳立卿	陈立卿
41	Chen, Li-Qing	Ch'ên, Li-Ch'ing	陳立卿	陈立卿
42	Chen, Min	Chen, Men	陳　敏	陈　敏
43	Chen, Ming-Zhe	Chen, Ming-Che	陳明哲	陈明哲
44	Chen, Qing-Cheng	Ch'ên, Ch'ing-Cheng	陳慶誠	陈庆诚
45	Chen, Qing-Xia	Chen, Ching-Hsia	陳擎霞	陈擎霞
46	Chen, Ren-Jun	Chun, Ren-Jun	陳仁鈞	陈仁钧
47	Chen, Rong	Ch'ên, Yung	陳　嶸	陈　嵘
47	Chen, Rong	Chen, Yung	陳　嶸	陈　嵘
48	Chen, Rong-Rui	Chen, Yung-Reui	陳榮銳	陈荣锐
49	Chen, Rui-Qing	Chen, Zuei-Ching	陳瑞青	陈瑞青
50	Chen, Shan-Ming	Ch'ên, Shan-Ming	陳善銘	陈善铭
51	Chen, Sheng-Zhen	Chen, Sen-Jen	陳升振	陈升振
52	Chen, Shi-Hui	Chen, Shih-Huei	陳世輝	陈世辉
53	Chen, Shi-Xiang	Chen, Si-Cien	陳世驤	陈世骧
54	Chen, Shu-Hua	Chen, Su-Hwa	陳淑華	陈淑华
55	Chen, Shu-Zhen	Ch'ên, Shu-Chên	陳淑珍	陈淑珍
55	Chen, Shu-Zhen	Chun, Shu-Chên	陳淑珍	陈淑珍
56	Chen, Si-Yi	Ch'ên, Ssŭ-I	陳思義	陈思义
57	Chen, Wei-Qiu	Chen, Wei-Chiu	陳偉球	陈伟球
58	Chen, Wei-Xin	Chen, Wei-Sin	陳維新	陈维新
59	Chen, Xiang-Fa	Ch'ên, Hsiang-Fa	陳祥發	陈祥发
60	Chen, Xi-Kai	Chen, Hsi-Kai	陳希凱	陈希凯
60	Chen, Xi-Kai	Ch'ên, Hsi-K'ai	陳希凱	陈希凯
61	Chen, Xin-Qi	Chen, Sing-Chi	陳心啓	陈心启
62	Chen, Xiu-Ying	Ch'en, Hsiu-Ying	陳秀英	陈秀英
63	Chen, Yan-Zhuo	Ch'ên, Yen-Cho	陳彥卓	陈彦卓
64	Chen, Ya-Qi	Chen, Ya-Chi	陳雅琦	陈雅琦
65	Chen, Yi-Ming	Chen, I-Ming	陳益明	陈益明
66	Chen, Ying-Huang	Ch'ên, Ying-Huang	陳映璜	陈映璜
67	Chen, You-Jun	Chen, Yo-Jiun	陳又君	陈又君
68	Chen, Zhi-Xiong	Chen, Chih-Hsiung	陳志雄	陈志雄
69	Chen, Zhuo	Chen, Cho	陳　倬	陈　倬
69	Chen, Zhuo	Ch'ên, Cho	陳　倬	陈　倬
70	Chen, Zong-Lian	Chen, Chung-Lien	陳宗蓮	陈宗莲
71	Chen, Zu-Keng	Chen, Tsu-Keng	陳祖鏗	陈祖铿
72	Cheng, Jing-Rong	Cheng, Ching-Yung	誠靜容	诚静容
73	Cheng, Yong-Qian	Tseng, Yong-Qian	程用謙	程用谦
73	Cheng, Yong-Qian	Tseng, Yung-Chien	程用謙	程用谦
74	Cui, Bo-Tang	Ts'ui, Po-T'ang	崔伯棠	崔伯棠
75	Cui, Hong-Bin	Tsui, Hung-Pin	崔鴻賓	崔鸿宾

75	Cui, Hong-Bin	Ts'ui, Hung-Pin	崔鴻賓	崔鸿宾
76	Cui, You-Wen	Tsui, You-Wen	崔友文	崔友文
76	Cui, You-Wen	Tsui, Yu-Wen	崔友文	崔友文
76	Cui, You-Wen	Ts'ui, Yu-Wên	崔友文	崔友文
77	Dai, Fang-Lan	Tai, Fang-Lan	戴芳瀾	戴芳澜
77	Dai, Fang-Lan	Tai, Fung-Lan	戴芳瀾	戴芳澜
78	Dai, Fan-Jin	Tai, Fan-Chin	戴蕃瑨	戴蕃瑨
78	Dai, Fan-Jin	Tai, Fan-Tsien	戴蕃瑨	戴蕃瑨
79	Dai, Lun-Yan	Tai, Lun-Yen	戴倫焰	戴伦焰
80	Dai, Ming-Jie	Tai, Ming-Chieh	戴銘傑	戴铭杰
81	Deng, Gui-Ling	Têng, Kuei-Ling	鄧桂玲	邓桂玲
82	Deng, Mei-Zhi	Deng, Mei-Jih	鄧美智	邓美智
83	Deng, Shu-Qun	Têng, Shu-Ch'ün	鄧叔羣	邓叔群
84	Ding, Guang-Qi	Ting, Kuang-Ch'i	丁廣奇	丁广奇
84	Ding, Guang-Qi	Ting, Kuang-Chi	丁廣奇	丁广奇
85	Ding, Su	Ting, Sü	丁 驌	丁 骕
86	Ding, Ying	Ting, Ying	丁 穎	丁 颖
87	Ding, Yuan	Ting, Yüan	丁 源	丁 源
88	Ding, Zhi-Zun	Ting, Chih-Chi	丁志遵	丁志遵
88	Ding, Zhi-Zun	Ting, Chih-Tsun	丁志遵	丁志遵
89	Dong, Chang-Chun	Tung, Ch'ang-Ch'un	董長椿	董长椿
90	Dong, Gui-Yang	Tong, Koe-Yang	董貴陽	董贵阳
91	Dong, Hui-Min	Tung, Hui-Min	董惠民	董惠民
91	Dong, Hui-Min	Tung, Hwei-Nin	董惠民	董惠民
92	Dong, Shi-Lin	Tung, Shih-Lin	董世林	董世林
92	Dong, Shi-Lin	Tung, Shi-Lin	董世林	董世林
93	Dong, Shuang-Qiu	Tung, Shuang-Ch'iu	董爽秋	董爽秋
94	Du, Fu	Tu, Fu	杜 複	杜 复
95	Du, Nai-Qiu	Tu, Nai-Chiu	杜乃秋	杜乃秋
96	Du, Ya-Quan	Tu, Ya-Tsuan	杜亞泉	杜亚泉
97	Duan, Pei-Qin	Tuan, Pei-Chen	段佩琴	段佩琴
98	Duanmu, Xin	Duanmu, Shing	端木炘	端木炘
99	Fan, Gong-Ju	Fan, Kung-Chü	樊恭炬	樊恭炬
100	Fan, Qing-Sheng	Fan, Ch'ing-Sêng	樊慶生	樊庆生
101	Fan, Qing-Sheng	Fan, Ch'ing-Shêng	樊慶生	樊庆生
102	Fang, Rui-Zheng	Fang, Rhui-Cheng	方瑞徵	方瑞征
103	Fang, Wen-Pei	Fang, Wên-P'ei	方文培	方文培
104	Fang, Xi-Chen	Fang, Hsi-Ch'ên	方錫琛	方锡琛
104	Fang, Xi-Chen	Fang, Si-Tzen	方錫琛	方锡琛
105	Fang, Yong-Huai	Fang, Yung-Huai	方永槐	方永槐
106	Fang, Yong-Xin	Fang, Yong-Sing	方永鑫	方永鑫

107	Fang, Zhen-Fu	Fang, Cheng-Fu	方振富	方振富
108	Feng, Guo-Mei	Feng, Kuo-Mei	馮國楣	冯国楣
108	Feng, Guo-Mei	Fêng, Kuo-Mei	馮國楣	冯国楣
109	Feng, Xue-Lin	Fung, Hok-Lam	馮學琳	冯学琳
110	Feng, Yao-Zong	Fêng, Yao-Tsung	馮耀宗	冯耀宗
111	Feng, Ze-Fang	Fêng, Tsê-Fang	馮澤芳	冯泽芳
112	Feng, Ze-Feng	Fêng, Tse-Fêng	豐澤豐	丰泽丰
113	Feng, Zhao-Nan	Fung, Chao-Nan	馮肇南	冯肇南
114	Feng, Zhao-Xun	Feung, Chao-Shieung	馮朝勛	冯朝勋
115	Fu, Kun-Jun	Fu, Kun-Tsun	傅坤俊	傅坤俊
116	Fu, Li-Guo	Fu, Li-Kuo	傅立國	傅立国
117	Fu, Shu-Xia	Fu, Shu-Hsia	傅書遐	傅书遐
118	Fu, Xiang-Qian	Fu, Hiang-Chian	富象乾	富象乾
119	Fu, Yu-Qin	Fu, Yu-Chin	符毓秦	符毓秦
120	Gao, Bao-Chun	Kao, Pao-Chung	高寶蒓	高宝莼
121	Gao, De-Pei	Kao, Teh-P'ei	高德培	高德培
121	Gao, De-Pei	Kao, Tê-P'ei	高德培	高德培
122	Gao, Ling	Kao, Ling	高　嶺	高　岭
123	Gao, Mu-Cun	Kao, Muh-Tsuen	高木村	高木村
124	Gao, Qian	Gao, Chien	高　謙	高　谦
124	Gao, Qian	Ko, Chien	高　謙	高　谦
125	Gao, Yun-Zhang	Kao, Yün-Chang	高藴璋	高蕴璋
125	Gao, Yun-Zhang	Ko, Wan-Chang	高藴璋	高蕴璋
125	Gao, Yun-Zhang	Ko, Wan-Cheung	高藴璋	高蕴璋
126	Gao, Zuo-Jing	Kao, Tso-Ching	高作經	高作经
127	Ge, Ding-Bang	Ko, Ting-Pang	戈定邦	戈定邦
128	Ge, Ming-Yu	Ko, Ming-Yü	葛明裕	葛明裕
129	Geng, Bai-Jie	Keng, Pai-Chieh	耿伯介	耿伯介
129	Geng, Bai-Jie	Kêng, Pai-Chieh	耿伯介	耿伯介
130	Geng, Kuan-Hou	Kêng, Kwan-Hou	耿寬厚	耿宽厚
131	Geng, Qing-Han	Kêng, Ch'ing-Han	耿慶漢	耿庆汉
132	Geng, Xuan	Keng, Hsuan	耿　煊	耿　煊
132	Geng, Xuan	Keng, Hsüan	耿　煊	耿　煊
132	Geng, Xuan	Keng, Hsiuan	耿　煊	耿　煊
133	Geng, Yi-Li	Kêng, I-Li	耿以禮	耿以礼
133	Geng, Yi-Li	Keng, Yi-Li	耿以禮	耿以礼
134	Geng, Zuo-Lin	Kêng, Tso-Lin	耿作霖	耿作霖
135	GengLin, Ruo-Xiu	KengLing, Ro-Siu	耿林若琇	耿林若琇
136	Gong, Qi-Wu	Kung, Ch'i-Wu	龔啓鋈	龚启鋈
137	Gu, Cui-Zhi	Ku, Tsue-Chih	谷粹芝	谷粹芝
138	Gu, Guan	Ku, Kuan	古　灌	古　灌

139	Gu, Qian-Ji	Ku, Ch'ien-Chi	顧謙吉	顾谦吉
140	Guan, Ke-Jian	Kuan, Ke-Chien	關克儉	关克俭
140	Guan, Ke-Jian	Kuan, K'o-Chien	關克儉	关克俭
141	Guan, Yu-Ying	Kuan, Yü-Ying	關毓英	关毓英
142	Guan, Zhong-Tian	Kuan, Chung-Tian	管中天	管中天
143	Gui, Yao-Lin	Kwei, Yao-Lin	桂耀林	桂耀林
144	Guo, Bao-Zhang	Kuo, Pao-Chang	郭寶章	郭宝章
145	Guo, Ben-Zhao	Kuo, Pen-Chao	郭本兆	郭本兆
145	Guo, Ben-Zhao	Kuo, Pung-Chao	郭本兆	郭本兆
146	Guo, Chang-Sheng	Kuoh, Chang-Sheng	郭長生	郭长生
147	Guo, Chao-Sheng	Kuo, Ch'ao-Shêng	郭朝勝	郭朝胜
148	Guo, Cheng-Meng	Kuo, Chen-Meng	郭城孟	郭城孟
149	Guo, Jian-Mao	Kuo, Chien-Mao	過鑒楙	过鉴楙
150	Guo, Ji-Fan	Kuo, Chi-Fan	郭紀凡	郭纪凡
151	Guo, Qiu-Cheng	Kuo, Chin-Chen	郭秋成	郭秋成
152	Guo, Xing-Rong	Kuo, Shing-Rong	郭幸榮	郭幸荣
153	Guo, Yu-Jie	Kuo, Yü-Chieh	郭玉潔	郭玉洁
154	Han, Shu-Jin	Han, Shu-Chih	韓樹金	韩树金
155	Hang, Jin-Xin	Hang, King-Hsing	杭金欣	杭金欣
156	Hao, Jing-Sheng	Hao, Ching-Sheng	郝景盛	郝景盛
156	Hao, Jing-Sheng	Hao, Kin-Shen	郝景盛	郝景盛
156	Hao, Jing-Sheng	Ho, Ching-Shêng	郝景盛	郝景盛
157	He, Chun-Nian	Ho, Chun-Nien	何椿年	何椿年
157	He, Chun-Nian	Ho, Ch'un-Nien	何椿年	何椿年
158	He, Feng-Ji	Ho, Feng-Chi	何豐吉	何丰吉
159	He, Feng-Ren	Ho, Feng-Jen	何鳳仁	何凤仁
160	He, Jing	Ho, Chin	何 景	何 景
160	He, Jing	Ho, Ching	何 景	何 景
160	He, Jing	Hoo, Gin	何 景	何 景
161	He, Jun-Feng	Ho, Hsün-Fêng	賀峻峯	贺峻峰
161	He, Jun-Feng	Ho, Tsun-Fung	賀峻峯	贺峻峰
162	He, Kai	Hô, K'ai	賀 闓	贺 闿
162	He, Kai	Ho, Kan	賀 闓	贺 闿
163	He, Nai-Wei	Ho, Lei-Wei	何乃維	何乃维
163	He, Nai-Wei	Ho, Nai-Wei	何乃維	何乃维
164	He, Shi-Yuan	Ho, Shih-Yuen	賀士元	贺士元
165	He, Tian-Xiang	Ho, T'ien-Hsiang	何天相	何天相
165	He, Tian-Xiang	Ho, Tien-Hsiang	何天相	何天相
166	He, Ting-Nong	Ho, Ting-Nung	何廷農	何廷农
167	He, Xian-Zhang	Ho, Hsien-Chang	何憲章	何宪章
167	He, Xian-Zhang	Hoh, Hin-Cheung	何憲章	何宪章

168	He, Ye-Qi	Ho, Ye-Chi	何業琪	何业琪
168	He, Ye-Qi	Hou, Yeh-Chi	何業琪	何业琪
169	He, Zhen-Xi	Ho, Chên-Hsi	賀貞熙	贺贞熙
170	He, Zhu	Ho, Chu	何 鑄	何 铸
171	Hong, Zhang-Xun	Hung, Chang-Hsun	洪章訓	洪章训
171	Hong, Zhang-Xun	Hung, Chang-Hsün	洪章訓	洪章训
172	Hou, Ding	Hou, Ting	侯 定	侯 定
173	Hou, Hui-Zong	Hou, Hui-Tsung	侯惠宗	侯惠宗
174	Hou, Kuan-Zhao	Hou, K'uan-Chao	侯寬昭	侯宽昭
174	Hou, Kuan-Zhao	How, Foon-Chew	侯寬昭	侯宽昭
175	Hu, Da-Wei	Hu, Ta-Wei	胡大維	胡大维
176	Hu, Gong-De	Hu, Kung-Tê	胡公德	胡公德
177	Hu, Jia-Qi	Hu, Chia-Chi	胡嘉琪	胡嘉琪
178	Hu, Jia-Ying	Hu, Chia-Ying	胡嘉穎	胡嘉颖
179	Hu, Jing-Hua	Hu, Chin-Hwa	胡敬華	胡敬华
180	Hu, Ji-Sheng	Hu, Tsi-Sun	胡濟生	胡济生
181	Hu, Lin-Zhen	Hu, Lin-Cheng	胡琳貞	胡琳贞
182	Hu, Qi-Ming	Hu, Chi-Ming	胡啓明	胡启明
183	Hu, Shi-Yi	Hu, Shih-I	胡適宜	胡适宜
183	Hu, Shi-Yi	Hu, Shih-Yi	胡適宜	胡适宜
184	Hu, Wen-Guang	Hu, Wen-Kwang	胡文光	胡文光
184	Hu, Wen-Guang	Hu, Wên-Kwang	胡文光	胡文光
185	Hu, Xian-Su	Hu, Hsen-Hsu	胡先驌	胡先骕
186	Hu, Xian-Yuan	Hu, Hsian-Yuan	胡顯源	胡显源
187	Hu, Xing-Zong	Hu, Hsing-Tsung	胡興宗	胡兴宗
187	Hu, Xing-Zong	Hu, Hsin-Tsung	胡興宗	胡兴宗
188	Hu, Xiu-Ying	Hu, Hsiu-Ying	胡秀英	胡秀英
188	Hu, Xiu-Ying	Hu, Shiu-Ying	胡秀英	胡秀英
189	Hu, Yu-Xi	Hu, Yu-Shi	胡玉熹	胡玉熹
190	Hu, Zhao-Hua	Hu, Chao-Hwa	胡兆華	胡兆华
191	Hua, Ru-Cheng	Hua, Ju-Ch'êng	華汝成	华汝成
191	Hua, Ru-Cheng	Hua, Ju-Cheng	華汝成	华汝成
192	Huang, Cheng-Jiu	Huang, Ch'êng-Chiu	黄成就	黄成就
192	Huang, Cheng-Jiu	Huang, Ching-Chieu	黄成就	黄成就
193	Huang, Cui-Mei	Huang, Ts'ui-Mei	黄翠梅	黄翠梅
193	Huang, Cui-Mei	Hwang, Ts'ui-Mei	黄翠梅	黄翠梅
194	Huang, Da-Zhang	Huang, Ta-Chang	黄達章	黄达章
195	Huang, Feng-Yuan	Huang, Fêng-Yüan	黄逢源	黄逢源
195	Huang, Feng-Yuan	Hwang, Fung-Yuan	黄逢源	黄逢源
196	Huang, Ji-Zhuang	Huang, Chi-Chuang	黄季莊	黄季庄
197	Huang, Pu-Hua	Huang, Pu-Hwa	黄普華	黄普华

198	Huang, Shao-Xu	Huang, Shao-Hsü	黄紹緒	黄绍绪
199	Huang, Sheng	Huang, Shong	黄　生	黄　生
200	Huang, Shou-Xian	Huang, Shou-Hsien	黄守先	黄守先
201	Huang, Shu-Mei	Hwang, Shu-Mei	黄淑美	黄淑美
202	Huang, Shu-Qiong	Huang, Shu-Chung	黄蜀瓊	黄蜀琼
203	Huang, Shu-Wei	Hwang, Shu-Wei	黄淑煒	黄淑炜
204	Huang, Song-Gen	Huang, Sang-Gen	黄松根	黄松根
204	Huang, Song-Gen	Huang, Sung-K'ên	黄松根	黄松根
205	Huang, Xie-Cai	Huang, Se-Zei	黄燮才	黄燮才
206	Huang, Xing-Fan	Huang, Shing-Fan	黄星凡	黄星凡
207	Huang, Xi-Wang	Huang, Ch'i-Wang	黄溪旺	黄溪旺
208	Huang, Yong-Qin	Huang, Yong-Chin	黄咏琴	黄咏琴
209	Huang, You-Ru	Huang, Yu-Ju	黄友儒	黄友儒
209	Huang, You-Ru	Hwang, You-Ru	黄友儒	黄友儒
210	Huang, Yun-Hui	Huang, Yun-Huei	黄雲輝	黄云辉
211	Huang, Zeng-Quan	Huang, Tseng-Chieng	黄增泉	黄增泉
212	Huang, Zhi	Wang, Chi	黄　志	黄　志
213	Huang, Zuo-Jie	Hwang, Tso-Chie	黄作傑	黄作杰
214	Huo, Shu-Hua	Hao, Shu-Hwa	火樹華	火树华
215	Ji, Han	Chi, Han	稽　含	稽　含
216	Ji, Jing-Yuan	Chi, Ching-Yuan	季景元	季景元
217	Ji, Jun-Mian	Chi, Chün-Mien	季君勉	季君勉
218	Ji, Zhan-He	Tsi, Zhan-Huo	吉占和	吉占和
219	Jia, Cheng-Zhang	Chia, Ch'êng-Chang	賈成章	贾成章
220	Jia, Liang-Zhi	Chia, Liang-Chih	賈良智	贾良智
221	Jia, Shen-Xiu	Chia, Shen-Hsiu	賈慎修	贾慎修
222	Jia, Zu-Shan	Chia, Tsu-San	賈祖珊	贾祖珊
223	Jia, Zu-Zhang	Chia, Tsu-Chang	賈祖璋	贾祖璋
224	Jian, Zhuo-Po	Tsien, Cho-Po	簡焯坡	简焯坡
225	Jiang, Cheng-Ji	Chiang, Ch'êng-Chi	蔣承基	蒋承基
225	Jiang, Cheng-Ji	Chiang, Cheng-Chi	蔣承基	蒋承基
226	Jiang, Han-Qiao	Chiang, Han-Ch'iao	姜漢僑	姜汉侨
227	Jiang, Pu	Chiang, P'u	蔣　溥	蒋　溥
228	Jiang, Xing-Lin	Tsiang, Hsing-Ling	蔣興麟	蒋兴麟
229	Jiang, Ying	Tsiang, Ying	蔣　英	蒋　英
229	Jiang, Ying	Ch'iang, Ying	蔣　英	蒋　英
230	Jiang, You-Long	Chiang, You-Long	江有龍	江有龙
231	Jiang, Zhen-Yu	Chiang, Tzen-Yuh	蔣鎮宇	蒋镇宇
232	JiangCai, Shu-Hua	ChiangTsai, Su-Hwa	江蔡淑華	江蔡淑华
233	Jiao, Qi-Yuan	Chiao, Chi-Yuan	焦啓源	焦启源
233	Jiao, Qi-Yuan	Chiao, Ch'i-Yüan	焦啓源	焦启源

233	Jiao, Qi-Yuan	Chiao, Chi-Yuen	焦啓源	焦启源
234	Jin, Cun-Li	Chin, Tsen-Li	金存禮	金存礼
235	Jin, De-Xiang	Chin, Tê-Hsiang	金德祥	金德祥
236	Jin, Duo-Han	Chin, To-Han	金多翰	金多翰
237	Jin, Wei-Jian	Chen, Wei-Chien	金維堅	金维坚
237	Jin, Wei-Jian	King, Wei-Jean	金維堅	金维坚
238	Jin, Yue-Xing	Chin, Yu-Xing	金岳杏	金岳杏
239	Jin, Zhen-Zhou	Chin, Chên-Chou	金振州	金振州
240	Kong, Qing-Lai	Kung, Ch'ing-Lai	孔慶萊	孔庆莱
240	Kong, Qing-Lai	K'ung, Ch'ing-Lai	孔慶萊	孔庆莱
241	Kong, Shi-Pei	Kung, Shi-Pei	孔世培	孔世培
242	Kong, Xian-Wu	Kung, Hsien-Wu	孔憲武	孔宪武
242	Kong, Xian-Wu	K'ung, Hsien-Wu	孔憲武	孔宪武
243	Kong, Xian-Xu	Kung, Hsian-Shiu	孔憲需	孔宪需
244	Kuang, Ke-Ren	K'uang, K'o-Jên	匡可任	匡可任
244	Kuang, Ke-Ren	Kuang, Ko-Zen	匡可任	匡可任
245	Lai, Ming-Zhou	Lai, Ming-Jou	赖明洲	赖明洲
246	Lang, Kai-Yong	Lang, Kai-Yung	郎楷永	郎楷永
247	Le, Tian-Yu	Lo, T'ien-Yü	樂天宇	乐天宇
248	Li, An-Ren	Lee, An-Jên	李安仁	李安仁
248	Li, An-Ren	Li, An-Jen	李安仁	李安仁
248	Li, An-Ren	Li, An-Jên	李安仁	李安仁
249	Li, Bing-Tao	Li, Ping-T'ao	李秉滔	李秉滔
250	Li, Cai-Qi	Le, Ts'ae-Ch'i	李彩祺	李彩祺
250	Li, Cai-Qi	Lee, Ts'ai-Ch'i	李彩祺	李彩祺
251	Li, Chao-Luan	Li, Chao-Luang	李朝鑾	李朝銮
252	Li, Dai-Fang	Li, Tai-Fang	李代芳	李代芳
253	Li, Dan-Qing	Lee, Dan-Ching	李淡清	李淡清
254	Li, De-Lin	Lee, Der-Lin	李德霖	李德霖
254	Li, De-Lin	Li, Tê-Lin	李德霖	李德霖
255	Li, Dong-Sheng	Li, Tung-Sheng	李東生	李东生
255	Li, Dong-Sheng	Li, Tung-Shêng	李東生	李东生
256	Li, Gong-De	Lee, Kong-Teh	黎功德	黎功德
256	Li, Gong-De	Li, Kung-Tê	黎功德	黎功德
256	Li, Gong-De	Li, Kung-Teh	黎功德	黎功德
257	Li, Gou-Tang	Li, Ko-T'ang	李構堂	李构堂
258	Li, Guang-Min	Lee, Kong-Ming	李廣民	李广民
258	Li, Guang-Min	Li, Kuang-Ming	李廣民	李广民
259	Li, Guo-Ren	Lee, Kwok-Yan	李國仁	李国仁
260	Li, Han-Zhang	Li, Han-Chang	李含章	李含章
261	Li, Jia-Wei	Li, Chia-Wei	李家維	李家维

262	Li, Jia-Wen	Li, Chia-Wen	李家文	李家文
263	Li, Jia-Xin	Li, Chia-Hsin	李嘉馨	李嘉馨
264	Li, Ji-Tong	Li, Chi-Tung	李繼侗	李继侗
264	Li, Ji-Tong	Li, Chi-T'ung	李繼侗	李继侗
265	Li, Ji-Yun	Li, Chi-Yun	李冀雲	李冀云
266	Li, Liang-Qing	Li, Liang-Chin	李良慶	李良庆
266	Li, Liang-Qing	Li, Liang-Ch'ing	李良慶	李良庆
267	Li, Pei-Qiong	Li, Pei-Chun	李沛瓊	李沛琼
268	Li, Peng-Fei	Lei, P'aang-Fei	李鵬飛	李鹏飞
268	Li, Peng-Fei	Li, P'êng-Fei	李鵬飛	李鹏飞
269	Li, Qian	Li, Chien	李　乾	李　乾
270	Li, Qing-Tao	Li, Ching-Tao	李清濤	李清涛
270	Li, Qing-Tao	Li, Ch'ing-T'ao	李清濤	李清涛
271	Li, Shang-Gui	Li, Shang-Kuei	李尚癸	李尚癸
272	Li, Shang-Hao	Ley, Shang-Hao	黎尚豪	黎尚豪
272	Li, Shang-Hao	Li, Shang-Hao	黎尚豪	黎尚豪
273	Li, Shu-Gang	Lee, Shu-Kang	李樹剛	李树刚
273	Li, Shu-Gang	Li, Shu-Kang	李樹剛	李树刚
274	Li, Shun-Qing	Lee, Shun-Ching	李順卿	李顺卿
274	Li, Shun-Qing	Lee, Shun-Ch'ing	李順卿	李顺卿
274	Li, Shun-Qing	Li, Shun-Ch'ing	李順卿	李顺卿
275	Li, Shu-Xuan	Lee, Shu-Hsien	李曙軒	李曙轩
275	Li, Shu-Xuan	Li, Shu-Hsien	李曙軒	李曙轩
276	Li, Song-Bo	Li, Sung-Po	李松柏	李松柏
277	Li, Xing-Hua	Li, Hsing-Hua	李興華	李兴华
278	Li, Xing-Jiang	Lee, Shin-Chiang	黎興江	黎兴江
278	Li, Xing-Jiang	Li, Hsin-Kiang	黎興江	黎兴江
278	Li, Xing-Jiang	Li, Shin-Chiang	黎興江	黎兴江
279	Li, Xi-Wen	Li, Hsi-Wen	李錫文	李锡文
280	Li, Yao-Ying	Lee, Yao-Yin	李堯英	李尧英
281	Li, Yin-Gong	Li, Yin-Kung	李寅恭	李寅恭
282	Li, Zong-Dao	Li, Tsung-Tao	李宗道	李宗道
283	Li, Zong-Ying	Li, Tsung-Ying	李宗英	李宗英
284	Lian, Wen-Yan	Lien, Wen-Yen	連文琰	连文琰
285	Lian, Yong-Shan	Lian, Yung-Shan	廉永善	廉永善
286	Liang, Bao-Han	Liang, Pao-Han	梁寶漢	梁宝汉
287	Liang, Hui-Zhou	Leong, Wai-Chao	梁慧舟	梁慧舟
288	Liang, Kui	Liang, Kuei	梁　葵	梁　葵
288	Liang, Kui	Liang, K'uei	梁　葵	梁　葵
289	Liang, Sheng-Ye	Liang, Sheng-Yeh	梁盛業	梁盛业
290	Liang, Tai-Ran	Liang, T'ai-Jan	梁泰然	梁泰然

291	Liang, Xi	Liang, Hsi	梁 希	梁 希
292	Liao, Ri-Jing	Liao, Jih-Ching	廖日京	廖日京
293	Lin, Ai-Jin	Lin, Ai-Chin	林艾進	林艾进
294	Lin, Bang-Juan	Lin, Pang-Jun	林邦娟	林邦娟
294	Lin, Bang-Juan	Lin, Pan-Juan	林邦娟	林邦娟
295	Lin, Bao-Shu	Lin, Pao-Shu	林寶樹	林宝树
296	Lin, Chong-Zhi	Lin, Ch'ung-Chih	林崇智	林崇智
296	Lin, Chong-Zhi	Lin, Tsung-Chih	林崇智	林崇智
297	Lin, Han-Xi	Lin, Han-Shi	林涵錫	林涵锡
298	Lin, Hou-Xuan	Lin, Hou-Hsüan	林厚萱	林厚萱
299	Lin, Lai-Guan	Ling, Lai-Kuan	林來官	林来官
300	Lin, Liang-Dong	Lin, Liang-Tung	林亮東	林亮东
301	Lin, Qing	Lin, Ching	林 青	林 青
301	Lin, Qing	Lin, Ch'ing	林 青	林 青
302	Lin, Quan	Ling, Chüan	林 泉	林 泉
302	Lin, Quan	Ling, Chuan	林 泉	林 泉
303	Lin, Rong	Lin, Jung	林 鎔	林 镕
303	Lin, Rong	Ling, Yong	林 鎔	林 镕
303	Lin, Rong	Ling, Yung	林 鎔	林 镕
304	Lin, Ruo-Xiu	Lin, Jo-Hsiu	林若琇	林若琇
304	Lin, Ruo-Xiu	Lin, Ro-Siu	林若琇	林若琇
305	Lin, Ru-Yao	Lin, Ju-Yao	林汝瑤	林汝瑶
306	Lin, Shan-Xiong	Lin, Shan-Hsiung	林善雄	林善雄
307	Lin, Shi-Rong	Lin, Shih-Yung	林仕榕	林仕榕
308	Lin, Shu-Qian	Lin, Shu-Chien	林叔芊	林叔芊
309	Lin, Wan-Tao	Lin, Wan-T'ao	林萬濤	林万涛
310	Lin, Wei-Zhi	Lin, Wei-Ch'ih	林維治	林维治
311	Lin, Wei-Zhi	Lin, Wei-Chih	林維治	林维治
312	Lin, Xie	Lin, Hsieh	林 協	林 协
312	Lin, Xie	Ling, Hsieh	林 協	林 协
313	Lin, Xin-Zi	Lin, Hsin-Tzu	林信子	林信子
314	Lin, Xiong-Xiang	Lin, Hsiung-Hsiang	林熊祥	林熊祥
315	Lin, You-Run	Ling, Yeou-Ruenn	林有潤	林有润
316	Lin, Zan-Biao	Lin, Tsan-Piao	林讃標	林赞标
317	Lin, Ze-Tong	Lin, Tzer-Tong	林則桐	林则桐
318	Lin, Zhong-Gang	Lin, Chung-Kang	林仲剛	林仲刚
319	Ling, Yuan-Jie	Ling, Yuan-Chi	凌元潔	凌元洁
319	Ling, Yuan-Jie	Ling, Yuan-Jue	凌元潔	凌元洁
320	Liu, Bao-Shan	Liu, Pao-Shan	劉寶山	刘宝山
321	Liu, Bo	Liu, Po	劉 波	刘 波
322	Liu, Chen	Liu, Ch'ên	劉 晨	刘 晨

323	Liu, Chong-Sheng	Leou, Chong-Sheng	柳重勝	柳重胜
324	Liu, Fang-Xun	Liu, Fang-Hsün	劉昉勛	刘昉勋
325	Liu, He-Yi	Liu, Ho-Yih	劉和義	刘和义
326	Liu, Hong-E	Liu, Hung-Eh	劉洪諤	刘洪谔
327	Liu, Hou	Liu, Ho	劉　厚	刘　厚
327	Liu, Hou	Liou, Ho	劉　厚	刘　厚
328	Liu, Hua-Xiang	Liew, Fah-Seong	劉華祥	刘华祥
329	Liu, Ji-Meng	Liu, Ki-Mon	劉繼孟	刘继孟
330	Liu, Jing-Fen	Liu, Ching-Fên	劉經芬	刘经芬
331	Liu, Jin-Hui	Liu, Chin-Hui	劉錦惠	刘锦惠
332	Liu, Li	Liu, Lee	劉　利	刘　利
333	Liu, Liang	Liou, Liang	劉　亮	刘　亮
334	Liu, Ming-Yuan	Liou, Ming-Yuan	劉鳴遠	刘鸣远
334	Liu, Ming-Yuan	Liou, Min-Yuan	劉鳴遠	刘鸣远
335	Liu, Ru-Qiang	Liu, Ju-Ch'iang	劉汝強	刘汝强
336	Liu, Shen-E	Liou, Tchen-Ngo	劉慎諤	刘慎谔
336	Liu, Shen-E	Liu, Shên-O	劉慎諤	刘慎谔
337	Liu, Shou-Lu	Liou, Sheo-Lu	劉守爐	刘守炉
337	Liu, Shou-Lu	Liou, Shou-Lu	劉守爐	刘守炉
338	Liu, Shou-Ren	Liu, Shou-Jên	劉守仁	刘守仁
338	Liu, Shou-Ren	Lau, Shau-Yan	劉守仁	刘守仁
339	Liu, Tang-Rui	Liu, T’ang-Shui	劉棠瑞	刘棠瑞
339	Liu, Tang-Rui	Liu, T'ang-Jui	劉棠瑞	刘棠瑞
339	Liu, Tang-Rui	Liu, Tang-Shui	劉棠瑞	刘棠瑞
339	Liu, Tang-Rui	Liu, Tung-Shui	劉棠瑞	刘棠瑞
340	Liu, Tong-Sheng	Liu, T’ung-Shêng	劉同生	刘同生
341	Liu, Xian-Qi	Liou, Sham-Che	劉獻奇	刘献奇
342	Liu, Xi-Jin	Liu, Hsi-Shin	劉錫進	刘锡进
342	Liu, Xi-Jin	Liu, Hsi-Tsing	劉錫進	刘锡进
343	Liu, Ye-Jing	Liu, Yeh-Ching	劉業經	刘业经
344	Liu, Ying-Xin	Liou, Ying-Sin	劉媖心	刘媖心
344	Liu, Ying-Xin	Liou, Ying-Xin	劉媖心	刘媖心
344	Liu, Ying-Xin	Liu, Ying-Hsin	劉媖心	刘媖心
345	Liu, You-Dong	Liu, Yu-Tung	劉有棟	刘有栋
346	Liu, Yu-Hu	Law, Yuh-Wu	劉玉壺	刘玉壶
346	Liu, Yu-Hu	Liu, Yü-Hu	劉玉壺	刘玉壶
347	Liu, Zi-Ming	Leou, Tze-Ming	柳子明	柳子明
347	Liu, Zi-Ming	Liu, Tsu-Ming	柳子明	柳子明
348	Lou, Zhi-Cen	Lou, Chih-Chen	樓之岑	楼之岑
348	Lou, Zhi-Cen	Lou, Chih-Ch'ên	樓之岑	楼之岑
349	Lü, Chang-Ze	Lu, Chang-Tze	呂長澤	吕长泽

350	Lü, Fu-Yuan	Lu, Fu-Yuen	呂福源	吕福源
351	Lü, Jin-Luo	Lü, Chin-Lo	呂金蘿	吕金萝
352	Lu, Jin-Yi	Lu, Chin-Yi	陸錦一	陆锦一
353	Lu, Kai-Yun	Lu, K'ai-Yün	盧開運	卢开运
354	Lu, Ling-Di	Lu, Ling-Ti	陸玲娣	陆玲娣
355	Lü, Sheng-You	Lu, Sheng-Yu	呂勝由	吕胜由
356	Lü, Wen-Bin	Leu, Wen-Pen	呂文賓	吕文宾
357	Lu, Zhi-Hua	Lu, Chi-Hua	陸志華	陆志华
358	Lu, An-Min	Lu, An-Ming	路安民	路安民
359	Luo, Han-Qiang	Lo, Hang-Chiang	羅漢強	罗汉强
360	Luo, Jian-Xin	Luo, Jian-Shing	羅健馨	罗健馨
361	Luo, Shi-Jiong	Lo, Shih-Chiung	羅士炯	罗士炯
362	Luo, Xian-Rui	Lo, Hsien-Shui	羅獻瑞	罗献瑞
362	Luo, Xian-Rui	Luo, Hsien-Shui	羅獻瑞	罗献瑞
363	Ma, Cheng-Gong	Ma, Cheng-Gung	馬成功	马成功
364	Ma, Da-Pu	Ma, Ta-P'u	馬大浦	马大浦
365	Ma, Ji	Ma, Chi	馬　驥	马　骥
366	Ma, Qi-Yun	Ma, Chi-Yun	馬其雲	马其云
367	Ma, Yong-Gui	Ma, Yong-Kui	馬永貴	马永贵
368	Ma, Yu-Quan	Ma, Yu-Chuan	馬毓泉	马毓泉
368	Ma, Yu-Quan	Ma, Yü-Ch'üan	馬毓泉	马毓泉
369	Mao, Hua-Xun	Mao, Hua-Hsün	毛華訓	毛华训
369	Mao, Hua-Xun	Mao, Hwa-Shing	毛華訓	毛华训
370	Mao, Pin-Yi	Mao, Ping-I	毛品一	毛品一
370	Mao, Pin-Yi	Mao, Pin-I	毛品一	毛品一
371	Mao, Xiu-Sheng	Mao, Hsiu-Sheng	茅秀生	茅秀生
372	Mao, Yong	Mao, Yung	毛　雍	毛　雍
373	Mao, Zong-Liang	Mao, Chung-Liang	毛宗良	毛宗良
373	Mao, Zong-Liang	Mao, Ts'ung-Liang	毛宗良	毛宗良
373	Mao, Zong-Liang	Mao, Tsung-Liang	毛宗良	毛宗良
374	Meng, Fan-Song	Men, Fan-Son	孟繁松	孟繁松
375	Meng, Ren-Xian	Meng, Jen-Hsien	蒙仁憲	蒙仁宪
376	Min, Tian-Lu	Ming, Tien-Lu	閔天祿	闵天禄
377	Mo, Xin-Li	Mo, Sin-Li	莫新禮	莫新礼
378	Mu, Chun-Tao	Moo, Chuen-Tau	穆椿濤	穆椿涛
379	Mu, Zong-Tao	Moo, Chuen-Tau	穆宗濤	穆宗涛
380	Ni, Zhi-Cheng	Ni, Chi-Cheng	倪志誠	倪志诚
381	Nie, Shao-Quan	Nie, Shou-Chuan	聶紹荃	聂绍荃
382	Ou, Chen-Xiong	Au, Chen-Hsiung	歐辰雄	欧辰雄
383	Ou, Chi-Nan	Au, Chih-Nan	歐熾南	欧炽南
383	Ou, Chi-Nan	Au, Ch'ih-Nan	歐熾南	欧炽南

384	Ou, Shi-Huang	Ou, Shih-Huang	歐世璜	欧世璜
385	Pan, Guo-Ying	Pan, Kwo-Ying	潘國瑛	潘国瑛
386	Pang, Xin-Min	P'ang, Hsin-Min	龐新民	庞新民
387	Pei, Jian	Pei, Chien	裴 鑒	裴 鉴
387	Pei, Jian	P'ei, Chien	裴 鑒	裴 鉴
388	Peng, Guang-Qin	P'êng, Kuang-Ch'in	彭光欽	彭光钦
388	Peng, Guang-Qin	P'êng, Kwang-Ch'ing	彭光欽	彭光钦
389	Peng, Jing-Yi	Peng, Ching-I	彭鏡毅	彭镜毅
390	Peng, Ren-Jie	Peng, Jen-Jye	彭仁傑	彭仁杰
391	Peng, Shi-Fang	P'êng, Shih-Fang	彭世芳	彭世芳
392	Peng, Ze-Xiang	Pen, Tse-Hsiang	彭澤祥	彭泽祥
393	Peng, Zuo-Quan	P'êng, Tso-Ch'üan	彭佐權	彭佐权
394	Pu, Fa-Ding	Pu, Fa-Ting	溥發鼎	溥发鼎
395	Qi, Jing-Wen	Chi, Chin-Wen	戚經文	戚经文
396	Qi, Pei-Kun	Chi, Pai-Kuen	戚佩坤	戚佩坤
397	Qian, Chong-Shu	Ch'ien, Ch'ung-Shu	錢崇澍	钱崇澍
397	Qian, Chong-Shu	Chien, Sung-Shu	錢崇澍	钱崇澍
397	Qian, Chong-Shu	Ch'ien, Sung-Shu	錢崇澍	钱崇澍
398	Qian, Jia-Ju	Chien, Chia-Chu	錢家駒	钱家驹
398	Qian, Jia-Ju	Ch'ien, Chia-Chü	錢家駒	钱家驹
398	Qian, Jia-Ju	Chien, Jia-Ju	錢家駒	钱家驹
399	Qian, Nan-Fen	Ch'ien, Nan-Fên	錢南芬	钱南芬
400	Qian, Xiao-Hu	Chien, Hsiao-Hu	錢嘯虎	钱啸虎
400	Qian, Xiao-Hu	Ch'ien, Hsiao-Hu	錢嘯虎	钱啸虎
401	Qiao, Zeng-Jian	Chiao, Tseng-Chien	喬曾鑒	乔曾鉴
401	Qiao, Zeng-Jian	Chiao, Tsêng-Chien	喬曾鑒	乔曾鉴
402	Qin, Hui-Zhen	Chin, Hui-Chen	秦慧貞	秦慧贞
402	Qin, Hui-Zhen	Chin, Hui-Zhen	秦慧貞	秦慧贞
403	Qin, Ren-Chang	Ch'in, Jên-Ch'ang	秦仁昌	秦仁昌
403	Qin, Ren-Chang	Ching, Ren-Chang	秦仁昌	秦仁昌
404	Qiu, Bao-Lin	Chiu, Pao-Lin	裘寶林	裘宝林
404	Qiu, Bao-Lin	Chiu, Pao-Ling	裘寶林	裘宝林
405	Qiu, Gui-Yuan	Ch'iu, Kuei-Yüan	裘桂元	裘桂元
406	Qiu, Hua-Xing	Chiu, Hua-Hsing	丘華興	丘华兴
406	Qiu, Hua-Xing	Ch'iu, Hua-Hsing	丘華興	丘华兴
406	Qiu, Hua-Xing	Kiu, Hua-Sing	丘華興	丘华兴
406	Qiu, Hua-Xing	Kiu, Hua-Xing	丘華興	丘华兴
407	Qiu, Lian-Qing	Chiu, Lien-Ching	邱蓮卿	邱莲卿
407	Qiu, Lian-Qing	Ch'iu, Lien-Ch'ing	邱蓮卿	邱莲卿
408	Qiu, Nian-Yong	Chiu, Nien-Yung	邱年永	邱年永
409	Qiu, Pei-Xi	Chiu, Pei-Hsi	裘佩熹	裘佩熹

409	Qiu, Pei-Xi	Chiu, Pei-Shi	裘佩熹	裘佩熹
409	Qiu, Pei-Xi	Chiu, Pei-Xi	裘佩熹	裘佩熹
410	Qiu, Shao-Ting	Chiu, Shau-Ting	邱少婷	邱少婷
411	Qiu, Wen-Liang	Chiou, Wen-Liang	邱文良	邱文良
412	Qu, Gui-Ling	Chu, Kuei-Ling	曲桂齡	曲桂龄
412	Qu, Gui-Ling	Ch'ü, Kuei-Ling	曲桂齡	曲桂龄
413	Qu, Shi-Zeng	Chu, Shin-Tseng	曲式曾	曲式曾
414	Qu, Zhong-Xiang	Chu, Chung-Hsiang	曲仲湘	曲仲湘
414	Qu, Zhong-Xiang	Ch'ü, Chung-Hsiang	曲仲湘	曲仲湘
415	Rao, Qin-Zhi	Jao, Chin-Chih	饒欽止	饶钦止
415	Rao, Qin-Zhi	Jao, Ch'in-Chih	饒欽止	饶钦止
416	Ren, Wei	Jên, Wei	任　瑋	任　玮
417	Ren, Xian-Wei	Jen, Hsien-Wei	任憲威	任宪威
418	Ruan, Xi-Zhong	Yuan, Hsi-Chung	阮希中	阮希中
419	Shan, Ren-Hua	Shan, Jên-Hua	單人驊	单人骅
419	Shan, Ren-Hua	Shan, Ren-Hwa	單人驊	单人骅
420	Shao, Yao-Nian	Shao, Yao-Nien	邵堯年	邵尧年
420	Shao, Yao-Nian	Shiu, Iu-Nin	邵堯年	邵尧年
421	She, Meng-Lan	Sheh, Meng-Lan	佘孟蘭	佘孟兰
422	Shen, Guan-Mian	Shen, Kuan-Mien	沈觀冕	沈观冕
423	Shen, Jun	Shên, Tsun	沈　雋	沈　隽
424	Shen, Lian-De	Shen, Lian-Te	沈聯德	沈联德
425	Shen, Yu-Feng	Shen, Yü-Fêng	沈毓鳳	沈毓凤
426	Shen, Zhong-Fu	Shen, Chung-Fu	沈中桴	沈中桴
427	Sheng, Cheng-Gui	Sheng, Cheng-Kuei	盛誠桂	盛诚桂
427	Sheng, Cheng-Gui	Shêng, Ch'êng-Kuei	盛誠桂	盛诚桂
428	Shi, Bing-Lin	Shih, Bing-Ling	施炳霖	施炳霖
429	Shi, Jiu-Zhuang	Shih, Chiu-Chuang	史久莊	史久庄
430	Shi, Wen-Liang	Shih, Wen-Liang	施文良	施文良
431	Shi, Xing-Hua	Shih, Hsin-Hua	施興華	施兴华
432	Shi, Xue-Xi	Shi, Hak-Shu	施學習	施学习
432	Shi, Xue-Xi	Shih, Hsüeh-Hsi	施學習	施学习
433	Shi, Zhu	Shih, Chu	石　鑄	石　铸
434	Shi, Zi-Xing	Shih, Tzu-Hsing	石子興	石子兴
435	Shu, Zui-Xiang	Shu, Tsui-Hsiang	舒醉香	舒醉香
436	Si, Xing-Jian	Ssu, Hsing-Chien	斯行健	斯行健
436	Si, Xing-Jian	Sze, Hsing-Chien	斯行健	斯行健
437	Song, Wan-Zhi	Sung, Wan-Chih	宋萬志	宋万志
437	Song, Wan-Zhi	Sung, Wang-Chih	宋萬志	宋万志
438	Song, Zi-Pu	Soong, Tse-Pu	宋滋圃	宋滋圃
439	Su, He-Yi	Su, Ho-Yi	粟和毅	粟和毅

440	Su, Hong-Jie	Su, Horng-Jye	蘇鴻傑	苏鸿杰
441	Su, Mei-Ling	So, May-Ling	蘇美靈	苏美灵
442	Sun, Bi-Xing	Sun, Bi-Sin	孫必興	孙必兴
443	Sun, Dai-Yang	Sun, Tai-Yang	孫岱陽	孙岱阳
444	Sun, Feng-Ji	Sun, Fêng-Chi	孫逢吉	孙逢吉
445	Sun, Jian-Zhang	Su, Chien-Chang	孫建璋	孙建璋
446	Sun, Long-Ji	Sun, Lung-Chi	孫隆吉	孙隆吉
446	Sun, Long-Ji	Sun, Von-Gee	孫隆吉	孙隆吉
447	Sun, Qi-Shi	Sun, Chi-Shi	孫啓時	孙启时
448	Sun, Shu-Xian	Sun, Hsu-Hsien	孫淑賢	孙淑贤
448	Sun, Shu-Xian	Sun, Shu-Hsien	孫淑賢	孙淑贤
449	Sun, Xiang-Zhong	Sun, Hsiang-Chung	孫祥鐘	孙祥钟
450	Sun, Xiong-Cai	Sun, Hsiung-Tsai	孫雄才	孙雄才
450	Sun, Xiong-Cai	Sun, Hsiung-Ts'ai	孫雄才	孙雄才
450	Sun, Xiong-Cai	Sun, Yon-Zai	孫雄才	孙雄才
451	Tan, Pei-Xiang	Tam, Pui-Cheung	譚沛祥	谭沛祥
452	Tan, Zhong-Ming	Tan, Chung-Ming	譚仲明	谭仲明
453	Tang, Jin	Tang, Tsin	唐　進	唐　进
453	Tang, Jin	T'ang, Chin	唐　進	唐　进
454	Tang, Jue	T'ang, Chio	唐　覺	唐　觉
455	Tang, Teng-Han	T'ang, T'êng-Han	湯滕漢	汤滕汉
456	Tang, Wei-Xin	Tang, Wei-Shin	湯惟新	汤惟新
457	Tang, Wen-Tong	T'ang, Wên-T'ung	湯文通	汤文通
458	Tang, Xin-Yao	Tang, Sin-Yao	唐心曜	唐心曜
459	Tang, Yan-Cheng	Tang, Yen-Cheng	湯彥承	汤彦承
459	Tang, Yan-Cheng	T'ang, Yen-Ch'êng	湯彥承	汤彦承
460	Tang, Yao	T'ang, Yao	唐　耀	唐　耀
461	Tang, Zhen-Zi	Tang, Chen-Zi	唐振緇	唐振缁
462	Tao, Yan-Qiao	T'ao, Yen-Ch'iao	陶延橋	陶延桥
463	Teng, Yong-Yan	Têng, Yung-Yen	滕咏延	滕咏延
463	Teng, Yong-Yan	Terng, Yeong-Yan	滕咏延	滕咏延
464	Tian, Jing-Quan	Tian, Kin-Tsuan	田景全	田景全
465	Tong, Zhi-Guo	Tung, Chi-Kuo	仝治國	仝治国
466	Tu, Xiao-Zuo	T'u, Hsiao-Tsu	塗孝祚	涂孝祚
467	Wan, Guo-Ding	Wan, Kuo-Ting	萬國鼎	万国鼎
467	Wan, Guo-Ding	Wan, Kwoh-Ting	萬國鼎	万国鼎
468	Wan, Zong-Ling	Wan, Thung-Ling	萬宗玲	万宗玲
468	Wan, Zong-Ling	Wan, Tsung-Ling	萬宗玲	万宗玲
468	Wan, Zong-Ling	Wane, Thung-Ling	萬宗玲	万宗玲
469	Wang, Bai-Sun	Wang, Pai-Sun	王伯蓀	王伯荪
470	Wang, Bing-Quan	Wang, Ping-Chuan	汪秉全	汪秉全

471	Wang, Cheng-Yin	Ouang, Teng-Ying	汪呈因	汪呈因
471	Wang, Cheng-Yin	Wang, Ch'êng-Yin	汪呈因	汪呈因
472	Wang, Fan-Long	Wang, Fan-Lung	王繁隆	王繁隆
473	Wang, Fa-Zuan	Wang, Fa-Tsuan	汪發纘	汪发缵
474	Wang, Fu-Xiong	Wang, Fu-Hsiung	王伏雄	王伏雄
475	Wang, Guang-Yu	Wang, Kuang-Yuh	王光育	王光育
476	Wang, Hao-Zhen	Wang, Hao-Chên	王浩真	王浩真
476	Wang, Hao-Zhen	Wong, Hao-Chên	王浩真	王浩真
477	Wang, Jin-Ting	Wang, Chin-Ting	王金亭	王金亭
477	Wang, Jin-Ting	Wang, Chin-T'ing	王金亭	王金亭
478	Wang, Ju-Yuan	Wang, Chü-Yüan	汪菊淵	汪菊渊
479	Wang, Ming-De	Wang, Ming-Tê	王明德	王明德
480	Wang, Ming-Jin	Wang, Ming-Chin	王名金	王名金
480	Wang, Ming-Jin	Wang, Ming-Kin	王名金	王名金
481	Wang, Qing-Rui	Wang, Ching-Jui	王慶瑞	王庆瑞
481	Wang, Qing-Rui	Wang, Ch'ing-Jui	王慶瑞	王庆瑞
482	Wang, Qiu-Mei	Wang, Chiou-Mei	王秋美	王秋美
483	Wang, Qi-Wu	Wang, Chi-Wu	王啓無	王启无
483	Wang, Qi-Wu	Wang, Ch'i-Wu	王啓無	王启无
484	Wang, Wen-Cai	Wang, Wen-Tsai	王文采	王文采
484	Wang, Wen-Cai	Wang, Wên-Ts'ai	王文采	王文采
485	Wang, Xian-Zhi	Wang, Hsien-Chih	王先志	王先志
486	Wang, Xian-Zhi	Wang, Hsien-Chih	王顯智	王显智
487	Wang, Yan-Jie	Wang, Yen-Chieh	汪燕傑	汪燕杰
488	Wang, Yan-Zu	Wang, Yen-Tsu	王彥祖	王彦祖
489	Wang, Yao-Yu	Wang, Yo-Yü	王藥雨	王药雨
490	Wang, Ye-Zhen	Wang, Yei-Zeing	王也珍	王也珍
491	Wang, Yi-Gui	Wang, I-Kuei	王一桂	王一桂
492	Wang, Yong-Kui	Wang, Yung-Kwei	王永魁	王永魁
493	Wang, Yun-Zhang	Wang, Yun-Chang	王雲章	王云章
494	Wang, Ze-Jun	Wang, Tze-Chun	王澤鋆	王泽鋆
494	Wang, Ze-Jun	Wong, Tzê-Chun	王澤鋆	王泽鋆
495	Wang, Zhan	Wang, Chan	王 戰	王 战
496	Wang, Zhao-Feng	Wan, Chao-Feng	王兆鳳	王兆凤
496	Wang, Zhao-Feng	Wang, Chao-Fêng	王兆鳳	王兆凤
497	Wang, Zhao-Wu	Wang, Chao-Wu	汪昭武	汪昭武
498	Wang, Zheng-Ping	Wang, Cheng-Ping	王正平	王正平
499	Wang, Zhen-Hua	Wang, Chên-Hua	王振華	王振华
499	Wang, Zhen-Hua	Wang, Chen-Hwa	王振華	王振华
500	Wang, Zhen-Ru	Wang, Chen-Ju	汪振儒	汪振儒
501	Wang, Zhen-Zhe	Wang, Jenn-Che	王震哲	王震哲

502	Wang, Zhi-Jia	Wang, Chih-Chia	王志稼	王志稼
502	Wang, Zhi-Jia	Wang, Chu-Chia	王志稼	王志稼
503	Wang, Zhong-Kui	Wang, Chung-Kuei	王忠魁	王忠魁
503	Wang, Zhong-Kui	Wang, Chung-K'uei	王忠魁	王忠魁
504	Wang, Zhong-Xin	Wang, Chung-Hsin	王忠信	王忠信
505	Wang, Zhu-Hao	Wang, Chu-Hao	王鑄豪	王铸豪
506	Wang, Zong-Xin	Wang, Chung-Hsin	王宗信	王宗信
507	Wang, Zong-Xun	Wang, Tsung-Hsuin	王宗訓	王宗训
507	Wang, Zong-Xun	Wang, Tsung-Hsün	王宗訓	王宗训
508	Wang, Zuo-Bin	Wang, Tso-Pin	王作賓	王作宾
508	Wang, Zuo-Bin	Wang, Tso-Ping	王作賓	王作宾
509	Wei, Jing-Chao	Wei, C. T.	魏景超	魏景超
510	Wei, Yu-Zong	Wei, Yue-Tsung	韋裕宗	韦裕宗
510	Wei, Yu-Zong	Wei, Yu-Tsung	韋裕宗	韦裕宗
511	Wei, Zhao-Fen	Wei, Chao-Fen	衛兆芬	卫兆芬
512	Wen, Hong-Han	Wên, Hung-Han	聞洪漢	闻洪汉
513	Wu, De-Lin	Wu, Te-Lin	吴德鄰	吴德邻
513	Wu, De-Lin	Wu, Te-Ling	吴德鄰	吴德邻
514	Wu, Geng-Min	Wu, Kêng-Min	吴耕民	吴耕民
515	Wu, Guo-Fang	Wu, Kuo-Fang	吴國芳	吴国芳
516	Wu, Hui-Min	Wu, Fei-Man	伍輝民	伍辉民
517	Wu, Ji-Zhi	Go, Kei-Shi	吴繼志	吴继志
517	Wu, Ji-Zhi	Go, Shi-Zen	吴繼志	吴继志
517	Wu, Ji-Zhi	Wu, Chi-Chih	吴繼志	吴继志
518	Wu, Jun-Zong	Wu, Jiunn-Tsong	吴俊宗	吴俊宗
518	Wu, Jun-Zong	Wu, Jiunn-Tzong	吴俊宗	吴俊宗
519	Wu, Ming-Xiang	Wu, Ming-Hsiang	吴明翔	吴明翔
520	Wu, Ming-Zhou	Wu, Ming-Jou	吴明洲	吴明洲
521	Wu, Peng-Cheng	Wu, Pan-Cheng	吴鵬程	吴鹏程
521	Wu, Peng-Cheng	Wu, Pang-Cheng	吴鵬程	吴鹏程
522	Wu, Qi-Jun	Wu, Ch'i-Chün	吴其浚	吴其浚
523	Wu, Qing-Ru	Wu, Ching-Ju	吴慶如	吴庆如
524	Wu, Rong-Fen	Wu, Yeng-Fen	吴容芬	吴容芬
524	Wu, Rong-Fen	Wu, Yong-Fen	吴容芬	吴容芬
524	Wu, Rong-Fen	Wu, Young-Fen	吴容芬	吴容芬
525	Wu, Su-Gong	Wu, Su-Kung	武素功	武素功
526	Wu, Wen-Xiang	Wu, Wen-Hsiang	鄔文祥	邬文祥
527	Wu, Wen-Zhen	Wu, Wen-Chên	吴韞珍	吴韫珍
527	Wu, Wen-Zhen	Wu, Yün-Chên	吴韞珍	吴韫珍
528	Wu, Xin-Gan	Wu, Hsin-Kan	吴信淦	吴信淦
529	Wu, Yin-Chan	Wu, Yin-Ch'an	吴印禪	吴印禅

530	Wu, Ying-Xiang	Wu, Ying-Siang	吳應祥	吴应祥
531	Wu, Yuan-Di	Wu, Yuan-Ti	吳元滌	吴元涤
532	Wu, Zhao-Hong	Wu, Shiew-Hung	吳兆洪	吴兆洪
533	Wu, Zheng-Yi	Wu, Cheng-I	吳徵鎰	吴征镒
533	Wu, Zheng-Yi	Wu, Cheng-Yih	吳徵鎰	吴征镒
534	Wu, Zhong-Lun	Wu, Chung-Luen	吳中倫	吴中伦
534	Wu, Zhong-Lun	Wu, Chung-Lun	吳中倫	吴中伦
534	Wu, Zhong-Lun	Wu, Chung-Lwen	吳中倫	吴中伦
535	Xia, Guang-Cheng	Hsia, Kuang-Cheng	夏光成	夏光成
536	Xia, Wei-Kun	Hsia, Wei-Kun	夏緯琨	夏纬琨
536	Xia, Wei-Kun	Hsia, Wei-K'un	夏緯琨	夏纬琨
537	Xia, Wei-Ying	Hsia, Wei-Ying	夏緯瑛	夏纬瑛
538	Xiang, Gong-Chuan	Hsiang, Kung-Ch'uang	項公傳	项公传
539	Xiang, Qi-Bai	Shang, Chih-Bei	向其柏	向其柏
540	Xiao, Jin-Long	Hsiao, Jin-Long	蕭錦隆	萧锦隆
541	Xiao, Pei-Gen	Hsiao, Pei-Ken	肖培根	肖培根
542	Xiao, Rong	Shiao, Yung	肖 溶	肖 溶
543	Xiao, Ru-Ying	Hsiao, Ju-Ying	肖如英	肖如英
544	Xie, Chang-Fu	Hsieh, Chang-Fu	謝長富	谢长富
545	Xie, Cheng-Ke	Hsieh, Chen-Ko	謝成科	谢成科
546	Xie, Cheng-Zhang	Hsieh, Cheng-Chang	謝成章	谢成章
547	Xie, Ming-Ke	Hsieh, Ming-Ke	謝鳴珂	谢鸣珂
548	Xie, Wan-Quan	Shieh, Wang-Chueng	謝萬權	谢万权
549	Xie, Yin-Tang	Hsieh, Yin-Tang	謝寅堂	谢寅堂
550	Xie, Zong-Wan	Hsieh, Tsung-Wan	謝宗萬	谢宗万
551	Xie, Zong-Xin	Hsieh, Tsung-Hsin	謝宗欣	谢宗欣
552	Xin, Shu-Zhi	Hsin, Shu-Ch'ih	辛樹幟	辛树帜
552	Xin, Shu-Zhi	Sin, S. S.	辛樹幟	辛树帜
553	Xing, Gong-Xia	Shing, Gung-Hsia	邢公俠	邢公侠
553	Xing, Gong-Xia	Shing, Kung-Hsia	邢公俠	邢公侠
554	Xing, Xi-Yong	Hsing, Hsi-Yung	邢錫永	邢锡永
555	Xiong, Wen-Yu	Hsiung, Wen-Yue	熊文愈	熊文愈
556	Xiong, Yao-Guo	Hsiung, Yao-Kuo	熊耀國	熊耀国
557	Xu, Bing-Sheng	Hsu, Ping-Sheng	徐炳聲	徐炳声
558	Xu, Cheng-Wen	Hsü, Ch'êng-Wên	許成文	许成文
559	Xu, De-Xi	Hsu, De-Xi	徐德喜	徐德喜
560	Xu, Guo-Shi	Hsu, Kuo-Shih	徐國士	徐国士
561	Xu, Ji	Hsü, Chi	許 基	许 基
562	Xu, Jian-Chang	Hsu, Chien-Chang	許建昌	许建昌
562	Xu, Jian-Chang	Hsu, Chien-Ch'ang	許建昌	许建昌
562	Xu, Jian-Chang	Hsu, Chien-Chieng	許建昌	许建昌

562	Xu, Jian-Chang	Shü, Chien-Tsang	許建昌	许建昌
563	Xu, Lang-Ran	Shu, Lon-Rhan	徐朗然	徐朗然
564	Xu, Ren	Hsu, Jen	徐 仁	徐 仁
564	Xu, Ren	Hsü, Jên	徐 仁	徐 仁
565	Xu, Rong-Zhang	Hsu, Yong-Zhang	徐榮章	徐荣章
566	Xu, Ting-Zhi	Hsu, Ting-Zhi	徐廷志	徐廷志
567	Xu, Wei-Dong	Hsü, Wei-Tung	徐緯東	徐纬东
568	Xu, Wei-Ying	Hsu, Wei-Ying	徐偉英	徐伟英
569	Xu, Wei-Ying	Hsü, Wei-Ying	徐緯英	徐纬英
570	Xu, Xiang-Hao	Hsu, Hsiang-Hao	徐祥浩	徐祥浩
570	Xu, Xiang-Hao	Hsü, Hsiang-Hao	徐祥浩	徐祥浩
570	Xu, Xiang-Hao	Hsue, Hsiang-Hao	徐祥浩	徐祥浩
570	Xu, Xiang-Hao	Hsue, Hsiang-How	徐祥浩	徐祥浩
571	Xu, Yang-Peng	Hsu, Yang-Peng	徐養鵬	徐养鹏
571	Xu, Yang-Peng	Hsu, Yang-Pong	徐養鵬	徐养鹏
572	Xu, Yan-Qian	Hsü, Yen-Ch'ien	徐燕千	徐燕千
573	Xu, Yin	Hsu, Yin	徐 垠	徐 垠
574	Xu, Yong-Chun	Hsü, Yun-Chun	徐永椿	徐永椿
574	Xu, Yong-Chun	Hsu, Yung-Chun	徐永椿	徐永椿
574	Xu, Yong-Chun	Hsü, Yung-Ch'un	徐永椿	徐永椿
575	Xu, Zai-Wen	Hsu, Tsai-Wen	許再文	许再文
576	Xu, Zhao-Peng	Xu, Chao-Peng	徐兆鵬	徐兆鹏
577	Xuan, Shu-Jie	Hsuan, Shwe-Jye	宣淑潔	宣淑洁
577	Xuan, Shu-Jie	Hsuan, Shwu-Jue	宣淑潔	宣淑洁
578	Xue, Cheng-Jian	Hsüeh, Ch'êng-Chien	薛承健	薛承健
579	Xue, Ji-Ru	Hsueh, Chi-Ju	薛紀如	薛纪如
580	Yan, Chu-Jiang	Yen, Tsu-Kiang	嚴楚江	严楚江
581	Yan, Xun-Chu	Yen, Sun-Ch'u	閻遜初	阎逊初
582	Yan, Zhen-Long	Yen, Chen-Lung	閻振龍	阎振龙
583	Yang, Cai-Yong	Yang, Tsai-Yeog	楊彩勇	杨彩勇
584	Yang, Cheng-Yuan	Yang, Ch'êng-Yüan	楊承元	杨承元
585	Yang, Chong-Ren	Yang, Tsung-Ren	楊崇仁	杨崇仁
586	Yang, Guo-Zhen	Yang, Kuoh-Cheng	楊國禎	杨国祯
587	Yang, Han-Bi	Yang, Han-Pi	楊漢碧	杨汉碧
588	Yang, Qin-Zhou	Yang, Chien-Chow	楊欽周	杨钦周
588	Yang, Qin-Zhou	Yang, Ching-Chow	楊欽周	杨钦周
589	Yang, Rong-Qi	Yang, Yung-Ch'i	楊榮啓	杨荣启
590	Yang, Shang-Guang	Yang, Shan-Kwan	楊上洸	杨上洸
590	Yang, Shang-Guang	Young, Shan-Kwan	楊上洸	杨上洸
591	Yang, Sheng-Ren	Yang, Sheng-Zehn	楊勝任	杨胜任
592	Yang, Sun-Liu	Yang, Suen-Liu	楊孫鎏	杨孙鎏

592	Yang, Sun-Liu	Yang, Sun-Liu	楊孫鎏	杨孙鎏
593	Yang, Xian-Jin	Yang, Yen-Chin	楊銜晉	杨衔晋
594	Yang, Xian-Jin	Yang, Hsien-Chin	楊銜晉	杨衔晋
595	Yang, Xi-Lin	Yang, Hsi-Ling	楊喜林	杨喜林
596	Yang, Xin-Mei	Yang, Hsin-Mei	楊新美	杨新美
597	Yang, Yong-Chang	Yang, Yung-Chang	楊永昌	杨永昌
598	Yang, Yuan-Bo	Yang, Yuen-Po	楊遠波	杨远波
599	Yang, Yu-Po	Yang, Yü-P'o	楊玉坡	杨玉坡
600	Yang, Zheng-Zhong	Yang, Jeng-Jung	楊正仲	杨正仲
601	Yang, Zong-Yu	Yang, Tsung-Yu	楊宗愈	杨宗愈
602	Yao, He-Nian	Yao, Hao-Nien	姚鶴年	姚鹤年
602	Yao, He-Nian	Yao, Ho-Nien	姚鶴年	姚鹤年
603	Ye, Kai-Wen	Yeh, Kai-Yun	葉開溫	叶开温
604	Ye, Ya-Ge	Ip, Nga-Kok	葉雅各	叶雅各
604	Ye, Ya-Ge	Yeh, Ya-Ko	葉雅各	叶雅各
605	Ye, Zhao-Qing	Yeh, Chao-Ch'ing	葉兆慶	叶兆庆
606	Yin, Zhao-Pei	Yin, Chao-P’ei	尹兆培	尹兆培
607	Yin, Zu-Tang	Yen, Tsu-Tang	尹祖棠	尹祖棠
608	Ying, Jun-Sheng	Ying, Tsun-Shen	應俊生	应俊生
609	Yu, Da-Fu	Yü, Ta-Fu	俞大紱	俞大绂
609	Yu, Da-Fu	Yu, Ta-Fuh	俞大紱	俞大绂
610	Yu, De-Jun	Yü, Tê-Chün	俞德浚	俞德浚
610	Yu, De-Jun	Yu, Te-Tsun	俞德浚	俞德浚
610	Yu, De-Jun	Yü, Tê-Chun	俞德浚	俞德浚
611	Yu, Jing-Rang	Yu, Ching-Jang	于景讓	于景让
611	Yu, Jing-Rang	Yü, Ching-Yang	于景讓	于景让
612	Yu, Qi-Bao	Yü, Ch'i-Pao	俞啓葆	俞启葆
613	Yu, Yong-Nian	Yu, Yung-Nien	余永年	余永年
614	Yu, Zong-Xiong	Yu, Tsung-Hsiung	郁宗雄	郁宗雄
615	Yu, Zuo-Ji	Yü, Tso-Chi	俞作楫	俞作楫
616	Yuan, Chang-Qi	Yuan, Ch'ang-Chi	袁昌齊	袁昌齐
617	Yue, Chong-Xi	Yueh, Chung-Hsi	樂崇熙	乐崇熙
617	Yue, Chong-Xi	Yueh, Chun-Hsi	樂崇熙	乐崇熙
618	Yue, Song-Jian	Yueh, Song-Chien	嶽松健	岳松健
618	Yue, Song-Jian	Yueh, Sung-Chien	嶽松健	岳松健
619	Zeng, Cang-Jiang	Tseng, Chang-Jiang	曾滄江	曾沧江
620	Zeng, Cheng-Kui	Tsêng, Ch'eng-K'uei	曾呈奎	曾呈奎
620	Zeng, Cheng-Kui	Tsêng, Chêng-K'uei	曾呈奎	曾呈奎
620	Zeng, Cheng-Kui	Tseng, Cheng-Kwei	曾呈奎	曾呈奎
621	Zeng, Huai-De	Tsang, Wai-Tak	曾懷德	曾怀德
622	Zeng, Ji-Kuan	Tsêng, Chi-K'uan	曾濟寬	曾济宽

623	Zeng, Mian-Zhi	Tsen, Mill	曾勉之	曾勉之
623	Zeng, Mian-Zhi	Tsêng, Mien-Chih	曾勉之	曾勉之
624	Zeng, Qiao	Tseng, Chiao	曾　樵	曾　樵
624	Zeng, Qiao	Tsêng, Ch'iao	曾　樵	曾　樵
624	Zeng, Qiao	Tseng, Charles Chiao	曾　樵	曾　樵
625	Zeng, Shu-Ying	Tseng, Shu-Ying	曾淑英	曾淑英
626	Zeng, Xian-Feng	Cheng, Xian-Feng	曾憲鋒	曾宪锋
627	Zeng, Yan-Xue	Tseng, Yen-Hsueh	曾彥學	曾彦学
628	Zeng, Zhao-Ran	Ts'êng, Chao-Jan	曾昭然	曾昭然
629	Zeng, Zhe-Ru	Tsêng, Che-Ju	曾哲如	曾哲如
629	Zeng, Zhe-Ru	Tseng, Chie-Ju	曾哲如	曾哲如
630	Zha, Ru-Qiang	Cha, Ju-Chiang	查汝強	查汝强
631	Zhan, Ying-Xian	Chan, Ying-Hsien	詹英賢	詹英贤
632	Zhang, Ben-Neng	Chang, Ben-Neng	張本能	张本能
633	Zhang, Bo-Liang	Chang, Po-Liang	張伯良	张伯良
634	Zhang, Chang-Bi	Chang, Tsang-Pi	張蒼碧	张苍碧
635	Zhang, Chao-Chang	Chang, Ch'ao-Ch'ang	張超常	张超常
636	Zhang, Chu-Bao	Chang, Ch'u-Pao	張楚寶	张楚宝
637	Zhang, Da-Cheng	Chang, Da-Cheng	張大成	张大成
638	Zhang, De-Rui	Chang, Tê-Shui	張德瑞	张德瑞
639	Zhang, Dian-Min	Chang, Dian-Mine	張佃民	张佃民
640	Zhang, Fang-Ci	Zhang, Fang-Szu	張芳賜	张芳赐
641	Zhang, Guang-Chu	Chang, Kuang-Chu	張光初	张光初
641	Zhang, Guang-Chu	Chang, Kwang-Chu	張光初	张光初
642	Zhang, Gui-Yi	Chang, Gui-Yi	張貴一	张贵一
643	Zhang, Guo-Liang	Chang, Kuo-Liang	張國梁	张国梁
644	Zhang, Guo-Wei	Chang, Kuo-Wei	張國維	张国维
645	Zhang, Hai-Dao	Zhang, Hai-Tau	張海道	张海道
646	Zhang, Hao	Chang, Hao	張　灝	张　灏
647	Zhang, He-Cen	Chang, Ho-Ch'ên	張和岑	张和岑
648	Zhang, He-Xi	Chang, Ho-Shii	張和喜	张和喜
649	Zhang, He-Zeng	Chang, Ho-Tseng	張盍曾	张盍曾
650	Zhang, Hong-Da	Chang, Hung-Ta	張宏達	张宏达
651	Zhang, Hui-Zhu	Chang, Huey-Ju	張惠珠	张惠珠
652	Zhang, Ji	Chang, Chi	張　伋	张　伋
653	Zhang, Jing-Yue	Chang, Chin-Yueh	張景鉞	张景钺
653	Zhang, Jing-Yue	Chang, Chin-Yüeh	張景鉞	张景钺
654	Zhang, Jin-Tan	Chang, Chin-T'an	張金談	张金谈
654	Zhang, Jin-Tan	Chang, King-Tang	張金談	张金谈
655	Zhang, Jun-Fu	Chang, Tsun-Fu	張峻甫	张峻甫
656	Zhang, Kui	Chang, K.	張　奎	张　奎

657	Zhang, Le-Min	Chang, Lo-Min	章樂民	章乐民
658	Zhang, Man-Xiang	Chang, Man-Hsiang	張滿祥	张满祥
658	Zhang, Man-Xiang	Chang, Man-Siang	張滿祥	张满祥
659	Zhang, Mei-Zhen	Chang, Mei-Chen	張美珍	张美珍
660	Zhang, Pei-Gao	Chang, Pei-Kao	張培杲	张培杲
661	Zhang, Peng-Yun	Chang, Pan-Yung	張鵬雲	张鹏云
662	Zhang, Qing-En	Chang, Ching-En	張慶恩	张庆恩
662	Zhang, Qing-En	Chang, Ch'ing-Nen	張慶恩	张庆恩
663	Zhang, Ruo-Hui	Chang, Roh-Hwei	張若蕙	张若蕙
664	Zhang, Shan-Zhen	Chang, Shan-Jean	張善楨	张善桢
665	Zhang, Shao-Yao	Chang, Shao-Yao	章紹堯	章绍尧
666	Zhang, Shao-Yun	Chang, Shao-Yun	張紹雲	张绍云
667	Zhang, Shu-Feng	Chang, Shu-Fêng	章樹楓	章树枫
668	Zhang, Shu-Lin	Chang, Shu-Lin	張樹林	张树林
669	Zhang, Ting-Yu	Chang, T'ing-Yü	張廷玉	张廷玉
670	Zhang, Ting-Zhen	Chang, Ting-Chien	張廷楨	张廷桢
671	Zhang, Xian-Wu	Chang, Hsien-Wu	張憲武	张宪武
672	Zhang, Xiao-Fang	Chang, Hsiao-Fang	張斅方	张敩方
673	Zhang, Xiao-Tai	Chang, Hsiao-Tai	張曉臺	张晓台
673	Zhang, Xiao-Tai	Chang, Hsiao-T'ai	張曉臺	张晓台
674	Zhang, Xiu-Jun	Chang, Hsiu-Chün	張秀俊	张秀俊
675	Zhang, Xiu-Shi	Chang, Siu-Shih	張秀實	张秀实
676	Zhang, Xue-Zhong	Chang, Hsuch-Chung	張學忠	张学忠
677	Zhang, Yan-Fu	Chang, Yan-Fu	張彥福	张彦福
678	Zhang, Yao-Zong	Chang, Yao-Tsung	張耀宗	张耀宗
678	Zhang, Yao-Zong	Chung, Yao-Tsung	張耀宗	张耀宗
679	Zhang, Yi	Chang, Yi	張　依	张　依
680	Zhang, Yong-Tian	Chang, Yong-Tian	張永田	张永田
681	Zhang, Yu-Liang	Chang, Yui-Liang	張玉良	张玉良
681	Zhang, Yu-Liang	Chang, Yu-Liang	張玉良	张玉良
681	Zhang, Yu-Liang	Chang, Yü-Liang	張玉良	张玉良
682	Zhang, Yu-Long	Chang, Yu-Lung	張玉龍	张玉龙
683	Zhang, Ze-Rong	Chang, Che-Yung	張澤榮	张泽荣
683	Zhang, Ze-Rong	Chang, Chieh-Yun	張澤榮	张泽荣
683	Zhang, Ze-Rong	Chang, Tseh-Yung	張澤榮	张泽荣
684	Zhang, Zhao-Qian	Chang, Chao-Chieh	張肇騫	张肇骞
684	Zhang, Zhao-Qian	Chang, Chao-Ch'ien	張肇騫	张肇骞
685	Zhang, Zhen-Qing	Chang, Chen-Ching	張振清	张振清
686	Zhang, Zhen-Wan	Chang, Chen-Wan	張振萬	张振万
687	Zhang, Zhe-Seng	Chang, Che-Tseng	張哲僧	张哲僧
688	Zhang, Zhi-Song	Chang, Che-Sung	張志松	张志松

689	Zhang, Zhi-Yu	Chang, Chih-Yu	張芝玉	张芝玉
690	Zhao, Ai-Zhen	Chao, Ai-Chên	趙愛真	赵爱真
690	Zhao, Ai-Zhen	Chao, Ai-Chêng	趙愛真	赵爱真
691	Zhao, Da-Chang	Chao, Ta-Ch'ang	趙大昌	赵大昌
692	Zhao, Hong-Ji	Chao, Hung-Chi	趙鴻基	赵鸿基
693	Zhao, Huai-Qing	Chao, Huai-Qing	趙懷青	赵怀青
694	Zhao, Ju-Huang	Chao, Chü-Huang	趙橘黄	赵橘黄
695	Zhao, Liang-Neng	Chao, Liang-Neng	趙良能	赵良能
695	Zhao, Neng	Chao, Neng	趙　能	赵　能
696	Zhao, Pu	Chao, Pu	趙　璞	赵　璞
697	Zhao, Qing-Sheng	Cao, Ching-Sen	趙清盛	赵清盛
698	Zhao, Qi-Seng	Chao, Chi-Son	趙奇僧	赵奇僧
699	Zhao, Shu-Miao	Chaw, Shu-Miaw	趙淑妙	赵淑妙
700	Zhao, Tian-Bang	Chao, Tien-Bang	趙天榜	赵天榜
700	Zhao, Tian-Bang	Chao, Tien-Bung	趙天榜	赵天榜
701	Zhao, Xiu-Qian	Chao, Hsiu-Chien	趙修謙	赵修谦
701	Zhao, Xiu-Qian	Chao, Hsiu-Ch'ien	趙修謙	赵修谦
702	Zhao, You-Wei	Chao, Yu-Wei	趙有爲	赵有为
703	Zhao, Zhe-Ming	Chao, Jew-Ming	趙哲明	赵哲明
704	Zheng, Chao-Zong	Cheng, Chao-Tsung	鄭朝宗	郑朝宗
704	Zheng, Chao-Zong	Cheng, Chao-Zong	鄭朝宗	郑朝宗
705	Zheng, Ji-Xi	Chêng, Chi-Hsi	鄭稷熙	郑稷熙
706	Zheng, Mian	Chêng, Mien	鄭　勉	郑　勉
706	Zheng, Mian	Cheng, Mien	鄭　勉	郑　勉
707	Zheng, Ru-Yong	Cheng, Ju-Yung	鄭儒永	郑儒永
708	Zheng, Si-Xu	Cheng, Sze-Hsue	鄭斯緒	郑斯绪
709	Zheng, Wan-Jun	Cheng, Wan-Chun	鄭萬鈞	郑万钧
709	Zheng, Wan-Jun	Chêng, Wan-Chün	鄭萬鈞	郑万钧
710	Zheng, Wu-Can	Cheng, Wu-Tsang	鄭武燦	郑武灿
711	Zheng, Zhong	Chêng, Chung	鄭　重	郑　重
712	Zhong, Bu-Qin	Chung, Pu-Chin	鐘補勤	钟补勤
712	Zhong, Bu-Qin	Chung, Pu-Ch'in	鐘補勤	钟补勤
712	Zhong, Bu-Qin	Tsoong, Pu-Ch'in	鐘補勤	钟补勤
713	Zhong, Bu-Qiu	Chung, Pu-Ch'iu	鐘補求	钟补求
713	Zhong, Bu-Qiu	Tsoong, Pu-Chiu	鐘補求	钟补求
713	Zhong, Bu-Qiu	Tsoong, Pu-Ch'iu	鐘補求	钟补求
714	Zhong, Chong-Xin	Chung, Chung-Hsin	仲崇信	仲崇信
715	Zhong, Guan-Guang	Chung, Kuan-Kuang	鐘觀光	钟观光
715	Zhong, Guan-Guang	Tsoong, Kuan-Kwang	鐘觀光	钟观光
716	Zhong, Guo-Fang	Chung, Kuo-Fang	鐘國芳	钟国芳
717	Zhong, Ji-Xin	Chung, Chi-Hsing	鐘濟新	钟济新

717	Zhong, Ji-Xin	Tsoong, Chi-Hsin	鐘濟新	钟济新
718	Zhong, Nian-Jun	Chung, Nian-June	鐘年鈞	钟年钧
719	Zhong, Shi-Wen	Chung, Shih-Wen	鐘詩文	钟诗文
720	Zhong, Xin-Xuan	Chung, Hsin-Hsuan	鐘心煊	钟心煊
720	Zhong, Xin-Xuan	Chung, Hsin-Hsüan	鐘心煊	钟心煊
721	Zhong, Yi	Chung, I	鐘　義	钟　义
722	Zhou, Chong-Guang	Chou, Chung-Kuang	周重光	周重光
722	Zhou, Chong-Guang	Chou, Ts'ung-Huang	周重光	周重光
722	Zhou, Chong-Guang	Chow, Chung-Huang	周重光	周重光
722	Zhou, Chong-Guang	Chow, Chung-Kuang	周重光	周重光
723	Zhou, Guang-Rong	Chou, Kuang-Yung	周光榮	周光荣
723	Zhou, Guang-Rong	Chow, Kuang-Yung	周光榮	周光荣
724	Zhou, Guang-Yu	Chou, Kuang-Yu	周光裕	周光裕
724	Zhou, Guang-Yu	Chou, Kuang-Yü	周光裕	周光裕
725	Zhou, Han-Fan	Chou, Han-Fan	周漢藩	周汉藩
725	Zhou, Han-Fan	Chow, Hang-Fan	周漢藩	周汉藩
726	Zhou, He-Chang	Chow, Ho-Chang	周鶴昌	周鹤昌
727	Zhou, Hong-Fu	Chow, Hong-Fu	周洪富	周洪富
727	Zhou, Hong-Fu	Chow, Hung-Fu	周洪富	周洪富
728	Zhou, Jian-Ren	Chou, Chien-Jen	周建人	周建人
729	Zhou, Jia-Qi	Chow, Chia-Chih	周家琪	周家琪
730	Zhou, Ji-Lun	Chou, Chi-Lun	周紀倫	周纪伦
731	Zhou, Tai-Xuan	Chou, T'ai-Hsüan	周太玄	周太玄
732	Zhou, Tai-Yan	Cheo, Tai-Yien	周太炎	周太炎
732	Zhou, Tai-Yan	Chou, T'ai-Yen	周太炎	周太炎
732	Zhou, Tai-Yan	Chou, Tai-Yen	周太炎	周太炎
733	Zhou, Xuan	Chow, Shuan	周　鉉	周　铉
734	Zhou, Xu-Yuan	Cheo, Shu-Yuen	周蓄源	周蓄源
734	Zhou, Xu-Yuan	Chou, Hsu-Yuan	周蓄源	周蓄源
734	Zhou, Xu-Yuan	Chou, Hsü-Yüan	周蓄源	周蓄源
735	Zhou, Yi-Liang	Chou, I-Liang	周以良	周以良
735	Zhou, Yi-Liang	Chou, Yi-Liang	周以良	周以良
735	Zhou, Yi-Liang	Chou, Yu-Liang	周以良	周以良
736	Zhou, Yin	Chou, Yin	周　崟	周　崟
737	Zhou, Ying	Chou, Ying	周　瑛	周　瑛
738	Zhou, Ying-Chang	Cheo, Ying-Ch'ang	周映昌	周映昌
738	Zhou, Ying-Chang	Chou, Ying-Ch'ang	周映昌	周映昌
741	Zhou, Zhen	Chou, Chên	周　楨	周　桢
741	Zhou, Zhen	Chow, Cheng	周　鎮	周　镇
740	Zhou, Zhen-Ying	Chou, Chen-Ying	周貞英	周贞英
740	Zhou, Zhen-Ying	Chou, Chên-Ying	周貞英	周贞英

741	Zhou, Zhong-Xuan	Chow, Chung-Hsuan	周鐘瑄	周钟瑄
742	Zhou, Zong-Huang	Chou, Tsung-Huang	周宗璜	周宗璜
742	Zhou, Zong-Huang	Chow, Chung-Hwang	周宗璜	周宗璜
742	Zhou, Zong-Huang	Chow, Tsung-Huang	周宗璜	周宗璜
743	Zhou, Zu-Ying	Chou, Chu-Ying	周祖英	周祖英
743	Zhou, Zu-Ying	Chou, Tsu-Ying	周祖英	周祖英
744	Zhu, Ge-Lin	Chu, Ge-Lin	朱格麟	朱格麟
744	Zhu, Ge-Lin	Chu, Ge-Ling	朱格麟	朱格麟
745	Zhu, Hao-Ran	Chu, Hao-Jan	朱浩然	朱浩然
746	Zhu, Hua	Chu, Hua	朱　華	朱　华
747	Zhu, Hui-Fang	Chu, Huei-Fang	朱會芳	朱会芳
748	Zhu, Jia-Ran	Chu, Chia-Nan	朱家柟	朱家柟
749	Zhu, Ji-Fan	Chu, Chi-Fan	朱濟凡	朱济凡
750	Zhu, Mao-Shun	Chu, Mao-Shun	朱懋順	朱懋顺
751	Zhu, Shu-Ping	Chu, Shu-P'ing	朱樹屏	朱树屏
752	Zhu, Ting-Cheng	Chu, T’ing-Ch'êng	祝廷成	祝廷成
753	Zhu, Wan-Jia	Chu, Wan-Chia	朱婉嘉	朱婉嘉
754	Zhu, Wei-Ming	Chu, Wei-Ming	朱維明	朱维明
755	Zhu, Xiang-San	Chu, Hsiang-San	朱象三	朱象三
756	Zhu, Yan-Cheng	Chu, Yen-Ch'êng	朱彥丞	朱彦丞
756	Zhu, Yan-Cheng	Tchou, Yen-Tch'êng	朱彥丞	朱彦丞
757	Zhu, You-Chang	Chu, You-Chang	朱有昌	朱有昌
758	Zhu, Zhao-Yi	Chu, Chao-I	朱兆儀	朱兆仪
758	Zhu, Zhao-Yi	Chu, Chao-Yi	朱兆儀	朱兆仪
759	Zhu, Zheng-De	Chu, Cheng-De	朱政德	朱政德
760	Zhu, Zhong-Han	Chu, Chung-Han	朱中翰	朱中翰
761	Zhu, Zong-Yuan	Chu, Zong-Yuan	朱宗元	朱宗元
762	Zhuang, Can-Yang	Chuang, Tsan-Iang	莊燦暘	庄灿旸
762	Zhuang, Can-Yang	Chuang, Tsan-Yang	莊燦暘	庄灿旸
762	Zhuang, Can-Yang	Chuang, Ts'an-Yang	莊燦暘	庄灿旸
763	Zhuang, Qing-Zhang	Chuang, Ching-Chang	莊清漳	庄清漳
764	Zhuang, Xuan	Chuang, Hsuan	莊　璇	庄　璇
765	Zhuo, Ren-Song	Cho, Jên-Sung	卓仁松	卓仁松
766	Zou, Bing-Wen	Tsou, Ping-Wên	鄒秉文	邹秉文
767	Zou, Zhi-Hua	Tsou, Chih-Hua	鄒稚華	邹稚华
768	Zou, Zhong-Lin	Tsou, Chung-Lin	鄒鐘琳	邹钟琳
769	Zuo, Da-Xun	Tso, Ta-Hsun	左大勛	左大勋
770	Zuo, Jing-Lie	Tso, Ching-Lieh	左景烈	左景烈

表 2：中国植物分类学者以老式名字的字母拼写为序
Table 2, Chinese plant taxonomists alphabetically by older transliterations of their names

Number	Old Spelling	Pinyin Spelling	Traditional Chinese	Simple Chinese
1	An, Cheng-Hsi	An, Zheng-Xi	安爭夕	安争夕
2	Aou, Lian-Do	Ao, Liang-De	敖良德	敖良德
2	Aou, Liang-Teh	Ao, Liang-De	敖良德	敖良德
382	Au, Chen-Hsiung	Ou, Chen-Xiong	歐辰雄	欧辰雄
383	Au, Chih-Nan	Ou, Chi-Nan	歐熾南	欧炽南
383	Au, Ch'ih-Nan	Ou, Chi-Nan	歐熾南	欧炽南
3	Aur, Chih-Wen	Ao, Zhi-Wen	敖志文	敖志文
3	Aur, Chin-Wen	Ao, Zhi-Wen	敖志文	敖志文
7	Bi, Lie-Jiu	Bi, Lie-Jue	畢列爵	毕列爵
11	Bu, Yu-Tze	Bu, Yu-Zhi	步毓芝	步毓芝
8	But, Paul Pui-Hay	Bi, Pei-Xi	畢培曦	毕培曦
697	Cao, Ching-Sen	Zhao, Qing-Sheng	趙清盛	赵清盛
229	Ch'iang, Ying	Jiang, Ying	蔣 英	蒋 英
630	Cha, Ju-Chiang	Zha, Ru-Qiang	查汝強	查汝强
631	Chan, Ying-Hsien	Zhan, Ying-Xian	詹英賢	詹英贤
632	Chang, Ben-Neng	Zhang, Ben-Neng	張本能	张本能
635	Chang, Ch'ao-Ch'ang	Zhang, Chao-Chang	張超常	张超常
684	Chang, Chao-Chieh	Zhang, Zhao-Qian	張肇騫	张肇骞
684	Chang, Chao-Ch'ien	Zhang, Zhao-Qian	張肇騫	张肇骞
685	Chang, Chen-Ching	Zhang, Zhen-Qing	張振清	张振清
686	Chang, Chen-Wan	Zhang, Zhen-Wan	張振萬	张振万
688	Chang, Che-Sung	Zhang, Zhi-Song	張志松	张志松
687	Chang, Che-Tseng	Zhang, Zhe-Seng	張哲僧	张哲僧
683	Chang, Che-Yung	Zhang, Ze-Rong	張澤榮	张泽荣
652	Chang, Chi	Zhang, Ji	張 伋	张 伋
683	Chang, Chieh-Yun	Zhang, Ze-Rong	張澤榮	张泽荣
689	Chang, Chih-Yu	Zhang, Zhi-Yu	張芝玉	张芝玉
662	Chang, Ching-En	Zhang, Qing-En	張慶恩	张庆恩
662	Chang, Ch'ing-Nen	Zhang, Qing-En	張慶恩	张庆恩
654	Chang, Chin-T'an	Zhang, Jin-Tan	張金談	张金谈
653	Chang, Chin-Yueh	Zhang, Jing-Yue	張景鉞	张景钺
653	Chang, Chin-Yüeh	Zhang, Jing-Yue	張景鉞	张景钺
636	Chang, Ch'u-Pao	Zhang, Chu-Bao	張楚寶	张楚宝
637	Chang, Da-Cheng	Zhang, Da-Cheng	張大成	张大成

639	Chang, Dian-Mine	Zhang, Dian-Min	張佃民	张佃民
642	Chang, Gui-Yi	Zhang, Gui-Yi	張貴一	张贵一
646	Chang, Hao	Zhang, Hao	張　灝	张　灏
647	Chang, Ho-Ch'ên	Zhang, He-Cen	張和岑	张和岑
648	Chang, Ho-Shii	Zhang, He-Xi	張和喜	张和喜
649	Chang, Ho-Tseng	Zhang, He-Zeng	張盍曾	张盍曾
672	Chang, Hsiao-Fang	Zhang, Xiao-Fang	張斅方	张敩方
673	Chang, Hsiao-Tai	Zhang, Xiao-Tai	張曉臺	张晓台
673	Chang, Hsiao-T'ai	Zhang, Xiao-Tai	張曉臺	张晓台
671	Chang, Hsien-Wu	Zhang, Xian-Wu	張憲武	张宪武
20	Chang, Hsing-Jo	Chang, Xing-Ruo	暢行若	畅行若
674	Chang, Hsiu-Chün	Zhang, Xiu-Jun	張秀俊	张秀俊
676	Chang, Hsuch-Chung	Zhang, Xue-Zhong	張學忠	张学忠
651	Chang, Huey-Ju	Zhang, Hui-Zhu	張惠珠	张惠珠
650	Chang, Hung-Ta	Zhang, Hong-Da	張宏達	张宏达
656	Chang, K.	Zhang, Kui	張　奎	张　奎
654	Chang, King-Tang	Zhang, Jin-Tan	張金談	张金谈
641	Chang, Kuang-Chu	Zhang, Guang-Chu	張光初	张光初
643	Chang, Kuo-Liang	Zhang, Guo-Liang	張國樑	张国梁
644	Chang, Kuo-Wei	Zhang, Guo-Wei	張國維	张国维
641	Chang, Kwang-Chu	Zhang, Guang-Chu	張光初	张光初
657	Chang, Lo-Min	Zhang, Le-Min	章樂民	章乐民
658	Chang, Man-Hsiang	Zhang, Man-Xiang	張滿祥	张满祥
658	Chang, Man-Siang	Zhang, Man-Xiang	張滿祥	张满祥
659	Chang, Mei-Chen	Zhang, Mei-Zhen	張美珍	张美珍
661	Chang, Pan-Yung	Zhang, Peng-Yun	張鵬雲	张鹏云
660	Chang, Pei-Kao	Zhang, Pei-Gao	張培杲	张培杲
633	Chang, Po-Liang	Zhang, Bo-Liang	張伯良	张伯良
663	Chang, Roh-Hwei	Zhang, Ruo-Hui	張若蕙	张若蕙
664	Chang, Shan-Jean	Zhang, Shan-Zhen	張善楨	张善桢
665	Chang, Shao-Yao	Zhang, Shao-Yao	章紹堯	章绍尧
666	Chang, Shao-Yun	Zhang, Shao-Yun	張紹雲	张绍云
667	Chang, Shu-Fêng	Zhang, Shu-Feng	章樹楓	章树枫
668	Chang, Shu-Lin	Zhang, Shu-Lin	張樹林	张树林
675	Chang, Siu-Shih	Zhang, Xiu-Shi	張秀實	张秀实
669	Chang, T'ing-Yü	Zhang, Ting-Yu	張廷玉	张廷玉
638	Chang, Tê-Shui	Zhang, De-Rui	張德瑞	张德瑞
670	Chang, Ting-Chien	Zhang, Ting-Zhen	張廷楨	张廷桢
634	Chang, Tsang-Pi	Zhang, Chang-Bi	張蒼碧	张苍碧
683	Chang, Tseh-Yung	Zhang, Ze-Rong	張澤榮	张泽荣
655	Chang, Tsun-Fu	Zhang, Jun-Fu	張峻甫	张峻甫

677	Chang, Yan-Fu	Zhang, Yan-Fu	張彥福	张彦福
678	Chang, Yao-Tsung	Zhang, Yao-Zong	張耀宗	张耀宗
679	Chang, Yi	Zhang, Yi	張 依	张 依
680	Chang, Yong-Tian	Zhang, Yong-Tian	張永田	张永田
681	Chang, Yui-Liang	Zhang, Yu-Liang	張玉良	张玉良
681	Chang, Yu-Liang	Zhang, Yu-Liang	張玉良	张玉良
681	Chang, Yü-Liang	Zhang, Yu-Liang	張玉良	张玉良
682	Chang, Yu-Lung	Zhang, Yu-Long	張玉龍	张玉龙
690	Chao, Ai-Chên	Zhao, Ai-Zhen	趙愛真	赵爱真
690	Chao, Ai-Chêng	Zhao, Ai-Zhen	趙愛真	赵爱真
698	Chao, Chi-Son	Zhao, Qi-Seng	趙奇僧	赵奇僧
694	Chao, Chü-Huang	Zhao, Ju-Huang	趙橘黃	赵橘黄
701	Chao, Hsiu-Chien	Zhao, Xiu-Qian	趙修謙	赵修谦
701	Chao, Hsiu-Ch'ien	Zhao, Xiu-Qian	趙修謙	赵修谦
693	Chao, Huai-Qing	Zhao, Huai-Qing	趙懷青	赵怀青
692	Chao, Hung-Chi	Zhao, Hong-Ji	趙鴻基	赵鸿基
703	Chao, Jew-Ming	Zhao, Zhe-Ming	趙哲明	赵哲明
695	Chao, Liang-Neng	Zhao, Liang-Neng	趙良能	赵良能
695	Chao, Neng	Zhao, Neng	趙 能	赵 能
696	Chao, Pu	Zhao, Pu	趙 璞	赵 璞
691	Chao, Ta-Ch'ang	Zhao, Da-Chang	趙大昌	赵大昌
700	Chao, Tien-Bang	Zhao, Tian-Bang	趙天榜	赵天榜
700	Chao, Tien-Bung	Zhao, Tian-Bang	趙天榜	赵天榜
702	Chao, Yu-Wei	Zhao, You-Wei	趙有爲	赵有为
699	Chaw, Shu-Miaw	Zhao, Shu-Miao	趙淑妙	赵淑妙
21	Ch'ên, An-Chi	Chen, An-Ji	陳安集	陈安集
37	Chen, Cheih	Chen, Jie	陳 介	陈 介
36	Chen, Chia-Jue	Chen, Jia-Rui	陳家瑞	陈家瑞
36	Chen, Chia-Jui	Chen, Jia-Rui	陳家瑞	陈家瑞
39	Ch'ên, Chien-Jên	Chen, Juan-Ren	陳雋人	陈隽人
68	Chen, Chih-Hsiung	Chen, Zhi-Xiong	陳志雄	陈志雄
44	Ch'ên, Ch'ing-Cheng	Chen, Qing-Cheng	陳慶誠	陈庆诚
45	Chen, Ching-Hsia	Chen, Qing-Xia	陳擎霞	陈擎霞
38	Chen, Chiung-Sung	Chen, Jiong-Song	陳炯松	陈炯松
38	Ch'ên, Chiung-Sung	Chen, Jiong-Song	陳炯松	陈炯松
69	Chen, Cho	Chen, Zhuo	陳 倬	陈 倬
69	Ch'ên, Cho	Chen, Zhuo	陳 倬	陈 倬
70	Chen, Chung-Lien	Chen, Zong-Lian	陳宗蓮	陈宗莲
25	Chen, Chung-Ming	Chen, Chong-Ming	陳重明	陈重明
26	Ch'ên, Ch'ung-Shou	Chen, Chun-Shou	陳椿壽	陈椿寿
26	Ch'ên, Ch'un-Shou	Chen, Chun-Shou	陳椿壽	陈椿寿

40	Chen, Chun-Yu	Chen, Jun-Yu	陳俊愉	陈俊愉
40	Chen, Chün-Yü	Chen, Jun-Yu	陳俊愉	陈俊愉
40	Ch'ên, Chün-Yü	Chen, Jun-Yu	陳俊愉	陈俊愉
27	Chen, Dao-Men	Chen, De-Mao	陳德懋	陈德懋
30	Ch'ên, Fêng-Huai	Chen, Feng-Huai	陳封懷	陈封怀
30	Chen, Feng-Hwai	Chen, Feng-Huai	陳封懷	陈封怀
33	Chen, Hao-Tzu	Chen, Hao-Zi	陳淏子	陈淏子
59	Ch'ên, Hsiang-Fa	Chen, Xiang-Fa	陳祥發	陈祥发
60	Chen, Hsi-Kai	Chen, Xi-Kai	陳希凱	陈希凯
60	Ch'ên, Hsi-K'ai	Chen, Xi-Kai	陳希凱	陈希凯
62	Ch'en, Hsiu-Ying	Chen, Xiu-Ying	陳秀英	陈秀英
34	Chen, Hua-Kwei	Chen, Hua-Gui	陳華癸	陈华癸
35	Ch'ên, Huan-Yung	Chen, Huan-Yong	陳煥鏞	陈焕镛
65	Chen, I-Ming	Chen, Yi-Ming	陳益明	陈益明
31	Ch'ên, Kuan-Yu	Chen, Guan-You	陳冠友	陈冠友
32	Ch'ên, Kuei	Chen, Gui	陳　貴	陈　贵
41	Chen, Li-Ching	Chen, Li-Qing	陳立卿	陈立卿
41	Ch'ên, Li-Ch'ing	Chen, Li-Qing	陳立卿	陈立卿
42	Chen, Men	Chen, Min	陳　敏	陈　敏
42	Chen, Ming-Che	Chen, Ming-Zhe	陳明哲	陈明哲
22	Chen, Pan-Chieh	Chen, Bang-Jie	陳邦傑	陈邦杰
22	Chen, Pang-Chieh	Chen, Bang-Jie	陳邦傑	陈邦杰
22	Ch'ên, Pang-Chieh	Chen, Bang-Jie	陳邦傑	陈邦杰
23	Chen, Pang-Yu	Chen, Bang-Yu	陳邦餘	陈邦余
24	Ch'ên, Pê-Ch'uan	Chen, Bo-Chuan	陳伯川	陈伯川
51	Chen, Sen-Jen	Chen, Sheng-Zhen	陳升振	陈升振
50	Ch'ên, Shan-Ming	Chen, Shan-Ming	陳善銘	陈善铭
52	Chen, Shih-Huei	Chen, Shi-Hui	陳世輝	陈世辉
55	Ch'ên, Shu-Chên	Chen, Shu-Zhen	陳淑珍	陈淑珍
53	Chen, Si-Cien	Chen, Shi-Xiang	陳世驤	陈世骧
61	Chen, Sing-Chi	Chen, Xin-Qi	陳心啓	陈心启
56	Ch'ên, Ssŭ-I	Chen, Si-Yi	陳思義	陈思义
54	Chen, Su-Hwa	Chen, Shu-Hua	陳淑華	陈淑华
28	Chen, Te-Chao	Chen, De-Zhao	陳德昭	陈德昭
28	Ch'ên, Tê-Chao	Chen, De-Zhao	陳德昭	陈德昭
29	Chen, Ti	Chen, Di	陳　蒂	陈　蒂
71	Chen, Tsu-Keng	Chen, Zu-Keng	陳祖鏗	陈祖铿
237	Chen, Wei-Chien	Jin, Wei-Jian	金維堅	金维坚
56	Chen, Wei-Chiu	Chen, Wei-Qiu	陳偉球	陈伟球
57	Chen, Wei-Sin	Chen, Wei-Xin	陳維新	陈维新
63	Chen, Ya-Chi	Chen, Ya-Qi	陳雅琦	陈雅琦

63	Ch'ên, Yen-Cho	Chen, Yan-Zhuo	陳彥卓	陈彦卓
66	Ch'ên, Ying-Huang	Chen, Ying-Huang	陳映璜	陈映璜
67	Chen, Yo-Jiun	Chen, You-Jun	陳又君	陈又君
47	Chen, Yung	Chen, Rong	陳　嶸	陈　嵘
47	Ch'ên, Yung	Chen, Rong	陳　嶸	陈　嵘
48	Chen, Yung-Reui	Chen, Rong-Rui	陳榮銳	陈荣锐
49	Chen, Zuei-Ching	Chen, Rui-Qing	陳瑞青	陈瑞青
704	Cheng, Chao-Tsung	Zheng, Chao-Zong	鄭朝宗	郑朝宗
704	Cheng, Chao-Zong	Zheng, Chao-Zong	鄭朝宗	郑朝宗
705	Chêng, Chi-Hsi	Zheng, Ji-Xi	鄭稷熙	郑稷熙
72	Cheng, Ching-Yung	Cheng, Jing-Rong	誠靜容	诚静容
711	Chêng, Chung	Zheng, Zhong	鄭　重	郑　重
707	Cheng, Ju-Yung	Zheng, Ru-Yong	鄭儒永	郑儒永
706	Cheng, Mien	Zheng, Mian	鄭　勉	郑　勉
706	Chêng, Mien	Zheng, Mian	鄭　勉	郑　勉
708	Cheng, Sze-Hsue	Zheng, Si-Xu	鄭斯緒	郑斯绪
709	Cheng, Wan-Chun	Zheng, Wan-Jun	鄭萬鈞	郑万钧
709	Chêng, Wan-Chün	Zheng, Wan-Jun	鄭萬鈞	郑万钧
710	Cheng, Wu-Tsang	Zheng, Wu-Can	鄭武燦	郑武灿
626	Cheng, Xian-Feng	Zeng, Xian-Feng	曾憲鋒	曾宪锋
734	Cheo, Shu-Yuen	Zhou, Xu-Yuan	周蓄源	周蓄源
732	Cheo, Tai-Yien	Zhou, Tai-Yan	周太炎	周太炎
738	Cheo, Ying-Ch'ang	Zhou, Ying-Chang	周映昌	周映昌
216	Chi, Ching-Yuan	Ji, Jing-Yuan	季景元	季景元
395	Chi, Chin-Wen	Qi, Jing-Wen	戚經文	戚经文
217	Chi, Chün-Mien	Ji, Jun-Mian	季君勉	季君勉
215	Chi, Han	Ji, Han	稽　含	稽　含
396	Chi, Pai-Kuen	Qi, Pei-Kun	戚佩坤	戚佩坤
219	Chia, Ch'êng-Chang	Jia, Cheng-Zhang	賈成章	贾成章
220	Chia, Liang-Chih	Jia, Liang-Zhi	賈良智	贾良智
221	Chia, Shen-Hsiu	Jia, Shen-Xiu	賈慎修	贾慎修
223	Chia, Tsu-Chang	Jia, Zu-Zhang	賈祖璋	贾祖璋
222	Chia, Tsu-San	Jia, Zu-Shan	賈祖珊	贾祖珊
225	Chiang, Ch'êng-Chi	Jiang, Cheng-Ji	蔣承基	蒋承基
225	Chiang, Cheng-Chi	Jiang, Cheng-Ji	蔣承基	蒋承基
226	Chiang, Han-Ch'iao	Jiang, Han-Qiao	姜漢僑	姜汉侨
227	Chiang, P'u	Jiang, Pu	蔣　溥	蒋　溥
231	Chiang, Tzen-Yuh	Jiang, Zhen-Yu	蔣鎮宇	蒋镇宇
230	Chiang, You-Long	Jiang, You-Long	江有龍	江有龙
232	ChiangTsai, Su-Hwa	JiangCai, Shu-Hua	江蔡淑華	江蔡淑华
233	Chiao, Chi-Yuan	Jiao, Qi-Yuan	焦啓源	焦启源

233	Chiao, Ch'i-Yüan	Jiao, Qi-Yuan	焦啓源	焦启源
233	Chiao, Chi-Yuen	Jiao, Qi-Yuan	焦啓源	焦启源
401	Chiao, Tseng-Chien	Qiao, Zeng-Jian	喬曾鑒	乔曾鉴
401	Chiao, Tsêng-Chien	Qiao, Zeng-Jian	喬曾鑒	乔曾鉴
398	Chien, Chia-Chu	Qian, Jia-Ju	錢家駒	钱家驹
398	Ch'ien, Chia-Chü	Qian, Jia-Ju	錢家駒	钱家驹
397	Ch'ien, Ch'ung-Shu	Qian, Chong-Shu	錢崇澍	钱崇澍
400	Chien, Hsiao-Hu	Qian, Xiao-Hu	錢嘯虎	钱啸虎
400	Ch'ien, Hsiao-Hu	Qian, Xiao-Hu	錢嘯虎	钱啸虎
398	Chien, Jia-Ju	Qian, Jia-Ju	錢家駒	钱家驹
399	Ch'ien, Nan-Fên	Qian, Nan-Fen	錢南芬	钱南芬
397	Chien, Sung-Shu	Qian, Chong-Shu	錢崇澍	钱崇澍
397	Ch'ien, Sung-Shu	Qian, Chong-Shu	錢崇澍	钱崇澍
239	Chin, Chên-Chou	Jin, Zhen-Zhou	金振州	金振州
402	Chin, Hui-Chen	Qin, Hui-Zhen	秦慧貞	秦慧贞
402	Chin, Hui-Zhen	Qin, Hui-Zhen	秦慧貞	秦慧贞
403	Ch'in, Jên-Ch'ang	Qin, Ren-Chang	秦仁昌	秦仁昌
235	Chin, Tê-Hsiang	Jin, De-Xiang	金德祥	金德祥
236	Chin, To-Han	Jin, Duo-Han	金多翰	金多翰
234	Chin, Tsen-Li	Jin, Cun-Li	金存禮	金存礼
238	Chin, Yu-Xing	Jin, Yue-Xing	金岳杏	金岳杏
403	Ching, Ren-Chang	Qin, Ren-Chang	秦仁昌	秦仁昌
411	Chiou, Wen-Liang	Qiu, Wen-Liang	邱文良	邱文良
406	Chiu, Hua-Hsing	Qiu, Hua-Xing	丘華興	丘华兴
406	Ch'iu, Hua-Hsing	Qiu, Hua-Xing	丘華興	丘华兴
405	Ch'iu, Kuei-Yüan	Qiu, Gui-Yuan	裘桂元	裘桂元
407	Chiu, Lien-Ching	Qiu, Lian-Qing	邱蓮卿	邱莲卿
407	Ch'iu, Lien-Ch'ing	Qiu, Lian-Qing	邱蓮卿	邱莲卿
408	Chiu, Nien-Yung	Qiu, Nian-Yong	邱年永	邱年永
404	Chiu, Pao-Lin	Qiu, Bao-Lin	裘寶林	裘宝林
404	Chiu, Pao-Ling	Qiu, Bao-Lin	裘寶林	裘宝林
409	Chiu, Pei-Hsi	Qiu, Pei-Xi	裘佩熹	裘佩熹
409	Chiu, Pei-Shi	Qiu, Pei-Xi	裘佩熹	裘佩熹
409	Chiu, Pei-Xi	Qiu, Pei-Xi	裘佩熹	裘佩熹
401	Chiu, Shau-Ting	Qiu, Shao-Ting	邱少婷	邱少婷
765	Cho, Jên-Sung	Zhuo, Ren-Song	卓仁松	卓仁松
13	Choi, P. S.	Cai, Pan-Sheng	蔡盤生	蔡盘生
739	Chou, Chên	Zhou, Zhen	周　楨	周　桢
740	Chou, Chen-Ying	Zhou, Zhen-Ying	周貞英	周贞英
740	Chou, Chên-Ying	Zhou, Zhen-Ying	周貞英	周贞英
728	Chou, Chien-Jen	Zhou, Jian-Ren	周建人	周建人

730	Chou, Chi-Lun	Zhou, Ji-Lun	周紀倫	周纪伦
722	Chou, Chung-Kuang	Zhou, Chong-Guang	周重光	周重光
743	Chou, Chu-Ying	Zhou, Zu-Ying	周祖英	周祖英
725	Chou, Han-Fan	Zhou, Han-Fan	周漢藩	周汉藩
734	Chou, Hsu-Yuan	Zhou, Xu-Yuan	周蓄源	周蓄源
734	Chou, Hsü-Yüan	Zhou, Xu-Yuan	周蓄源	周蓄源
735	Chou, I-Liang	Zhou, Yi-Liang	周以良	周以良
724	Chou, Kuang-Yu	Zhou, Guang-Yu	周光裕	周光裕
724	Chou, Kuang-Yü	Zhou, Guang-Yu	周光裕	周光裕
723	Chou, Kuang-Yung	Zhou, Guang-Rong	周光榮	周光荣
731	Chou, T'ai-Hsüan	Zhou, Tai-Xuan	周太玄	周太玄
732	Chou, T'ai-Yen	Zhou, Tai-Yan	周太炎	周太炎
732	Chou, Tai-Yen	Zhou, Tai-Yan	周太炎	周太炎
722	Chou, Ts'ung-Huang	Zhou, Chong-Guang	周重光	周重光
742	Chou, Tsung-Huang	Zhou, Zong-Huang	周宗璜	周宗璜
743	Chou, Tsu-Ying	Zhou, Zu-Ying	周祖英	周祖英
735	Chou, Yi-Liang	Zhou, Yi-Liang	周以良	周以良
736	Chou, Yin	Zhou, Yin	周 崟	周 崟
737	Chou, Ying	Zhou, Ying	周 瑛	周 瑛
738	Chou, Ying-Ch'ang	Zhou, Ying-Chang	周映昌	周映昌
735	Chou, Yu-Liang	Zhou, Yi-Liang	周以良	周以良
739	Chow, Cheng	Zhou, Zhen	周 鎮	周 镇
729	Chow, Chia-Chih	Zhou, Jia-Qi	周家琪	周家琪
741	Chow, Chung-Hsuan	Zhou, Zhong-Xuan	周鐘瑄	周钟瑄
722	Chow, Chung-Huang	Zhou, Chong-Guang	周重光	周重光
742	Chow, Chung-Hwang	Zhou, Zong-Huang	周宗璜	周宗璜
722	Chow, Chung-Kuang	Zhou, Chong-Guang	周重光	周重光
725	Chow, Hang-Fan	Zhou, Han-Fan	周漢藩	周汉藩
726	Chow, Ho-Chang	Zhou, He-Chang	周鶴昌	周鹤昌
727	Chow, Hong-Fu	Zhou, Hong-Fu	周洪富	周洪富
727	Chow, Hung-Fu	Zhou, Hong-Fu	周洪富	周洪富
723	Chow, Kuang-Yung	Zhou, Guang-Rong	周光榮	周光荣
733	Chow, Shuan	Zhou, Xuan	周 鉉	周 铉
742	Chow, Tsung-Huang	Zhou, Zong-Huang	周宗璜	周宗璜
13	Choy, Pun-Sang	Cai, Pan-Sheng	蔡盤生	蔡盘生
758	Chu, Chao-I	Zhu, Zhao-Yi	朱兆儀	朱兆仪
758	Chu, Chao-Yi	Zhu, Zhao-Yi	朱兆儀	朱兆仪
759	Chu, Cheng-De	Zhu, Zheng-De	朱政德	朱政德
748	Chu, Chia-Nan	Zhu, Jia-Ran	朱家柟	朱家柟
749	Chu, Chi-Fan	Zhu, Ji-Fan	朱濟凡	朱济凡
760	Chu, Chung-Han	Zhu, Zhong-Han	朱中翰	朱中翰

414	Chu, Chung-Hsiang	Qu, Zhong-Xiang	曲仲湘	曲仲湘
414	Ch'ü, Chung-Hsiang	Qu, Zhong-Xiang	曲仲湘	曲仲湘
744	Chu, Ge-Lin	Zhu, Ge-Lin	朱格麟	朱格麟
744	Chu, Ge-Ling	Zhu, Ge-Lin	朱格麟	朱格麟
745	Chu, Hao-Jan	Zhu, Hao-Ran	朱浩然	朱浩然
755	Chu, Hsiang-San	Zhu, Xiang-San	朱象三	朱象三
746	Chu, Hua	Zhu, Hua	朱　華	朱　华
747	Chu, Huei-Fang	Zhu, Hui-Fang	朱會芳	朱会芳
412	Chu, Kuei-Ling	Qu, Gui-Ling	曲桂齡	曲桂龄
412	Ch'ü, Kuei-Ling	Qu, Gui-Ling	曲桂齡	曲桂龄
750	Chu, Mao-Shun	Zhu, Mao-Shun	朱懋順	朱懋顺
413	Chu, Shin-Tseng	Qu, Shi-Zeng	曲式曾	曲式曾
751	Chu, Shu-P'ing	Zhu, Shu-Ping	朱樹屏	朱树屏
752	Chu, T'ing-Ch'êng	Zhu, Ting-Cheng	祝廷成	祝廷成
753	Chu, Wan-Chia	Zhu, Wan-Jia	朱婉嘉	朱婉嘉
754	Chu, Wei-Ming	Zhu, Wei-Ming	朱維明	朱维明
756	Chu, Yen-Ch'êng	Zhu, Yan-Cheng	朱彦丞	朱彦丞
757	Chu, You-Chang	Zhu, You-Chang	朱有昌	朱有昌
761	Chu, Zong-Yuan	Zhu, Zong-Yuan	朱宗元	朱宗元
763	Chuang, Ching-Chang	Zhuang, Qing-Zhang	莊清漳	庄清漳
764	Chuang, Hsuan	Zhuang, Xuan	莊　璇	庄　璇
762	Chuang, Tsan-Iang	Zhuang, Can-Yang	莊燦暘	庄灿旸
762	Chuang, Tsan-Yang	Zhuang, Can-Yang	莊燦暘	庄灿旸
762	Chuang, Ts'an-Yang	Zhuang, Can-Yang	莊燦暘	庄灿旸
46	Chun, Ren-Jun	Chen, Ren-Jun	陳仁鈞	陈仁钧
55	Chun, Shu-Chên	Chen, Shu-Zhen	陳淑珍	陈淑珍
35	Chun, Woon-Young	Chen, Huan-Yong	陳煥鏞	陈焕镛
35	Chun, Woon-Yung	Chen, Huan-Yong	陳煥鏞	陈焕镛
717	Chung, Chi-Hsing	Zhong, Ji-Xin	鐘濟新	钟济新
714	Chung, Chung-Hsin	Zhong, Chong-Xin	仲崇信	仲崇信
720	Chung, Hsin-Hsuan	Zhong, Xin-Xuan	鐘心煊	钟心煊
720	Chung, Hsin-Hsüan	Zhong, Xin-Xuan	鐘心煊	钟心煊
721	Chung, I	Zhong, Yi	鐘　義	钟　义
715	Chung, Kuan-Kuang	Zhong, Guan-Guang	鐘觀光	钟观光
716	Chung, Kuo-Fang	Zhong, Guo-Fang	鐘國芳	钟国芳
718	Chung, Nian-June	Zhong, Nian-Jun	鐘年鈞	钟年钧
712	Chung, Pu-Chin	Zhong, Bu-Qin	鐘補勤	钟补勤
712	Chung, Pu-Ch'in	Zhong, Bu-Qin	鐘補勤	钟补勤
713	Chung, Pu-Ch'iu	Zhong, Bu-Qiu	鐘補求	钟补求
719	Chung, Shih-Wen	Zhong, Shi-Wen	鐘詩文	钟诗文
678	Chung, Yao-Tsung	Zhang, Yao-Zong	張耀宗	张耀宗

82	Deng, Mei-Jih	Deng, Mei-Zhi	鄧美智	邓美智
98	Duanmu, Shing	Duanmu, Xin	端木炘	端木炘
100	Fan, Ch'ing-Sêng	Fan, Qing-Sheng	樊慶生	樊庆生
101	Fan, Ch'ing-Shêng	Fan, Qing-Sheng	樊慶生	樊庆生
99	Fan, Kung-Chü	Fan, Gong-Ju	樊恭炬	樊恭炬
107	Fang, Cheng-Fu	Fang, Zhen-Fu	方振富	方振富
104	Fang, Hsi-Ch'ên	Fang, Xi-Chen	方錫琛	方锡琛
102	Fang, Rhui-Cheng	Fang, Rui-Zheng	方瑞徵	方瑞征
104	Fang, Si-Tzen	Fang, Xi-Chen	方錫琛	方锡琛
103	Fang, Wên-P'ei	Fang, Wen-Pei	方文培	方文培
106	Fang, Yong-Sing	Fang, Yong-Xin	方永鑫	方永鑫
105	Fang, Yung-Huai	Fang, Yong-Huai	方永槐	方永槐
108	Feng, Kuo-Mei	Feng, Guo-Mei	馮國楣	冯国楣
108	Fêng, Kuo-Mei	Feng, Guo-Mei	馮國楣	冯国楣
111	Fêng, Tsê-Fang	Feng, Ze-Fang	馮澤芳	冯泽芳
112	Fêng, Tse-Fêng	Feng, Ze-Feng	豐澤豐	丰泽丰
110	Fêng, Yao-Tsung	Feng, Yao-Zong	馮耀宗	冯耀宗
114	Feung, Chao-Shieung	Feng, Zhao-Xun	馮朝勛	冯朝勛
118	Fu, Hiang-Chian	Fu, Xiang-Qian	富象乾	富象乾
115	Fu, Kun-Tsun	Fu, Kun-Jun	傅坤俊	傅坤俊
116	Fu, Li-Kuo	Fu, Li-Guo	傅立國	傅立国
117	Fu, Shu-Hsia	Fu, Shu-Xia	傅書遐	傅书遐
119	Fu, Yu-Chin	Fu, Yu-Qin	符毓秦	符毓秦
113	Fung, Chao-Nan	Feng, Zhao-Nan	馮肇南	冯肇南
109	Fung, Hok-Lam	Feng, Xue-Lin	馮學琳	冯学琳
124	Gao, Chien	Gao, Qian	高　謙	高　谦
517	Go, Kei-Shi	Wu, Ji-Zhi	吳繼志	吴继志
517	Go, Shi-Zen	Wu, Ji-Zhi	吳繼志	吴继志
154	Han, Shu-Chih	Han, Shu-Jin	韓樹金	韩树金
155	Hang, King-Hsing	Hang, Jin-Xin	杭金欣	杭金欣
156	Hao, Ching-Sheng	Hao, Jing-Sheng	郝景盛	郝景盛
156	Hao, Kin-Shen	Hao, Jing-Sheng	郝景盛	郝景盛
214	Hao, Shu-Hwa	Huo, Shu-Hua	火樹華	火树华
169	Ho, Chên-Hsi	He, Zhen-Xi	賀貞熙	贺贞熙
160	Ho, Chin	He, Jing	何　景	何　景
160	Ho, Ching	He, Jing	何　景	何　景
156	Ho, Ching-Shêng	Hao, Jing-Sheng	郝景盛	郝景盛
170	Ho, Chu	He, Zhu	何　鑄	何　铸
157	Ho, Chun-Nien	He, Chun-Nian	何椿年	何椿年
157	Ho, Ch'un-Nien	He, Chun-Nian	何椿年	何椿年
158	Ho, Feng-Chi	He, Feng-Ji	何豐吉	何丰吉

159	Ho, Feng-Jen	He, Feng-Ren	何鳳仁	何凤仁
167	Ho, Hsien-Chang	He, Xian-Zhang	何憲章	何宪章
161	Ho, Hsün-Fêng	He, Jun-Feng	賀峻峯	贺峻峰
162	Hô, K'ai	He, Kai	賀 闓	贺 闿
162	Ho, Kan	He, Kai	賀 闓	贺 闿
163	Ho, Lei-Wei	He, Nai-Wei	何乃維	何乃维
163	Ho, Nai-Wei	He, Nai-Wei	何乃維	何乃维
164	Ho, Shih-Yuen	He, Shi-Yuan	賀士元	贺士元
165	Ho, T'ien-Hsiang	He, Tian-Xiang	何天相	何天相
165	Ho, Tien-Hsiang	He, Tian-Xiang	何天相	何天相
166	Ho, Ting-Nung	He, Ting-Nong	何廷農	何廷农
168	Ho, Ye-Chi	He, Ye-Qi	何業琪	何业琪
161	Ho, Tsun-Fung	He, Jun-Feng	賀峻峯	贺峻峰
167	Hoh, Hin-Cheung	He, Xian-Zhang	何憲章	何宪章
160	Hoo, Gin	He, Jing	何 景	何 景
173	Hou, Hui-Tsung	Hou, Hui-Zong	侯惠宗	侯惠宗
174	Hou, K'uan-Chao	Hou, Kuan-Zhao	侯寬昭	侯宽昭
172	Hou, Ting	Hou, Ding	侯 定	侯 定
168	Hou, Yeh-Chi	He, Ye-Qi	何業琪	何业琪
174	How, Foon-Chew	Hou, Kuan-Zhao	侯寬昭	侯宽昭
535	Hsia, Kuang-Cheng	Xia, Guang-Cheng	夏光成	夏光成
536	Hsia, Wei-Kun	Xia, Wei-Kun	夏緯琨	夏纬琨
536	Hsia, Wei-K'un	Xia, Wei-Kun	夏緯琨	夏纬琨
537	Hsia, Wei-Ying	Xia, Wei-Ying	夏緯瑛	夏纬瑛
538	Hsiang, Kung-Ch'uang	Xiang, Gong-Chuan	項公傳	项公传
540	Hsiao, Jin-Long	Xiao, Jin-Long	蕭錦隆	萧锦隆
543	Hsiao, Ju-Ying	Xiao, Ru-Ying	肖如英	肖如英
541	Hsiao, Pei-Ken	Xiao, Pei-Gen	肖培根	肖培根
544	Hsieh, Chang-Fu	Xie, Chang-Fu	謝長富	谢长富
546	Hsieh, Cheng-Chang	Xie, Cheng-Zhang	謝成章	谢成章
545	Hsieh, Chen-Ko	Xie, Cheng-Ke	謝成科	谢成科
547	Hsieh, Ming-Ke	Xie, Ming-Ke	謝鳴珂	谢鸣珂
551	Hsieh, Tsung-Hsin	Xie, Zong-Xin	謝宗欣	谢宗欣
550	Hsieh, Tsung-Wan	Xie, Zong-Wan	謝宗萬	谢宗万
549	Hsieh, Yin-Tang	Xie, Yin-Tang	謝寅堂	谢寅堂
552	Hsin, Shu-Ch'ih	Xin, Shu-Zhi	辛樹幟	辛树帜
554	Hsing, Hsi-Yung	Xing, Xi-Yong	邢錫永	邢锡永
555	Hsiung, Wen-Yue	Xiong, Wen-Yu	熊文愈	熊文愈
556	Hsiung, Yao-Kuo	Xiong, Yao-Guo	熊耀國	熊耀国
558	Hsü, Ch'êng-Wên	Xu, Cheng-Wen	許成文	许成文
561	Hsü, Chi	Xu, Ji	許 基	许 基

562	Hsu, Chien-Chang	Xu, Jian-Chang	許建昌	许建昌
562	Hsu, Chien-Ch'ang	Xu, Jian-Chang	許建昌	许建昌
562	Hsu, Chien-Chieng	Xu, Jian-Chang	許建昌	许建昌
559	Hsu, De-Xi	Xu, De-Xi	徐德喜	徐德喜
570	Hsu, Hsiang-Hao	Xu, Xiang-Hao	徐祥浩	徐祥浩
570	Hsü, Hsiang-Hao	Xu, Xiang-Hao	徐祥浩	徐祥浩
564	Hsu, Jen	Xu, Ren	徐　仁	徐　仁
564	Hsü, Jên	Xu, Ren	徐　仁	徐　仁
560	Hsu, Kuo-Shih	Xu, Guo-Shi	徐國士	徐国士
557	Hsu, Ping-Sheng	Xu, Bing-Sheng	徐炳聲	徐炳声
566	Hsu, Ting-Zhi	Xu, Ting-Zhi	徐廷志	徐廷志
575	Hsu, Tsai-Wen	Xu, Zai-Wen	許再文	许再文
567	Hsü, Wei-Tung	Xu, Wei-Dong	徐緯東	徐纬东
568	Hsu, Wei-Ying	Xu, Wei-Ying	徐偉英	徐伟英
569	Hsü, Wei-Ying	Xu, Wei-Ying	徐緯英	徐纬英
571	Hsu, Yang-Peng	Xu, Yang-Peng	徐養鵬	徐养鹏
571	Hsu, Yang-Pong	Xu, Yang-Peng	徐養鵬	徐养鹏
572	Hsü, Yen-Ch'ien	Xu, Yan-Qian	徐燕千	徐燕千
573	Hsu, Yin	Xu, Yin	徐　垠	徐　垠
565	Hsu, Yong-Zhang	Xu, Rong-Zhang	徐榮章	徐荣章
574	Hsü, Yun-Chun	Xu, Yong-Chun	徐永椿	徐永椿
574	Hsu, Yung-Chun	Xu, Yong-Chun	徐永椿	徐永椿
574	Hsü, Yung-Ch'un	Xu, Yong-Chun	徐永椿	徐永椿
577	Hsuan, Shwe-Jye	Xuan, Shu-Jie	宣淑潔	宣淑洁
577	Hsuan, Shwu-Jue	Xuan, Shu-Jie	宣淑潔	宣淑洁
570	Hsue, Hsiang-Hao	Xu, Xiang-Hao	徐祥浩	徐祥浩
570	Hsue, Hsiang-How	Xu, Xiang-Hao	徐祥浩	徐祥浩
578	Hsüeh, Ch'êng-Chien	Xue, Cheng-Jian	薛承健	薛承健
579	Hsueh, Chi-Ju	Xue, Ji-Ru	薛紀如	薛纪如
190	Hu, Chao-Hwa	Hu, Zhao-Hua	胡兆華	胡兆华
177	Hu, Chia-Chi	Hu, Jia-Qi	胡嘉琪	胡嘉琪
178	Hu, Chia-Ying	Hu, Jia-Ying	胡嘉穎	胡嘉颖
182	Hu, Chi-Ming	Hu, Qi-Ming	胡啓明	胡启明
179	Hu, Chin-Hwa	Hu, Jing-Hua	胡敬華	胡敬华
185	Hu, Hsen-Hsu	Hu, Xian-Su	胡先驌	胡先骕
186	Hu, Hsian-Yuan	Hu, Xian-Yuan	胡顯源	胡显源
187	Hu, Hsing-Tsung	Hu, Xing-Zong	胡興宗	胡兴宗
187	Hu, Hsin-Tsung	Hu, Xing-Zong	胡興宗	胡兴宗
188	Hu, Hsiu-Ying	Hu, Xiu-Ying	胡秀英	胡秀英
176	Hu, Kung-Tê	Hu, Gong-De	胡公德	胡公德
181	Hu, Lin-Cheng	Hu, Lin-Zhen	胡琳貞	胡琳贞

183	Hu, Shih-I	Hu, Shi-Yi	胡適宜	胡适宜
183	Hu, Shih-Yi	Hu, Shi-Yi	胡適宜	胡适宜
188	Hu, Shiu-Ying	Hu, Xiu-Ying	胡秀英	胡秀英
175	Hu, Ta-Wei	Hu, Da-Wei	胡大維	胡大维
180	Hu, Tsi-Sun	Hu, Ji-Sheng	胡濟生	胡济生
184	Hu, Wen-Kwang	Hu, Wen-Guang	胡文光	胡文光
184	Hu, Wên-Kwang	Hu, Wen-Guang	胡文光	胡文光
189	Hu, Yu-Shi	Hu, Yu-Xi	胡玉熹	胡玉熹
191	Hua, Ju-Cheng	Hua, Ru-Cheng	華汝成	华汝成
191	Hua, Ju-Ch'êng	Hua, Ru-Cheng	華汝成	华汝成
207	Huang, Ch'i-Wang	Huang, Xi-Wang	黃溪旺	黄溪旺
192	Huang, Ch'êng-Chiu	Huang, Cheng-Jiu	黃成就	黄成就
196	Huang, Chi-Chuang	Huang, Ji-Zhuang	黃季莊	黄季庄
192	Huang, Ching-Chieu	Huang, Cheng-Jiu	黃成就	黄成就
195	Huang, Fêng-Yüan	Huang, Feng-Yuan	黃逢源	黄逢源
197	Huang, Pu-Hwa	Huang, Pu-Hua	黃普華	黄普华
204	Huang, Sang-Gen	Huang, Song-Gen	黃松根	黄松根
205	Huang, Se-Zei	Huang, Xie-Cai	黃燮才	黄燮才
198	Huang, Shao-Hsü	Huang, Shao-Xu	黃紹緒	黄绍绪
206	Huang, Shing-Fan	Huang, Xing-Fan	黃星凡	黄星凡
199	Huang, Shong	Huang, Sheng	黃　生	黄　生
200	Huang, Shou-Hsien	Huang, Shou-Xian	黃守先	黄守先
202	Huang, Shu-Chung	Huang, Shu-Qiong	黃蜀瓊	黄蜀琼
204	Huang, Sung-K'ên	Huang, Song-Gen	黃松根	黄松根
194	Huang, Ta-Chang	Huang, Da-Zhang	黃達章	黄达章
211	Huang, Tseng-Chieng	Huang, Zeng-Quan	黃增泉	黄增泉
193	Huang, Ts'ui-Mei	Huang, Cui-Mei	黃翠梅	黄翠梅
208	Huang, Yong-Chin	Huang, Yong-Qin	黃咏琴	黄咏琴
209	Huang, Yu-Ju	Huang, You-Ru	黃友儒	黄友儒
210	Huang, Yun-Huei	Huang, Yun-Hui	黃雲輝	黄云辉
171	Hung, Chang-Hsun	Hong, Zhang-Xun	洪章訓	洪章训
171	Hung, Chang-Hsün	Hong, Zhang-Xun	洪章訓	洪章训
195	Hwang, Fung-Yuan	Huang, Feng-Yuan	黃逢源	黄逢源
201	Hwang, Shu-Mei	Huang, Shu-Mei	黃淑美	黄淑美
203	Hwang, Shu-Wei	Huang, Shu-Wei	黃淑煒	黄淑炜
213	Hwang, Tso-Chie	Huang, Zuo-Jie	黃作傑	黄作杰
193	Hwang, Ts'ui-Mei	Huang, Cui-Mei	黃翠梅	黄翠梅
209	Hwang, You-Ru	Huang, You-Ru	黃友儒	黄友儒
604	Ip, Nga-Kok	Ye, Ya-Ge	葉雅各	叶雅各
415	Jao, Chin-Chih	Rao, Qin-Zhi	饒欽止	饶钦止
415	Jao, Ch'in-Chih	Rao, Qin-Zhi	饒欽止	饶钦止

417	Jen, Hsien-Wei	Ren, Xian-Wei	任憲威	任宪威
416	Jên, Wei	Ren, Wei	任　瑋	任　玮
122	Kao, Ling	Gao, Ling	高　嶺	高　岭
123	Kao, Muh-Tsuen	Gao, Mu-Cun	高木村	高木村
120	Kao, Pao-Chung	Gao, Bao-Chun	高寶蒓	高宝莼
121	Kao, Teh-P'ei	Gao, De-Pei	高德培	高德培
121	Kao, Tê-P'ei	Gao, De-Pei	高德培	高德培
126	Kao, Tso-Ching	Gao, Zuo-Jing	高作經	高作经
125	Kao, Yün-Chang	Gao, Yun-Zhang	高蘊璋	高蕴璋
131	Kêng, Ch'ing-Han	Geng, Qing-Han	耿慶漢	耿庆汉
132	Keng, Hsiuan	Geng, Xuan	耿　煊	耿　煊
132	Keng, Hsuan	Geng, Xuan	耿　煊	耿　煊
132	Keng, Hsüan	Geng, Xuan	耿　煊	耿　煊
133	Kêng, I-Li	Geng, Yi-Li	耿以禮	耿以礼
130	Kêng, Kwan-Hou	Geng, Kuan-Hou	耿寬厚	耿宽厚
129	Keng, Pai-Chieh	Geng, Bai-Jie	耿伯介	耿伯介
129	Kêng, Pai-Chieh	Geng, Bai-Jie	耿伯介	耿伯介
134	Kêng, Tso-Lin	Geng, Zuo-Lin	耿作霖	耿作霖
133	Keng, Yi-Li	Geng, Yi-Li	耿以禮	耿以礼
135	KengLing, Ro-Siu	GengLin, Ruo-Xiu	耿林若琇	耿林若琇
237	King, Wei-Jean	Jin, Wei-Jian	金維堅	金维坚
406	Kiu, Hua-Sing	Qiu, Hua-Xing	丘華興	丘华兴
406	Kiu, Hua-Xing	Qiu, Hua-Xing	丘華興	丘华兴
124	Ko, Chien	Gao, Qian	高　謙	高　谦
128	Ko, Ming-Yü	Ge, Ming-Yu	葛明裕	葛明裕
127	Ko, Ting-Pang	Ge, Ding-Bang	戈定邦	戈定邦
125	Ko, Wan-Chang	Gao, Yun-Zhang	高蘊璋	高蕴璋
125	Ko, Wan-Cheung	Gao, Yun-Zhang	高蘊璋	高蕴璋
139	Ku, Ch'ien-Chi	Gu, Qian-Ji	顧謙吉	顾谦吉
138	Ku, Kuan	Gu, Guan	古　灌	古　灌
137	Ku, Tsue-Chih	Gu, Cui-Zhi	谷粹芝	谷粹芝
142	Kuan, Chung-Tian	Guan, Zhong-Tian	管中天	管中天
140	Kuan, Ke-Chien	Guan, Ke-Jian	關克儉	关克俭
140	Kuan, K'o-Chien	Guan, Ke-Jian	關克儉	关克俭
141	Kuan, Yü-Ying	Guan, Yu-Ying	關毓英	关毓英
244	K'uang, K'o-Jên	Kuang, Ke-Ren	匡可任	匡可任
244	Kuang, Ko-Zen	Kuang, Ke-Ren	匡可任	匡可任
240	Kung, Ch'ing-Lai	Kong, Qing-Lai	孔慶萊	孔庆莱
240	K'ung, Ch'ing-Lai	Kong, Qing-Lai	孔慶萊	孔庆莱
136	Kung, Ch'i-Wu	Gong, Qi-Wu	龔啓鋈	龚启鋈
243	Kung, Hsian-Shiu	Kong, Xian-Xu	孔憲需	孔宪需

242	Kung, Hsien-Wu	Kong, Xian-Wu	孔憲武	孔宪武
242	K'ung, Hsien-Wu	Kong, Xian-Wu	孔憲武	孔宪武
241	Kung, Shi-Pei	Kong, Shi-Pei	孔世培	孔世培
147	Kuo, Ch'ao-Shêng	Guo, Chao-Sheng	郭朝勝	郭朝胜
148	Kuo, Chen-Meng	Guo, Cheng-Meng	郭城孟	郭城孟
149	Kuo, Chien-Mao	Guo, Jian-Mao	過鑒楙	过鉴楙
150	Kuo, Chi-Fan	Guo, Ji-Fan	郭紀凡	郭纪凡
151	Kuo, Chin-Chen	Guo, Qiu-Cheng	郭秋成	郭秋成
144	Kuo, Pao-Chang	Guo, Bao-Zhang	郭寶章	郭宝章
145	Kuo, Pen-Chao	Guo, Ben-Zhao	郭本兆	郭本兆
145	Kuo, Pung-Chao	Guo, Ben-Zhao	郭本兆	郭本兆
152	Kuo, Shing-Rong	Guo, Xing-Rong	郭幸榮	郭幸荣
153	Kuo, Yü-Chieh	Guo, Yu-Jie	郭玉潔	郭玉洁
146	Kuoh, Chang-Sheng	Guo, Chang-Sheng	郭長生	郭长生
143	Kwei, Yao-Lin	Gui, Yao Lin	桂耀林	桂耀林
245	Lai, Ming-Jou	Lai, Ming-Zhou	賴明洲	赖明洲
246	Lang, Kai-Yung	Lang, Kai-Yong	郎楷永	郎楷永
338	Lau, Shau-Yan	Liu, Shou-Ren	劉守仁	刘守仁
346	Law, Yuh-Wu	Liu, Yu-Hu	劉玉壺	刘玉壶
250	Le, Ts'ae-Ch'i	Li, Cai-Qi	李彩祺	李彩祺
248	Lee, An-Jên	Li, An-Ren	李安仁	李安仁
253	Lee, Dan-Ching	Li, Dan-Qing	李淡清	李淡清
254	Lee, Der-Lin	Li, De-Lin	李德霖	李德霖
258	Lee, Kong-Ming	Li, Guang-Min	李廣民	李广民
256	Lee, Kong-Teh	Li, Gong-De	黎功德	黎功德
259	Lee, Kwok-Yan	Li, Guo-Ren	李國仁	李国仁
278	Lee, Shin-Chiang	Li, Xing-Jiang	黎興江	黎兴江
275	Lee, Shu-Hsien	Li, Shu-Xuan	李曙軒	李曙轩
273	Lee, Shu-Kang	Li, Shu-Gang	李樹剛	李树刚
274	Lee, Shun-Ching	Li, Shun-Qing	李順卿	李顺卿
274	Lee, Shun-Ch'ing	Li, Shun-Qing	李順卿	李顺卿
250	Lee, Ts'ai-Ch'i	Li, Cai-Qi	李彩祺	李彩祺
280	Lee, Yao-Yin	Li, Yao-Ying	李堯英	李尧英
268	Lei, P'aang-Fei	Li, Peng-Fei	李鵬飛	李鹏飞
287	Leong, Wai-Chao	Liang, Hui-Zhou	梁慧舟	梁慧舟
323	Leou, Chong-Sheng	Liu, Chong-Sheng	柳重勝	柳重胜
347	Leou, Tze-Ming	Liu, Zi-Ming	柳子明	柳子明
356	Leu, Wen-Pen	Lü, Wen-Bin	呂文賓	吕文宾
272	Ley, Shang-Hao	Li, Shang-Hao	黎尚豪	黎尚豪
248	Li, An-Jen	Li, An-Ren	李安仁	李安仁
248	Li, An-Jên	Li, An-Ren	李安仁	李安仁

251	Li, Chao-Luang	Li, Chao-Luan	李朝鑾	李朝銮
263	Li, Chia-Hsin	Li, Jia-Xin	李嘉馨	李嘉馨
261	Li, Chia-Wei	Li, Jia-Wei	李家維	李家维
262	Li, Chia-Wen	Li, Jia-Wen	李家文	李家文
269	Li, Chien	Li, Qian	李　乾	李　乾
270	Li, Ching-Tao	Li, Qing-Tao	李清濤	李清涛
270	Li, Ch'ing-T'ao	Li, Qing-Tao	李清濤	李清涛
264	Li, Chi-Tung	Li, Ji-Tong	李繼侗	李继侗
264	Li, Chi-T'ung	Li, Ji-Tong	李繼侗	李继侗
265	Li, Chi-Yun	Li, Ji-Yun	李冀雲	李冀云
260	Li, Han-Chang	Li, Han-Zhang	李含章	李含章
277	Li, Hsing-Hua	Li, Xing-Hua	李興華	李兴华
278	Li, Hsin-Kiang	Li, Xing-Jiang	黎興江	黎兴江
279	Li, Hsi-Wen	Li, Xi-Wen	李錫文	李锡文
257	Li, Ko-T'ang	Li, Gou-Tang	李構堂	李构堂
258	Li, Kuang-Ming	Li, Guang-Min	李廣民	李广民
256	Li, Kung-Tê	Li, Gong-De	黎功德	黎功德
256	Li, Kung-Teh	Li, Gong-De	黎功德	黎功德
266	Li, Liang-Chin	Li, Liang-Qing	李良慶	李良庆
266	Li, Liang-Ch'ing	Li, Liang-Qing	李良慶	李良庆
267	Li, Pei-Chun	Li, Pei-Qiong	李沛瓊	李沛琼
268	Li, P'êng-Fei	Li, Peng-Fei	李鵬飛	李鹏飞
249	Li, Ping-T'ao	Li, Bing-Tao	李秉滔	李秉滔
272	Li, Shang-Hao	Li, Shang-Hao	黎尚豪	黎尚豪
271	Li, Shang-Kuei	Li, Shang-Gui	李尚癸	李尚癸
278	Li, Shin-Chiang	Li, Xing-Jiang	黎興江	黎兴江
275	Li, Shu-Hsien	Li, Shu-Xuan	李曙軒	李曙轩
273	Li, Shu-Kang	Li, Shu-Gang	李樹剛	李树刚
274	Li, Shun-Ch'ing	Li, Shun-Qing	李順卿	李顺卿
276	Li, Sung-Po	Li, Song-Bo	李松柏	李松柏
252	Li, Tai-Fang	Li, Dai-Fang	李代芳	李代芳
254	Li, Tê-Lin	Li, De-Lin	李德霖	李德霖
282	Li, Tsung-Tao	Li, Zong-Dao	李宗道	李宗道
283	Li, Tsung-Ying	Li, Zong-Ying	李宗英	李宗英
255	Li, Tung-Sheng	Li, Dong-Sheng	李東生	李东生
255	Li, Tung-Shêng	Li, Dong-Sheng	李東生	李东生
281	Li, Yin-Kung	Li, Yin-Gong	李寅恭	李寅恭
285	Lian, Yung-Shan	Lian, Yong-Shan	廉永善	廉永善
291	Liang, Hsi	Liang, Xi	梁　希	梁　希
288	Liang, Kuei	Liang, Kui	梁　葵	梁　葵
288	Liang, K'uei	Liang, Kui	梁　葵	梁　葵

286	Liang, Pao-Han	Liang, Bao-Han	梁寶漢	梁宝汉
289	Liang, Sheng-Yeh	Liang, Sheng-Ye	梁盛業	梁盛业
290	Liang, T'ai-Jan	Liang, Tai-Ran	梁泰然	梁泰然
292	Liao, Jih-Ching	Liao, Ri-Jing	廖日京	廖日京
284	Lien, Wen-Yen	Lian, Wen-Yan	連文琰	连文琰
328	Liew, Fah-Seong	Liu, Hua-Xiang	劉華祥	刘华祥
293	Lin, Ai-Chin	Lin, Ai-Jin	林艾進	林艾进
301	Lin, Ching	Lin, Qing	林 青	林 青
301	Lin, Ch'ing	Lin, Qing	林 青	林 青
296	Lin, Ch'ung-Chih	Lin, Chong-Zhi	林崇智	林崇智
318	Lin, Chung-Kang	Lin, Zhong-Gang	林仲剛	林仲刚
297	Lin, Han-Shi	Lin, Han-Xi	林涵錫	林涵锡
298	Lin, Hou-Hsüan	Lin, Hou-Xuan	林厚萱	林厚萱
312	Lin, Hsieh	Lin, Xie	林 協	林 协
313	Lin, Hsin-Tzu	Lin, Xin-Zi	林信子	林信子
314	Lin, Hsiung-Hsiang	Lin, Xiong-Xiang	林熊祥	林熊祥
304	Lin, Jo-Hsiu	Lin, Ruo-Xiu	林若琇	林若琇
303	Lin, Jung	Lin, Rong	林 鎔	林 镕
305	Lin, Ju-Yao	Lin, Ru-Yao	林汝瑤	林汝瑶
300	Lin, Liang-Tung	Lin, Liang-Dong	林亮東	林亮东
294	Lin, Pang-Jun	Lin, Bang-Juan	林邦娟	林邦娟
294	Lin, Pan-Juan	Lin, Bang-Juan	林邦娟	林邦娟
295	Lin, Pao-Shu	Lin, Bao-Shu	林寶樹	林宝树
304	Lin, Ro-Siu	Lin, Ruo-Xiu	林若琇	林若琇
306	Lin, Shan-Hsiung	Lin, Shan-Xiong	林善雄	林善雄
307	Lin, Shih-Yung	Lin, Shi-Rong	林仕榕	林仕榕
308	Lin, Shu-Chien	Lin, Shu-Qian	林叔芊	林叔芊
316	Lin, Tsan-Piao	Lin, Zan-Biao	林讚標	林赞标
296	Lin, Tsung-Chih	Lin, Chong-Zhi	林崇智	林崇智
317	Lin, Tzer-Tong	Lin, Ze-Tong	林則桐	林则桐
309	Lin, Wan-T'ao	Lin, Wan-Tao	林萬濤	林万涛
310	Lin, Wei-Ch'ih	Lin, Wei-Zhi	林維治	林维治
311	Lin, Wei-Chih	Lin, Wei-Zhi	林維治	林维治
302	Ling, Chuan	Lin, Quan	林 泉	林 泉
302	Ling, Chüan	Lin, Quan	林 泉	林 泉
312	Ling, Hsieh	Lin, Xie	林 協	林 协
299	Ling, Lai-Kuan	Lin, Lai-Guan	林來官	林来官
315	Ling, Yeou-Ruenn	Lin, You-Run	林有潤	林有润
303	Ling, Yong	Lin, Rong	林 鎔	林 镕
319	Ling, Yuan-Chi	Ling, Yuan-Jie	凌元潔	凌元洁
319	Ling, Yuan-Jue	Ling, Yuan-Jie	凌元潔	凌元洁

303	Ling, Yung	Lin, Rong	林　鎔	林　镕
327	Liou, Ho	Liu, Hou	劉　厚	刘　厚
333	Liou, Liang	Liu, Liang	劉　亮	刘　亮
334	Liou, Ming-Yuan	Liu, Ming-Yuan	劉鳴遠	刘鸣远
334	Liou, Min-Yuan	Liu, Ming-Yuan	劉鳴遠	刘鸣远
341	Liou, Sham-Che	Liu, Xian-Qi	劉獻奇	刘献奇
337	Liou, Sheo-Lu	Liu, Shou-Lu	劉守爐	刘守炉
337	Liou, Shou-Lu	Liu, Shou-Lu	劉守爐	刘守炉
336	Liou, Tchen-Ngo	Liu, Shen-E	劉慎諤	刘慎谔
344	Liou, Ying-Sin	Liu, Ying-Xin	劉媖心	刘媖心
344	Liou, Ying-Xin	Liu, Ying-Xin	劉媖心	刘媖心
322	Liu, Ch'ên	Liu, Chen	劉　晨	刘　晨
330	Liu, Ching-Fên	Liu, Jing-Fen	劉經芬	刘经芬
331	Liu, Chin-Hui	Liu, Jin-Hui	劉錦惠	刘锦惠
324	Liu, Fang-Hsün	Liu, Fang-Xun	劉昉勛	刘昉勋
327	Liu, Ho	Liu, Hou	劉　厚	刘　厚
325	Liu, Ho-Yih	Liu, He-Yi	劉和義	刘和义
342	Liu, Hsi-Shin	Liu, Xi-Jin	劉錫進	刘锡进
342	Liu, Hsi-Tsing	Liu, Xi-Jin	劉錫進	刘锡进
326	Liu, Hung-Eh	Liu, Hong-E	劉洪諤	刘洪谔
335	Liu, Ju-Ch'iang	Liu, Ru-Qiang	劉汝強	刘汝强
329	Liu, Ki-Mon	Liu, Ji-Meng	劉繼孟	刘继孟
332	Liu, Lee	Liu, Li	劉　利	刘　利
320	Liu, Pao-Shan	Liu, Bao-Shan	劉寶山	刘宝山
321	Liu, Po	Liu, Bo	劉　波	刘　波
336	Liu, Shên-O	Liu, Shen-E	劉慎諤	刘慎谔
338	Liu, Shou-Jên	Liu, Shou-Ren	劉守仁	刘守仁
339	Liu, T'ang-Shui	Liu, Tang-Rui	劉棠瑞	刘棠瑞
340	Liu, T'ung-Shêng	Liu, Tong-Sheng	劉同生	刘同生
339	Liu, T'ang-Jui	Liu, Tang-Rui	劉棠瑞	刘棠瑞
339	Liu, Tang-Shui	Liu, Tang-Rui	劉棠瑞	刘棠瑞
347	Liu, Tsu-Ming	Liu, Zi-Ming	柳子明	柳子明
339	Liu, Tung-Shui	Liu, Tang-Rui	劉棠瑞	刘棠瑞
343	Liu, Yeh-Ching	Liu, Ye-Jing	劉業經	刘业经
344	Liu, Ying-Hsin	Liu, Ying-Xin	劉媖心	刘媖心
346	Liu, Yü-Hu	Liu, Yu-Hu	劉玉壺	刘玉壶
345	Liu, Yu-Tung	Liu, You-Dong	劉有棟	刘有栋
359	Lo, Hang-Chiang	Luo, Han-Qiang	羅漢強	罗汉强
362	Lo, Hsien-Shui	Luo, Xian-Rui	羅獻瑞	罗献瑞
361	Lo, Shih-Chiung	Luo, Shi-Jiong	羅士炯	罗士炯
247	Lo, T'ien-Yü	Le, Tian-Yu	樂天宇	乐天宇

348	Lou, Chih-Chen	Lou, Zhi-Cen	樓之岑	楼之岑
348	Lou, Chih-Ch'ên	Lou, Zhi-Cen	樓之岑	楼之岑
358	Lu, An-Ming	Lu, An-Min	路安民	路安民
349	Lu, Chang-Tze	Lü, Chang-Ze	吕長澤	吕长泽
357	Lu, Chi-Hua	Lu, Zhi-Hua	陸志華	陆志华
351	Lü, Chin-Lo	Lü, Jin-Luo	吕金蘿	吕金萝
352	Lu, Chin-Yi	Lu, Jin-Yi	陸錦一	陆锦一
350	Lu, Fu-Yuen	Lü, Fu-Yuan	吕福源	吕福源
353	Lu, K'ai-Yün	Lu, Kai-Yun	盧開運	卢开运
354	Lu, Ling-Ti	Lu, Ling-Di	陸玲娣	陆玲娣
355	Lu, Sheng-Yu	Lü, Sheng-You	吕勝由	吕胜由
362	Luo, Hsien-Shui	Luo, Xian-Rui	羅獻瑞	罗献瑞
360	Luo, Jian-Shing	Luo, Jian-Xin	羅健馨	罗健馨
363	Ma, Cheng-Gung	Ma, Cheng-Gong	馬成功	马成功
365	Ma, Chi	Ma, Ji	馬　驥	马　骥
366	Ma, Chi-Yun	Ma, Qi-Yun	馬其雲	马其云
364	Ma, Ta-P'u	Ma, Da-Pu	馬大浦	马大浦
367	Ma, Yong-Kui	Ma, Yong-Gui	馬永貴	马永贵
368	Ma, Yu-Chuan	Ma, Yu-Quan	馬毓泉	马毓泉
368	Ma, Yü-Ch'üan	Ma, Yu-Quan	馬毓泉	马毓泉
373	Mao, Chung-Liang	Mao, Zong-Liang	毛宗良	毛宗良
371	Mao, Hsiu-Sheng	Mao, Xiu-Sheng	茅秀生	茅秀生
369	Mao, Hua-Hsün	Mao, Hua-Xun	毛華訓	毛华训
369	Mao, Hwa-Shing	Mao, Hua-Xun	毛華訓	毛华训
370	Mao, Ping-I	Mao, Pin-Yi	毛品一	毛品一
370	Mao, Pin-I	Mao, Pin-Yi	毛品一	毛品一
373	Mao, Ts'ung-Liang	Mao, Zong-Liang	毛宗良	毛宗良
373	Mao, Tsung-Liang	Mao, Zong-Liang	毛宗良	毛宗良
372	Mao, Yung	Mao, Yong	毛　雍	毛　雍
374	Men, Fan-Son	Meng, Fan-Song	孟繁松	孟繁松
375	Meng, Jen-Hsien	Meng, Ren-Xian	蒙仁憲	蒙仁宪
376	Ming, Tien-Lu	Min, Tian-Lu	閔天祿	闵天禄
377	Mo, Sin-Li	Mo, Xin-Li	莫新禮	莫新礼
378	Moo, Chuen-Tau	Mu, Chun-Tao	穆椿濤	穆椿涛
379	Moo, Chuen-Tau	Mu, Zong-Tao	穆宗濤	穆宗涛
380	Ni, Chi-Cheng	Ni, Zhi-Cheng	倪志誠	倪志诚
381	Nie, Shou-Chuan	Nie, Shao-Quan	聶紹荃	聂绍荃
384	Ou, Shih-Huang	Ou, Shi-Huang	歐世璜	欧世璜
473	Ouang, Teng-Ying	Wang, Cheng-Yin	汪呈因	汪呈因
5	Pai, Pei-Yu	Bai, Pei-Yu	白佩瑜	白佩瑜
4	Pai, Ts'ai	Bai, Cai	白　采	白　采

6	Pai, Yin-Yuan	Bai, Yin-Yuan	白蔭元	白荫元
6	Pai, Yin-Yüan	Bai, Yin-Yuan	白蔭元	白荫元
385	Pan, Kwo-Ying	Pan, Guo-Ying	潘國瑛	潘国瑛
386	P'ang, Hsin-Min	Pang, Xin-Min	龐新民	庞新民
387	Pei, Chien	Pei, Jian	裴　鑒	裴　鉴
387	P'ei, Chien	Pei, Jian	裴　鑒	裴　鉴
392	Pen, Tse-Hsiang	Peng, Ze-Xiang	彭澤祥	彭泽祥
389	Peng, Ching-I	Peng, Jing-Yi	彭鏡毅	彭镜毅
390	Peng, Jen-Jye	Peng, Ren-Jie	彭仁傑	彭仁杰
388	P'êng, Kuang-Ch'in	Peng, Guang-Qin	彭光欽	彭光钦
388	P'êng, Kwang-Ch'ing	Peng, Guang-Qin	彭光欽	彭光钦
391	P'êng, Shih-Fang	Peng, Shi-Fang	彭世芳	彭世芳
393	P'êng, Tso-Ch'üan	Peng, Zuo-Quan	彭佐權	彭佐权
9	Pi, Chung-Pen	Bi, Zhong-Ben	畢中本	毕中本
10	Pi, T. K.	Bi, Zu-Gao	畢祖高	毕祖高
394	Pu, Fa-Ting	Pu, Fa-Ding	溥發鼎	溥发鼎
11	Pu, Yu-Chih	Bu, Yu-Zhi	步毓芝	步毓芝
419	Shan, Jên-Hua	Shan, Ren-Hua	單人驊	单人骅
419	Shan, Ren-Hwa	Shan, Ren-Hua	單人驊	单人骅
539	Shang, Chih-Bei	Xiang, Qi-Bai	向其柏	向其柏
420	Shao, Yao-Nien	Shao, Yao-Nian	邵堯年	邵尧年
421	Sheh, Meng-Lan	She, Meng-Lan	佘孟蘭	佘孟兰
426	Shen, Chung-Fu	Shen, Zhong-Fu	沈中桴	沈中桴
422	Shen, Kuan-Mien	Shen, Guan-Mian	沈觀冕	沈观冕
424	Shen, Lian-Te	Shen, Lian-De	沈聯德	沈联德
423	Shên, Tsun	Shen, Jun	沈　雋	沈　隽
425	Shen, Yü-Fêng	Shen, Yu-Feng	沈毓鳳	沈毓凤
427	Sheng, Cheng-Kuei	Sheng, Cheng-Gui	盛誠桂	盛诚桂
427	Shêng, Ch'êng-Kuei	Sheng, Cheng-Gui	盛誠桂	盛诚桂
432	Shi, Hak-Shu	Shi, Xue-Xi	施學習	施学习
542	Shiao, Yung	Xiao, Rong	肖　溶	肖　溶
548	Shieh, Wang-Chueng	Xie, Wan-Quan	謝萬權	谢万权
428	Shih, Bing-Ling	Shi, Bing-Lin	施炳霖	施炳霖
429	Shih, Chiu-Chuang	Shi, Jiu-Zhuang	史久莊	史久庄
433	Shih, Chu	Shi, Zhu	石　鑄	石　铸
431	Shih, Hsin-Hua	Shi, Xing-Hua	施興華	施兴华
432	Shih, Hsüeh-Hsi	Shi, Xue-Xi	施學習	施学习
434	Shih, Tzu-Hsing	Shi, Zi-Xing	石子興	石子兴
430	Shih, Wen-Liang	Shi, Wen-Liang	施文良	施文良
553	Shing, Gung-Hsia	Xing, Gong-Xia	邢公俠	邢公侠
553	Shing, Kung-Hsia	Xing, Gong-Xia	邢公俠	邢公侠

420	Shiu, Iu-Nin	Shao, Yao-Nian	邵堯年	邵尧年
562	Shü, Chien-Tsang	Xu, Jian-Chang	許建昌	许建昌
563	Shu, Lon-Rhan	Xu, Lang-Ran	徐朗然	徐朗然
435	Shu, Tsui-Hsiang	Shu, Zui-Xiang	舒醉香	舒醉香
552	Sin, S. S.	Xin, Shu-Zhi	辛樹幟	辛树帜
441	So, May-Ling	Su, Mei-Ling	蘇美靈	苏美灵
438	Soong, Tse-Pu	Song, Zi-Pu	宋滋圃	宋滋圃
436	Ssu, Hsing-Chien	Si, Xing-Jian	斯行健	斯行健
445	Su, Chien-Chang	Sun, Jian-Zhang	孫建璋	孙建璋
440	Su, Horng-Jye	Su, Hong-Jie	蘇鴻傑	苏鸿杰
439	Su, Ho-Yi	Su, He-Yi	粟和毅	粟和毅
442	Sun, Bi-Sin	Sun, Bi-Xing	孫必興	孙必兴
447	Sun, Chi-Shi	Sun, Qi-Shi	孫啓時	孙启时
444	Sun, Fêng-Chi	Sun, Feng-Ji	孫逢吉	孙逢吉
449	Sun, Hsiang Chung	Sun, Xiang-Zhong	孫祥鐘	孙祥钟
450	Sun, Hsiung-Tsai	Sun, Xiong-Cai	孫雄才	孙雄才
450	Sun, Hsiung-Ts'ai	Sun, Xiong-Cai	孫雄才	孙雄才
448	Sun, Hsu-Hsien	Sun, Shu-Xian	孫淑賢	孙淑贤
446	Sun, Lung-Chi	Sun, Long-Ji	孫隆吉	孙隆吉
448	Sun, Shu-Hsien	Sun, Shu-Xian	孫淑賢	孙淑贤
443	Sun, Tai-Yang	Sun, Dai-Yang	孫岱陽	孙岱阳
446	Sun, Von-Gee	Sun, Long-Ji	孫隆吉	孙隆吉
450	Sun, Yon-Zai	Sun, Xiong-Cai	孫雄才	孙雄才
437	Sung, Wan-Chih	Song, Wan-Zhi	宋萬志	宋万志
437	Sung, Wang-Chih	Song, Wan-Zhi	宋萬志	宋万志
436	Sze, Hsing-Chien	Si, Xing-Jian	斯行健	斯行健
78	Tai, Fan-Chin	Dai, Fan-Jin	戴蕃瑨	戴蕃瑨
77	Tai, Fang-Lan	Dai, Fang-Lan	戴芳瀾	戴芳澜
78	Tai, Fan-Tsien	Dai, Fan-Jin	戴蕃瑨	戴蕃瑨
77	Tai, Fung-Lan	Dai, Fang-Lan	戴芳瀾	戴芳澜
79	Tai, Lun-Yen	Dai, Lun-Yan	戴倫焰	戴伦焰
80	Tai, Ming-Chieh	Dai, Ming-Jie	戴銘傑	戴铭杰
451	Tam, Pui-Cheung	Tan, Pei-Xiang	譚沛祥	谭沛祥
452	Tan, Chung-Ming	Tan, Zhong-Ming	譚仲明	谭仲明
461	Tang, Chen-Zi	Tang, Zhen-Zi	唐振緇	唐振缁
453	T'ang, Chin	Tang, Jin	唐 進	唐 进
454	T'ang, Chio	Tang, Jue	唐 覺	唐 觉
458	Tang, Sin-Yao	Tang, Xin-Yao	唐心曜	唐心曜
455	T'ang, T'êng-Han	Tang, Teng-Han	湯滕漢	汤滕汉
453	Tang, Tsin	Tang, Jin	唐 進	唐 进
456	Tang, Wei-Shin	Tang, Wei-Xin	湯惟新	汤惟新

457	T'ang, Wên-T'ung	Tang, Wen-Tong	湯文通	汤文通
460	T'ang, Yao	Tang, Yao	唐 耀	唐 耀
459	Tang, Yen-Cheng	Tang, Yan-Cheng	湯彥承	汤彦承
459	T'ang, Yen-Ch'êng	Tang, Yan-Cheng	湯彥承	汤彦承
462	T'ao, Yen-Ch'iao	Tao, Yan-Qiao	陶延橋	陶延桥
24	Tchen, Te-Tchouan	Chen, Bo-Chuan	陳伯川	陈伯川
756	Tchou, Yen-Tch'êng	Zhu, Yan-Cheng	朱彥丞	朱彦丞
81	Têng, Kuei-Ling	Deng, Gui-Ling	鄧桂玲	邓桂玲
83	Têng, Shu-Ch'ün	Deng, Shu-Qun	鄧叔羣	邓叔群
463	Têng, Yung-Yen	Teng, Yong-Yan	滕咏延	滕咏延
463	Terng, Yeong-Yan	Teng, Yong-Yan	滕咏延	滕咏延
464	Tian, Kin-Tsuan	Tian, Jing-Quan	田景全	田景全
88	Ting, Chih-Chi	Ding, Zhi-Zun	丁志遵	丁志遵
88	Ting, Chih-Tsun	Ding, Zhi-Zun	丁志遵	丁志遵
84	Ting, Kuang-Ch'i	Ding, Guang-Qi	丁廣奇	丁广奇
84	Ting, Kuang-Chi	Ding, Guang-Qi	丁廣奇	丁广奇
85	Ting, Sü	Ding, Su	丁 驌	丁 骕
86	Ting, Ying	Ding, Ying	丁 穎	丁 颖
87	Ting, Yüan	Ding, Yuan	丁 源	丁 源
90	Tong, Koe-Yang	Dong, Gui-Yang	董貴陽	董贵阳
628	Ts'êng, Chao-Jan	Zeng, Zhao-Ran	曾昭然	曾昭然
16	Tsai, Hse-Tao	Cai, Xi-Tao	蔡希陶	蔡希陶
16	Tsai, Hsi-Tao	Cai, Xi-Tao	蔡希陶	蔡希陶
16	Ts'ai, Hsi-T'ao	Cai, Xi-Tao	蔡希陶	蔡希陶
12	Tsai, Jenn-Lai	Cai, Jin-Lai	蔡進來	蔡进来
12	Tsai, Jinn-Lai	Cai, Jin-Lai	蔡進來	蔡进来
14	Ts'ai, shao-Lan	Cai, Shao-Lan	蔡少蘭	蔡少兰
15	Tsai, Su-Hwa	Cai, Shu-Hua	蔡淑華	蔡淑华
17	Tsai, Tseng-Chun	Cai, Zhen-Cong	蔡振聰	蔡振聪
621	Tsang, Wai-Tak	Zeng, Huai-De	曾懷德	曾怀德
18	Tsao, Kan	Cao, Kan	曹 侃	曹 侃
18	Ts'ao, K'an	Cao, Kan	曹 侃	曹 侃
623	Tsen, Mill	Zeng, Mian-Zhi	曾勉之	曾勉之
19	Ts'ên, Ming-Shu	Cen, Ming-Shu	岑銘恕	岑铭恕
619	Tseng, Chang-Jiang	Zeng, Cang-Jiang	曾滄江	曾沧江
624	Tseng, Charles Chiao	Zeng, Qiao	曾 樵	曾 樵
629	Tsêng, Che-Ju	Zeng, Zhe-Ru	曾哲如	曾哲如
620	Tsêng, Ch'eng-K'uei	Zeng, Cheng-Kui	曾呈奎	曾呈奎
620	Tsêng, Chêng-K'uei	Zeng, Cheng-Kui	曾呈奎	曾呈奎
620	Tseng, Cheng-Kwei	Zeng, Cheng-Kui	曾呈奎	曾呈奎
624	Tseng, Chiao	Zeng, Qiao	曾 樵	曾 樵

624	Tsêng, Ch'iao	Zeng, Qiao	曾 樵	曾 樵
629	Tseng, Chie-Ju	Zeng, Zhe-Ru	曾哲如	曾哲如
622	Tsêng, Chi-K'uan	Zeng, Ji-Kuan	曾濟寬	曾济宽
623	Tsêng, Mien-Chih	Zeng, Mian-Zhi	曾勉之	曾勉之
625	Tseng, Shu-Ying	Zeng, Shu-Ying	曾淑英	曾淑英
627	Tseng, Yen-Hsueh	Zeng, Yan-Xue	曾彥學	曾彦学
73	Tseng, Yong-Qian	Cheng, Yong-Qian	程用謙	程用谦
73	Tseng, Yung-Chien	Cheng, Yong-Qian	程用謙	程用谦
218	Tsi, Zhan-Huo	Ji, Zhan-He	吉占和	吉占和
228	Tsiang, Hsing-Ling	Jiang, Xing-Lin	蔣興麟	蒋兴麟
229	Tsiang, Ying	Jiang, Ying	蔣 英	蒋 英
224	Tsien, Cho-Po	Jian, Zhuo-Po	簡焯坡	简焯坡
770	Tso, Ching-Lieh	Zuo, Jing-Lie	左景烈	左景烈
769	Tso, Ta-Hsun	Zuo, Da-Xun	左大勛	左大勋
717	Tsoong, Chi-Hsin	Zhong, Ji-Xin	鐘濟新	钟济新
715	Tsoong, Kuan-Kwang	Zhong, Guan-Guang	鐘觀光	钟观光
712	Tsoong, Pu-Ch'in	Zhong, Bu-Qin	鐘補勤	钟补勤
713	Tsoong, Pu-Chiu	Zhong, Bu-Qiu	鐘補求	钟补求
713	Tsoong, Pu-Ch'iu	Zhong, Bu-Qiu	鐘補求	钟补求
767	Tsou, Chih-Hua	Zou, Zhi-Hua	鄒稚華	邹稚华
768	Tsou, Chung-Lin	Zou, Zhong-Lin	鄒鐘琳	邹钟琳
766	Tsou, Ping-Wên	Zou, Bing-Wen	鄒秉文	邹秉文
75	Tsui, Hung-Pin	Cui, Hong-Bin	崔鴻賓	崔鸿宾
75	Ts'ui, Hung-Pin	Cui, Hong-Bin	崔鴻賓	崔鸿宾
74	Ts'ui, Po-T'ang	Cui, Bo-Tang	崔伯棠	崔伯棠
76	Tsui, You-Wen	Cui, You-Wen	崔友文	崔友文
76	Tsui, Yu-Wen	Cui, You-Wen	崔友文	崔友文
76	Ts'ui, Yu-Wên	Cui, You-Wen	崔友文	崔友文
94	Tu, Fu	Du, Fu	杜 複	杜 复
466	T'u, Hsiao-Tsu	Tu, Xiao-Zuo	塗孝祚	涂孝祚
95	Tu, Nai-Chiu	Du, Nai-Qiu	杜乃秋	杜乃秋
96	Tu, Ya-Tsuan	Du, Ya-Quan	杜亞泉	杜亚泉
97	Tuan, Pei-Chen	Duan, Pei-Qin	段佩琴	段佩琴
89	Tung, Ch'ang-Ch'un	Dong, Chang-Chun	董長椿	董长椿
465	Tung, Chi-Kuo	Tong, Zhi-Guo	仝治國	仝治国
91	Tung, Hui-Min	Dong, Hui-Min	董惠民	董惠民
92	Tung, Shih-Lin	Dong, Shi-Lin	董世林	董世林
92	Tung, Shi-Lin	Dong, Shi-Lin	董世林	董世林
93	Tung, Shuang-Ch'iu	Dong, Shuang-Qiu	董爽秋	董爽秋
91	Tung, Hwei-Nin	Dong, Hui-Min	董惠民	董惠民
496	Wan, Chao-Feng	Wang, Zhao-Feng	王兆鳳	王兆凤

467	Wan, Kuo-Ting	Wan, Guo-Ding	萬國鼎	万国鼎
467	Wan, Kwoh-Ting	Wan, Guo-Ding	萬國鼎	万国鼎
468	Wan, Thung-Ling	Wan, Zong-Ling	萬宗玲	万宗玲
468	Wan, Tsung-Ling	Wan, Zong-Ling	萬宗玲	万宗玲
468	Wane, Thung-Ling	Wan, Zong-Ling	萬宗玲	万宗玲
495	Wang, Chan	Wang, Zhan	王　戰	王　战
496	Wang, Chao-Fêng	Wang, Zhao-Feng	王兆鳳	王兆凤
497	Wang, Chao-Wu	Wang, Zhao-Wu	汪昭武	汪昭武
498	Wang, Cheng-Ping	Wang, Zheng-Ping	王正平	王正平
471	Wang, Ch'êng-Yin	Wang, Cheng-Yin	汪呈因	汪呈因
499	Wang, Chên-Hua	Wang, Zhen-Hua	王振華	王振华
499	Wang, Chen-Hwa	Wang, Zhen-Hua	王振華	王振华
500	Wang, Chen-Ju	Wang, Zhen-Ru	汪振儒	汪振儒
212	Wang, Chi	Huang, Zhi	黄　志	黄　志
502	Wang, Chih-Chia	Wang, Zhi-Jia	王志稼	王志稼
481	Wang, Ching-Jui	Wang, Qing-Rui	王慶瑞	王庆瑞
481	Wang, Ch'ing-Jui	Wang, Qing-Rui	王慶瑞	王庆瑞
477	Wang, Chin-Ting	Wang, Jin-Ting	王金亭	王金亭
477	Wang, Chin-T'ing	Wang, Jin-Ting	王金亭	王金亭
482	Wang, Chiou-Mei	Wang, Qiu-Mei	王秋美	王秋美
483	Wang, Chi-Wu	Wang, Qi-Wu	王啓無	王启无
483	Wang, Ch'i-Wu	Wang, Qi-Wu	王啓無	王启无
502	Wang, Chu-Chia	Wang, Zhi-Jia	王志稼	王志稼
505	Wang, Chu-Hao	Wang, Zhu-Hao	王鑄豪	王铸豪
504	Wang, Chung-Hsin	Wang, Zhong-Xin	王忠信	王忠信
506	Wang, Chung-Hsin	Wang, Zong-Xin	王宗信	王宗信
503	Wang, Chung-Kuei	Wang, Zhong-Kui	王忠魁	王忠魁
503	Wang, Chung-K'uei	Wang, Zhong-Kui	王忠魁	王忠魁
478	Wang, Chü-Yüan	Wang, Ju-Yuan	汪菊淵	汪菊渊
472	Wang, Fan-Lung	Wang, Fan-Long	王繁隆	王繁隆
473	Wang, Fa-Tsuan	Wang, Fa-Zuan	汪發纘	汪发缵
474	Wang, Fu-Hsiung	Wang, Fu-Xiong	王伏雄	王伏雄
476	Wang, Hao-Chên	Wang, Hao-Zhen	王浩真	王浩真
485	Wang, Hsien-Chih	Wang, Xian-Zhi	王先志	王先志
486	Wang, Hsien-Chih	Wang, Xian-Zhi	王顯智	王显智
491	Wang, I-Kuei	Wang, Yi-Gui	王一桂	王一桂
501	Wang, Jenn-Che	Wang, Zhen-Zhe	王震哲	王震哲
475	Wang, Kuang-Yuh	Wang, Guang-Yu	王光育	王光育
480	Wang, Ming-Chin	Wang, Ming-Jin	王名金	王名金
480	Wang, Ming-Kin	Wang, Ming-Jin	王名金	王名金
479	Wang, Ming-Tê	Wang, Ming-De	王明德	王明德

469	Wang, Pai-Sun	Wang, Bai-Sun	王伯蓀	王伯荪
470	Wang, Ping-Chuan	Wang, Bing-Quan	汪秉全	汪秉全
508	Wang, Tso-Pin	Wang, Zuo-Bin	王作賓	王作宾
508	Wang, Tso-Ping	Wang, Zuo-Bin	王作賓	王作宾
507	Wang, Tsung-Hsuin	Wang, Zong-Xun	王宗訓	王宗训
507	Wang, Tsung-Hsün	Wang, Zong-Xun	王宗訓	王宗训
494	Wang, Tze-Chun	Wang, Ze-Jun	王澤鋆	王泽鋆
484	Wang, Wen-Tsai	Wang, Wen-Cai	王文采	王文采
484	Wang, Wên-Ts'ai	Wang, Wen-Cai	王文采	王文采
490	Wang, Yei-Zeing	Wang, Ye-Zhen	王也珍	王也珍
487	Wang, Yen-Chieh	Wang, Yan-Jie	汪燕傑	汪燕杰
488	Wang, Yen-Tsu	Wang, Yan-Zu	王彦祖	王彦祖
489	Wang, Yo-Yü	Wang, Yao-Yu	王藥雨	王药雨
493	Wang, Yun-Chang	Wang, Yun-Zhang	王雲章	王云章
492	Wang, Yung-Kwei	Wang, Yong-Kui	王永魁	王永魁
509	Wei, C. T.	Wei, Jing-Chao	魏景超	魏景超
511	Wei, Chao-Fen	Wei, Zhao-Fen	衛兆芬	卫兆芬
510	Wei, Yue-Tsung	Wei, Yu-Zong	韋裕宗	韦裕宗
510	Wei, Yu-Tsung	Wei, Yu-Zong	韋裕宗	韦裕宗
512	Wên, Hung-Han	Wen, Hong-Han	聞洪漢	闻洪汉
476	Wong, Hao-Chên	Wang, Hao-Zhen	王浩真	王浩真
494	Wong, Tzê-Chun	Wang, Ze-Jun	王澤鋆	王泽鋆
533	Wu, Cheng-I	Wu, Zheng-Yi	吳徵鎰	吴征镒
533	Wu, Cheng-Yih	Wu, Zheng-Yi	吳徵鎰	吴征镒
517	Wu, Chi-Chih	Wu, Ji-Zhi	吳繼志	吴继志
522	Wu, Ch'i-Chün	Wu, Qi-Jun	吳其濬	吴其浚
523	Wu, Ching-Ju	Wu, Qing-Ru	吳慶如	吴庆如
534	Wu, Chung-Luen	Wu, Zhong-Lun	吳中倫	吴中伦
534	Wu, Chung-Lun	Wu, Zhong-Lun	吳中倫	吴中伦
534	Wu, Chung-Lwen	Wu, Zhong-Lun	吳中倫	吴中伦
516	Wu, Fei-Man	Wu, Hui-Min	伍輝民	伍辉民
528	Wu, Hsin-Kan	Wu, Xin-Gan	吳信淦	吴信淦
518	Wu, Jiunn-Tsong	Wu, Jun-Zong	吳俊宗	吴俊宗
518	Wu, Jiunn-Tzong	Wu, Jun-Zong	吳俊宗	吴俊宗
514	Wu, Kêng-Min	Wu, Geng-Min	吳耕民	吴耕民
515	Wu, Kuo-Fang	Wu, Guo-Fang	吳國芳	吴国芳
519	Wu, Ming-Hsiang	Wu, Ming-Xiang	吳明翔	吴明翔
520	Wu, Ming-Jou	Wu, Ming-Zhou	吳明洲	吴明洲
521	Wu, Pan-Cheng	Wu, Peng-Cheng	吳鵬程	吴鹏程
521	Wu, Pang-Cheng	Wu, Peng-Cheng	吳鵬程	吴鹏程
532	Wu, Shiew-Hung	Wu, Zhao-Hong	吳兆洪	吴兆洪

525	Wu, Su-Kung	Wu, Su-Gong	武素功	武素功
513	Wu, Te-Lin	Wu, De-Lin	吳德鄰	吴德邻
513	Wu, Te-Ling	Wu, De-Lin	吳德鄰	吴德邻
527	Wu, Wen-Chên	Wu, Wen-Zhen	吳韞珍	吴韫珍
526	Wu, Wen-Hsiang	Wu, Wen-Xiang	鄔文祥	邬文祥
524	Wu, Yeng-Fen	Wu, Rong-Fen	吳容芬	吴容芬
529	Wu, Yin-Ch'an	Wu, Yin-Chan	吳印禪	吴印禅
530	Wu, Ying-Siang	Wu, Ying-Xiang	吳應祥	吴应祥
524	Wu, Yong-Fen	Wu, Rong-Fen	吳容芬	吴容芬
524	Wu, Young-Fen	Wu, Rong-Fen	吳容芬	吴容芬
531	Wu, Yuan-Ti	Wu, Yuan-Di	吳元滌	吴元涤
527	Wu, Yün-Chên	Wu, Wen-Zhen	吳韞珍	吴韫珍
576	Xu, Chao-Peng	Xu, Zhao-Peng	徐兆鵬	徐兆鹏
584	Yang, Ch'êng-Yüan	Yang, Cheng-Yuan	楊承元	杨承元
588	Yang, Chien-Chow	Yang, Qin-Zhou	楊欽周	杨钦周
588	Yang, Ching-Chow	Yang, Qin-Zhou	楊欽周	杨钦周
587	Yang, Han-Pi	Yang, Han-Bi	楊漢碧	杨汉碧
594	Yang, Hsien-Chin	Yang, Xian-Jin	楊銜晉	杨衔晋
595	Yang, Hsi-Ling	Yang, Xi-Lin	楊喜林	杨喜林
596	Yang, Hsin-Mei	Yang, Xin-Mei	楊新美	杨新美
600	Yang, Jeng-Jung	Yang, Zheng-Zhong	楊正仲	杨正仲
586	Yang, Kuoh-Cheng	Yang, Guo-Zhen	楊國禎	杨国祯
590	Yang, Shan-Kwan	Yang, Shang-Guang	楊上洸	杨上洸
591	Yang, Sheng-Zehn	Yang, Sheng-Ren	楊勝任	杨胜任
592	Yang, Suen-Liu	Yang, Sun-Liu	楊孫鎏	杨孙鎏
592	Yang, Sun-Liu	Yang, Sun-Liu	楊孫鎏	杨孙鎏
583	Yang, Tsai-Yeog	Yang, Cai-Yong	楊彩勇	杨彩勇
585	Yang, Tsung-Ren	Yang, Chong-Ren	楊崇仁	杨崇仁
601	Yang, Tsung-Yu	Yang, Zong-Yu	楊宗愈	杨宗愈
593	Yang, Yen-Chin	Yang, Xian-Jin	楊銜晉	杨衔晋
598	Yang, Yuen-Po	Yang, Yuan-Bo	楊遠波	杨远波
597	Yang, Yung-Chang	Yang, Yong-Chang	楊永昌	杨永昌
589	Yang, Yung-Ch'i	Yang, Rong-Qi	楊榮啓	杨荣启
599	Yang, Yü-P'o	Yang, Yu-Po	楊玉坡	杨玉坡
602	Yao, Hao-Nien	Yao, He-Nian	姚鶴年	姚鹤年
602	Yao, Ho-Nien	Yao, He-Nian	姚鶴年	姚鹤年
605	Yeh, Chao-Ch'ing	Ye, Zhao-Qing	葉兆慶	叶兆庆
603	Yeh, Kai-Yun	Ye, Kai-Wen	葉開溫	叶开温
604	Yeh, Ya-Ko	Ye, Ya-Ge	葉雅各	叶雅各
582	Yen, Chen-Lung	Yan, Zhen-Long	閻振龍	阎振龙
581	Yen, Sun-Ch'u	Yan, Xun-Chu	閻遜初	阎逊初

580	Yen, Tsu-Kiang	Yan, Chu-Jiang	嚴楚江	严楚江
607	Yen, Tsu-Tang	Yin, Zu-Tang	尹祖棠	尹祖棠
606	Yin, Chao-P'ei	Yin, Zhao-Pei	尹兆培	尹兆培
608	Ying, Tsun-Shen	Ying, Jun-Sheng	應俊生	应俊生
590	Young, Shan-Kwan	Yang, Shang-Guang	楊上洸	杨上洸
611	Yu, Ching-Jang	Yu, Jing-Rang	于景讓	于景让
611	Yü, Ching-Yang	Yu, Jing-Rang	于景讓	于景让
612	Yü, Ch'i-Pao	Yu, Qi-Bao	俞啓葆	俞启葆
609	Yü, Ta-Fu	Yu, Da-Fu	俞大紱	俞大绂
609	Yu, Ta-Fuh	Yu, Da-Fu	俞大紱	俞大绂
610	Yü, Tê-Chun	Yu, De-Jun	俞德浚	俞德浚
610	Yü, Tê-Chün	Yu, De-Jun	俞德浚	俞德浚
610	Yu, Te-Tsun	Yu, De-Jun	俞德浚	俞德浚
615	Yü, Tso-Chi	Yu, Zuo-Ji	俞作楫	俞作楫
614	Yu, Tsung-Hsiung	Yu, Zong-Xiong	郁宗雄	郁宗雄
613	Yu, Yung-Nien	Yu, Yong-Nian	余永年	余永年
616	Yuan, Ch'ang-Chi	Yuan, Chang-Qi	袁昌齊	袁昌齐
418	Yuan, Hsi-Chung	Ruan, Xi-Zhong	阮希中	阮希中
617	Yueh, Chung-Hsi	Yue, Chong-Xi	樂崇熙	乐崇熙
617	Yueh, Chun-Hsi	Yue, Chong-Xi	樂崇熙	乐崇熙
618	Yueh, Song-Chien	Yue, Song-Jian	嶽松健	岳松健
618	Yueh, Sung-Chien	Yue, Song-Jian	嶽松健	岳松健
640	Zhang, Fang-Szu	Zhang, Fang-Ci	張芳賜	张芳赐
645	Zhang, Hai-Tau	Zhang, Hai-Dao	張海道	张海道

索引
Indices

1 中文人名索引

1 Index of Authors in Chinese

F

G

H

J

K

L

T

W

X

Y

Z

2 西文人名索引

2 Index of Authors

D

I

J

K

L

M

N

O

P

Q

R

S

T

V

W

Z

3 植物中文名索引

3 Index of Chinese Plant Names

4 植物学名索引

4 Index of Scientific Plant Names

5 中文期刊和图书索引

5 Index of Journals and Books in Chinese

E

F

G

H

J

K

L

M

N

T

W

X

Y

Z

6 西文期刊和图书索引

6 Index of Journals and Books

B

C

G

H

K

L

M

N

O

P

Q

R

S

T

后记

撰写本书的想法始于 2011 年，当澳大利亚墨尔本举办的第十八届国际植物学大会确定了第十九届国际植物学大会将于 2017 年在中国深圳举行时，笔者欣喜之余，也想做点贡献，共襄盛举。尤其东道国——中国被誉为"园林之母"，大会似不应仅限为具体的学术活动。如果届时能有一部关于中国植物分类学基本情况的出版物，不仅能回顾东道主的植物分类学足迹，同时也能向国内外同仁展示植物分类学界的中国历史。

本书动笔始于 2015 年新年；原计划 2017 年春天交稿，在夏天植物学大会开始前出版。怎奈临近 2017 年深圳国际植物学大会时发现，稿子还有诸多缺漏；如果勉强投入印刷，不但难以向后人交代，更觉得愧对先辈。不得已，放弃了原有的出版计划，希望尽最大努力完善书稿，哪怕出版时间需要延后一些。当初未曾想到，这一改又是两年半之多，直到 2019 年夏才交付排版，之后历经三审三校、四次更改；加之其它因素掣肘，最后 2019 年秋天才付梓。

史海苍茫，收集资料实属不易，考证更是倍加艰辛。时间仓促，精力有限，就此付印难免遗漏甚至错误，敬请各位老友新朋批评指正。感谢四位合作者各方面的支持与合作；特别是哈佛大学植物标本馆鲍棣伟博士润色英文、提供采集信息、给与的建议与讨论、提供照片，庐山植物园胡宗刚先生参与讨论与确认内容、提供资料与图片，原助理、现华东师范大学在读博士生廖帅收集相关文献与校对文本，中国科学院华南植物园叶文博士核对苔藓资料与润色中文以及翻译。没有他们的帮助，本书不可能完成。然而，本书中任何错误、遗漏或不当记载，均由笔者负责。

中国植物分类学纪事为中国植物分类学的历史变迁，尤其是过去百年间的历史提供了一个新的视角。然而，希望读者的探索与思考绝不止于此！

永久邮箱：jinshuangma@gmail.com

2019 年 11 月 5 日星期二

Postscript

Since 2011, at the XVIII International Botanical Congress in Melbourne, Australia, when it was confirmed that the XIX International Botanical Congress in 2017 would take place in Shenzhen, China, I thought about what I could contribute. Since China has been called 'Mother of Gardens,' it should be more than an account of specific academic activities. It could be a publication on Chinese plant taxonomy that not only reviews taxonomic activities in the host country, but also informs colleagues, both domestic and foreign, on the history of Chinese activities in plant systematics.

Thus, I began compiling information for such a book at the beginning of 2015, initially thinking that it could be completed and submitted by the spring of 2017 and published before the 19th IBC in mid summer of that year. I certainly did not expect that the manuscript would not be completed before the Congress in Shenzhen. I found that although it was possible to print what I had, I could certainly not hold my head up before future generations, let alone our ancestors! At last, I had to give up the original plan to publish before the International Botanical Congress and to recommit to doing as much as I could and the best that I could, but would need more time. Unexpectedly, this work has continued for more than another two and a half years, until the summer of 2019, when the manuscript was sent to press. After many reviews and edits due to various factors, it was finally printed in the autumn of 2019!

The sea of history is so vast, the collection of data not easy, and the research even more difficult. However, since time and energy are always limited, it is inevitable that some facts will be missed and even some mistakes made. I welcome criticism and corrections from friends old and new. Many thanks to the support and cooperation of my four partners. In particular, I would like to note the polishing of the English, collection information, friendly suggestions and discussions on the manuscript and photos provided by David Boufford, Harvard University Herbaria, kindly discussions and confirmation of the contents, information and photos provided by Mr. Zong Gang HU, Lushan Botanical Garden, collecting of the literature and proofreading by Shuai LIAO, my former assistant and now a doctoral student at East China Normal University, and the bryophyte information and translation and polishing of the Chinese by Dr. Wen YE, South China Botanical Garden, Chinese Academy of Sciences. Without their help, it would have been impossible to complete such a book. However, any mistakes, omissions or improper records are surely my responsibility.

A Chronicle of Plant Taxonomy in China provides a new perspective by systematically displaying the historical changes in Chinese plant taxonomy, particularly during the past 100 years. However, I hope that the reader's exploration and thinking will go beyond what is presented here!

Jinshuang Ma

Permanent email: jinshuangma@gmail.com

November 5, 2019, Tuesday

作者
Authors

马金双　Jin-Shuang MA

胡宗刚　Zong-Gang HU

廖帅　Shuai LIAO

叶文　Wen YE

鲍棣伟　David E. Boufford